Exploring Engineering

Exploring Engineering
An Introduction to Engineering and Design

Sixth Edition

Philip Kosky

Robert Balmer

William Keat

George Wise

ACADEMIC PRESS

An imprint of Elsevier

ELSEVIER

Academic Press is an imprint of Elsevier
125 London Wall, London EC2Y 5AS, United Kingdom
525 B Street, Suite 1650, San Diego, CA 92101, United States
50 Hampshire Street, 5th Floor, Cambridge, MA 02139, United States

Notices

Knowledge and best practice in this field are constantly changing. As new research and experience broaden our understanding, changes in research methods, professional practices, or medical treatment may become necessary.

Practitioners and researchers must always rely on their own experience and knowledge in evaluating and using any information, methods, compounds, or experiments described herein. In using such information or methods they should be mindful of their own safety and the safety of others, including parties for whom they have a professional responsibility.

To the fullest extent of the law, neither the Publisher nor the authors, contributors, or editors, assume any liability for any injury and/or damage to persons or property as a matter of products liability, negligence or otherwise, or from any use or operation of any methods, products, instructions, or ideas contained in the material herein.

ISBN: 978-0-443-13541-5

For information on all Academic Press publications visit our website at
https://www.elsevier.com/books-and-journals

Publisher: Peter Linsley
Acquisitions Editor: Stephen Merken
Editorial Project Manager: Joshua Mearns
Production Project Manager: Sharmila Kirouchenadassou
Cover Designer: Greg Harris

Typeset by TNQ Technologies

Working together
to grow libraries in
developing countries

www.elsevier.com • www.bookaid.org

Contents

Preface

Young men and women exploring engineering as a career are excited about the future—*their future*—and about the engineering challenges 10–20 years from now when they are in the spring and summer of their careers. The National Academy of Engineering proposed the following 14 **Grand Challenges for Engineering in the 21st century**. In this text, we have chosen to include material that engages many of these topics because they represent the future of engineering creativity.

1) Make solar energy economical
2) Provide energy from fusion
3) Develop carbon sequestration methods
4) Manage the nitrogen cycle
5) Provide access to clean water
6) Restore and improve urban infrastructure
7) Advance health informatics
8) Engineer better medicines
9) Reverse-engineer the brain
10) Prevent nuclear terror
11) Secure cyberspace
12) Enhance virtual reality
13) Advance personalized learning
14) Engineer the tools of scientific discovery

The 21st century will be filled with many exciting challenges for engineers, architects, physicians, sociologists, and politicians. Fig. 1 illustrates an enhanced set of future challenges as envisioned by Joseph Bordogna, former deputy director and chief operating officer of the National Science Foundation.

The structure of this text

We have tried to provide an exciting introduction to the engineering profession. Between its covers, you will find material on classical engineering fields as well as introductory material for new interdisciplinary 21st century engineering fields.

- **Part 1**, which we call **Lead-On**, emphasizes what we consider the fundamentals that thread through virtually all the branches of engineering. There are just six chapters in Part 1: What Do Engineers Do, Engineering Ethics, Elements of Engineering Analysis, Force and Motion, Energy, and Engineering Economics. Most chapters in Part 1 are organized around just one or two principles, have several worked examples, and include end-of-chapter exercises with several levels of difficulty. Occasionally, answers are given to selected exercises to encourage students to work toward self-proficiency. Solutions are available online for instructors at the publisher's website. *We recommend all the first five chapters be used in first-year engineering courses.*

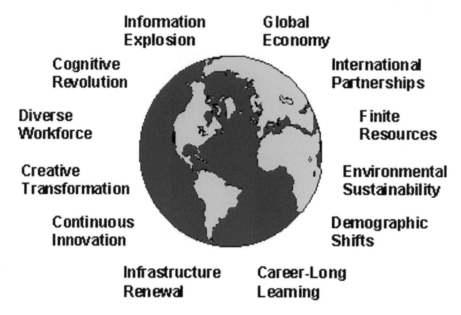

FIGURE 1

Future Trajectories in Science, Engineering, and Technology.

Source: nsf.gov/news/speeches/bordogna/jb98_nrl/sld002.htm.

- **Part 2**, which we call **Minds-On,** covers introductory material explicitly from the following well-established engineering disciplines: aeronautical, chemical, civil, computer, electrical, industrial, manufacturing, materials, mechanical, and nuclear. There is some minor overlap because some engineering fields *do* overlap. In addition, five more chapters are devoted to new interdisciplinary engineering fields: bioengineering, electrochemical engineering, environmental engineering "green" energy engineering, and an introduction to mechatronics and physical computing. We have tried to keep the level of this material commiserate with the background of first-year students, though with some occasional challenge. Our recommendation to instructors is to select appropriate discipline-specific chapters.
- **Part 3,** which we call **Hands-On,** is just as essential and challenging as the engineering aspects covered in Parts 1 and 2, and for most students *it is a lot more fun.* It provides the content for a **design studio** and contains eight specific design steps that should be covered in full. Few things are more satisfying than seeing a machine, an electronic device, or a computer program you designed, built, and tested doing exactly what you intended it to do.

Key elements of this book

The approach taken in this text is unique, because of the character of authorship. Two of the authors have *industrial* backgrounds, mostly at the GE Research Center in Niskayuna, New York. The other two authors followed more traditional academic paths and have the experience with first-year students. Specifically, this textbook reflects the combined author's experience that engineering is not a "spectator sport." Here are some of our approaches to familiar engineering topics:

1) The **Accreditation Board for Engineering and Technology** (ABET) requires engineering students to have "an understanding of professional and ethical responsibility." To help instructors meet this requirement, we have an early chapter on engineering ethics that includes a pedagogical tool called the **Engineering Ethics Decision Matrix.** The rows of the matrix are the canons of engineering ethics, and the columns are possible ways to resolve an ethical problem.

2) We also introduce **spreadsheets** and **MATLAB** as calculation tools early in the text because they are commonly used today by working engineers.

3) We rigorously try to use the appropriate number of **significant figures** in examples and exercises throughout the text. For example, we always try to differentiate between 60. and 60 (notice the decimal point or its absence).

4) We introduce an effective problem-solving technique called the **Need-Know-How-Solve** method. This technique is designed to prevent students from blindly plugging numbers into equations and writing down meaningless answers. Initially this might be clumsy to implement, but it is invaluable in training young engineers to leave a clear audit trail of their work now and in the future.

5) While the metric unit system of measurement is rapidly gaining popularity in the United States, we recognize that the old **Engineering English** unit system is still in use today. In the solutions to our example problems, we use a clear methodology to track units in all the calculations to arrive at the correct units (and significant figures) in the answer.

6) Pacing of hands-on projects in Part 3 is accomplished through **design milestones.** These are well-known design steps that are powerful tools in motivating first-year students toward engineering success.

7) The **design project examples** in Part 3 were selected from past student projects to be readily grasped by the beginning engineering student. The culmination of these designs may lead to head-to-head **team competition.**

WDK at Union College, Schenectady, New York
RTB at the University of Wisconsin—Milwaukee
A companion student web site for this textbook is available at
https://www.elsevier.com/books-and-journals/book-companion/9780443135415
For instructors, a solution manual, design contest material, and Power Point lecture slides are available by registering at https://educate.elsevier.com/book/details/9780443135415. It contains solutions to the exercises using the Need-Know-How-Solve technique introduced in this text.

Acknowledgments

This textbook covers introductory material for many engineering disciplines. We want to thank those proficient in those disciplines and who gave us some of their precious time. First, we acknowledge help, suggestions, and advice from our Union College colleagues and especially from instructors in the Union College freshman engineering course: Professors Brad Bruno, James Hedrick, David Hodgson (who wrote the mechatronics chapter), Thomas Jewell, John Spinelli, Cherrice Travers, and Frank Wicks. In addition, we have received advice and council from Professor Nicholas Krouglicof of Memorial University, Newfoundland, Canada. Other assistance came from Dr. John Rogers, Mechanical Engineering Division, West Point, and Dr. Andrew Wolfe, Civil Engineering Technology, SUNY Institute of Technology, Utica.

We thank the following faculty who provided valuable revision feedback:

Jamil AL-Nouman, SENMC formerly NMSU Carlsbad
Stephanie Adams, Virginia Tech
Stacy Birmingham, Grove City College
Aaron Budge, Minnesota State University
Ray Burynski, SUNY Polytechnic Institute
Mauro Caputi, Hofstra University
Kelly Crittenden, Louisiana Technical University
Brian DeJong, Central Michigan University
Mebougna Drabo, Alabama A&M University
Yalcin M Ertekin, Drexel University
Jessica Furrer, Benedict College
Haitham Abu Ghazaleh, Tarleton State University
Darin Gray, University of Southern California
Michael Gregg, Virginia Polytechnic Institute
Daniel Guino, Ohio University
Jerry Hamann, University of Wyoming
Alfonso Juan Hinojosa, Laredo College
Robert Krchnavek, Rowan University
Ken Manning, SUNY Adirondack
Steven McIntosh, University of Virginia
Samuel Morton, James Madison University
Francelina Neto, California State Polytechnic University
Jin Park, Minnesota State University
James Riddell, Baker College

The competition-based hands-on approach to teaching engineering design used in this book was influenced by Michael Larson and Daniel Retajczyk. When Professors Keat and Larson[1] were at MIT, they were impressed by the undergraduate design course of Prof. Woody Flowers. Later, when Prof. Keat was at Clarkson University, Daniel Retajczyk (then a graduate student there) helped to improve the design competition format.

First year Union College student Craig Ferguson developed the student design for the "A Bridge Too Far" example in Part 3, and first year Union College students Marcello Agnitti, James Cooper, and Justin Gold developed the Automatic Air Freshener design solution presented in Part 3.

Finally, we thank Mr. Dan Beller (University of Wisconsin-Milwaukee), for his assistance in developing the material on drawing and sketching that appears in the "COMPANION MATERIALS" section of the books Elsevier web site.

Many of the graphic illustrations were produced by Theodore A. Balmer at March Twenty Productions (http://marchtwenty.com) in Milwaukee Wisconsin.

---

[1]Currently Professor of Engineering and Innovation at the University of Colorado at Colorado Springs.

Lead-on

Source: ArcMann/Shutterstock

What engineers do

Source: Antonis Papantoniou/iStockphoto.com

Exploring Engineering. https://doi.org/10.1016/B978-0-443-13541-5.00018-0

1.1 Introduction

What is an engineer, and what does he or she do? You can get a good answer to this question by just looking at the word itself. The word **engine** comes from the Latin **ingenerare**, meaning "to create." About 2000 years ago, the Latin word *ingenium* ("the product of genius") was used to describe the design of a new machine. Soon after, the word **ingen** was used to describe all machines. In English, *ingen* was spelled "engine," and people who designed creative things were known as "engine-**ers**." In French, German, and Spanish today, the word for *engineer* is *ingenieur*, and in Italian it is *ingegnere*.

So, what does the word "engineer" really mean?

The word engineer *refers to someone who is a creative, ingenious person who finds solutions to practical problems.*

Today the word **engineer** refers to people who use creative design and analysis processes that incorporate energy, materials, motion, and information to serve human needs in innovative ways. Engineers express knowledge in the form of **variables**, **numbers**, and **units of measurement** (e.g., meters, kilograms, seconds, etc.,). There are many kinds of engineers, but all share the concepts and methods introduced in this book.

1.2 What is engineering?

While engineers use science to help them solve work-related problems, there are some key differences between the two when it comes to their responsibilities. Engineers often use science and math to research more about the world and make discoveries that could benefit humanity. The late scientist and science fiction writer Isaac Asimov once said that "Science can amuse and fascinate us all, but it is engineering that changes the world."[1] Almost everything you see around you has been touched by an engineer. Engineers are creative people who use mathematics, scientific principles, material properties, and computer methods to design new products and to solve human problems. Engineers can and do just about anything, designing and building roads, bridges, cars, planes, space stations, cell phones, computers, medical equipment, and much more.

Engineers can be classified according to the kind of work they do—administration, construction, consulting, design, development, teaching, planning (also called *applications engineers*), production, research, sales, service, and test engineers. Because engineering deals with the world around us, the number of engineering disciplines is very large. Table 1.1 lists some of the many engineering fields.

---

[1] Isaac Asimov's *Book of Science and Nature Quotations* (New York: Simon & Schuster, 1970).

Table 1.1 A few of the many engineering fields available today.

Aerospace	Biomedical	Ecological	Marine	Ocean
Aeronautical	Ceramic	Electrical	Materials	Petroleum
Agricultural	Chemical	Environmental	Mechanical	Sanitary
Architectural	Civil	Geological	Mining	Textiles
Automotive	Computer	Manufacturing	Nuclear	Transportation

1.3 What do engineers do?

Most engineers have a bachelor's degree in engineering from a program accredited by the **Accreditation Board for Engineering and Technology** (ABET). This nonprofit organization assures confidence in training programs that use science, technology, engineering, and math. By taking courses approved by ABET, you are prepared to enter the global engineering workforce.

Most engineers specialize in a specific field of engineering. The following list contains information on a few of the engineering fields in the Federal Government's Standard Occupational Classification (SOC) system.[2] Note that some of the engineering fields may have several subdivisions. For example, Civil Engineering includes Structural and Transportation Engineering, and Materials Engineering includes Ceramic, Metallurgical, Polymer and Nanotechnology Engineering.

- **Aerospace engineers**[3] apply scientific and technological principles to research, design, develop, maintain, and test the performance of civil and military aircraft, missiles, weapons systems, satellites, and space vehicles. They also work on the different components that make up these aircraft and systems.
- **Biomedical engineers** develop devices and procedures that solve medical and health-related problems by combining biology and medicine with engineering principles. Many biomedical engineers develop and evaluate systems and products such as artificial limbs and organs, instrumentation, and health management and care delivery systems.
- **Chemical engineers** apply the principles of chemistry and engineering to solve problems involving the production or use of chemicals, fuels, drugs, food, and plastics. They design equipment and processes for chemical manufacturing, petroleum refining, byproduct treatment, and supervise production.
- **Civil engineers** design and supervise the construction of roads, buildings, airports, tunnels, dams, bridges, and water supply and sewage systems. Civil

---

[2] Abstracted from the Bureau of Labor Statistics (http://www.bls.gov/oco/ocos027.htm).
[3] www.prospects.ac.uk

engineering is one of the oldest engineering disciplines[4] and encompasses many specialties. The major ones are structural, water resources, construction, transportation, and geotechnical engineering.

- **Computer engineers** design, develop, test, and oversee the manufacture and installation of computer hardware, including computer chips, circuit boards, computer systems, and related equipment, such as keyboards, routers, and printers. Computer engineers may also design and develop the software systems that control computers.

- **Electrical engineers** design, develop, test, and supervise the manufacture of electrical equipment. Some of this equipment includes electric motors; machinery controls, lighting, and wiring in buildings; radar and navigation systems; communications systems; power generation, and transmission devices used by electric utilities.

- **Environmental engineers** use the principles of biology, chemistry, and engineering to develop solutions to environmental problems. They are involved in water and air pollution control, recycling, waste disposal, and public health issues. Environmental engineers conduct hazardous waste management studies in which they evaluate the significance of potential hazards and develop regulations to prevent mishaps.

- **Industrial and manufacturing engineers** determine the most effective ways to use the basic items of production—people, machines, materials, information, and energy—to make a product or provide a service. They are concerned with increasing productivity through the management of people, methods of business organization, and technology. These engineers study product requirements and then design manufacturing systems to meet those requirements.

- **Materials engineers** are involved in the development, processing, and testing of the materials used to create a range of products, from computer chips and aircraft wings to golf clubs and snow skis. They work with metals, ceramics, plastics, semiconductors, and composites to create new materials that meet certain mechanical, electrical, and chemical requirements.

- **Mechanical engineers** design, develop, manufacture, and test all types of mechanical devices. Mechanical Engineering is one of the broadest engineering disciplines. Mechanical engineers work on power-producing machines such as electric generators, internal combustion engines, and steam and gas turbines; they also work on power-using machines such as refrigeration and air-conditioning equipment, machine tools, material-handling systems, and robots.

- **Nuclear engineers** develop the processes, instruments, and systems used to derive benefits from nuclear energy and radiation. They design, develop, monitor, and operate nuclear plants to generate electric power. They may work on the nuclear

---

[4] The oldest type of engineering is military engineering. The first nonmilitary engineers were called "civilian" engineers, which later became shortened to civil engineers.

fuel cycle—the production, handling, and use of nuclear fuel and the safe disposal of nuclear waste.

You can find more about what today's engineers do within their specialties by searching the Internet. A few of the engineering societies that represent different engineering fields are:

- AIAA (aeronautical engineering)
- AIChE (chemical engineers)
- ANS (nuclear engineering)
- ASCE (civil engineers)
- ASME (mechanical engineers)
- ASTM (materials and testing engineers)
- BMES (biomedical engineering)
- IEEE (electrical engineers)

You will discover that your basic college engineering courses have much in common with all engineering disciplines. They cover scientific principles, application of logical problem-solving processes, principles of design, and the value of teamwork. If you are considering an engineering career, we highly recommend you consult web resources to refine your understanding of the various fields of engineering.

1.4 Where do engineers work?

Most engineers work in office buildings, laboratories, or industrial plants. Others may spend time outdoors at construction sites and oil and gas exploration and production sites, where they monitor or direct operations or solve onsite problems. Some engineers travel extensively to plants or worksites here and abroad. Engineers are salaried, not hourly, employees and typically work a nominal 40-hour week. At times, deadlines or design standards may bring extra pressure to a job, requiring engineers to work longer hours.

Engineers usually work in teams. Sometimes, the team has only two or three engineers, but in large companies, engineering teams can have hundreds of people working on a single project (for example, in the design and manufacture of a large aircraft). Engineers are capable of designing the processes and equipment needed for a project, and that sometimes involves inventing new technologies. Engineers must also test their work carefully before it is used by trying to anticipate all of the things that could go wrong and make sure that their products perform safely and effectively.

More than 1.6 million engineers work in the United States today, making engineering the nation's second-largest profession. According to the National Association of Colleges and Employers, **engineering majors typically have the highest 4-year baccalaureate degree starting salaries**, averaging around $70,000 per year in 2023.

An engineering degree also opens doors to other careers. Engineering graduates can move into other professions, such as medicine, law, and business, where their engineering problem-solving ability is a valuable asset. A list[5] is available of famous engineers who became American Presidents, Nobel Prize winners, astronauts, corporate presidents, entertainers, inventors, and scientists.

In the United States, distinguished engineers may be elected to the National Academy of Engineering (NAE); it is the highest national honor for engineers. In many countries, there are parallel organizations (e.g., The Royal Academy of Engineering in the United Kingdom).

1.5 What is the difference between engineering and engineering technology?

Students who obtain a Bachelor of Science in **Engineering** begin their careers as entry-level **engineers**. Graduates with a 4-year **Engineering Technology** degree are called "technologists" and students who complete 2-year **engineering technology** programs are called "technicians."

The following definition of engineering technology was approved by the Engineering Technology Council of the American Society for Engineering Education.

Engineering technology is the profession in which a knowledge of mathematics and natural sciences gained by higher education, experience, and practice is devoted primarily to the implementation and extension of existing technology for the benefit of humanity.

Engineering technologists work closely with engineers in coordinating people, material, and machinery to achieve the specific goals of a particular project. Many engineering technicians work in quality control, inspecting products and processes, conducting tests, or collecting data. In manufacturing, they may assist engineers in product design, development, and production.

1.6 What makes a "good" engineer?

This is actually a difficult question to answer because the knowledge and skills required to be a good engineer (i.e., to create ingenious solutions) is a moving target. Generally speaking, a good engineer should have strong technical knowledge and experience using various software design and analysis programs. They should also have strong creative thinking and problem-solving skills to design and build

---

[5] See http://www.sinc.sunysb.edu/Stu/hnaseer/interest.htm

products that are valuable and practical. Most engineers work on projects that have fixed deadlines, so they also need effective time-management and organization skills. A key characteristic of a successful 21st-century engineer is her/his ability to utilize multidisciplinary engineering approaches to produce competitive solutions to societal needs.

So just what *does* the 21st century hold for the young engineer? It will be characterized by the *convergence* of many technologies and engineering systems (such as mechanical and electrical engineering converging into the new field of "*mechatronics*"). The products of today and tomorrow will be faster and smarter. The incorporation of computers, sensors, controls, modern alloys, and plastics are as important as continuing expertise in the traditional engineering disciplines.

1.7 Keys to success as an engineering student

Students who have had considerable success in high school often struggle in college. Here are five key steps to successfully study engineering.

(1) **Read the book before each class.** The course instructor cannot discuss every detail that needs to be covered in an engineering course.

(2) **Do all the homework.** To do well in sports, music, art, or engineering, you must practice, practice, practice. Homework is your "engineering practice." Just as watching a soccer game will not make you a good soccer player, not doing your engineering homework (or just copying someone else's homework) will not help you learn the material.

(3) **Attend class, every class.** You are paying a considerable amount of money for college courses, so be an active participant in class—ask questions about things you don't understand.

(4) **Treat College like a job.** Expect to spend 2—3 hours of homework for every hour of lecture. Treat a 3-credit engineering course as an 8 or 10 hour a week job.

(5) **It is YOUR responsibility to learn the material.** The faculty are responsible for guiding you through the process, but you are ultimately responsible for your success.

- Prioritize your study time. There are only so many hours in a day you can devote to studying so it is important that you divide study time between subjects and focus on upcoming exams.
- Dedicate a specific time and place to study.
- Start with the most difficult subject you need to study. You are more effective with difficult material at the beginning of your study session than later when you are tired.

1.8 Engineering is not an spectator sport

Engineering is not a spectator sport! It is a profession that involves design, implementation, and leadership skills to solve complex technological problems. In this book, we have included a **Hands-On** section that is intended to be both challenging and *fun!* In this section, you will design, build, and test increasingly complex engineering systems, starting with the tallest tower that can be made from a single sheet of paper and ending with a computer-controlled device that combines many parts into a system aimed at achieving a specific goal.

You will have to integrate skills learned in construction, electrical circuits, logic, and computers in building your device (which could be a model car, robot, boat, bridge, or anything else appropriate to your course). It may have to compete against similar devices built by other students in your class whose motivation may be to stop your device from achieving its goals! You will learn how to organize data and the vital importance of good communication skills. You will also be required to present your ideas and your designs both orally and in written reports.

As a start to that portion of this book, can you mentally take apart and put back together an imaginary automobile, toaster, computer, or bicycle? Instead of using wrenches and screwdrivers, you will be using mental and computer tools for engineering thought.

Example 1.1

Fig. 1.1 shows a generic car with numbered parts. Without looking at the footnote below, can you fill in the correct number corresponding to the parts in each of the blanks.[6]

[6] Answer: 1—distributor, 2—transmission, 3—spare tire, 4—muffler, 5—gas tank, 6—starter motor, 7—exhaust manifold, 8—oil filter, 9—radiator, 10—alternator, 11—battery.

As visually appealing as Fig. 1.1 is, an engineer would consider it inadequate because it fails to express the functional connections among the various parts. Expressing in visual form the basic elements involved in a problem is a crucial tool of engineering, called a **conceptual sketch**. The first step in an engineer's approach to a problem is to draw a conceptual sketch of the problem. Artistic talent is not an issue nor is graphic accuracy. The engineer's conceptual sketch may not look at all like the thing it portrays. Rather, it is intended to (1) help the engineer identify the elements in a problem, (2) see how groups of elements are connected to form subsystems, and (3) understand how all those subsystems work together to create a working system.

Example 1.2

On a piece of paper, draw a conceptual sketch of what happens when you push on the pedal of a bicycle. Before you begin here are some questions you should think about.

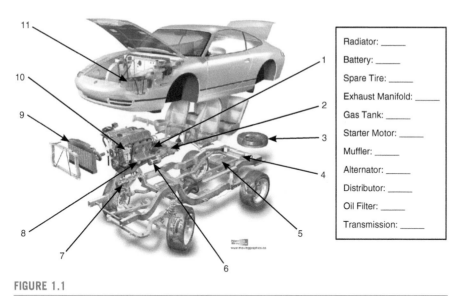

Radiator: _____

Battery: _____

Spare Tire: _____

Exhaust Manifold: _____

Gas Tank: _____

Starter Motor: _____

Muffler: _____

Alternator: _____

Distributor: _____

Oil Filter: _____

Transmission: _____

FIGURE 1.1

An exploded view of a modern automobile.

Source: © Moving Graphics.

- What are the key components that connect the pedal to the wheel?
- Which components are connected to each other?
- How does doing something to one of the components affect the others?
- What do those connections and changes have to do with accomplishing the task of accelerating the bicycle?

Solution Fig. 1.2 shows what your sketch probably looks like. But this is just the final form of the bicycle; it does not give much insight into what was needed to design and to build it. It's the idealized representation in the block diagram that will clarify the functions needed in its design.

For any engineering concept, many different conceptual sketches are possible. You are encouraged to draw conceptual sketches of each of the key points in the chapters of this book.

1.8.1 Lead-on skills

Many engineers become leaders in their field. They rise to positions of managers, division heads, executives, and company presidents. You may even start your own company and will need specialized leadership skills to be successful.

Leadership skills can help you in all aspects of your career, from applying for jobs to seeking career advancement. They help you encourage and organize people to reach a shared goal and motivate others. Leadership is not just one skill but a combination of several different skills.

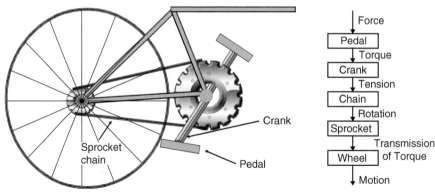

FIGURE 1.2

Bicycle transmission.

1.8.1.1 Seven important lead-on leadership skills

(1) Decisiveness—The ability to make timely decisions is a valuable leadership trait that demonstrates your capacity to think objectively and weigh different options.

(2) Integrity—Integrity is the act of behaving honorably and following moral and ethical principles in all aspects of life.

(3) Creativity—Leaders with creativity promote a free exchange of new ideas and inspire innovation and collaboration in the workplace.

(4) Flexibility—A flexible leader can accept last-minute changes or new issues and be open to suggestions and feedback.

(5) Positive Attitude—Good leaders work to create a positive work environment, even during stressful periods, because employees are motivated to do their best when they are happy and feel valued.

(6) Communication—Good leaders can clearly explain important project issues because open communication between team members promotes a comfortable atmosphere and transparency.

(7) Relationship-Building—Leadership requires effective task delegation and conflict resolution to build and maintain a strong, collaborative team working toward the same goal.

Summary

Engineering is about changing the world by creating solutions to society's problems. What is common to the branches of engineering is their use of fundamental ideas involving **variables**, **numbers**, and **units of measurement** and the creative use of energy, materials, motion, and information. Engineering is a lead-on, hands-on, and minds-on profession. In the **Hands-On** section of this book, you will use good design practice to construct a "*device*" of your choosing or one to compete against similar devices built by other students. You will learn how to keep design records and how to protect your designs. You will use conceptual sketches and engineering drawings to advance your designs. And, of course, you will become a leader in the organizations of your future.

Exercises

1. Draw a conceptual sketch of your computer. Identify the keyboard, screen, power source, and information storage devices using arrows and labels.
2. Draw a conceptual sketch of a football and identify all the components (stitches, cover, bladder, inflation port, etc.) using arrows and numbers as in Fig. 1.1.
3. Draw a conceptual sketch of your ballpoint pen. Identify all the components with arrows and labels as in Fig. 1.2.
4. Check this exploded view of a table. Identify and label all the components.

5. Check this exploded view of a box with a hinged top. Identify and label all the components.

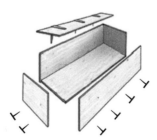

6. Search Wikipedia or the Internet for information on **three** of the professional engineering associations in the following table that are of interest to you and write a short paragraph on each about their purpose and history.

1	**American Association of Aeronautical and Astronautics (AIAA)**
2	**American Institute of Chemical Engineers** (AIChE)
3	**American Society of Civil engineers** (ASCE)
4	**American Institute of Electrical Engineers** (AIEE)
5	**American Society of Mechanical Engineers** (ASME)
6	**American Nuclear Society** (ANS)
7	**American Society of Naval Engineers** (ASNE)
8	**American Society of Testing and Materials** (ASTM)
9	**Biomedical Engineering Society** (BMSE)
10	**Institute of Electrical and Electronic Engineers** (IEEE)
11	**National Society of Black Engineers** (NSBE)
12	**Society of Automotive Engineers** (SAE)
13	**Society of Manufacturing Engineers** (SME)
14	**Institute of Transportation Engineers** (ITE)
15	**Society of Women Engineers** (SWE)

In problems 7 through 12 describe which leadership skills (there may be more than one) described in the Lead-On Leadership Skills section is missing.

7. At my first job, my manager was new and did things herself rather than empowering her employees. Initially, I liked her micromanagement since at that point I was new to the job. After finally working up the confidence to lighten the mood in a meeting I told a joke. I watched as my manager immediately made a painful overdramatic cringe in response. My stomach dropped, and so did my excitement for the job.

Missing Lead-On skills _____

8. I once had a boss who, in weekly team meetings, would treat one person like a hero and another person like a loser who was not doing anything right and was going to bring the team or the whole business down. You never knew who was going to be the victim, so we all dreaded the meeting. Not surprisingly, there was not much substance behind why an individual was being praised or condemned—it was just his "management" technique.

Missing Lead-On skills _____

9. I once had a boss who decided that, despite a very strong performance, they did not like someone on our team and wanted to see them gone. They did not have the ability to fire him, so they cut the employee out of emails and meetings, talked poorly about him behind his back, and disregarded any good performance from him. It made getting our team's work done way, way harder.

Missing Lead-On skills _____

10. I once had a boss who, while I was replying to a question addressed to me by her boss in a meeting, put her hand in front of my face to silence me so that she could answer instead.

Missing Lead-On skills _____

11. I once had a boss who nearly always multitasked in meetings by being on the phone while in the meeting. In group settings, she would shift her attention constantly from the speaker to her phone—back and forth for the entire time. At first, I just thought she was extremely busy, and it was the only way for her to get everything done—until 1 day during a meeting, I caught her playing solitaire on her phone.

Missing Lead-On skills _____

12. Although I had gotten approval to work from home 1 day a week, if I took my eyes off my computer screen (we used video conferencing to communicate throughout the day) for 3 min, my manager berated me. He told me I would lose my remote privileges unless I started letting him know when I was getting up to check on my baby, make a cup of coffee, or use the bathroom.

    Missing Lead-On skills _____

Engineering ethics

2.1 Introduction

Four days into her maiden voyage the luxury British steamship **RMS *Titanic*** struck an iceberg and sank on April 15, 1912 in less than 3 hours. At that time, a ship's lifeboat system only needed to ferry passengers to nearby rescue vessels, not to hold everyone on board. Titanic only had 20 lifeboats on board with space for about 1200 people. Because this was an inaugural voyage, there were more than 2200 passengers and crew on board. Unfortunately, the extra lifeboats had been removed because the ship owners were worried that it made the ship look unsafe. Consequently, more than 1500 people died.

Ethics is based on well-founded standards of right and wrong that prescribe what humans ought to do, usually in terms of rights, obligations, benefits to society,

Exploring Engineering. https://doi.org/10.1016/B978-0-443-13541-5.00035-0

fairness, or specific virtues. **Engineering Ethics** is the study of decisions, policies, and values that are morally desirable in engineering practice and research.

The word "**Ethics**" originates from the Greek word "*ethos*" meaning "character." Ethics are a set of rules or principles that are generally considered as standards of good and bad or right and wrong, which are usually imposed by an external group or a society or a profession. Ethics can be understood as the rules of conduct of a society or a particular group or culture often vary from one culture to another.

The word "**Morals**" originates from the Latin word "*mos*" meaning "custom." Morals are the societal customs that define what is right or wrong in one's conduct. Morals are what you have been taught or feel is good or bad. They can change because they depend on your ongoing experience with social issues in life.

Both morals and ethics have to do with differentiating between "**good** and **bad**" or "**right** and **wrong**." However, morality is usually governed by personal behavior whereas ethics are rules defined by a community.

An engineer deals with ethical and moral issues that may involve theft of confidential information, fraud, negligence, misrepresenting a product's performance, ignoring safety concerns, and so on. Sometimes these concerns lead to physical or economic disasters.

2.2 What are *personal* ethics

The word "ethics" can have several meanings. It can mean *personal ethics* (or morals) which involves your behavior relative to others or it can mean *professional ethics* where the limits of right and wrong behavior are defined by society or a profession.

Personal ethics are the standards of human behavior that individuals of different cultures have constructed to make moral judgments about personal or group situations. Ethical principles developed as people reflected on the intentions and consequences of their acts. Naturally, they vary over time and from culture to culture, resulting in conflict when what is acceptable in one culture is not in another. For example, the notion of privacy in the United States culture is very strong, and a desk in an office is considered an extension of that privacy, whereas in another culture, such as Japan, office space is open and one's desk is considered public domain.

Suppose you are a passenger in a car driven by a close friend. The friend is exceeding the speed limit and has an accident. There are no witnesses, and his lawyer tells you that if you testify that your friend was not exceeding the speed limit, it will save him from a jail sentence. What do you do?

Lying is often more acceptable in cultures that stress human relationships, but it is less accepted in cultures that stress laws. People in cultures that emphasize human relationships would most likely lie to protect the relationship, whereas people in cultures that put greater value on laws would lie less to obey the law.

How do you reconcile a belief in certain moral absolutes such as "I will not kill anyone" with the reality that, in some circumstances (e.g., war), it might be necessary to endanger or kill innocent people for the greater good? This issue gets particularly difficult since the prevailing morality that most of us would describe as "good" is to extend tolerance to others.

Finally, students today are faced with a new ethical crisis. Artificial intelligence (AI) and the Internet have evolved to the point where there are textbook solution manuals online and dozens of websites where you can pay someone to do your homework for you (e.g., Chegg, ChatGPT, Finish My Math, Course Hero, and so forth.). Should you use them? Getting someone to do your homework for you has traditionally been treated as cheating and can get you a failing grade or even expelled. On the other hand, getting someone to help you understand the course material without doing your homework for you is generally acceptable. This is an issue of your personal ethics, but clearly if someone else does your work (in this case homework) for you then you will never be successful in your career.

2.2.1 Five cornerstones of personal ethical behavior

Here are some examples of codes of personal ethics. At this point, you might want to compare your own personal code of ethics with the ones listed here.[1]

(1) Do what you say you will do.
(2) Never divulge information given to you in confidence.
(3) Accept responsibility for your actions (and mistakes).
(4) Never become involved in a lie.
(5) Never accept gifts that reward you for ignoring another persons' mistake.

2.2.2 Top ten questions you should ask yourself when making a personal ethical decision[2]

10. Could the decision become habit forming? – *If so, don't do it.*
9. Is it legal?—*If it isn't, don't do it.*
8. Is it safe?—*If it isn't, don't do it.*
7. Is it the right thing to do?—*If it isn't, don't do it.*
6. Will this stand the test of public scrutiny?—*If it won't, don't do it.*
5. If something terrible happened, could I defend my actions?—*If you can't, don't do it.*
4. Is it moral, balanced, and fair?—*If it isn't, don't do it.*
3. How will it make me feel about myself?—*If it feels lousy, don't do it.*

---

[1] Manske, F.A., Jr., *Secrets of Effective Leadership*, (Columbia, SC: Leadership Education and Development, Inc., 1999).
[2] From: http://www.cs.bgsu.edu/maner/heuristics/1990Taylor.htm

2. Does this choice lead to the greatest good for the greatest number?—*If it doesn't, don't do it.*

And the number one question you should ask yourself when making an ethical decision.

1. Would I do this in front of my mother?—*If you wouldn't, don't do it.*

2.3 What are *professional* ethics?

A *professional code of ethics* has the goal of ensuring that a profession serves the legitimate goals of *all* its constituencies: self, employer, profession, and public. The code protects the members of the profession from undesired consequences of competition (for example, the pressure to cut corners to save money) while leaving the members of the profession free to benefit from the desirable consequences of competition (such as invention and innovation).

Having a code of ethics enables an engineer to resist administrative pressure to approve inferior work by saying, *"As a professional, I cannot ethically put business concerns ahead of professional ethics."* It also enables the engineer to similarly resist pressure to allow concerns such as personal desires, greed, ideology, religion, or politics to override professional ethics.

2.3.1 National Society of Professional Engineers code of ethics for engineers

Engineering is an important and learned profession. As members of this profession, engineers are expected to exhibit the highest standards of honesty and integrity. Engineering has a direct and vital impact on the quality of life for all people. Accordingly, the services provided by engineers require honesty, impartiality, fairness, and equity and must be dedicated to the protection of public health, safety, and welfare. Engineers must perform under a standard of professional behavior that requires adherence to the highest principles of ethical conduct.[3]

2.3.2 Fundamental canons[4]

Engineers, in the fulfillment of their professional duties, shall

- Hold paramount the safety, health, and welfare of the public.
- Perform services only in areas of their competence.
- Issue public statements only in an objective and truthful manner.

---

[3] See http://www.nspe.org/ethics/eh1-code.asp
[4] Canons were originally church laws; the word has come to mean rules of acceptable behavior for specific groups.

- Act for each employer or client as faithful agents or trustees.
- Avoid deceptive acts.
- Conduct themselves honorably, responsibly, ethically, and lawfully so as to enhance the honor, reputation, and usefulness of the profession.

Example 2.1: An ethical dilemma

The following scenario is a common situation faced by engineering students. Read it and think about how you would respond. What are your ethical responsibilities?

You and your roommate are both enrolled in the same engineering class. Your roommate spent the weekend partying and did not do the homework that is due on Monday. You did your homework, and your roommate asked to see it. You are afraid he/ she will just copy it and turn it in as his/her own work. What are you ethically obligated to do?

(a) Show your roommate the homework?
(b) Show the homework but ask your roommate not to copy it?
(c) Show the homework and tell the roommate that if the homework is copied, you will tell the professor?
(d) Refuse to show the homework?
(e) Refuse to show the homework but offer to spend time helping the roommate?

Solution For the purposes of this course, the answer to an ethics question consists of appropriately *applying a code of ethics.*[5] In this example, the Five Cornerstones of Ethical Behavior will be used since they are familiar to you in one form or another. Let us see which of the Five Cornerstones apply here.

1. **Do what you say you will do.** If the teacher has made it clear that this is an individual assignment, then by participating in the assignment, you have implicitly agreed to keep your individual effort private. Allowing one's homework to be copied means going back on this implicit promise. This implies that answer **d** or **e**, "Refuse to show the homework" is at least part of the right answer.
2. **Never divulge information given to you in confidence.** Again, homework is implicitly confidential communication between an individual student and a teacher. By solving the problem, you have created confidential communication with the teacher. This is more support for choice **d** or **e**.
3. **Accept responsibility for your actions.** Sharing your homework enables your roommate to evade this standard. Being an accomplice in the violation of standards by others is itself an ethical violation. This is further support for choice **d** or **e**.
4. **Never become involved in a lie.** Allowing your homework to be copied is participating in a lie: that the work the roommate turns in is his or her own work. This further supports choice **d** or **e**.
5. **Never accept gifts that compromise your ability to perform in the best interests of your organization.** Since the roommate has not offered anything in exchange for help, this standard appears not to apply in this case.

Four of the five cornerstones endorse choice **d** or **e**, refuse to show the homework, while the fifth cornerstone is silent. These results indicate that your ethical obligation under this particular code of personal ethics is to refuse to show your homework.

---

[5] Also be aware that each college or university has its own code of ethics or honor code that would apply here.

Many people find the Five Cornerstones to be incomplete because they lack a canon common to most of the world's ethical codes: the **Golden Rule**.[6] Including the Golden Rule would create the additional obligation to show some empathy for your roommate's plight, just as you would hope to receive such empathy if you were in a similar situation. This suggests the appropriateness of choice **(e)**, offering to help the roommate understand how to do the homework. In much the same way, in subsequent exercises, you may feel the need to supplement the Code of Ethics for Engineers with elements from your own personal code of ethics. However, this must not take the form of *replacing* an element in the Code of Ethics for Engineers with a personal preference.

In subsequent chapters, the National Society of Professional Engineers (NSPE) Code of Ethics for Engineers will be used as a reminder that you must accept that code in your professional dealings if you want to be a professional engineer.

2.4 Engineering ethics decision matrix

To help you deal with some of the ethical issues in your engineering career, we provide a very useful tool: The **Engineering Ethics Decision Matrix**. This tool presents a simple way of applying the canons of engineering ethics and to evaluate the spectrum of responses that might apply in any given situation. It should give you pause **not** to accept the first simple **do** or **do not** response that comes to mind.

In Table 2.1, the rows of the matrix are the canons of engineering ethics (the NSPE set) and the columns are possible ways to resolve the problem. Each box of the matrix must be filled with a very brief answer to the question: "Does this particular solution meet this particular canon?" Like other engineering tools, the Ethical Decision Matrix is a way to divide and conquer a problem rather than trying to address all its dimensions simultaneously.

---

[6] There are many versions of the Golden Rule in the world's major religions. Here's one attributed to Confucius: "Do not do to others what you would not like yourself."

Table 2.1 The engineering ethics decision matrix.

Options → NSPE Canons ↓	Go along with the decision	Appeal to higher management	Quit your job	Write your state representative	Call a newspaper reporter
Hold paramount the safety, health, and welfare of the public					
Perform services only in the area of your competence					
Issue public statements only in an objective and truthful manner					
Act for each employer or client as faithful agents or trustees					
Avoid deceptive acts					
Conduct themselves honorably					

Example 2.2: A budget crisis

You are a civil engineer on a team designing a bridge for a state government. Your team submits what you believe to be the best design by all criteria, at a cost that is within the limits originally set. However, some months later the state undergoes a budget crisis and cuts your funds. Your supervisor, also a qualified civil engineer, makes design changes to achieve cost reduction that he/she believes will not compromise the safety of the bridge. However, you cannot conclusively demonstrate that the new design does not contain a safety hazard. You request that a new safety analysis be carried out. Your supervisor denies your request on the grounds of time and limited budget. What do you do?

Solution Table 2.2 shows a typical set of student responses. How would *you* fill out this table?

Table 2.2 Student responses to the ethical scenario.

Options → Canons ↓	Go along with the decision	Appeal to higher management	Quit your job	Write your state representative	Call a newspaper reporter
Hold paramount the safety, health, and welfare of the public.	**No.** Total assent may put public at risk.	**Maybe.** Addresses risk, but boss may bury issue.	**No.** If you just quit, the risk is less likely to be addressed.	**Yes.** Potential safety risk will be put before public.	**Yes.** Potential safety risk will be put before public.
Perform services only in the area of your competence.	**Yes.** You are not a safety expert.	**Yes.** Though not a safety expert, you are competent to surface an issue.	**Maybe.**	**No.** You are not an expert in government relations.	**No.** You are not an expert in press relations.
Issue public statements only in an objective and truthful manner	**No.** Silence may seem untruthful agreement.	**Maybe.** You are publicly silent but have registered dissent.	**No.** Quitting to avoid the issue is being untruthful.	**Maybe.** Your personal involvement may hurt your objectivity.	**No.** The press is likely to sensationalize what is as yet only a potential safety issue.
Act for each employer or client as faithful agents or trustees	**Yes.** As an agent, you are expected to follow orders.	**Yes.** As an agent, you are expected to alert management to potential problems.	**Maybe.** Quitting a job is not bad faith.	**No.** As an agent or trustee, you may not make internal matters public without higher approval.	**No.** As an agent or trustee, you may not make internal matters public without higher approval.
Avoid deceptive acts	**No.** Assent to something you disagree with is deceptive.	**Yes.** You honestly reveal your disagreement.	**No.** Quitting to avoid responsibility is deceptive.	**Yes.** You honestly reveal your disagreement.	**Yes.** You honestly reveal your disagreement.
Conduct themselves honorably …	**No.** Deceptive assent dishonors the profession.	**Yes.** Honorable dissent is in accord with obligations.	**Maybe.**	**Yes.** Honorable dissent is in accord with obligations.	**Maybe.** Might be publicity seeking, not honorable dissent.
Totals (neglecting "maybe")	Yes = 2 No = 4	Yes = 4 No = 0	Yes = 0 No = 3	Yes = 3 No = 2	Yes = 2 No = 3

*Notice the multidimensional character of these answers. Here's one way to make some sense of your answer. Total the **yes's** and **no's** in each column (ignore **"maybe"**). By this analysis (4-Yes and 0-No), you should **appeal to higher management**, which of course might still ignore you. But that is the first action you should consider even though your boss may strenuously disagree with you.*

You now have a powerful ally in the **Engineering Ethics Decision Matrix** to persuade others to your point of view. Some Engineering Ethics Decision Matrix analysis may have just one overwhelming criterion that negates all other ethical responses on your part; if so, you must follow that path. Sometimes the Engineering Ethics Decision Matrix may contain multiple conflicting factors. All you should expect from the matrix is that it will stimulate most or all the relevant items you should consider and help you avoid immediately accepting the first thought that enters your head.

2.4.1 The ethics challenge featuring Dilbert and Dogbert

In 1997 the Lockheed Martin Corporation developed **The Ethics Challenge** game based on the Dilbert comic strip. It was designed to help their employees understand the ethics of the company in an entertaining way. Players were divided into teams, and a simple ethical situation is presented along with four possible responses. Their responses earned the team points that moved their Dilbert characters around the board (Fig. 2.1).

FIGURE 2.1

Lockheed Martin ethics challenge game.

Today Lockheed Martin sponsors a national ethics in engineering competition for undergraduate engineering students. For more information, search for **Lockheed Martin Ethics in Engineering Competition** on the internet.

Summary

As a student you should explore your personal ethics to better understand your decision-making opinions in everyday life. The five cornerstones of personal ethics and the 10 questions you should ask yourself in making an ethically correct decision discussed in this chapter is a good place to start. Always remember that engineering is a profession and, in your work, (even as a student) you must strive to meet our professional ethics standards.

Exercises

1. Repeat Example 2.1 using the NSPE Code of Ethics for Engineers. Solve using the Engineering Ethics Decision Matrix.
2. Repeat Example 2.2 using the Five Cornerstones of Ethical Behavior. Solve using the Engineering Ethics Decision Matrix.
3. The Lockheed Martin Ethics Challenge game (see Fig. 2.1) example #1: You overhear your supervisor ask his secretary to find out from a female interviewing for a job at your company if she is married and if she has any children. What should you do:
   a. Ignore the comment because your supervisor is probably just interested in learning about all prospective employees.
   b. Tell the secretary she should do what the boss asks.
   c. Tell your supervisor that it was an inappropriate request because it may violate federal and state laws.
   d. Tell your supervisor his request is OK if he shares the information with everyone else.
4. The Lockheed Martin Ethics Challenge game (see Fig. 2.1) example #2: A lazy coworker in your department is married to the boss's daughter. You've noticed that he has a new car and is bragging about a recent salary increase. What should you do?
   a. Ignore it and just mind your own business for a change.
   b. Contact the Human Relations Depart and ask them to investigate your boss.
   c. Go to your boss's boss and complain.
   d. Ask your boss if he has another daughter you can marry.
5. It is the last semester of your senior year, and you are anxious to get an exciting electrical engineering position in a major company. You accept a position from Company A early in the recruiting process but continue to interview hoping for a better offer. Then, your dream job offer comes along from Company B. More salary, better company, more options for advancement, it is just what you have been looking for. What should you do about the offer you have already accepted from Company A?
   a. Just don't show up for work at Company A.
   b. Send a letter to Company A retracting your job acceptance with them.
   c. Ask Company B to contact Company A and tell them you won't be working for them.
   d. Reject the offer from Company B and go to work for Company A as agreed.
6. A company purchased an expensive computer program for your summer job with them. The license agreement states that you can make a backup copy, but you can use the program on only one computer at a time and your company is currently using that copy. Your senior design course professor would like you to use the program for your senior design project. What should you do?
   a. Give the program to your professor and let him/her worry about the consequences.
   b. Copy the program and use it because no one will know.
   c. Ask your supervisor at the company that purchased the program if you can use it at school on your senior project.

  d. Ask your professor to contact the company and ask for permission to use the program at school.
7. You are attending a regional student XYZ conference along with five other students from your institution. The night before the group is scheduled to return to campus, one of the students is arrested for public intoxication and is jailed. Neither he nor the other students have enough cash for bail, and he doesn't want his parents to know. He asks you to lend him money from the XYZ's emergency cash fund so that he doesn't have to spend the night in jail; he'll repay you as soon as his parents send the money. What should you do?
  a. Lend him the money, since his parents are wealthy, and you know he can repay it quickly.
  b. Tell him to contact his parents now and ask them for help.
  c. Give him the money but ask him to write and sign a note to repay it.
  d. Tell him to call a lawyer since it's not your problem.
8. You are testing motorcycle helmets manufactured by a variety of your competitors. Your company has developed an inexpensive helmet with a liner that will withstand multiple impacts but is much less effective on the initial impact than your competitor's. The vice president for sales is anxious to get this new helmet on the market and is threatening to fire you if you do not release it to the manufacturing division. What should you do?
  a. Follow the vice president's orders since he/she will ultimately be responsible for the decision.
  b. Call a newspaper to "blow the whistle" on the unsafe company policies.
  c. Refuse to release the product because it is unsafe and take your chances on being fired.
  d. Stall the vice president while you look for a job at a different company.
9. Paul Ledbetter is employed at Bluestone Ltd. as a manufacturing engineer. He regularly meets with vendors who offer to supply Bluestone with needed services and parts. Paul discovers that one of the vendors, Duncan Mackey, like Paul, is an avid golfer. They begin by comparing notes about their favorite golf courses. Paul says he's always wanted to play at the Cherry Orchard Country Club, but since it is a private club, he's never had the opportunity. Duncan says he's been a member there for several years and that he's sure he can arrange a guest visit for Paul. What should Paul do?[7]
  a. Paul should accept the invitation since he has always wanted to play there.
  b. Paul should reject the invitation since it might adversely affect his business relationship with Duncan.
  c. Paul should ask Duncan to nominate him for membership in the club.
  d. Paul should ask his supervisor if it's OK to accept Duncan's invitation.
10. Some American companies have refused to promote women into positions of high authority in their international operations in Asia, the Middle East, and South America. Their rationale is that business will be hurt because some foreign customers do not wish to deal with women. It might be contended that this practice is justified out of respect for the customs of countries that discourage women from entering business and the professions.

Some people feel that such practices are wrong, and that gender should not be used in formulating job qualifications. Further, they believe that customer preferences should not justify gender discrimination. Present and defend your views to your class or course instructor on whether or not this discrimination is justified.
11. Marvin Johnson is an environmental engineer at a local manufacturing plant whose waste water discharges into a lake in a tourist area. Included in Marvin's responsibilities is the monitoring of water and air discharges at his plant and periodically preparing reports to be submitted to the Department of Natural Resources.

Marvin just prepared a report that indicates that the level of pollution in the plant's water discharge slightly exceeds the legal limitations. However, there is little reason to believe that this excessive amount poses any danger to people in the area. At worst, it will endanger a small number of fish. On the other hand, solving this problem will cost the plant more than $200,000.

Marvin's supervisor says the excess pollution should be regarded as a mere "technicality" and he asks Marvin to "adjust" the data so that the plant appears to be in compliance. He explains: "We can't afford the $200,000. It would set us behind our competitors. Besides the bad publicity we'd get, it might scare off some of the tourist industry." How do you think Marvin should respond to Edgar's request?

12. Derek Evans used to work for **Company A**, which is a small computer firm that specializes in developing software for management tasks. Derek was a primary contributor in designing an innovative software system for customer services. This software system is essentially the "lifeblood" of the firm. The small computer firm never asked Derek to sign an agreement that software designed during his employment there becomes the property of the company.

    Derek is now working for **Company B**, which is a much larger computer firm. Derek's job is in the customer service area, and he spends most of his time on the telephone talking with customers having systems problems. This requires him to cross-reference large amounts of information. It now occurs to him that by making a few minor alterations in the innovative software system he helped design at the small computer firm, **Company A**, the task of cross-referencing can be greatly simplified.

    On Friday, Derek decides he will make the adaptation early Monday morning. However, on Saturday evening, he attends a party with two of his old friends from **Company A**, you and Horace Jones. Since it has been some time since you have seen each other, you spend some time discussing what you have been doing recently. Derek mentions his plan to adapt the software system on Monday. Horace asks, "Isn't that unethical? That system is really the property of **Company A**." "But" Derek replies, "I'm just trying to make my work more efficient. I'm not selling the system to anyone, or anything like that. It's just for my own use—and, after all, I did help design it. Besides, it's not exactly the same system—I've made a few changes." What should Derek do?[8]

13. Janet, a professional engineer on unpaid leave, is a graduate student at a small private university and is enrolled in a research class for credit taught by Professor Dimanro, at the university. Part of the research being performed by Janet involves the use of innovative geothermal technology.

    The university is in the process of enlarging its facilities, and Professor Dimanro, a member of the university's building committee, has responsibility for developing a request to solicit interested engineering firms. Professor Dimanro plans to incorporate an application of Janet's innovative geothermal technology into the new facility. Professor Dimanro asks Janet to serve as a paid consultant to the university's building committee in reviewing proposals. Janet's employer will not be submitting a proposal and is not averse to having Janet work on the proposal reviews. Janet agrees to serve as a paid consultant.

    Is it a conflict of interest for Janet to be enrolled in a class for credit at the university and at the same time serve as a consultant to the university?[9]

14. During a construction project, a young engineer working on the project found a structure that he/she felt was unsafe. He/she informed a superior who said to stay calm and not to discuss it with anyone until next year's budget was passed to get the funds to repair the unsafe structure. What should the engineer do?

15. What should an engineer do who observes a colleague copying company confidential information? If she/he stops the colleague without reporting it, the colleague could repeat this behavior in the future. If she/he chooses to report it to management, the colleague could lose their job. What is the morally correct action?

16. An engineer developed a successful working prototype for an important engineering project that was to be presented to upper management. But just before the presentation occurred the engineer discovered that the prototype stopped working. Is it morally correct for the engineer to just claim the prototype operated correctly and that upper management should invest a substantial sum of money to move on with the project? What should the engineer do?

17. You are browsing the Internet and find some units conversion software that may be useful in this course. You would like to download the software at school and use it in this course. What do you do? Use the Fundamental NSPE Canons and fill in an Engineering Ethics Matrix.
    a. Check with the Internet site to make sure this software is freeware for your use in this course.
    b. Just download the software and use it because no one will know.
    c. Download the software at home and bring it to school.
    d. Never use software found on the Internet.
18. On December 11, 1998, the Mars *Climate Orbiter* was launched on a 760-million-mile journey to the Red Planet. On September 23, 1999, a final rocket firing was to put the spacecraft into orbit, but it disappeared. An investigation board concluded that NASA engineers failed to convert the rocket's thrust from pounds force to newtons (the unit used in the guidance software), causing the spacecraft to miss its intended 140−150 km altitude above Mars during orbit insertion, instead entering the Martian atmosphere at about 57 km. The spacecraft was then destroyed by atmospheric stresses and friction at this low altitude. As chief NASA engineer on this mission, how do you react to the national outcry for such a foolish mistake? Use the Fundamental Canons and fill in an Engineering Ethics Matrix.
    a. Take all the blame yourself and resign.
    b. Find the person responsible, and fire, demote, or penalize that person.
    c. Make sure it doesn't happen again by conducting a software audit for specification compliance on all data transferred between working groups.
    d. Verify the consistent use of units throughout the spacecraft design and operations.
19. You email a classmate in this course for some information about a spreadsheet homework problem. In addition to answering your question, your classmate also attaches a spreadsheet solution to the homework. What do you do?
    a. Delete the spreadsheet without looking at it.
    b. Look at the spreadsheet to make sure your classmate did it correctly.
    c. Copy the spreadsheet into your homework and change the formatting so that it doesn't look like the original.
    d. d, email the spreadsheet to all your friends so that they can have the solution, too.
20. Stephanie knew Adam, the environmental manager, would not be pleased with her report on the chemical spill. The data clearly indicated that the spill was large enough that regulations required it to be reported to the state. When Stephanie presented her report to Adam, he lost his temper. "A few gallons over the limit isn't worth the time it's going to take to fill out those damned forms. Go back to your desk and rework those numbers until it comes out right." What should Stephanie do?[10]
    a. Tell Adam that she will not knowingly violate state law and threaten to quit.
    b. Comply with Adam's request since he is in charge and will suffer any consequences.
    c. Send an anonymous report to the state documenting the violation.
    d. Go over Adam's head and speak to his supervisor about the problem.

---

[7] Extracted from *Teaching Engineering Ethics, A Case Study Approach*, Michael S. Pritchard, editor, Center for the Study of Ethics in Society Western Michigan University: http://ethics.tamu.edu/pritchar/golfing.htm

[8] Adapted from: http://ethics.tamu.edu/pritchar/property.htm

[9] Adapted from NSPE Board of Ethical Review Case No. 91-5.

[10] Abstracted from *Engineering Ethics: Concepts and Cases* at http://wadsworth.com/philosophy_d/templates/student_resources/0534605796_harris/cases/Cases.htm

Elements of engineering analysis

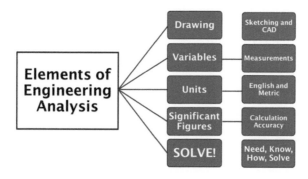

3.1 Introduction

How should an engineering problem be described so it can be analyzed and solved? Often there is an initial sketch that shows how things operate. Sometimes these sketches become drawings, drawings that obey a number of graphical rules that make spatial ideas clear. However, many engineering problems can be solved with mathematical analysis alone, and they may or may not be accompanied by any sketches or drawings. All engineering analysis methods share some key elements. One of them, a **numerical value**, is familiar to you. Answering a numerical question requires coming up with the right number. However, in engineering that is only one part of answering a question.

This chapter introduces other core elements of engineering analysis: **engineering drawing and sketching, engineering variables, dimensions, engineering units of measurement, significant figures**, and **computer analysis** as well as a fail-safe method of dealing with units and dimensions.

We also use a self-prompting memory aid called the **Need−Know−How−Solve** method to set up and solve engineering problems. It accomplishes two things: (1) it

Exploring Engineering. https://doi.org/10.1016/B978-0-443-13541-5.00013-1

leaves an "audit trail" of how you solved the problem and (2) it provides a methodology that sorts out and simplifies the problem in a formal way.

The result of an engineering analysis must involve the appropriate variables; it must be expressed in the appropriate units of measurement; it must express the numerical value with the appropriate number of digits (i.e., significant figures); and it must be accompanied by an explicit method so that others can understand and evaluate your analysis.

In addition to the preceding concepts, modern engineers have computerized tools at their fingertips. Because these tools can significantly enhance an engineer's productivity, it is necessary for the beginning engineer to learn them as soon as possible in his or her career. Today, all engineering reports and presentations are prepared on a computer. However, there is another computer tool that all engineers use: **spreadsheets.** This tool is another computer program that the engineer must master. It is used in this chapter so you can get experience in the use of this tool. We also use spreadsheet techniques throughout this book in the relevant exercise sections.

3.2 Engineering drawing and sketching

What are "engineering drawings" and why do we engineers need them? Engineering drawings are used to *communicate* design ideas and technical information to engineers and other professionals throughout the design process. An engineering drawing represents a complex three-dimensional object on a two-dimensional piece of paper or computer screen by a process called *projection*. The most common types of engineering drawing projections are shown in Table 3.1.

Table 3.1 Types of drawing projections.

Drawing type	Example
Isometric—The isometric projection is the basis for a three-dimensional engineering sketch. The three axes of the isometric drawing form 120-degree angles with each other. Isometric grids are a convenient aid in sketching isometric drawings with both straight edges and circular features.	

Table 3.1 Types of drawing projections.—*cont'd*

Drawing type	Example
Axonometric—The axonometric drawing is a pictorial representation in which both the axes' directions and scales can vary depending upon the application. The angular direction of each axis can also vary, so an axonometric sketch is more complex than an isometric sketch.	
Oblique—The three axes of an oblique drawing are drawn horizontal, vertical, and at a receding angle that can vary from 30 to 60 degrees. The main advantage of an oblique drawing is that circles parallel to the front plane of the projection are drawn true size and shape.	
Perspective—A perspective drawing adds realism to the representation because objects appear the way the human eye would see them. In perspective drawings, parallel lines converge to a single point (called the vanishing point) at the horizon.	
Orthographic—Orthographic drawings are commonly used in engineering because they contain multiview images. In an orthographic drawing, the object appears to be placed inside a "glass box" with each face of the object projected onto its side of the box. Only three views (front, top, and right side) are needed to convey the geometric data of the object.	

3.2.1 Drawing scale and dimensioning

Quite often you will not be able to create a drawing of an object at its actual size. The "scale" of a drawing is the ratio of the dimensional size on the drawing to the actual size of the object. So if a drawing has a scale of 1:2, the drawing is ½ the size of the actual object.

When you add dimensions to a drawing, put in only as many dimensions as necessary for a person to understand or manufacture the object. Repeatedly measuring from one point to another on the object will lead to confusion and inaccuracies. It is better to measure from one side of the object to various relevant points (center of a hole, another side, etc.). Try to place the dimensions in a way that will help a machinist create the object as in Fig. 3.1.

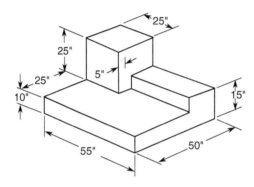

FIGURE 3.1

A properly dimensioned isometric engineering drawing.

3.2.2 Computer-aided design software[1]

Today, engineering drawings are produced on computers with computer-aided design (CAD) software such as SolidWorks, AutoCAD, Pro/ENGINEER, and SketchUp[2]. Every engineering college or school includes training in one or more of these sophisticated computer programs. Not only can you accurately represent a three-dimensional image of an object, but you can also carry out complex engineering stress-strain, fluid flow, and heat transfer analysis for various loading conditions on the object (Fig. 3.2). Also, the resulting computer files could then be downloaded to a **rapid prototyping machine** that will manufacture the object to scale.

---

[1] A free trial for AutoCAD can be found at https://www.autodesk.com/products/autocad/free-trial
[2] SketchUp is a basic 3-D CAD program that is available for free at http://www.sketchup.com

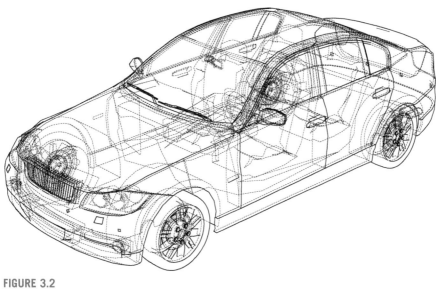

FIGURE 3.2

A 3-D CAD file.

3.2.3 Engineering Sketches

An engineering sketch is a free-hand drawing used to get ideas on paper quickly. A sketch does not need to be made to an exact scale, but it should be roughly proportional to the actual object (see Fig. 3.3).

How you prepare a sketch depends on who else will be viewing the sketch. A pictorial sketch works best when communicating design concepts with other engineers. When communicating technical details to a technician, a multiview (orthographic) sketch would be the appropriate choice.

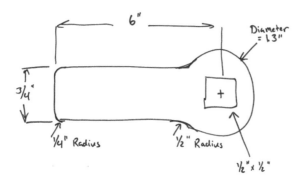

FIGURE 3.3

A typical engineering sketch.

Just because a sketch is a "free hand" drawing, it should still be relatively neat. However, sketching does not require any real "artistic" skill. Lines drawn quickly and confidently look better. The skills you will need are: (1) sketching straight lines and (2) sketching arcs.

Straight lines are easy to sketch if you keep your eye on where you want to draw to, not on the end of the pencil. Sketch horizontal lines away from your body (otherwise you tend to have an up-curve at their end). Sketch vertical lines downward, otherwise they will tend to curve in the direction of the hand holding the pencil. For inclined lines, turn the paper until it is either horizontal or vertical.

Lightly mark points at each end of the intended line. Move your pencil between the points a couple of times, with the pencil point not touching the paper, to "teach" yourself to follow the route. Draw several light guidelines between the points and then go over the "best line" more firmly.

With practice, sketching arcs and circles is relatively easy. To sketch a complete circle, start by sketching a square box that will enclose the circle. Mark the centers of the sides of the square and then sketch arcs that are tangent to the sides of the square. For larger diameters, you can add diagonals to the square by marking the circle's radius on the diagonals to get additional points to join with a smooth curve. Do not try to draw the circle in one motion; instead, draw it as a series of connected arcs. To draw an ellipse or oval, sketch it inside a rectangular box. For more complex curves, join dots along the curve.

Lastly, even though a sketch is defined as a free-hand drawing, here are some common "aids" to help you sketch:

- Graph paper—helps to maintain proportion and keep straight lines
- Scraps of paper—good for transferring distances
- Coins—make good circle templates
- Credit cards—make nice straight edges

3.3 Engineering variables

Engineers typically seek answers to such questions as "How hot will this get?" "How heavy will it be?" and "What's the voltage?" Each of these questions involves a variable. The question "How hot?" is answered with the variable *temperature*, "How heavy?" is answered with the variable *weight*, and "What voltage?" is answered with the variable *electric potential*.

For our purposes, variables almost always are defined in terms of measurements made with familiar instruments, such as thermometers, rulers, and clocks. Speed, for example, is defined as a ruler measurement of distance, divided by a clock measurement of time.

What makes an engineering calculation different from a purely mathematical calculation? For example, 7.35 is a legitimate answer to the mathematical question

"What is the value of 2.10 times 3.50?" However, expressing something in numbers is only the beginning of an engineering calculation. In addition to variables based on measurements and expressed as numbers, we require a second key element of engineering analysis: **units of measurement**.

3.4 Engineering units of measurement

What if you are stopped by the highway patrol on a Canadian highway and get a ticket saying you were driving at "100"? You would probably guess that the variable involved is speed. However, it would also be of interest to know if the claim was that you were traveling at 100 *miles* per hour or 100 kilometers per hour, knowing that 100 kph is only 62 mph. Units do make a difference!

In engineering a calculated quantity always has two parts: the numerical value *and* its associated units, if any.[3] Therefore, the result of any engineering calculation must always be correct in two separate categories: It must have the **correct numerical value**, and it must also have the **correct units.**

Units are a way of quantifying the underlying concept of dimensions. *Dimensions* are the fundamental quantities we perceive, such as mass, length, and time. Units provide us with a numerical scale whereby we can carry out a measurement of a quantity. They units are established quite arbitrarily and are codified by civil law or cultural custom. How the dimension of length ends up being measured in units of feet or meters has nothing to do with any physical law. It is solely dependent on the creativity and ingenuity of people. Therefore, the basic tenets of units systems are often grounded in the complex roots of past civilizations and cultures.

3.4.1 SI unit system

The International System of Units (in French, Le **Système International** d'Unités, which is where the "**SI**" comes from) is the metric standard of units and is based on **MKS** (meter, kilogram, second) units. The fundamental units in the SI system are as follows:

■ The meter (m), the fundamental unit of length.
■ The second (s), the fundamental unit of time.
■ The kilogram (kg), the fundamental unit of mass.
■ The degree kelvin (K), the fundamental unit of temperature.
■ The mole (mol), the fundamental unit of quantity of particles.
■ The ampere (A), the fundamental unit of electric current.

---

[3] Some engineering quantities legitimately have no associated units; for instance, a ratio of like quantities

3.4.2 Unit names and abbreviations

Table 3.2 lists several SI units that were named after the scientists who made discoveries in the fields in which these units are used.

Table 3.2 Some SI units and their abbreviations. Note that these unit names do not begin with a capital letter, but their abbreviation is capitalized (e.g., newton, N)

Famous SI Units and Their Abbreviation		
ampere (A)	joule (J)	siemens (S)
celsius (°C)	kelvin (K)	tesla (T)
coulomb (C)	newton (N)	volt (V)
farad (F)	ohm (Ω)	watt (W)
hertz (Hz)	pascal (Pa)	weber (Wb)

Here are the rules for writing units:

■ *All* unit names are written *without capitalization* (unless they appear at the beginning of a sentence), regardless of whether they were derived from proper names.

■ When the unit is to be abbreviated, the *abbreviation is capitalized if the unit was derived from a person's name*.

■ Unit abbreviations use two letters *only* when necessary to prevent them from being confused with other established unit abbreviations.[4] For example, the unit for the magnetic field is the *weber*, and we abbreviate it as Wb to distinguish it from the unit for power, the *watt*, which is abbreviated as W. Also, two letters are used to express prefixes such as kW for kilowatt.

■ A unit abbreviation is never pluralized, whereas the unit's name may be pluralized. For example, "kilograms" is abbreviated as kg, and *not* kgs, "newtons" as N and *not* as Ns, and the correct abbreviation of seconds is s, *not* sec. or secs.

■ Unit name abbreviations are never written with a terminal period unless they appear at the end of a sentence.

■ All other units whose names were not derived from the names of historically important people are both written and abbreviated with lowercase letters; for example, meter (m), kilogram (kg), and second (s).

---

[4] Non-SI unit systems do not generally follow this simple rule. For example, the English length unit, foot, could be abbreviated f rather than ft. However, ft is well established within society, and changing it at this time would only cause confusion

In the examples that follow in this text, we introduce a *fail-safe* method that will always allow you to develop the correct units in your calculations. A numerical value used in a calculation will always be followed with square brackets [...] that enclose the value's units. For example, if something has a length of 3 inches, it will be written in a calculation as **3 [in]**. In the case of unit conversion factors such as 12 inches = 1 foot we will write the conversion factor as **12 [in/ft]**, so if we wanted to convert 10 feet into inches, we will write (10 [ft]) × (12 [in/ft]) = 120 inches. **Note that we do not put brackets on the units of the final answer.**

Although this may seem ponderous in this example, in examples that are more complicated *it is essential* to follow this methodology. We will attempt to be consistent in presenting solutions and follow this technique throughout this text.

Example 3.1

The height of horses is still measured in the old unit of "hands." How many feet high is a horse that is 13 hands tall?

The conversion of old units to more familiar ones is of course known:

Unit	Equivalent	[Conversion]
2 hands	1 span	½ [span/hand]
2 spans	1 cubit	½ [cubit/span]
2 cubits	3 feet.	$^3/_2$ [feet/cubit]

The height of the horse = 13 [hands] × ½ [span/hand] × ½ [cubit/span] × $^3/_2$ [foot/cubit]

= 13 × ½ × ½ × $^3/_2$ [feet] = 4.9 feet

There are many SI units pertaining to different quantities being measured. Table 3.3 has value beyond merely listing these units: it relates the unit's name to the fundamental MKS units, that is, the fact that a frequency is expressed in hertz may not be as useful as the fact that a hertz is just the name of an inverse second, s^{-1}. In Table 3.4, multiples of these quantities are arranged in factors of 1000 for convenience for very large and very small multiples.[5]

---

[5] In recent years, a new subcategory of technology known as *nanotechnology* has arisen; it is so-called because it deals with materials whose characteristic size is in the nanometer or 10^{-9} m range.

Table 3.3 Some derived SI units. Note that these unit names do not begin with a capital letter, but their abbreviation is capitalized (e.g., newton, N)

Quantity	Name	Symbol	Formula	Fundamental units
Frequency	hertz	Hz	1/s	s^{-1}
Force	newton	N	$kg \cdot m/s^2$	$m \cdot kg \cdot s^{-2}$
Energy	joule	J	$N \cdot m$	$m^2 \cdot kg \cdot s^{-2}$
Power	watt	W	J/s	$m^2 \cdot kg \cdot s^{-3}$
Electric charge	coulomb	C	$A \cdot s$	$A \cdot s$
Electric potential	volt	V	W/A	$m^2 \cdot kg \cdot s^{-3} \cdot A^{-1}$
Electric resistance	ohm	Ω	V/A	$m^2 \cdot kg \cdot s^{-3} \cdot A^{-2}$
Electric capacitance	farad	F	C/V	$m^{-2} \cdot kg^{-1} \cdot s^4 \cdot A^2$

Table 3.4 SI unit prefixes.

Multiples	Prefixes	Symbols	Submultiples	Prefixes	Symbols
10^{18}	exa	E	10^{-1}	deci	d
10^{15}	peta	P	10^{-2}	centi	c
10^{12}	tera	T	10^{-3}	milli	m
10^{9}	giga	G	10^{-6}	micro	μ
10^{6}	mega	M	10^{-9}	nano	n
10^{3}	kilo	k	10^{-12}	pico	p
10^{2}	hecto	h	10^{-15}	femto	f
10^{1}	deka	da	10^{-18}	atto	a

3.5 Significant figures

Having defined your variable and specified its units, you are now ready to calculate its value. Your calculator will obediently spew out that value to as many digits as its display will hold. However, how many of those digits really matter? How many of those digits actually contribute toward achieving the purpose of engineering, which is to design useful objects and systems and to understand, predict, and control their function in useful ways? This question introduces into engineering analysis the concept of **significant figures**.[6]

[6] The use of significant figures in engineering calculations is a method to avoid mistakes like saying that 10/6 = 1.666666667 (as provided by an electronic calculator or spreadsheet), whereas the strict answer is 2 because both 6 and 10 have only one *significant* figure. In this sense, engineering numbers differ from pure mathematical numbers.

Definition

A significant figure[7] is any one of the digits 1, 2, 3, 4, 5, 6, 7, 8, 9, and 0. Note that zero is a significant figure except when it is used simply to fix the decimal point or to fill the places of unknown or discarded digits.

The number 234 has three significant figures, and the number 7305 has four significant figures since the zero within the number is a legitimate significant digit.

Leading zeroes before a decimal point are not significant. Thus, the number 0.000452 has three significant figures (4, 5, and 2), the leading zeroes (including the first one before the decimal point) being place markers rather than significant figures.

How about trailing zeroes? For example, the number 12,300 is indeed twelve thousand, three hundred, but written this way it has just three significant figures. It should be written as 1.2×10^4 if it has two significant figures, as 1.23×10^4 if it has three, as 1.230×10^4 if it has four, and as 1.2300×10^4 or 12,300. (**note the decimal point here**) if it has five.

Finally, **exact** integer numbers such as 1 foot = 12 inches, or numbers that come from counting, or in definitions such as a diameter = 2 × radius can be assumed to have an **infinite number** of significant figures. In addition, fractions such as 1/3 or 1/6 are integer fractions and can be assumed to have an **infinite number** of significant figures. When we want to use an integer or integer fraction as a value for a variable in an example or exercise at the end of a chapter, we insert the word **"exactly"** before the number. That means the number can be assumed to have an infinite number of significant figures. For example, if we say that the length, L, of a rectangle is *exactly* 3 feet, and the height of the rectangle is $H = 2.0$ feet, and you need to find the area of the rectangle, $A = L \times H$, the answer does not have just one significant figure as it would if we just said $L = 3$ feet. Since we said L is "exactly" 3 feet, then the 3 becomes an integer with an infinite number of significant figures, and the answer is $A = 6.0$ ft^2 (to two significant figures since H has two significant figures).

Rules for finding the number of significant figures

1. All **non-zero digits** are significant: 1, 2, 3, 4, 5, 6, 7, 8, and 9. The number 1.5 has two significant figures.
2. Zeros **between** non-zero digits are significant. For example, 105 has three significant figures.
3. Leading zeros are **never** significant. For example, 0.002 has just one significant figure.
4. Trailing zeros **are significant** when they are associated with a decimal point. For example, 85.00 and 8500. have four significant figures but 8500 has only two significant figures.
5. Exact numbers have an **INFINITE** number of significant figures. For example, 1 meter = 1000 mm, 12 inches = 1 foot, etc. all have an infinite number of significant figures.

---

[7] There are many good websites on the Internet that deal with this concept. The following site has tests that you can use to check your understanding of significant figures: http://science.widener.edu/svb/tutorial/sigfigures.html

Deciding how many significant figures are associated with a physical measurement or a dimension on an engineering drawing only comes through knowing how the measurement was made and how the part is to be machined. For example, if we measure the diameter of a shaft with a ruler, the result might be 3.5 inches (two significant figures), but if it is measured with a digital micrometer, Fig. 3.4, it might be 3.512 inches (four significant figures).

FIGURE 3.4

Digital micrometer.

Engineering calculations often deal with numbers having unequal numbers of significant figures. Several rules have been developed for various computations. These rules are the result of a strict mathematical understanding of the propagation of errors due to arithmetical operations such as addition, subtraction, multiplication, and division. Here are the rules for rounding off the results of a calculation to the correct number of significant figures:

Rule 1: Addition and Subtraction

The sum or difference of two values should contain no significant figures further to the right of the decimal place than occur in the <u>least precise</u> number in the operation.

For example, $113.2 + 1.43 = 114.63$, which must now be rounded to **114.6**. The least precise number in this operation is 113.2 (having only one place to the right of the decimal point), so the final result can have no more than one place to the right of the decimal point. Similarly, $113.2 - 1.43 = 111.77$ must now be rounded to **111.8**. This is vitally important when subtracting two numbers of similar magnitudes since their difference may be much less significant than the two numbers that were subtracted. For example, $113.212 - 113.0 = 0.2$ has only one significant figure even though the "measured" numbers each had four or more significant figures.

Rule 2: Multiplication and Division

The product or quotient should contain no more significant figures than are contained by the term with the <u>least number of significant figures</u> used in the operation.

For example, $(113.2) \times (1.43) = 161.876$, which must now be rounded to **162**, and $113.2/1.43 = 79.16$, which must now be rounded to **79.2** because 1.43 contains the least number of significant figures (i.e., three).

Rule 3: Rounding Numbers Up or Down

When the discarded part of the number is 0, 1, 2, 3, or 4, the last remaining digit should not be changed. When the discarded part of the number is 5, 6, 7, 8, or 9, then the last remaining digit should be increased by 1.

For example, if we were to round 113.2 to three significant figures, it would be **113**. If we were to round it further to two significant figures, it would be **110**, and if we were to round it to one significant figure, it would be **100** with the trailing zeroes only representing placeholders. As another example, 116.876 rounded to five significant figures is 116.88, which further rounded to four significant figures is 116.9, which further rounded to three significant figures is 117. As another example, 1.55 rounds to 1.6, but 1.54 rounds to 1.5.

Example 3.2

a. Determine the number of significant figures in 2.2900×10^7.
b. Determine the number of significant figures in 4.00×10^{-3}.
c. Determine the number of significant figures in 480.
d. What is the value of $345.678 - 345.912$?
e. What is the value of $345.678 - 345.9$?

   Solution
a. The number of significant figures in 2.2900×10^7 is **5**.
b. The number of significant figures in 4.00×10^{-3} is **3**.
c. The number of significant figures in 480 is **2**.
d. The value of $345.678 - 345.912 = -$**0.234**.
e. The value of $345.678 - 345.9 = -0.222 = -$**0.2**.

3.5.1 Only apply the significant figures rule to your *final answer*

Using the guidelines given above you might be tempted to convert the result of every intermediate calculation to its correct number of significant figures. Doing so actually introduces a small (but sometimes significant) amount of error when you have a complicated calculation involving many steps. This type of error is avoided when you keep all those intermediate digits until the final answer.

For example, suppose you need to calculate the product of a series of numbers like

$$X = 0.7083 \times 1.375 \times 0.3529 \times 1.7 = 0.5844$$

Which when then rounded to two significant figures (the term with the least number of significant figures is 1.7, which has only two significant figures) as $X =$ **0.58**.

However, if you round all the numbers to two significant figures before you calculate their product you get.

$$X = 0.71 \times 1.40 \times 0.36 \times 1.7 = 0.61$$

This may seem like a small difference, but engineering calculations are often made with many intermediate steps, and the accuracy of the final result can be substantial. This problem usually does not exist if you do your calculations on a computer with a spread sheet program like Excel (see Section 3.7 in this chapter) and only then correctly round the final numerical result.

3.6 The "Need—Know—How—Solve" method

In this book, we use a memory aid called the "**Need—Know—How—Solve**" (NKHS) method for setting up and solving problems. It has the powerful benefit of enabling you to attack complicated problems in a systematic manner, thereby making your success more likely. It may seem cumbersome at first for simple problems, but for those with an eye to your final course grades, it has the additional benefit of ensuring at least some partial credit if you do not get the correct answer! It also tracks an essential element of modern engineering practice by providing an "**audit trail**" by which past errors can be traced to their source.

In the Need—Know—How—Solve method, the **Need** is the variable or variables for which you are solving. It is the very first thing you should write down.[8] The **Know** is the quantities that are known, either through the explicit statement of the problem or through your background knowledge. Write these down next. They may be graphical sketches or schematic figures as well as numbers or principles. The **How** is the method you will use to solve the problem, typically expressed in an equation, although rough sketches or graphs may also be included. The **How** also includes the assumptions you make to solve the problem. Write it out in sentence or symbolic form before applying it. Only then, with **Need**, **Know**, and **How** explicitly laid out, should you proceed to **Solve** the problem.

The pitfall this method guards against is the normal human temptation to look at a problem, think you know the answer, and simply write it down. However, rather than gambling with this hit-and-miss method, the Need-Know-How-Solve scheme gives a logical development that shows your thought processes.

Many of the problems you solve in this book involve specific equations. However, do not be deceived into thinking that this method involves the blind plugging of numbers into equations if you do not understand where those equations came from. The essence of the method is to realize that, once you have correctly defined the variable you **Need** to solve for, you know a lot more about the problem than you thought you did. With the aid of that knowledge, there now is a method (**How**) to solve the problem to an appropriate level of accuracy. Sometimes, this method requires no equations at all.

When applying the method, it may be especially helpful, as part of the **How** step, to sketch the situation described in the problem. Many people are discouraged from sketching because they feel they lack artistic talent. Do not let this stop you. As the next example illustrates, even the crudest sketch can be illuminating.

---

[8] There is actually an important step before identifying the "Need" in a problem statement. It is **read the question and then reread the question again**! Few things are more frustrating than solving the wrong problem.

Example 3.3

A Texan wants to purchase the largest fenced-in square ranch she can afford. She has $320,000 available for the purchase. Fencing costs $10. $\times 10^3$ a mile, and land costs $10. $\times 10^4$ a square mile. (Note the decimal points here.) How large a ranch, as measured by the length of one side of the square, can she buy?

Need: The length of a side of the largest square of land the Texan can buy.

Know: Fencing costs $10. $\times 10^3$ a mile, and land costs $10. $\times 10^4$ a square mile. Our Texan has $320,000 to invest.

How: Let the unknown length $= X$ miles. It may not be immediately obvious how to write an equation to find X. Now sketch the ranch.

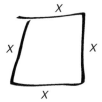

From the crude picture (and that is generally all you need), it is immediately obvious that the length of the fence surrounding the ranch is $4X$, and the area of the ranch is X^2. So, the cost of the ranch is about:

Length of fence [miles] $\times 10. \times 10^3$ [$/mile] + area of ranch [square miles] $\times 10. \times 10^4$ [$/sq. mi] = total cost of ranch = $320,000 or

$4X$ [miles] $\times$ $10. \times 10^3$ [$/mile] + X^2 [sq. miles] $\times$ $10. \times 10^4$ [$/sq. mi] = $320,000.

Solve: $(4X) \times (\$10. \times 10^3) + X^2 \times (\$10. \times 10^4) = 320,000.$

Therefore, $(\$10. \times 10^4)X^2 + 4X(\$10. \times 10^3) - 320,000 = 0$, which is a quadratic equation.

The solution is: $X = \textbf{1.6 miles}$ (to two significant figures).

There are actually two solutions here, $X = 1.6$ miles and $X = -2$ miles. Although it was not listed in the **Know** section, the length of the side of the ranch must be greater than zero (if for no other reason than to have room to put the cattle!). This yields the final answer. **The ranch has 1.6 miles on a side.**

When you do not have very accurate numbers to work with it is difficult to carry out an accurate analysis. The next example illustrates how to carry out an approximate analysis by just estimating how many significant figures should be in the result.

Example 3.4

Consider the following problem: How many barbershops are there in the city of Schenectady (population about 60,000 people)? Your first reaction may be "I haven't got a clue."

Need: The number of barbershops.

Know: There are about 60,000 people in the city of Schenectady, of whom about half are male. Assume the average male gets about 10 haircuts a year. A barber can probably do one haircut every half hour, or about 16 in the course of an 8-h day. There are about three barbers in a typical barbershop.

How: The number of haircuts given by the barbers must be equal to the number of haircuts received by the customers. So, if we calculate the number of haircuts per day received by all those 30,000 males, we can find the number of barbers needed to give those haircuts. Then, we can calculate the number of shops needed to hold those barbers. Assume that barbershops are open 300 days per year.

Solve: 30,000 males require about 10 [haircuts/male-year] × 30,000 [males]

$$= 300,000 \text{ haircuts/year.}$$

On a per day basis, this is about 300,000 [haircuts/year] × [1 year/300 days] = 1000 haircuts/day. This requires: 1000 [haircuts/day] × [1 barber-day/16 haircuts] = 62.5 barbers. At three barbers per shop, this means 62.5 [barbers] × [1 shop/3 barbers] = 20.8 shops.

So, the solution is **21 barbershops** (since surely not more than two of the digits used in the calculations are significant).

Looking at a recent yellow page phone directory and counting the number of barbershops in Schenectady gives the result of 23 barbershops. So, we are within about 10%, which is a very good answer, given the roughness of our estimates and the many likely sources of error. Notice also in this problem how the method of carrying the units in square brackets [..] helps your analysis of the problem and directs your thinking.

Haircuts and barbershops and Texas ranches are not among the variables or units used by engineers. However, applying common sense as well as equations is a crucial component of engineering analysis as demonstrated in the next example.

Example 3.5

A 2.00-meter-long steel wire is suspended from a hook in the ceiling. When a 10.0 kg mass is attached to its lower end the wire stretches by 15.0 mm. If the same mass is used to stretch a 4.00-meter-long piece of the same steel wire, how much will it stretch?

Need: Stretch = ____ mm for a 4.00 m piece of wire.

Know: 10.0 kg mass stretches a 2.00 m wire by 15.0 mm.

How: We need to deduce a possible law for extending a wire under load. Without experimentation, we cannot know if our "theoretical law" is correct, but a mixture of common sense and dimensional analysis can yield a plausible relationship. A longer wire should stretch further than a shorter one (a larger mass presumably also stretches the wire further).

Solve: A *plausible* model is therefore the extension X is proportional to the unstretched length of wire L (all other things such as the stretching force and the wire's cross-section being equal).

Then, under a fixed load: $X \propto L$ and the ratio between the two cases is

$$\frac{X_2}{X_1} = \frac{L_2}{L_1}$$

Therefore, the new extension $X_2 = 15.0 \times 4.00/2.00$ [mm][m/m] = **30.0 mm**.

The proportionality "law" used in this example is not necessarily correct. To guarantee that it is physically correct, we need to either understand the underlying principles of wire stretching or have good experimental data relating wire stretching to the wire's length. In fact, it is correct and is part of a physical law called *Hooke's Law*.

In some problems, the Need–Know–How–Solve method is decidedly clumsy in execution. Nevertheless, it is recommended that you use the basic method and learn how to modify it to suit the problem at hand. You will soon find there is ambiguity between the *Know, How*, and even *Solve* components of the method. Do not be concerned about this: the relevant information can sometimes go into any one of the baskets, just as long as it acts as a jog in the direction you need to solve the problem.

3.7 Spreadsheet analysis

Unfortunately, even though the Need-Know-How-Solve method gets you to an understanding of the answer, the subsequent mathematics may be too difficult to solve on a piece of paper, or you may need multiple cases, or graphical solutions, or there may be other complications that prevent you from easily working out the answer you are seeking. Today, an experienced engineer on a laptop computer can obtain solutions to problems that were daunting generations ago. In addition, spreadsheets do a superior job of displaying data—no small advantage if you want a busy person, perhaps your customer or your manager, to pay attention to the data.

Spreadsheets allow you to distinguish among three distinct concepts—*data, information,* and **knowledge**. First, what are data[9]?

What we call **data** is just a jumble of facts or numbers such 4, 1, 3, 1, 6. They become "information" when you sort them out in some way such as 3, 1, 4, 1, 6, and they become "knowledge" when the numerical sequence is interpreted as 3.1416 (i.e., π).

What makes spreadsheets useful is the visual way they can handle large arrays of numbers. These numbers are identified by their position in arrays by reading across their rows and down their columns. Interactions among these numbers are rapidly enhanced by providing sophisticated and complex mathematical functions that can manipulate the numbers in these arrays. Further, these arrays can be graphed so that a visual output is easily produced.

Fig. 3.5 is a reproduction of an Excel spreadsheet with some guidelines as to its key parameters. It is navigated by the intersection of rows and columns. The first cell is A1. Active cells, such as E10 in the example, are boxed with a bold border. You can scroll across and down the spreadsheet using the horizontal and vertical scroll bars. There is even the option to have multiple pages of spreadsheets that can be thought of as forming a "3D" spreadsheet.

[9] The singular form of data is *datum*, as in one datum point but five data points.

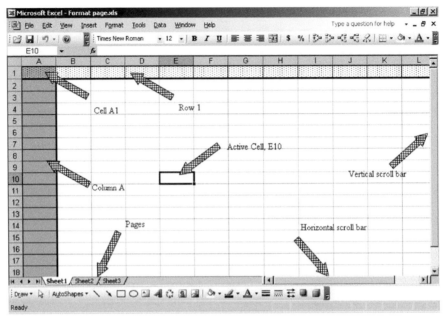

FIGURE 3.5

A typical spreadsheet.

3.7.1 Cell addressing modes

The window that you view is just a small fraction of the whole spreadsheet; the remaining virtual spreadsheet has thousands of rows and hundreds of columns of available space as well as many interactive sheets per workbook, enabling "3D addressing." Types of data that can be in a cell include:

Alphanumeric (text): Headings and labels.

Numeric: Numbers and references to the contents of other cells (using the cell address).

Equations: Formulae for a cell such as typing $= B1 + B4/H13$ in cell C3, which means adding the contents of cell B1 to the result of dividing the contents of cell B4 by the contents of cell H13 and putting the result in cell C3.

Predefined Functions: There are over 400 different function operations that you can use in Excel. Some of the basic ones are Sum, Average, Max, Min, Sin, Cos, Tan, Exp, and so forth.

You will need to explicitly tell the computer what data type is to be stored in a cell. The default cell entry is text containing numbers (i.e., an alphanumeric) and it is automatically left justified in the cell. But if numbers are entered, they are automatically right justified in the cell. If you put an equal sign (=) at the beginning of your cell entry, that means that an equation will follow (e.g., =SIN(D5). If you put a + or − at the beginning of your cell entry, that means a numeric will follow.[10]

Example 3.6

FleetsR'Us owns a fleet of rental cars; it wants to offer its customers an inclusive all-in-one-contract that takes account of the cost of the gasoline used. To estimate the cost and, therefore, pricing to their customers, they do a survey of a typical driver's journey and determine how many miles are driven. The raw data are given in Table 3.5.

Table 3.5 Input to a spreadsheet for **example 3.5**.

Driver	City miles	Suburban miles	Highway miles
Davis	16	28	79
Graham	10	31	112
Washington	4	7	158
Meyers	22	61	87
Richardson	12	56	198
Thomas	5	22	124
Williams	4	14	142

Using a spreadsheet, determine the total miles in each category of driving by each driver and the average miles driven in each category.

Need: Driver mileage and average miles driven in each segment of renter's journey.

Know: Mileage data in Table 3.5.

How: Total miles for each driver is the sum of the individual miles driven and the average mileage in each category of driving is the total miles divided by the number of drivers.

Solve: Use a spreadsheet to input the data and the indicated mathematical operations (see Table 3.6).

---

[10] However, "+ name" means a specially defined macro (a "macro" is just a type of program).

Table 3.6 Output of spreadsheet for **example 3.6**.

	A	B	C	D	E
1	Miles driven in various categories				
2					
3					
4	**Renter**	**City miles**	**Suburban miles**	**Highway miles**	**Total miles**
5	Davis	16	28	79	123
6	Graham	10	31	112	153
7	Washington	4	7	158	169
8	Meyers	22	61	87	170
9	Richardson	12	56	198	266
10	Thomas	5	22	124	151
11	Williams	4	14	142	160
12					
13	Average	10.4	31.3	128.6	170.3

We have added the sheet title in cell A1 (note that it actually sprawls across adjacent cells). It helps in organizing the spreadsheet so that its contents can be identified by you (or others) later. Descriptive labels for some columns have been added in row 4 as **A4:E4** (the colon in A4:E4 means it includes all the cells between A4 and E4). The labels may be simple, such as these are, or they may be informative, such as the units involved in the column (or row). Other descriptive labels that appear in this spreadsheet are in A5 to A11 (e.g., A5:A11) and A13. All of these are **text** statements.

The numerical input data for this example are given in the block of cells B5:D11. The results of some arithmetical results appear as row E5:E11 and as a block of cells B13: E13. Note that the variables are formatted to a reasonable number of significant figures. To get these results we type "=SUM(B5:D5)" in column E5, then copy and paste this cell into cells E6 through E11. The averages are obtained by typing "=AVERAGE(B5:B11)" in cell B12, then copying this cell and pasting it into C13 and D13.

Table 3.7 shows the mathematical operations behind Table 3.6 and what's in each cell. You get this version of the spreadsheet in Table 3.6 by *simultaneously* pressing the control key () and a tilde ($\sim$). This operation is written in shorthand as ^$\sim$.

Table 3.7 Expanded Spreadsheet of Table 3.6 using control-tilde (^~).

	A	B	C	D	E
1	**Miles driven**				
2					
3					
4	**Renter**	**City miles**	**Suburban miles**	**Highway miles**	**Total miles**
5	Davis	16	28	79	=SUM(B5:D5)
6	Graham	10	31	112	=SUM(B6:D6)
7	Washington	4	7	158	=SUM(B7:D7)
8	Meyers	22	61	87	=SUM(B8:D8)
9	Richardson	12	56	198	=SUM(B9:D9)
10	Thomas	5	22	124	=SUM(B10:D10)
11	Williams	4	14	142	=SUM(B11:D11)
12					
13	Average	=AVERAGE(B5:B11)	=AVERAGE(C5:C11)	=AVERAGE(D5:D11)	=AVERAGE(E5:E11)

FleetsR'Us now realizes that the gas mileage in each segment of the journey is different; the length of the journey is less important than the gallons used in each segment. It also realizes that gas mileage is a moving target. Vehicles are more or less fuel efficient depending on their size (a large SUV vs. a compact hybrid vehicle, for example). It needs to keep its options open and be able to update the spreadsheet when the market changes. Spreadsheets have a very good way of doing this. The tool is called **cell addressing.**

Excel normally uses what it calls **relative cell addressing**. When you copy a formula **across** the spreadsheet it increments the column letters in the formula. For example, copying the formula =Sum(B5:B11) across to column G will increment the column letter "B" to "G" giving =SUM(G5:G11).

However, when you copy a formula **down**, it increments the row numbers in the formula. For example, copying =B2/G2 down to row 4 will produce =B3/G3 and =B4/G4 in cells B3 and B4.

Mixed mode and absolute cell addressing is used when you want to lock a specific cell into a formula so that when you copy across or down the cell in the formula does not change. It is absolutely locked. For example, if you need the numerical value in cell B17 to be used in a formula used in other cells, then you need to use an absolute cell address for B17. An absolute cell address contains a dollar sign ($) before the row number to lock it in the row or before the column number to lock it in the column letter, or both as shown below.

$B17	Allows the row reference to change, but not the column reference.
B$17	Allows the column reference to change, but not the row reference.
$B$17	Allows neither the column nor the row reference to change.

Excel calls **$B17** and **B$17** mixed mode cell addressing, and it calls **$B$17** absolute cell addressing.

Example 3.7

FleetsR'Us wants to add the gas mileage to its spreadsheet in a flexible manner, so it can be updated later if needed. It wants to now know the gallons used per trip and the average used by their typical drivers. Here is the miles per gallon (mpg) by journey that need to be added to the spreadsheet:

	City	Suburbs	Highway
mpg	12.	18.	26.

Need: Average gallons per journey segment and average per driver.
Know: Mileage data from Example 3.6 plus mpg given above.
How: Divide mileage by mpg in Table 3.7 to get gallons used in each segment.
Solve: Table 3.8 shows several different ways we can program these changes using absolute cell referencing, (We are still looking at the ˆ~ spreadsheet here).

Table 3.8 MPG data added to Table 3.7 using different absolute cell referencing.

	A	B	C	D	E
4	**Renter**	**City gallons**	**Suburban gallons**	**Highway gallons**	**Total gallons**
5	Davis	=16/12	=28/18	=79/26	=SUM(B5:D5)
6	Graham	=10/B17	=31/$C$17	=112/D$17	=SUM(B6:D6)
7	Washington	=4/B17	=7/$C$17	=158/D$17	=SUM(B7:D7)
8	Meyers	=22/B17	=61/$C$17	=87/D$17	=SUM(B8:D8)
9	Richardson	=12/B17	=56/$C$17	=198/$D17	=SUM(B9:D9)
10	Thomas	=5/B17	=22/$C$17	=124/$D17	=SUM(B10:D10)
11	Williams	=4/B17	=14/$C$17	=142/$D17	=SUM(B11:D11)
12					
13	Average	=AVERAGE(B5: B11)	=AVERAGE(C5: C11)	=AVERAGE(D5: D11)	=AVERAGE(E5: E11)
14					
15					
16		**City**	**Suburbs**	**Highway**	
17	**mpg**	12	18	26	

Cells B5 through D5 in Table 3.8 do exactly what they are asked to do, divide one number by another number. However, they are *inflexible* should we later want to enter different mpg values. We would then need to input the new mpg values by hand into every affected cell (here there are just three cells, but in other problems there may be many more).

In Table 3.8, we used the contents of cell B17 to divide each cell in columns B6 through B11. Then, if we need to change the mpg for column B (city driving), all we need to do is change the number in cell B17. This is a powerful improvement over entering the data in individual cells.

In column C, we have fixed the contents of cell C17 by using absolute cell addressing $C$17 to modify cells C6:C11. We could also use either C$17 or $C17 as shown in column D to modify the cells in D6:D11. What does this mean and why do it? The importance of these operations is best explained by copying and pasting a section of the spreadsheet as in the next example.

Example 3.8

FleetsR'Us wants to copy part of its spreadsheet to another area on the spreadsheet to make it clearer for someone to read. Move the entire spreadsheet shown in Table 3.8 from A1 through E13 to a new origin at H1. Analyze what you see and report whether the relocated spreadsheet is correct or not; if not, explain what is wrong.

Need: Copied spreadsheet.

Know: You can copy and paste across the spreadsheet form.

How: Use the "Edit", "Copy", and "Paste" commands.

Solve: Suppose you want to copy the table to another part of the spreadsheet. All you need to do is highlight the block of cells containing the information that you want to move in A1 through E13, then go to "Edit," then "Copy." Move your cursor to the corner cell you want to move, cell H1, then go to "Edit" and then "Paste." Observe the effects after this is done in Table 3.9.

Table 3.9 Cells A1 through E13 in Table 3.8 moved to cells H1 through L13.

	H	I	J	K	L
4	**Renter**	**City gallons**	**Suburban gallons**	**Highway gallons**	**Total gallons**
5	Davis	=16/12	=28/18	=79/26	=SUM(I5:K5)
6	Graham	=10/I17	=31/$C17	=112/K$17	=SUM(I6:K6)
7	Washington	=4/I17	=7/$C17	=158/K$17	=SUM(I7:K7)
8	Meyers	=22/I17	=61/$C17	=87/K$17	=SUM(I8:K8)
9	Richardson	=12/I17	=56/$C17	=198/$D17	=SUM(I9:K9)
10	Thomas	=5/I17	=22/$C17	=124/$D17	=SUM(I1O:K1O)
11	Williams	=4/I17	=14/$C17	=142/$D17	=SUM(I11:K11)
12					
13	Average	=AVERAGE(I5: I11)	=AVERAGE(J5: J11)	=AVERAGE(K5: K11)	=AVERAGE(L5: L11)

Did we get the expected results? No, instead of modifying cells to give the correct answers, cells I6 through I11 and K6 through K8 give the wrong answers. What has happened? Use the toggle ˆ~ again and observe the numerical results in Table 3.10.

Table 3.10 Numerical contents of Table 3.9 after using ^ ~.

	H	I	J	K	L
1	**Miles driven in various categories**				
2					
3					
4	**Renter**	**City gallons**	**Suburban gallons**	**Highway gallons**	**Total gallons**
5	Davis	1.3	1.6	3.0	5.9
6	Graham	#DIV/0!	1.7	#DIV/0!	#DIV/0!
7	Washington	#DIV/0!	0.4	#DIV/0!	#DIV/0!
8	Meyers	#DIV/0!	3.4	#DIV/0!	#DIV/0!
9	Richardso	#DIV/0!	3.1	7.6	#DIV/0!
10	Thomas	#DIV/0!	1.2	4.8	#DIV/0!
11	Williams	#DIV/0!	0.8	5.5	#DIV/0!
12					
13	Average	#DIV/0!	1.7	#DIV/0!	#DIV/0!

Of course, the manually input cells I5 through K5 are correct, since they call for no information outside of these particular cells, but the entries in I6 through I11 are divided, not by B17, but by I17. In other words, when we translocated the cells, we also shifted B17 "relative" to the move (i.e., cell B17 was moved to I17 when cell A1 was moved to H1). And cell I17 (not shown) contains a default empty cell value of zero—hence, the apparent disaster that the gallons used in city driving are now all infinite (#DIV/0!).

Notice that cells J6 through J11 are multiplied by the contents of cell $C$17 as desired since we used the "$" designator for both row and column. Cell $C$17 has not relocated relative to the move because its address is locked and "absolute."

The results of the highway mileage in column K are explained by the "mixed" mode of addressing used there. Cells K6 through K8 now have been divided by K$17 since the column designator moved from column D to column K (again, just as column A did to H). Since we fixed the row designator at $17, it did not move to a new row. However, the cell K$17 contains a default value of zero and hence the cells K6 through K8, all of which contain this as a divisor, are each infinite.

However, cells K9 through K11 have the correct answers because they were divided by $D17. The $ fixed the column at column D. We did not choose to change the rows when we copied the block of cells from A1 to H1 and the divisor $D17 still refers to the original mpg in cell D17. We simply took advantage of our simple horizontal move from A1 to H1 so that it did not change the row locations. Had we also changed the row location—say, by pasting the block at A1 through E13 to H2 instead of H1—the rest of the Highway Gallons calculations would also have been in error due to an mpg call to cell $D18.

Sometimes we want to use relative, absolute, or mixed modes. It depends on what we want to do. With experience, your ability to do this will improve. One nice feature of spreadsheets is that your mistakes are usually immediately obvious. If, for example, you forget to fix an absolute cell address and add, multiply, divide, exponentiate, and so on by some unintended cell, the worksheet will complain as the target cell picks up whatever cell values were in the inadvertent cells.

Example 3.9

Correct Example 3.8 using correct absolute addressing modes.

Need: Corrected spreadsheet.

Know: How to use different absolute cell addressing modes.

How: Make sure that no cell addresses are moved to undesired cells.

Solve: Had all the cells in rows B, C, and D been divided by their respective absolute constants, $B$17 through $D$17, the final relocated result would have been as initially desired, as shown in Table 3.11.

Table 3.11 Results of Table 3.8 using correct absolute cell addressing.

	H	I	J	K	L
1	Miles driven in various categories				
2					
3					
4	Renter	City gallons	Suburban gallons	Highway gallons	Total gallons
5	Davis	1.33	1.56	3.04	5.93
6	Graham	0.833	1.72	4.31	6.86
7	Washington	0.333	0.389	6.08	6.80
8	Meyers	1.83	3.39	3.35	8.57
9	Richardson	1.00	3.11	7.62	11.7
10	Thomas	0.417	1.22	4.77	6.41
11	Williams	0.333	0.778	5.46	6.57
12					
13	Average	0.869	1.74	4.95	7.55

3.7.2 Graphing in spreadsheets

One of the big advantages of spread-sheeting is that we can display our results in ways that can visually convey a great deal of information. In a real sense, most of our "knowledge" is accessible only then, since the raw data, even if arranged as logical information in neat tables, may still be difficult to interpret.

We confine ourselves for the moment to simple Cartesian graphing, but you will soon recognize the pattern to follow if you want to use other modes of display.

Example 3.10

SpeedsR'Us wants to sell after-market performance booster kits for cars. They equipped a car with a booster kit and found that it had the performance on a level road shown in Table 3.12.

Table 3.12 Performance for a booster kit modified car.

Acceleration time (seconds, s)	Car speed (miles per hour, mph)
0.0	0.00
1.0	18.6
2.0	32.1
3.0	46.5
4.0	58.2
5.0	78.6
6.0	86.9
7.0	96.9
8.0	103

The car manufacturer says that the ***un*modified** car speed characteristic on a level road obeys the following equation: $mph = 136.5(1 - e^{-0.158t})$ with t in seconds. Is the modified car faster to 100 mph than the standard car?

Need: To compare the tabular data to the equation given.

Know: A spreadsheet provides multiple ways of looking at data.

How: Graph the tabular data and compare that data to the theoretical equation.

Solve: Set up the spreadsheet and show the mathematical operations (i.e., use ^ ~) as in Table 3.13.

Table 3.13 The $^\wedge\sim$ view of a spreadsheet for evaluating booster kit results.

	A	B	C	D
1	Compare measured car speed to manufacturer's specification			
2				
3	Speed in mph = 136.51 $(1 - e^{-0.158t})$ with t in seconds			
4				
5				
6	Time, s	Mph, actual	Mph, theory	
7	0	0	=136.5*(1 − EXP(−0.158*A7))	
8	=A7+1	18.6	=136.5*(1 − EXP(−0.158*A8))	
9	=A8+1	32.1	=136.5*(1 − EXP(−0.158*A9))	
10	=A9+1	46.5	=136.5*(1 − EXP(−0.158*A10))	
11	=A10+1	58.2	=136.5*(1 − EXP(−0.158*A11))	
12	=A11+1	78.6	=136.5*(1 − EXP(−0.158*A12))	
13	=A12+1	86.9	=136.5*(1 − EXP(−0.158*A13))	
14	=A13+1	96.9	=136.5*(1 − EXP(−0.158*A14))	
15	=A14+1	102.9	=136.5*(1 − EXP(−0.158*A15))	

Start with a spreadsheet title in cell A1. Make it descriptive. The cell labels in A6 through C6 should indicate what the columns under them contain. Of course, the numbers in columns A7 through C15 are just raw data—not easy to interpret by merely looking at them. Columns A, B, and C are the data you want to plot, that is, the value of the variable "mph" given as input data and calculated by the manufacturer's equation.

You do not have to enter too much to get a lot of information.[11] In cells A7 through B7, enter 0, and in C7, enter **=136.5*(1 − exp(−0.158*A7)**. The result in C7 is zero since $1 - e^0 = 1 - 1$ is zero. Now in A8 enter **= A7 + 1**. The value that appears in A8 is 1 since A7 is zero. Now copy cell A8 to cells A9 through A15, then copy cell C7 to cells C8 through C15. Now, you will have the manufacturer's function defined for each time increment.

You might want to graph your results, A7 through A15 versus B7 through B15 and A7 through A15 versus C7 through C15. Just highlight A7 through C15 and go to "Insert, Chart" in the header. Then choose the kind of graph you want. Let us choose a "**xy** *(Scatter)*" plot. You will see two sets of points corresponding to each column of speed data. Then, hit "Next" and fill in your preferred titles. Optionally, delete the "Legend" box and finish by copying and pasting the graph into your problem solution. You can also highlight either series of data and fit a continuous curve to the data, as in Fig. 3.6. On the basis of this graph, we see that the performance booster kit achieves its goal.

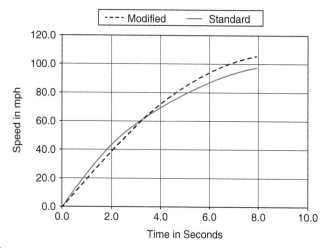

FIGURE 3.6

Graph of SpeedsR'Us vehicle performance with and without modification.

3.8 Computer programming with MATLAB

MATLAB is a computer programming language with special features designed to give engineers quick access to a wide variety of mathematical analysis tools, such as numerical methods, matrix computation, signal processing, and graphics.

A computer program (or code) can be written in any one of a multitude of readily available languages. These range from legacy languages such as Basic and Fortran to next-generation languages like C/C++ and Java and the more recent additions to the scene such as MATLAB and Python. Here, we will consider just the MATLAB programming language with the modest goal of introducing some of the fundamental concepts in a way that is both practical and accessible to first-year engineering students. We will begin by demonstrating how to formulate a short 3-line code and run it to solve a system of equations. The special functions and matrix operators available in MATLAB will then be discussed and applied to minimize and plot a 2D

---

[11] You can take advantage of pull-downs (called "smart tags") on the cells that automatically copy or increment adjacent cells. See the Help menu for your particular spreadsheet to see if this usage is supported.

mathematical function using only seven lines of executable code. Finally, the control statements **"if,"** **"for"** and **"while"** which are universal to most programming languages will be introduced. These are essential for implementing new computer programs for which existing special functions may not be available.

3.8.1 Representation of numbers

Numbers can be categorized by type as **integer, real or imaginary**. In some programming languages, every quantity represented by a variable must be classified as one of these types and their corresponding values must be represented in accordance with the rules governing those number types, for example, integers cannot have a decimal point while real numbers must have a decimal point to be recognized as such. However, MATLAB is different. There is no need to declare variable type and the numbers are handled the same way in calculations irrespective of whether they carry decimal points.

The use of scientific notation is always an option, especially for very small or very large numbers. In MATLAB, the exponent is identified by preceding it with an "**e**." For example, the number 1.05×10^7 is represented in MATLAB as **1.05e7**.

3.8.2 Variables, vectors, and arrays

Quantities that change value, either while running the code or from one run to the next, should be represented in the code by a **variable** name that is consistent with the naming conventions of the programming language. For example, in MATLAB, variable names must begin with a letter and utilize only letters, numbers, and the underscore character (-). Often, variable names are descriptive to remind the programmer of the quantity it represents.

Variable names can be associated with a single value or sets of values. If the latter, they are commonly referred to as **arrays**. A two-dimensional array, also known as a **matrix**, can be thought of as consisting of rows and columns of storage locations. Each storage location is identified by its row and column numbers (also referred to as indices), where the numbering of the rows and columns starts from the storage location in the upper left-hand corner of the matrix (see Fig. 3.7). For example, a list of five-point coordinates in three-dimensional space could be stored as an array named "A" with five rows (one for each point) and three columns (for storing the x, y, and z coordinates). A particular value in the array can be accessed by referring to it by its variable name followed by its row and column number separated by a comma both within parentheses. For example, A(4,2) would access the y-coordinate of the fourth spatial point stored in array A.

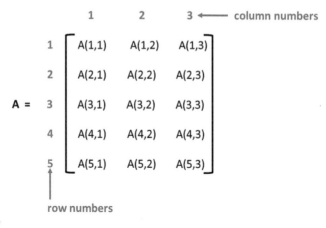

FIGURE 3.7

Storage locations for a 5 by 3 array.

A one-dimensional array with just one row or one column is referred to as a **vector**. A specific value in a vector can be accessed using a single index. For example, if the array LIST consists of 10 rows and one column of storage locations, then the values stored in the sixth row could be accessed using either LIST(6,1) or LIST(6), the latter form being more commonly used.

Fig. 3.8 illustrates how equations are used to assign initial values to single-valued variables, arrays, and vectors. Like all equations in a MATLAB code, the result is always stored in the quantity to the left of the "=" sign. Values in an array are defined row-wise with spaces (or commas) separating values in a row and a semicolon denoting the end of a row.

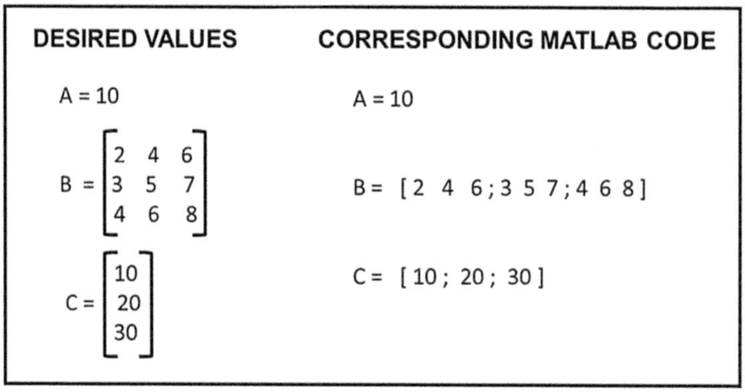

DESIRED VALUES	CORRESPONDING MATLAB CODE
$A = 10$	$A = 10$
$B = \begin{bmatrix} 2 & 4 & 6 \\ 3 & 5 & 7 \\ 4 & 6 & 8 \end{bmatrix}$	$B = [2\ 4\ 6; 3\ 5\ 7; 4\ 6\ 8]$
$C = \begin{bmatrix} 10 \\ 20 \\ 30 \end{bmatrix}$	$C = [10; 20; 30]$

FIGURE 3.8

MATLAB code for initializing variable A, array B, and vector C.

3.8.3 **Equations**

Each equation constitutes one line of code and can be identified by a single " $=$ ". All equations have the same basic structure—numeric values are defined or computed on the right side of the " $=$ " and the result is assigned to the variable name on the left side of the " $=$ ". MATLAB can interpret by itself whether the result is single valued, an array or a vector. The MATLAB code for a typical equation is presented below:

$$R = 5*(x - 4) \wedge 2 + 10/y;$$

in which a semicolon is used to suppress the printing of the result to the Command Window.

Both the use of parentheses and knowledge of the hierarchy of operations are required when coding an equation to ensure that mathematical operations are performed by MATLAB as intended. **Operations within the innermost parentheses are performed first**, with the order of operations within these parentheses being governed by the hierarchy of operations. Quantities with exponents " $\wedge$ " are evaluated first. "/" and "*" are next, and having equal priority is evaluated in the order in which they appear as read from left to right. The next highest priority is " $+$ " and " $-$ ", which also have equal priority and therefore are evaluated in the order in which they appear when in series. When all operations within the innermost parentheses have been completed, operations within the next innermost parentheses are performed and so on. Thus, for our typical equation, MATLAB will perform the operations in the following order:

$$R1 = x - 4, R2 = R 1 \wedge 2, R3 = 5*R2, R4 = 10/y, \text{ and finally, } R = R3 + R4.$$

In addition to the standard operators ($+$, $-$, *, /, and $\wedge$) used in equations, MATLAB also employs special operators to represent symbolically the coding used to carry out various matrix operations. For example, the backslash "\" when used in the right context is used to indicate the solution of simultaneous equations by Gaussian elimination. This is illustrated in the following example.

Example 3.10

Use MATLAB to simultaneously solve the three equations below using the special operator "\" in MATLAB.

$$2x + 3y + z - 12 = 0$$

$$-3x - y - 8z + 25 = 0$$

$$x + y + 2z - 9 = 0$$

Need: Solve three simultaneous equations.

Know: Equations are given.

How: Rewrite the equations in matrix form, put the result into MATLAB and then solve the equations by Gaussian elimination using the backslash "\" special operator available in MATLAB.

To **solve** these equations using MATLAB, first move the constants to the right-hand side.

$$2x + 3y + z = 12$$
$$-3x - y - 8z = -25$$
$$x + y + 2z = 9$$

Next, rewrite the equations in matrix form as follows:

$$\begin{bmatrix} 2 & 3 & 1 \\ -3 & -1 & -8 \\ 1 & 1 & 2 \end{bmatrix} \begin{Bmatrix} x \\ y \\ z \end{Bmatrix} = \begin{Bmatrix} 12 \\ -25 \\ 9 \end{Bmatrix}$$

where each row of the matrix contains the coefficients of x, y, and z for one of the equations and the column vector on the right-hand side of the "=" contains the corresponding constants that were moved to the right-hand side of the equations. The order of the coefficients within each row of the matrix must be consistent with the order of x, y, and z within the column vector just to the right of the matrix.

This matrix equation can be rewritten in shorthand notation as: $[A]\{\mathbf{x}\} = \{b\}$ in which $[A]$ is the 3 by 3 matrix of constant coefficients, $\{x\}$ represents the 3 by 1 column vector of unknowns, and $\{\mathbf{b}\}$ is a column vector containing the constants on the right-hand side. The MATLAB code in Fig. 3.9 can then be used to assign values to $[A]$ and $\{b\}$ and then solve the equations for the unknown values of x, y, and z. The expression "$A\backslash b$," which is unique to MATLAB, tells MATLAB to solve the equations by a method known as Gaussian elimination.

```
A = [ 2 3 1 ; -3 -1 -8 ; 1 1 2 ] ;
b = [ 12 ; -25 ; 9 ] ;
x = A\b
```

FIGURE 3.9

MATLAB code used to solve the equations in Example 3.10. yielding the result that **{x, y, z} = {−2.5, 4.5, 3.5}**.

3.8.4 Running the code in MATLAB

To run a MATLAB code, you must first open up the MATLAB computing environment. You can do this either by clicking on the MATLAB program icon (usually found on your desktop or in your computer's list of programs) or by clicking on an existing MATLAB script file that contains one of your MATLAB codes. A MATLAB script file can be recognized by its **.m** extension.

When opened up, the MATLAB computing environment should resemble the layout shown in Fig. 3.10. A toolbar is displayed at the top of the screen. The most commonly used toolbars are the EDITOR tool bar and the HOME toolbar. The layout of Fig. 3.10 has four open windows below the toolbar.

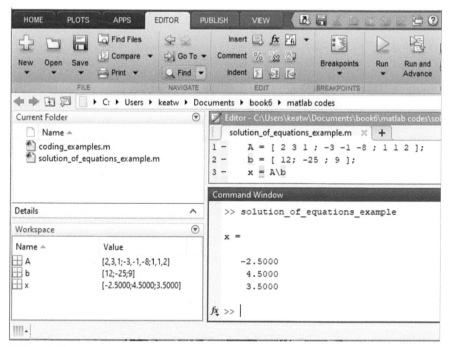

FIGURE 3.10

Screenshot of the MATLAB computing environment used to run the code in Example 3.10.

- The **Current Folder** window (top left) displays the files in the current folder, which by default is the folder holding the MATLAB script file that was opened up to start the session.
- The **Editor** window (top right) displays the text of the most recently opened script file, in this case, the script that contains the code for Example 3.10.
- The **Command Window** (bottom right) is where commands are typed into the right of the cursor symbol ">>" (e.g., to run the codes) and where output from the codes can be viewed.
- The **Workspace** window (bottom left) displays the final values of all of the variables and arrays used in the equations. If the array is too large to display in the format shown, its values can be viewed by clicking on the array name, which will open up a fifth window dedicated to displaying the array values.

Before running the code, the Workspace window should be cleared of all values by selecting the **Clear Workspace** tab on the HOME toolbar, as leaving the old values there could corrupt your new results. There are two ways to run a code using MATLAB:

1) Simplest is to **use MATLAB like a calculator,** that is, type the code into the Command window, line by line, being sure to hit the "enter" key on the keyboard after typing in each line of code. This method works well for very short codes like the code in Example 3.10.

2) The more common approach used with longer codes is to **first type the entire code into a MATLAB script file** where you will have full access to all of MATLAB's editing and debugging capabilities. Use the **New** tab on the Editor toolbar to start typing in your code and the Save tab to name the file and save it for future use. When finished, **run the code by typing the name of the script file** (without the .m extension) in the Command window after the cursor symbol, then hit "enter."

The **Command** window of Fig. 3.10 shows how this second approach was used to run the short code of Example 3.10. The name of the script file "solution_of_equations_example" was typed in after the cursor. Upon hitting the "enter" key, the solution was printed to the **Command Window** in the form of a column vector. The final values of A, b, and x can also be observed in the **Workspace**.

3.8.5 Functions

Some tasks best performed by using codes that have already been written and are stored in MATLAB function libraries which can be accessed by naming the function and providing the necessary inputs. The built-in MATLAB functions can be executed in a single line of code having the following general form:

$$[output1, output2, \ldots] = function\_name \text{ (input1, input2, } \ldots)$$

in which parentheses enclose the input variables and brackets enclose the output variables. The number of inputs and outputs can depend on the variable type (i.e., single valued, array, vector) and the desired features. Input and feature options are best learned by typing "**help**" followed by the function name into the command window as represented below:

$$\textbf{help } function\_name$$

To view the list of functions available in MATLAB, select the help button "**(?)**" on the HOME toolbar and then select "MATLAB" to view the standard built-in functions or select one of the available toolboxes (e.g., optimization) to view its available functions. Commonly used functions include **plot**, **max**, **min**, **linspace**, and the mathematics functions such as **exp**(x), **log**(x), **log10**(x), **abs**(x), **cos**(x), **acos**(x), and many others.

Example 3.11

A fourth-order polynomial with constraints on the range of the independent variable is given below:

$$y = x^4 + 3x^3 - 9x^2 - 23x - 12 \quad \text{where} \quad -4.2 \le x \le 3.2$$

We wish to complete the following sequence of tasks using MATLAB using the **NKHS** method:

1) Generate a vector of 1000 equally spaced x-values, spanning from $x = -4.2$ to $x = 3.2$.
2) Generate the corresponding vector of y-values by substituting the x-values into the polynomial.
3) Find the $(x.y)$ point on the polynomial curve corresponding to the minimum y-value.
4) Verify the result in step (**3**) by creating a 2D plot of x versus y.

Need: Using MATLAB, find vectors x and y defining the points along the polynomial curve, the location of the point with the smallest y-value and generate a 2D plot of the polynomial to verify the result.

Know: The polynomial and the range of x are given.

How: Use the MATLAB program functions.

Solve: Use the **linspace** function to calculate the x-values, the **min** function to determine the smallest y-value, and the **plot** function with the **xlabel** and **ylabel** functions to create the 2D plot.

The resulting code is presented in Fig. 3.11. **Comment statements**, identified by the "%" that precedes them, are there to remind the programmer of the tasks being performed by each section of the code. Inserting blank lines between each section of code further clarifies where one section ends and the next one begins. Note that each equation in Fig. 3.11 ends with a semicolon to suppress printing the results of those equations to the **Command Window** screen.

```
% generate 1000 equally-spaced x-values
x = linspace(-4.2,3.2,1000);

% sub into the polynomial to obtain the corresponding y-values
y = x.^4 + 3*x.^3 - 9*x.^2 - 23*x - 12;

% find the point with the minimum y-value
[ymin,i] = min(y);
xmin = x(i);

% print final values of xmin and ymin to Command Window
xmin
ymin

% plot x versus y
plot(x,y)
xlabel('x')
ylabel('y')
```

FIGURE 3.11

This concise MATLAB code solves Example 3.11 by utilizing the following built-in functions: **linspace**, **min**, **plot**, **xlabe**l, and **ylabel** as well as the special operator "**.^**".

Functions used in this code include:
- **linspace**—generates 1000 equal values between −4.2 and 3.2 and puts the result in the vector x.
- **min**—finds the smallest value in the vector y. Outputs are the smallest y-value ($ymin$) and the index (i) identifying its position within the vector y.

- **plot**—receives input from the x and y vectors and creates a 2D plot of the polynomial curve.
- **xlabel**—adds the label in single quotes to the horizontal axis of the 2D plot.
- **ylabel**—adds the label in single quotes to the vertical axis of the 2D plot.

The equation used to evaluate the polynomial employs the special operator (.^) because x is a vector of values and not just a single value. It tells MATLAB that every value of x in the vector should be raised to the indicated power. Other operations (*, /, +, −) between a vector and a single-valued quantity do not require a special operator. MATLAB automatically infers from the use of the vector x in the equation that y must also be a vector.

The results from running the code are displayed in Fig. 3.12. The 2D plot confirmed the computational result that appeared in the **Command** window as a result of listing $x$min and $y$min in the code without semicolons.

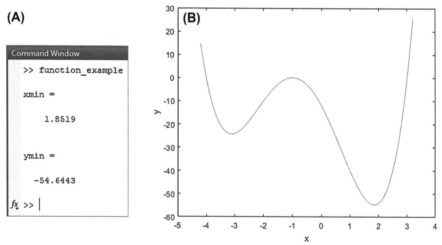

FIGURE 3.12

Results printed to the computer screen include (A) the values for $x$min and $y$min which appeared in the Command window and (B) the 2D plot.

3.8.6 Control statements

The codes presented so far progress sequentially line by line from the top of the code to the bottom of the code. This was possible in part because the functions and special operators simplified the coding requirements. In general, the order in which lines of code are executed is more complicated than this. For example, certain lines of code may have to undergo thousands of iterations, while other lines of code are skipped over under certain conditions.

To control the order in which lines of code are executed, we now introduce three **control statements** that are found in most programming languages—the "**if**" statement which skips lines of code if a condition is not satisfied, the "**for**" statement which repeats execution of lines of code a predefined number of times, and the "**while**" statement, which also repeats execution of lines of code but only as long as a certain condition is satisfied. The structure of these three multiline control statements is represented in Fig. 3.13. In other programming languages, the details of the syntax could differ slightly.

if x >= 0	**for i = i1 : iadd : i2**	**while x >= 0**
Execute these lines	*Repeat execution of*	*Repeat execution of*
of code only if x ≥ 0	*these lines of code*	*these lines of code*
end	*until i > i2*	*until it is no longer*
	end	*true that x ≥ 0*
		end
	Note: iadd defaults to 1 if	
	not included	

FIGURE 3.13

The structure of the **if**, **for**, and **while** control statements.

All three control statements begin with their identifying keyword and close with an "**end**" statement. In between are the isolated lines of code that require some degree of control. These lines of code are usually indented to help make the code more readable.

The first lines of "**if**" and "**while**" statements contain **logical expressions** that determine if code within these statements is to be executed or skipped. Logical expressions such as "$x \geq 0$" typically compare two algebraic quantities using one of the relational operators in Table 3.14.

Table 3.14 Relational operators used in logical expressions.

= =	Equal to
~ =	Not equal to
<	Less than
< =	Less than or equal to
>	Greater than
> =	Greater than or equal to

The result of the comparison is interpreted by MATLAB as being either TRUE or FALSE, depending on the current values of the variables appearing in the logical expression. When TRUE, the lines of code within the "**if**" or "**while**" statements are executed. When FALSE, the program skips these lines of code and moves to the next executable statement.

Both "**for**" and "**while**" statements can be employed to repeat the execution of specific lines of code. If the number of iterations is known, a "**for**" statement should be used. If instead, the required number of iterations depends on the results of calculations being performed inside the control statement, use a **while** statement. To ensure a finite number of iterations when using a **while** statement, its logical expression must be designed so that it ultimately changes in state from TRUE to FALSE due to calculations occurring within the **while** statement.

The "**for**" statement comes equipped with a counter that can be used, for example, to track the number of iterations or to identify storage locations within a vector or array. First time through the **for** statement, the counter "i" is set equal to its initial value "i1". Upon reaching the "**end**" statement, the current value of "i" is incremented by the amount "iadd" and then compared to the value of "i2", which represents the upper bound on "i". If $i \leq$ i2, the lines of code are executed once again, this time using the updated value of "i". The program will continue to execute the lines of code and update "i" in this manner until $i >$ i2.

Example 3.12

Rewrite the code for Example 3.11 using control statements instead of MATLAB's special functions. Then compare and contrast the programming details of these two approaches.

Recall that the polynomial function is defined by

$$y = x^4 + 3x^3 - 9x^2 - 23x - 12 \quad \text{where} \quad -4.2 \leq x \leq 3.2$$

The goals remain the same but the **How** step has changed.

Need: Using MATLAB, calculate the vectors x and y, find the smallest y-value, and plot the function.

How: Use a "**for**" statement to iterate over each point on the curve and an "**if**" statement to identify the smallest y-value. Retain the plot-related functions (**plot, xlabel, ylabel**) because replacing them with alternative code is not practical.

The new code developed using **for** and **if** statements is presented in Fig. 3.14. In this new version of the code, the **linspace** and **min** functions have been removed and the special operator "**.^**" has been replaced by the standard exponent symbol "**^**" The equivalent calculations are now performed inside a single **for** statement, which was selected over a **while** statement because the number of required iterations (1000) is known without performing any calculations inside the control statement.

```
% calculate distance between x-values
xinc = (3.2 - (-4.2)) / (1000-1) ;

% initialize ymin to a large value
ymin = 1.0e10 ;

% calculate x, y and ymin
for i=1:1000
    x(i) = -4.2 + (i-1)*xinc ;
    y(i) = x(i)^4 + 3*x(i)^3 - 9*x(i)^2 - 23*x(i) - 12 ;
    % check if ymin should be updated
    if y(i)<ymin
        ymin = y(i) ;
        xmin = x(i) ;
    end
end

% print final values of xmin and ymin to Command Window
xmin
ymin

% plot x versus y
plot(x,y)
xlabel('x')
ylabel('y')
```

FIGURE 3.14

Alternative code for Example 3.11 developed by using control statements.

Two variables are initialized at the top of the code. The variable "*x*inc" precedes the **for** statement because it only needs to be calculated once. An initial estimate of "*y*min" is needed so that *y*min has an assigned value for the first iteration ($i = 1$).

All calculations inside the **for** statement are focused on just one point per iteration as tracked by the counter "*i*" of the **for** statement. Therefore, all references to the vectors *x* and *y* in these equations are with respect to the *i*th elements of those vectors as denoted by $x(i)$ and $y(i)$. This contrasts with the original code (see Fig. 3.11) in which all references are to the vectors *x* and *y*, and not their individual elements.

An **if** statement embedded inside the **for** statement checks if the *i*th element of the vector *y* is smaller than the current best estimate of the smallest *y* (*y*min). If it is smaller, *y*min and its corresponding *x*-value are updated. The search for the smallest *y*-value continues until all 1000 points have been checked.

3.9 The Engineering Equation Solver

Perhaps the easiest computational program for a student to learn is the **Engineering Equation Solver** (EES) by **F-Chart Software**. It is a general equation-solving program that can numerically solve linear and nonlinear algebraic and differential equations automatically, which frees the user from having to follow a complicated programming structure. EES will also check the units of measurement of the values used in the equations, solve tables, and plot results as well as many other functions.

EES has the remarkable ability to allow the user to enter equations and variables in a random order rather than in a specific format, as required by MATLAB and most other programs. This is shown in Example 3.12.

Example 3.12

Find the acceleration required to cause an object to travel 10 m in 3 s from a standing start. The kinematics equations for velocity and position are $V = V_0 + at$ and $x = V_0 t + (\frac{1}{2})at^2$. Here, $V_0 = 0$ (a standing start), $x = 10$ m, and $t = 3$ s. Just type these equations (in any order) into EES along with their units, and then press F2 on your keyboard. The result shown in Fig. 3.15 then appears along with the solution for the acceleration as $a = 2.22$ m/s^2.

(A)   **(B)**

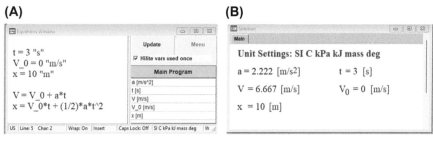

FIGURE 3.15
(A) EES screen where equations and variables are entered and (B) EES solution screen after F2 is pressed on the keyboard.

Summary

To summarize this chapter, the results of an engineering analysis must be correct in *four* ways.

1) It must involve the appropriate **variables.**
2) It must have the correct **numerical values.**
3) It must have the correct **units.**
4) it must have the correct number of **significant figures.**

Sketching an object you are analyzing can be helpful in focusing on the problem at hand. Some sketches are just that—crude unscaled drawings done with a pencil and paper. However, some are more formal and called "engineering drawings". Fortunately, there are computer programs known as CAD that are easy to use and will produce an accurate dimensioned drawing that can be given to fabricators for production.

You should carry **all** the digits you have available in a calculation until the very end, and only then modify your final answer to the correct number of **significant figures**. In other words, don't reduce the result of every intermediate calculation to its correct number of significant figures. Only apply the significant figures rules to the final answer.

A problem-solving method that guarantees you a path to the answer to engineering problems is a valuable addition to an engineer's tool kit. We use a self-prompting approach called the **Need—Know—How—Solve** (NKHS) method. It allows for a systematic approach to engineering analysis.

Many solutions to engineering problems require both complex analysis and repetitive solutions with graphical results. These situations are where spreadsheet analysis stands out. **Spread sheeting** allows you to organize, sort, and analyze the data to gain knowledge about the problem you are solving. Alternative computer programs now exist such as **MATLAB** and **EES** that allow you to rapidly analyze complex mathematical problems that occur in your education.

Exercises

You should use the *Need—Know—How—Solve* method in setting up and solving these problems. Make sure you are reporting the proper units and a number of significant figures.

1. Sketch the isometric plate below and add the following dimensions: height = 5.0 cm, length = 40.0 cm, and width = 20.0 cm. The diameter of the hole is 10.0 cm, and it is centered 15.0 cm from the front.

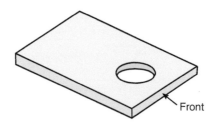

Front

2. Sketch and complete the orthographic drawing below.

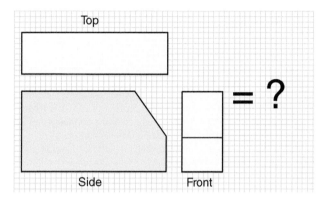

3. Sketch the top, left, and front views of the object below.

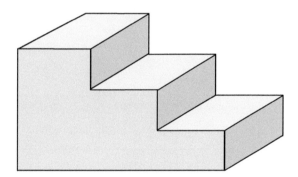

4. If a US gallon[12] has a volume of 0.134 ft^3 and a human mouth has a volume of 0.900 in^3, then how many mouthfuls of water are required to fill a 5.00 (US) gallon can? (**A: 1.29 × 10^3 mouthfuls.**)

5. Identify whether you would perform the following unit conversions by definition, by conversion factors, by geometry, or by scientific law.
   a. How many square miles in a square kilometer?
   b. How many microfarads in a farad?

---

[12] Distinguish US gallons from the old English measure of Imperial gallons; 1 Imp. gallon = 1. 20 U S. gallons.

    c. What is the weight on Earth in newtons of an object with a mass of 10 kg?

    d. How many square miles are on the surface of the Earth?

6. An acre was originally defined as the amount of area that an oxen team could plow in a day.[13] Suppose a team could plow 0.040 hectare per day, where a hectare is exactly 10^4 m². There are 1609 m in a mile. How many acres are there in a square mile?

7. There are 39 inches in a meter. What is the area in the SI system of the skin of a spherical orange that is 4.0 inches in diameter? (**A: 3.3 × 10^{-2} m²**)

8. There are 39 inches in a meter. What is the volume in the Engineering English system of a spherical apple that is 10. (**note the decimal point here**) cm in diameter?

9. Suppose the ranch in Example 3.3 was a circle instead of a square. Using the same financial information ($320,000 available funds, $10. × 10^3 a mile for fence, and $10. × 10^4 per square mile land cost), what would be the diameter of the ranch? **Note the decimal points here.**

10. The great physicist Enrico Fermi used to test the problem-solving ability of his students at the University of Chicago by giving them the following problem: Estimate the number of piano tuners in the city of Chicago. (Assume the population of Chicago is 5 million people.) (**Plausible answers 50–250 tuners.**[14])

11. Of all the rectangles that have an area of one square meter, what are the dimensions (length and width) of the one that has the smallest perimeter? Solve by graphing on a spreadsheet.

12. Suppose you want to make a cylindrical can that holds 0.01 m³ of soup. The sheet steel for the can costs $0.01/m². It costs $0.02/m to seal circular pieces to the top and bottom of the can and along the seam. What is the cost of the cylindrical can that is least expensive to make? (**A: 3 cents/can.**)

13. Suppose the mass used in Example 3.5 was increased from 10.0 to 20.0 kg, and the wire stretched by twice as much. If the 20.0 kg mass was then used to stretch a 4.00 m piece of the same steel wire, how much would it axially stretch?

14. Use a spreadsheet analysis to determine the total miles driven by each driver and the average miles driven in each category for the following car renters:

Renter	City miles	Suburban miles	Highway miles
Geske	35	57	93
Pollack	27	11	275
Loth	14	43	159
Sommerfeld	12	31	305
Thunes	22	16	132
Lu	5.0	21	417

15. Using the renter mileage information given in the previous exercise and the miles per gallon information given in Example 3.7, determine the average gallons per journey segment and average per driver for this set of drivers.

---

[13] The furlong or "furrow-long" was the length of a narow groove plowed by a team of oxen in a 10-acre farm field (220 yards or 1/8 of a mile). The usage goes back 1000 years and is still used in horse racing.

[14] See http://www.grc.nasa.gov/WWW/K12/Numbers/Math/Mathematical_Thinking/fermis_piano_tuner.htmS

16. Using the technique introduced in Example 3.10, create a spreadsheet graph of the following data for the median annual salaries in dollars for engineers based on years of experience, supervisory responsibility, and level of education.[15]

	Number of years after BS degree					
	0	**5**	**10**	**15**	**25**	**35**
Nonsupervisory						
B.S.	$55,341	$63,649	$73,162	$80,207	$85,116	$92,745
M.S.	—	$79,875	$86,868	$90,134	$97,463	$110,289
Ph.D.	—	—	$91,352	$98,053	$108,747	$122,886
Supervisory						
B.S.	—	—	$72,632	$80,739	$108,747	$122,886
M.S.	—	$99,367	$109,450	$110,360	$113,91	$117,146
Ph.D.	—	—	—	$110,877	$132,800	$147,517

Exercises 17–19 involve the following situation: Suppose that the weight of the gasoline in pounds-force [lbf] in a car's gas tank equaled the weight of the car in pounds-force [lbf] raised to 2/3 power. In other words, if G = gasoline weight in lbf and W = car weight in lbf, then

$$G = KW^{2/3} \text{ where K} = 1 \text{ } [\text{lbf}^{1/3}]$$

Assume that gasoline weighs 8.0 lbf/gal, and gas mileage, measured in miles per gallon [mpg], varies with weight according to the formula.

Gas mileage in mpg = (84,500 mpg • lbf) ÷ (Car's weight in lbf) − 3.0 mpg.

17. What is the fuel usage in miles per gallon of a 3.00×10^3 lbf car?
18. What is the heaviest car that can achieve a range of 600 miles? (**A: 3.7×10^3 lbf.**)
19. Suppose the formula for the weight of the gas was $G = W^b$, where b can be varied in the range 0.50–0.75. Graph the range of a 3.69×10^3 lbm car as a function of b.

Exercises 20–23: These exercises are concerned with bungee jumping as displayed in the figure. At full stretch, the elastic rope of original length L stretches to $L + X$. For a person whose weight is W lbf and a cord with a stiffness K lbf/ft, the extension x is given by the following formula:

$$X = \frac{W}{K} + \sqrt{\frac{W^2}{K^2} + \frac{2W \times L}{K}}$$

which can be written as a spreadsheet equation as: $X = W/K + \text{sqrt}(W^2/K^2 + 2 * W * L/K)$.

---

[15] These data are from the 2007 report of the Engineering Workforce Commission of the American Association of Engineering Societies (AAES).

20. If the height of the tower is 150.0 ft, $K = 6.25$ lbf/ft, $L = 40.0$ ft, and the person's weight is 150.0 lbf, will the person be able to bungee jump safely? Support your answer by giving the final value for length $= L + X$. (**A: 114 ft $<$ 150 ft, yes.**)

21. Americans are getting heavier. What is the jumper's weight limit for a 40.0-ft unstretched bungee with the stiffness of $K = 6.25$ lbf/ft? Graph final length $L + X$ versus W for weights from 100. lbf to 300. lbf in increments of 25 lbf. Print a warning if the jumper is too heavy for a 150.0 ft initial height. (**Hint:** Look up the application of the "IF" statement in your spreadsheet program.)

22. If the height of the tower is 200.0 ft, and the weight of the person is 150.0 lbf, and the unstretched length $L = 45.0$ ft, find a value of K that enables this person to stop exactly 5 feet above the ground. (**A: 2.60 lbf/ft.**)

23. By copying and pasting your spreadsheet from Exercise 22, find and plot the values of L needed (in ft) versus W, weight of jumper (in lbf) for successful bungee jumps (coming to a stop 5 ft above the ground) for $K = 6.25$ lbf/ft and from a tower of height 150. ft above the ground. The graph should cover weights from 100. lbf to 300. lbf in increments of 25 lbf. (**Hint:** The function "Goal seek" under "Tools" is one way to solve this exercise.)

Exercises 24–27: The fixed costs per year of operating an automobile is approximately 20.% of the initial price of the car. Therefore, the operating cost/mile $= 0.20$/yr $\times$ (purchase price of automobile)/(miles driven per year) + (price of gasoline/gallon) × (gallons used per mile). In the exercises that follow, assume that the automobile is driven 2.0×10^4 miles per year. Assume gasoline costs $5.00/gallon.

24. Estimate the operating cost per mile of an automobile with a price of $15,000 that gets 30.0 miles per gallon. (**A: $0.32/mile.**)

25. If one were to double the price of the automobile in Exercise 24, what would its gas mileage have to be to cost the same to operate per mile as the automobile in Exercise 24?

26. Suppose that the purchase price of automobiles varies with weight according to the formula that weight in lbf × $8.00, and gas mileage (miles/gallon) varies according to:

$$mpg = (84{,}500 \text{ mile-lbf/gallon})/W - 3.0 \text{ miles/gallon}$$

   Graph the cost per mile of operating a car as a function of the car's weight, in increments of 500. lbf from 2000. lbf to 5000. lbf. (**Partial A: 29 cents/mile for 2000. lbf cars and 76 cents/mile for 5000. lbf cars.**)

Exercises 27–29: In visiting stores, one finds the following prices for various things. Broccoli crowns cost $2.89 per pound. Soft drinks cost $2.00 per 2-L bottle. (A liter is 0.001 m^3) A new automobile weighs 2.50×10^3 lbf and costs 1.50×10^4. A dozen oranges, each of which is 0.06 m in diameter, cost $2.05. A 1.5-lb package of chicken thighs costs $5.35. A dictionary weighs 5.00 pounds and costs $20. A refrigerator weighs 200. lbf and costs $900.00 Assume that one cubic meter of any solid object or liquid weighs 1.00×10^4 N.

27. For the objects just listed, make a table and graph of the cost of objects in dollars as a function of their weight in newtons. It is suggested for this graph to use a *line* (not "scatter") graph with markers displayed at each data value. (Get rid of the unwanted line using the "Format series function.") The value of the line graph is that everything plotted is at the same horizontal displacement and is not dependent on its value.

28. What simple generalization about the cost of things might one make based on the table and graph of Exercise 27? (**Hint:** Does the comparative lack of spread in price surprise you?)

29. Name a product or group of products that does *not* fit the generalization you made in Exercise 28 and add and label the point on the graph in Exercise 28. To get a better perspective, use a *log scale* for the y-axis, $/N, since the ordinate should be much larger than the coordinates of those points from Exercise 28.

30. An unnamed country has obtained the population of passenger cars on its roads as determined by 250. kg mass differences. These data are shown in the table that follows. You have to make these data clear to the undersecretary of that country's transport minister. Plot these data by two methods: (1) as a "pie chart" and (2) as a histogram to show the distribution in an effective manner.

	Car mass in kg	% of all cars
1	1000.	12.1
2	1250.	13.1
3	1500.	15.4
4	1750.	18.6
5	2000.	14.8
6	2250.	9.2
7	2500.	7.5
8	2750.	6.3
9	3000.	2.0
10	3250.	1.0

Exercises 31—33 deal with Hubbert's Peak,[16] a model of the supply and demand for oil. It looks at the amount of available oil and its rate of consumption to draw conclusions about continuing the current course of our oil-based economy.

31. Suppose the world originally had 3 trillion (3000. $\times 10^9$) barrels of oil and exploration for oil began in 1850. Suppose 10.0% of the remaining *undiscovered* oil has been found in every quarter century since 1850. Call the *discovered*, but not yet consumed oil, *reserves*. Suppose oil consumption was 1000. $\times 10^5$ barrels in 1850, and further suppose oil consumption has grown by exactly a factor of 5 in every quarter century since 1850. When will the oil start to run out? (That is, when will the reserves become negative?) Give your answer to the nearest 25 years and provide a spreadsheet showing reserves and consumption as a factor. (**Ans. Year 2025.**)

32. Suppose the world originally had 10 trillion (1000. $\times 10^{10}$) barrels of oil. Use the data from Exercise 31 to predict when the oil will start to run out again.

33. Repeat Exercise 32, but instead of assuming the exponential growth in consumption continuing unabated by exactly a factor of 5 in every quarter century since 1850, curtail growth since the year 2000 and assume consumption has stayed constant since then. Predict when the oil will start to run out again.

34. Your friend tells you that the *Need—Know—How—Solve* problem-solving method seems overly complicated. He just wants to find the answer to the problem in the quickest possible way—say, by finding some formula in the text and plugging numbers into it. What do you tell him?
    a. Go ahead and do whatever you want, then you'll flunk out, and I'll survive.
    b. Talk to the instructor and have her explain why this methodology works.
    c. Find someone who has used this method and ask to copy his homework.
    d. Explain why this technique will lead to a fail-safe method of getting the correct answer.

35. Calculate with the correct significant figures: (a) $100/(2.0 \times 10^2)$, (b) $1.0 \times 10^2/(2.0 \times 10^2)$. (**A: 0.5, 0.50.**)

36. Calculate with the correct significant figures: (a) 10/6, (b) 10.0/6, (c) 10/6.0, (d) 10./6.0, and (e) 10.0/6.00.

37. What is 2.68×10^8 minus 2.33×10^3 to the correct significant figures? (**A: 2.68×10^8.**)

38. A machinist has a sophisticated micrometer that can measure the diameter of a drill bit to exactly 1/10,000 of an inch. What is the maximum number of significant figures that should be reported if the approximate diameter of the drill bit is (a) 0.0001 inches, (b) 0.1 inches, (c) 1 inch.

39. Round off to three significant places: (a) 1.53, (b) 15.345, (c) 16.67, (d) 102.04, (e) −124.7, and (f) 0.00123456.

40. Suppose you want to design a front door and doorway to fit snugly enough to keep out the drafts, yet still be easy to open. The dimensions are to be given in inches. To how many significant figures would you specify the length and width of the door and doorway?[17] Assume a standard door is 30.0" by 81.0".

41. Rewrite the four simultaneous equations given below in matrix form and then solve them in MATLAB using "$x = A\backslash b$".

---

[16] M. King Hubbert was a geologist with Shell Oil who, in the 1950s, correctly pointed out that the US supply of oil was going to fall short of demand by the 1970s. His methods have since been applied to world oil production and, based on demand exceeding production, predict an ongoing oil supply crisis. See http://www.hubbertpeak.com/hubbert/

[17] The subject of tolerancing is important in mass manufacturing to ensure a proper fit with one part and another, since each part is not exact, their combined tolerance determines how well, if at all, they fit together. It is the subject of significant statistical analysis. Applying the methods rigorously allowed the Japanese automotive industry to eclipse those of the rest of the world in terms of quality.

$$w + 6x + 5y + 2z - 27 = 0$$
$$3w + 2x + y + 4z - 33 = 0$$
$$5w + 8x + 7y + 3z - 23 = 0$$
$$4w + 16x + 8y + 7z - 48 = 3.0$$

(Ans. (w, x, y, z) = (-5.3765, -4.9176, 7.2000, 12.9412))

42. You are given the following function and constraints on x:

$$y = \sin(x) + \sin(10x / 3) \text{ where } -2.7 \le x \le 7.5$$

Develop and run a MATLAB script file to complete the following tasks:
   a. Use the **linspace** function to generate 1500 equally spaced x-values over the range of x and store the results in a vector x.
   b. Evaluate the given function at each of the generated x-values and store the results in a vector y.
   c. Use the **plot** function with functions **xlabel** and **ylabel** to generate a 2D plot of x versus y.
   d. Use the **min** function to find the smallest y-value and its corresponding x-value. (**Ans. (x,y) = (5.1456, -1.8996)**)
   e. Use the **max** function to find the largest y-value and its corresponding y-value. (**Ans. (x,y) = (2.2945, 1.7283)**)

   Programming Hint: Code, test and debug one task at a time.

43. The **Trapezoidal Rule** is a numerical method that is used to determine the area under a curve that has been plotted on x–y axes. The method is summarized in the figure below:

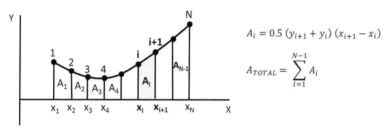

$$A_i = 0.5 \, (y_{i+1} + y_i) \, (x_{i+1} - x_i)$$

$$A_{TOTAL} = \sum_{i=1}^{N-1} A_i$$

in which $\sum_{i=1}^{N-1} A_i$ represents a summation of the areas A_i where the index "i" can vary from 1 to N-1, with N being the number of points used to represent the curve. When calculating area, the mathematical function is modeled as piecewise linear by calculating N number of points along the curve. The Trapezoidal Rule assumes the function varies linearly between these points. Total area is then obtained by summing the areas of the N-1 trapezoids located under the curve, where the area of each trapezoid is calculated as the product of its width and its average height. Goal of this problem is to use MATLAB with the Trapezoidal Rule to find the area under the curve defined by the following function:

$$y = 2 \sqrt{x} \text{ where } 0 \le x \le 10$$

To calculate the area, implement the following capabilities in a Matlab script file:
   a. Use the **linspace** function to generate 500 equally spaced x-values over the range of x and store the result in a vector x.
   b. Evaluate the given function at each of the generated x-values and store the results in a vector y.
   c. Use the **plot** function with functions **xlabel** and **ylabel** to generate a 2D plot of x versus y.

  d. Determine the total area under the curve using a single **for** statement to calculate and sum the areas according to the Trapezoidal Rule.

  (**Ans. Area = 42.1625**)

  e. Verify the result of (c) using the **trapz** function, which is available in MATLAB. **Hint:** Type "**help trapz**" in the Command Window to get input and output details.

44. When a physical experiment is repeated at the same settings of the control factors, some variation in performance as measured by the output variable x should still be expected as there are typically some variables that cannot be controlled. Therefore quality of the results is often measured in terms of two statistical parameters—the **mean** ($\bar{x}$) and the **standard deviation** (σ) which are calculated as follows: $\bar{x} = \dfrac{\sum\limits_{i=1}^{N} x_i}{N}; \sigma = \sqrt{\dfrac{\sum\limits_{i=1}^{N} (x_i - \bar{x})^2}{N-1}}$

  in which N is the number of experiments run at the same settings, x_i is the output associated with the ith experiment and $\Sigma_{i=1}^{N}$ indicates that the subscript "i" can vary from 1 to N in the terms immediately following and that all N terms thus identified should be summed. The mean represents the average output while the standard deviation provides a measure of the variability of the results around the mean. Let us now assume that 10 trials of an experiment run at the same settings yielded the following 10 values of the output variable x:

  $$x = 21.63\ 21.81\ 20.25\ 21.83\ 21.26\ 20.20\ 20.56\ 21.09\ 21.92\ 21.93$$

  Develop and run a MATLAB script file to analyze the data as follows:

  a. Calculate the mean ($\bar{x}$) and standard deviation (σ) of these 10 trials using **for** statements to perform the summations. **Hint:** The summation appearing in the definition of mean can be coded using a **for** statement as follows:

```
xsum = 0.
for i=1:10
    xsum = xsum + x(i);
end
```

  (**Ans. $\bar{x}$ = 21.2480, σ = 0.6920**)

  b. Verify the results calculated in (a) using the **mean** and **std** functions that are available in MATLAB. **Hint:** Type "help mean" and "help std" in the Command Window to get input and output details.

45. When plotting the results from a physical experiment for different values of the independent variable, it is often known or hypothesized that there is a linear relationship between the independent variable and the measured output variable. However, due to small errors associated with the experiments, the data will rarely fall exactly along a straight line. The **Method of Least Squares** fits a straight line of the form: $y = mx + b$ to experimental data by calculating the constants m and b which minimize the sum of the square of the residuals as defined by the quantity $(y_i - f(x_i))^2$, where y_i and $f(x_i)$ are the measured and best fit estimate of the output variable y. The resulting formulae for m and b are well known and given by:

$$m = \frac{N \sum\limits_{i=1}^{N} (x_i y_i) - \sum\limits_{i=1}^{N} x_i \sum\limits_{i=1}^{N} y_i}{N \sum\limits_{i=1}^{N} x_i^2 - \left(\sum\limits_{i=1}^{N} x_i\right)^2}; b = \frac{\sum\limits_{i=1}^{N} y_i - m \sum\limits_{i=1}^{N} x_i}{N}$$

  where N is the number of data points and x_i and y_i are the values of the independent and measured output variables for the ith data point, respectively. $\Sigma_{i=1}^{N}$ indicates that the subscript "i" can vary from 1 to N in the terms immediately following and that all N terms thus identified should be summed. All of the summations used to calculate m and b can be evaluated in this problem using a single **for** statement by building on the following code.

```
N = 10.
xsum = 0.
xysum= 0.
for I = 1:N
xsum = xsum + x(i);
xysum = xysum + x(i)*y(i);
end
```

The goal of this problem is to fit a straight line to the following 10 data points by using the Method of Least Squares.

Experimental Data:

$$x = 11.0 \ 12.0 \ 13.0 \ 14.0 \ 15.0 \ 16.0 \ 17.0 \ 18.0 \ 19.0 \ 20.0$$

$$y = 118.9 \ 121.2 \ 129.9 \ 135.3 \ 142.6 \ 154.7 \ 160.8 \ 165.2 \ 176.6 \ 177.2$$

where x is the independent variable, which was controlled in the experiment and y is the dependent, or output, variable. Generate and visualize a linear least-squares best fit in MATLAB by writing a script file that implements the following steps:

a. Calculate the constants m and b using the formulae and data provided.

   (Ans. m = 7.1370, b = 37.6170)

b. Use the **plot** function with functions **xlabel** and **ylabel** to graph the experimental data as points (represented by symbols) on x–y axes. The code for doing this is:

$$\texttt{plot(x,y,'o')}$$

c. Overlay the plot in (b) with the linear least squares fit by adding the following lines of code:

```
hold
fplot(@(x) m*x+b,[10,20])
```

where the **hold** command allows more than one plot function to be applied to the same graph. The **fplot** function plots the function "$m*x + b$" over the range of x defined in brackets, where "$@(x)$" assigns the letter x to represent the independent variable in the function.

Force and motion

graphicgeoff/Shutterstock.com

4.1 Introduction

The universe is full of motion, and it all happens because of forces. The Sun moves around our galaxy and the Earth moves around the Sun, and we move around on the Earth. This movement never stops.

Force and motion are part of all engineering disciplines. Civil engineering need to understand the forces and motions of buildings and bridges subject to high wind loads, mechanical engineers need to understand the forces and motions of automobiles, bioengineers study how the human body responds to forces and motions in normal and abnormal situations (such as in a fall), and so forth.

Statics is the study of forces on objects that do not experience acceleration. **Dynamics** is the study of forces on the motion of objects that are accelerating, while

Exploring Engineering. https://doi.org/10.1016/B978-0-443-13541-5.00003-9

kinematics is the study of the motion of objects without considering the mass or the forces that caused the motion.

4.2 What is a force?

The word *force* is used in many ways. The force of law, brute force, police force, driving force, nuclear force, electromagnetic force, and the force of gravity are all examples of its use. In engineering, a **force** is something that pushes or pulls and can cause an object to change speed, direction, or shape.

Newton's laws of force and motion are very important for engineers because they are involved in a wide variety of technologies. These laws tell us things like how cars work, how water flows, how planes fly, and how bridges are supported and basically describe the motion of everything around us. Newton's three physical laws can be summarized as follows.

1. **Newton's first law.** The speed v of a body remains constant unless the body is acted upon by an external force F.
2. **Newton's second law.** The acceleration a of a body is parallel to and directly proportional to the net force F applied to it and inversely proportional to its mass m (or $a \propto F/m$ where the symbol $\propto$ means "proportional to").
3. **Newton's third law.** The mutual forces of action and reaction between two bodies are equal, opposite, and collinear.

4.3 Newton's first law (statics)

Newton's first law of motion can be stated as

An object at rest stays at rest and an object in motion stays in motion with the same speed and in the same direction unless acted upon by an unbalanced force.

This can be written mathematically as:

$$\sum_{external} Forces = 0 \text{ implies that the speed } v = \text{constant}$$

where the Greek letter "Σ" means summation.
Consequently,

- An object at rest will stay at rest unless an unbalanced force acts upon it.
- An object in motion will not change its speed unless an unbalanced force acts upon it.

In systems where objects are moving at different speeds, it is impossible to determine which object is "in motion" and which object is "at rest." In other words, the laws of motion are the same in every frame of reference. This is illustrated in Example 4.1.

Example 4.1

When you are traveling slowly down a street in an open convertible at a constant speed, you can throw a heavy ball straight up in the air and catch it just as you would if the car was not moving. But what does a person standing on the sidewalk see as you pass by?

Need: What a stationary observer sees.

Know—How: If you are traveling in a car at a constant speed, the laws of motion are the same as when the car is at rest.

Solve: A person observing you from the sidewalk sees the ball follow a curve in the same direction as the motion of the car. From your perspective inside the car, the car and everything inside of it is at rest, it is the world outside of the car that is moving with a constant speed.

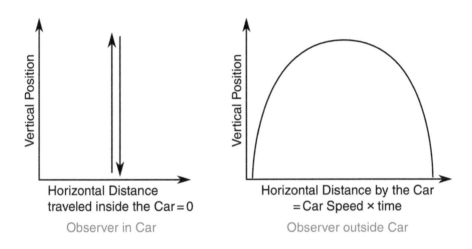

No experiment can tell whether it is the car that is at rest or the outside world that is at rest; the two situations are physically the same. Consequently, Newton's first law applies to any constant speed motion (even when the speed is zero).

4.4 Newton's second law (dynamics)

Newton's second law can be stated as

An object will accelerate if the net force on the object is not zero. The direction of the acceleration is the same as the direction of the net force. The magnitude of the acceleration is directly proportional to the magnitude of the net force, and inversely proportional to the mass of the object.

Central to any scientific set of units is the definition of **force**. According to Newton's second law, force is *proportional* to mass multiplied by acceleration and can be written mathematically as (remember that the letter "$\propto$", which means "is proportional to")

$$F \propto ma$$

where a is the **acceleration** of mass m. Acceleration is simply the rate of change of velocity (speed), so Newton's second law can be written for bodies of constant mass as $F \propto m(\Delta v/\Delta t)$ with Δv being the change in the speed and Δt being the change in the time during the change in speed. To convert the proportionality in Newton's second law into "equality," we need to introduce a **constant of proportionality**, K, such that

$$F = Kma \tag{4.1}$$

To evaluate the constant of proportionality in Eq. (4.1), suppose there exists a force F_1 that accelerates the mass m_1 by acceleration a_1. Then, Newton's second law can be written as

$$F_1 = Km_1a_1 \text{ so that } K = F_1/m_1a_1$$

We now eliminate the proportionality constant K in Eq. (4.1) to obtain

$$F = Kma = \left(\frac{F_1}{m_1a_1}\right)ma \tag{4.2}$$

Clearly, the fact that the proportionality constant is $K = F_1/m_1a_1$, the ratio of a specific force to a specific mass and acceleration is very important to the calculations made with this equation. It has become customary to give the proportionality constant the symbol $F_1/m_1a_1 = K = 1/g_c$, where $g_c \equiv m_1a_1/F_1$. Then, Newton's second law becomes

$$F = \left(\frac{F_1}{m_1a_1}\right)ma = \frac{ma}{\left(\dfrac{m_1a_1}{F_1}\right)} = \frac{ma}{g_c} \tag{4.3}$$

Any consistent set of units will satisfy this equation.

We must now choose both the magnitude and the dimensions of the proportionality constant g_c. This is illustrated in Example 4.2.

Example 4.2

Suppose we define a unit of mass called a "**slug**" that when exactly one slug is acted upon by exactly 1 pound force (lbf), it accelerates by exactly 1 ft/s^2. In this case, the proportionality constant in Eq. (4.2) is exactly one and dimensionless, so that the slug unit of mass is the same as the unit grouping lbf·ft/s^2. What force is then required to accelerate a 5.00 slug mass at 32.2 ft/s^2?

(Remember that when we say something is "exactly" a number, we mean that the number is to be treated as an integer with an infinite number of significant figures).

Need: Force needed to accelerate a 5.00 slug mass at 32.2 ft/s^2.

Know: The acceleration of the mass is 32.2 ft/s^2.

How: The first part of Eq. (4.3) can be used here

$$F = \left(\frac{F_1}{m_1 a_1}\right) ma, \text{ in which } \frac{F_1}{m_1 a_1} = \frac{1\,[\mathrm{lbf}]}{1\,[\mathrm{slug}] \times 1\,[\mathrm{ft/s^2}]} = 1\left[\frac{\mathrm{lbf} \times \mathrm{s^2}}{\mathrm{slug} \times \mathrm{ft}}\right]$$

Therefore,

$$F = \left(\frac{F_1}{m_1 a_1}\right) mg = \left\{1\left[\frac{\mathrm{lbf} \times \mathrm{s^2}}{\mathrm{slug} \times \mathrm{ft}}\right]\right\} \times \{5.00\,[\mathrm{slug}]\} \times \left\{32.2\left[\frac{\mathrm{ft}}{\mathrm{s^2}}\right]\right\} = \mathbf{161\ lbf}$$

Note that with this definition of force,

$$g_c = \left(\frac{m_1 a_1}{F_1}\right) = 1\left[\frac{\mathrm{slug} \times \mathrm{ft}}{\mathrm{lbf} \times \mathrm{s^2}}\right] = 1\,(\text{and dimensionless})$$

We can simplify things by giving our mysterious "slug" the units of $\mathrm{lbf} \times \mathrm{s^2/ft}$ and then $g_c = 1$ and is dimensionless.

In fact, the slug unit of mass in the "Technical English" unit system (see Table 4.1) is still preferred in the United States in some engineering fields. Its obvious advantage is that Newton's second law constant of proportionality g_c is exactly equal to one and is dimensionless, so Newton's second law can be written in the familiar form of $F = ma$, with the stipulation that mass m is in slugs and the acceleration is in $\mathrm{ft/s^2}$. Its disadvantage is that you probably have little feel for the size of a "slug" of mass (1 slug weighs 32.2 lbf at standard gravity).

Table 4.1 A comparison of various common unit systems using $F = ma/g_c$.

System name	Force unit	Mass unit	Length unit	Time unit	Value and units of g_c
SI (System International)	newton[a] [N]	kilogram, [kg]	meter [m]	second [s]	1 Dimensionless
Engineering English	pound force, [lbf]	pound mass [lbm]	foot [ft]	second [s]	32.2 [lbm·ft/lbf·s²]
Technical English	pound force [lbf]	slug[b] [slug]	foot [ft]	second [s]	1 Dimensionless

[a] Note that $1 = 1\,N = 1\,kg \cdot m/s^2$, which is useful when balancing units in an equation.
[b] A slug is a $lbf \cdot s^2/ft$ and since $g_c = 1$ and dimensionless here, 1 slug weighs 32.2 lbf at standard gravity.

The SI system is somewhat similar to the slug system with the choice for the constant of proportionality g_c being both easy and logical. In the SI system, the numerical value of the ratio $F_1/m_1 a_1$ was chosen to be exactly "1" and dimensionless. The unit of force must then simply be equivalent to the units of mass times the units of acceleration. It is convenient to give this set of units a unique new name, and in the SI system, the force unit is called the **newton** (abbreviated N), so $1\,N = 1\,kg \cdot m/s^2$.

In other words, a force F_1 of 1 newton (N) acting on a mass m_1 of exactly 1 kg causes the mass to accelerate a_1 at exactly 1 m/s². Thus, in the SI system, $g_c = m_1 a_1/F_1 \equiv 1$ (dimensionless), and Newton's second law can be written

$$F = ma/g_c = ma \text{ since } g_c = 1 \text{ and is dimensionless in SI units} \qquad (4.4)$$

This simplification led to the familiar form of Newton's law of motion that you have seen before. Since the conversion factor g_c here is [(kg·m/s²)/N], which is exactly **one** and is **dimensionless** (check this for yourself), we have dropped it out of the equation altogether.

Example 4.3

What is the force in newtons on a mass 1.00 kg that is accelerated by gravity at 9.81 m/s²?

Need: Force in newtons on a mass of 1.00 kg accelerated at 9.81 m/s².
Know: Newton's second law using SI units.
How: We use Eq. (4.4) where $g_c = 1$ and dimensionless in the SI system.
Solve: $F = ma = 1.00$ [kg] $\times 9.81$[m/s²] $= 9.81$ kg m/s² $=$ **9.81 N**.

These last two examples use an acceleration of special interest, that which is caused by Earth's gravity: $g = 32.2$ ft/s² $= 9.81$ m/s². Looking at Eq. (4.4), the SI force acting on exactly 1 kg mass due to gravity is not 1 N but 9.81 N, a fact that causes distress to newcomers to the subject. Weight is thus just a special kind of force—that is due to gravity. In this sense, Eq. (4.4) can be written with the acceleration of gravity to yield the force we call *weight*, **W**, by writing it as

$$Weight = W = mg/g_c \qquad (4.5)$$

Remember that the value and units of g_c depend on the unit system you are using.

Example 4.4

The *mass* of one small apple is 0.102 kg. What is its *weight* in newtons in the Earth's gravity of 9.81 m/s²?

Need: Weight in newtons of a 0.102 kg apple in Earth's gravity of 9.81 m/s².
Know: Newton's second law applies using SI units.
How: Eq. (4.5) with $g_c = 1$ and dimensionless (since we are using SI units).
Solve: The arithmetic is identical to that of Example 4.2—the force on 0.102 kg of mass is
$F = $ weight $= mg/g_c = (0.102$ kg) $(9.81$ m/s²)/1 $= 1.00$ kg m/s² $= 1.00$ N
In other words, 1 *newton is just about the weight of a small apple here on Earth!* Perhaps this will help you mentally visualize the magnitude of a force given in newtons (e.g., 100 N $\approx$ 100 apples).

In addition to SI units, you must master at least one other system of units that relate mass and force, one whose persistence in the United States is due more to custom than logic, the **Engineering English** system of units. In this system, the conversion factor g_c between force and mass $\times$ acceleration is *not* unity and is *not* dimensionless. Because of this, we must carry an explicit proportionality constant every time we use this unit system. In addition, because the proportionality factor is not unity, it has an explicit set of units as well.

The 2000-year-old Engineering English unit system evolved a rather unfortunate convention regarding the "pound" unit. When Newton's laws came into use, it was decided that the name *pound* would be used *both* for mass *and* for weight (i.e., force). As mass and force are distinctly different quantities, a modifier had to be added to the "pound" unit to indicate which (mass or weight) was being used. This was solved by simply using the phrase **pound mass** or **pound force** (with the associated abbreviations, lbm and lbf, respectively) to distinguish between them.

In the Engineering English system, it was also decided that one "pound mass" should weigh exactly one "pound force" at standard gravity, 32.2 ft/s^2. This was accomplished by setting the ratio $(F_1/m_1a_1) = 1/32.2$ [lbf·s^2/lbm·ft]. In other words, 1 lbf is defined as the force that will accelerate exactly 1 lbm by exactly 32.2 ft/s^2. The designers of the Engineering English system cleverly decided to define the numerical value of the proportionality constant as g_c, as

$$g_c \equiv 32.2 \frac{\text{lbm} \cdot \text{ft}}{\text{lbf} \cdot \text{s}^2} \tag{4.6}$$

This is awkward because it tends to make you think that g_c is the same as (i.e., equal to) the acceleration of local gravity, g, **which it definitely is not**. The constant g_c is nothing more than a proportionality constant with dimensions of [mass × length/(force × time^2)]. Because the use of g_c is so widespread today in the United States and because it is important that you are able to recognize the meaning of g_c when you see it elsewhere, it is used in the relevant equations in this book except when we are using the much more convenient SI units. Table 4.1 illustrates the values and units of g_c for various common unit systems.

In the English Engineering unit system, we henceforth write Newton's second law as

$$F = \frac{ma}{g_c} \text{ in Engineering English units} \tag{4.7}$$

The consequence of this choice is that you can easily calculate the force F in lbf corresponding to an acceleration a in ft/s^2 and a mass m in lbm when you use the g_c given in Eq. (4.6).

Example 4.5

What is the force necessary to accelerate a mass of 65.0 lbm at a rate of 15.0 ft/s^2?

Need: Force to accelerate mass of 65.0 lbm at 15.0 ft/s^2.

Know: Newton's second law in Engineering English units, in which

$$g_c = 32.2 \text{ lbm} \cdot \text{ft/(lbf} \cdot \text{s}^2).$$

How: Eq. (4.7) is the principle used here: $F = \frac{ma}{g_c}$

Solve: As the problem is stated in Engineering English units, the answer must also be in these units.

$$F = \frac{ma}{g_c} = \frac{65.0 \times 15.0}{32.2} = [\text{lbm}] \left[\frac{\text{ft}}{\text{s}^2} \right] \left[\frac{\text{lbf} \cdot \text{s}^2}{\text{lbm} \cdot \text{ft}} \right] = \textbf{30.3 lbf}$$

Notice how the **units** as well as the **value** of g_c enter the problem; without g_c our "force" would be in the nonsense units of lbm·ft/s^2 and our calculated numerical value in those nonsense units would be 975.

There are another series of consequences, too. In the next chapter, you will be introduced to the concept of **kinetic energy**. An engineer using the SI system defines kinetic energy as $\frac{1}{2}mv^2$ (remember, the v stands for speed). However, an engineer using the Engineering English system defines kinetic energy as $\frac{1}{2}\,(m/g_c)v^2$.

Of course, it is logically more elegant to show the units using the **[units]** bracket convention previously introduced. For example, if the grouping $\frac{1}{2}mv^2$ were expressed in Engineering English units, its units would be [lbm] × [ft/s]^2 if g_c were ignored. This is a meaningless collection of units! But, using the g_c proportionality factor, the definition of *kinetic energy* now becomes

$$kinetic\ energy \;=\; 1/2(mv^2/g_c) \;=\; \frac{[\text{lbm}] \times [\text{ft/s}]^2}{[\text{lbm}\cdot\text{ft}]/[\text{lbf/s}^2]} = [\text{ft}\cdot\text{lbf}]$$

which is a legitimate unit of energy in the Engineering English unit system.

Understand that the force conversion constant g_c is a universal constant that has the same value everywhere in the universe — the value of 1 and dimensionless in the SI system and 32.2 lbm·ft/(lbf·s^2) in the Engineering English system. In this regard, it should never be confused with the local acceleration of gravity g, which has different numerical values at different locations (as well as different dimensions from g_c).

The concept of weight always has the *local* gravity associated with it. The weight of a body of mass m at a point in space where the local acceleration of gravity is "g" is

$$\text{Weight} = W = \frac{mg}{g_c} \tag{4.8a}$$

where, on Earth, in the Engineering English units system, $g = 32.2$ ft/s^2 and $g_c = 32.2$ (lbm·ft/lbf·s^2), In the SI units system, $g = 9.81$ m/s^2 and $g_c = 1$ (and dimensionless), so in textbooks that use the SI system exclusively you will see Newton's second law written as:

$$\text{Weight} = W = mg \tag{4.8b}$$

A person on the International Space Station experiences microgravity, a much smaller g than we do on Earth; and a person on the Moon experiences about $^1/_6$ of g compared to a person on the surface of the Earth. However, a person who is an engineer using the Engineering English system of units must use the same numerical value for g_c wherever in the universe he or she may be.

Example 4.6

a. What is the weight on Earth in Engineering English units of a 10.0 lbm object?
b. What is the weight on Earth in SI units of a 10.0 kg mass?
c. What is the mass of a 10.0 lbm object on the Moon? Assume the Moon's local gravity is exactly 1/6 that of Earth's.
d. What is the weight of that 10.0 lbm object on the Moon?
   Solution: This is a straightforward unit conversion problem.
a. $W = mg/g_c$ in Engineering English units. Therefore,

$$W = (10.0 \times 32.2 / 32.2 \text{ [lbm][ft/s}^2\text{]/[lbm·ft/lbf·s}^2\text{]} = \textbf{10.0 lbf.}$$

b. $W = mg/g_c$ where $g_c = 1$ in SI units. Therefore,

$$W = 10.0 \times 9.81 \text{ [kg][m/s}^2\text{]/1} = 98.1 \text{ [kgm/s}^2\text{]} = \textbf{98.1 N.}$$

   (See Table 4.1 for the definition of newtons in terms of SI units).
c. Mass is a property of the material. Therefore, the object still has a mass of 10.0 lbm on Earth, on the Moon, or anywhere in the universe.
d. $W = mg/g_c$ in Engineering English units. On the Moon, $g = 32.2/6.00 = 5.37 \text{ ft/s}^2$. Therefore, on the Moon, $W = 10 \times (5.37)/32.2 \text{ [lbm][ft/s}^2\text{]/[lbm·ft/lbf·s}^2\text{]} = \textbf{1.67 lbf.}$

Until the mid-20th century, most English-speaking countries used one or more forms of the Engineering English unit system. But, because of world trade pressures and the worldwide acceptance of the SI system, many engineering textbooks today present examples and homework problems in both the Engineering English and the SI unit systems. The United States is *slowly* converting to the common use of the SI system. However, it appears likely that this conversion will take at least a significant fraction of your lifetime. So, to succeed as an engineer in the United States, you must learn the Engineering English system. Doing so will help you avoid future repetition of such disasters as NASA's embarrassing loss in 1999 of an expensive and scientifically important Mars Lander due to an improper conversion between Engineering English and SI units.[1]

Example 4.7

How much horizontal force is required to accelerate a 1000. kg car at *exactly* 5 m/s²?

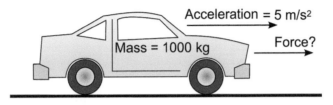

Need: The unbalanced force required to accelerate the car *exactly* 5 m/s².

---

[1] See http://www.cnn.com/TECH/space/9909/30/mars.metric for more information.

Know: The mass of the car is 1000. kg.

How: Newton's second law, in SI units.

Solve: Newton's second law relates an object's mass, the unbalanced force on it, and its acceleration: $F = ma/g_c$, and $g_c = 1$ and dimensionless in the SI system, so

$$F = ma/g_c = (1000. \times 5/1) \, [\text{kg}] \left[\frac{\text{m}}{\text{s}^2}\right] = 5000. \, \text{kg·m/s}^2 = \textbf{5000. newtons}$$

Note: Remember that when we say that a value is "**exactly**" some number, the number is to be treated as an integer (or integer fraction) and has an infinite number of significant figures.

4.5 Newton's third law

Newton's third law can be simply stated as:

For every action, there is an equal and opposite reaction.

For example, Newton's third law states that a force on object **A** that is due to the presence of a second object **B** is automatically accompanied by an equal and opposite force on object **B** due to the presence of object **A**. That is, any pair of forces between any two isolated objects does not cause the center of mass of these objects to accelerate. These objects can accelerate toward or away from each other, but their center of mass cannot accelerate.

A common demonstration of Newton's third law, called *Newton's Cradle*, is shown in Fig. 4.1. If one or more spheres are pulled away and then released, it strikes the next sphere in the series and comes to (nearly) a complete stop. The impact force is transmitted through the group of spheres until an equal number of spheres on the opposite end is lifted by the reaction force. More than half the spheres can be set in motion. For example, three out of five spheres result in the central sphere swinging without any apparent interruption.

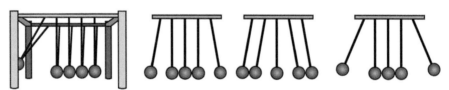

FIGURE 4.1

Newton's Cradle.

Example 4.8

The total weight of a new US space transport (vehicle and launch rockets) at liftoff is 4.40×10^6 lbf and the liftoff thrust is 6.78×10^6 lbf. What is the unbalanced force on the shuttle at the moment of liftoff? What is the rocket's initial acceleration?

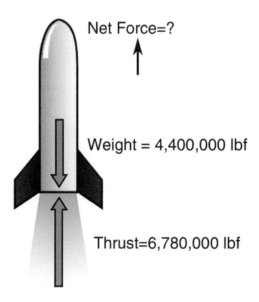

Net Force=?

Weight = 4,400,000 lbf

Thrust=6,780,000 lbf

Need: The unbalanced force on the space shuttle at the moment of liftoff.

Know–How: A rocket launch is a classic example of Newton's third law. A chemical reaction in the rocket engine generates gases that exert a downward force (thrust) at the bottom of the engine. This is the **action**. A force equal and opposite to the thrust exerts an upward force on the rocket pushing it in the upward direction. This is the **reaction**.

Solve: As for every action, there is an equal and opposite reaction, the net unbalanced force on the shuttle assembly is

$$F_{\text{thrust}} - F_{\text{weight}} = F_{\text{unbalanced}} = 6.78 \times 10^6 \text{ [lbf]} - 4.40 \times 10^6 \text{ [lbf]} = \mathbf{2.38 \times 10^6 \text{ lbf}}$$

This result can be used to calculate the initial acceleration of the vehicle as

$$a_{\text{initial}} = \frac{F_{\text{unblalanced}} \times g_c}{m} = \frac{2.38 \times 10^6}{4.40 \times 10^6} \times 32.2 \text{[lbf]} \left[\frac{\text{lbm} \cdot \text{ft}}{\text{lbf} \cdot \text{s}^2}\right] \left[\frac{1}{\text{lbm}}\right] = \mathbf{17.4 \text{ ft/s}^2}$$

which is about one-half the standard acceleration of gravity.

4.6 Free-body diagrams

A force acts in a specific direction with a magnitude that depends on the strength of the push or pull. Because of these characteristics, forces have both direction and magnitude. This means that forces follow a different set of mathematical rules than do quantities that do not have direction. For example, when determining what happens when two forces act on the same object, it is necessary to know both the magnitude and the direction of both forces to determine the resultant force.

Free-body diagrams can be used as a convenient way to keep track of forces acting on a system. Ideally, these diagrams are drawn with the angles and relative magnitudes of the forces listed so that graphical addition can be done to determine the resultant force (see Fig. 4.2).

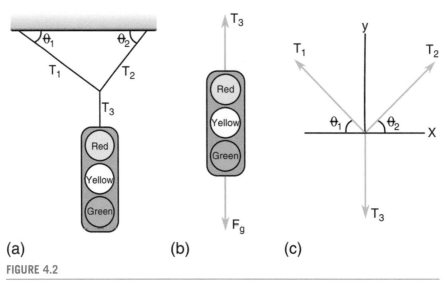

(a)   (b)   (c)

FIGURE 4.2

Free-body diagram of a traffic light suspension system: (a) cables holding the light, (b) resultant forces on the light, and (c) force diagram.

When two forces act on an object, the *resultant* force can be determined by what is called the *parallelogram rule of addition*. The addition of two forces represented by sides of a parallelogram gives a resultant force that is equal in magnitude and direction to the diagonal of the parallelogram. For example, in Fig. 4.2, the resultant force T_3 is equal to the vertical components of T_1 plus T_2, or

$$T_3 = T_1 \sin \theta_1 + T_2 \sin \theta_2$$

As well as being added, forces can also be resolved into independent coordinate axes components at right angles to each other. For many problems, these are simply horizontal and vertical directions. Resolving forces into x and y components is often an easier way to describe them than using their magnitudes and directions. Force equilibrium occurs when the resultant force acting on an object is zero. There are two kinds of equilibrium: static and dynamic.

4.6.1 Static equilibrium

The simplest case of static equilibrium occurs when two forces are equal in magnitude but opposite in direction. For example, an object on a level surface is pulled downward by the force of gravity (i.e., its weight). At the same time, the surface under the object resists the object's downward force with an equal upward force. This condition is one of zero net force and no acceleration.

Pushing horizontally on an object on a horizontal surface can result in a situation where the object does not move because the applied force is opposed by a force of static friction between the object and the surface. In this case, the static friction force balances the applied force resulting in no acceleration.

Static equilibrium between two forces is a common way of measuring the force of gravity (i.e., weight). This can be done using balance scales, as shown in Fig. 4.3.

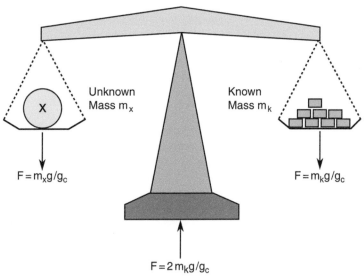

FIGURE 4.3

A balance scale force diagram illustrating static equilibrium.

4.6.2 **Dynamic equilibrium**

The definition of **dynamic equilibrium** is when an object moves at a constant speed, and all the forces on the object are balanced. A simple case of dynamic equilibrium occurs when a horizontal force is applied to an object causing it to move with a constant speed across a horizontal surface. In this case, the applied force is in the direction of motion, while a friction force in the opposite direction balances the applied force, producing no unbalanced force on the object. The object continues to move with a constant speed, as in Fig. 4.4.

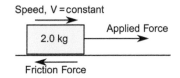

FIGURE 4.4

Dynamic equilibrium of a sliding object occurs when the applied force = friction force.

4.7 What is kinematics?

Kinematics is the study of how things move. Engineers need to understand the relationships among the important variables of distance, speed, acceleration, and time. These relationships are collectively called kinematics when they do not also involve the forces on objects or the inertia of the objects. Kinematics simply serves as a way to relate the effects of forces on a body to its subsequent motion without asking how we can determine the values of the forces.

Engineers are concerned with kinematic variables because they are the basic variables used to understand the motion of objects, such as the motion of cars in traffic. In the simplest cases, they are related to each other by geometric methods—in more complicated situations by the methods of calculus.

4.7.1 Distance, speed, and acceleration

In this book, only motion in a single direction (often called *one-dimensional*) is considered. For convenience, think of positive **distance** as from left to right, and negative distance as from right to left, just as in Cartesian geometry. **Speed** is a variable commonly measured in miles per hour (mph) in the United States but in kilometers per hour (kph) in most of the rest of the world. But, for engineering design, the SI units of meters per second (m/s) are often more convenient. Speed is calculated by dividing the distance traveled (Δx) by the time spent traveling (Δt). In mathematical terms, speed $v = \Delta x/\Delta t$ (the uppercase Greek delta Δ symbol means "difference"). In this case, speed means the final position minus the initial position divided by the final time minus the initial time (see Fig. 4.5).

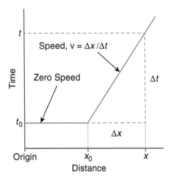

FIGURE 4.5

Constant speed occurs when $v = \Delta x/\Delta t =$ constant.

We use the symbol v (as in velocity) for speed. But, there is an important differentiation between the concepts of *speed* and *velocity*. Speed is the *magnitude* of velocity. Velocity may have a component in the y direction and another component in the x direction. Of course, in our one-dimensional analyses, speed and velocity are the same. For acceleration, we use a, except for that due to gravity where we use g.

Adding a word to *speed* gets us to the phrase *speed up*. We have already met the speed up variable, **acceleration**. It is defined as change of speed divided by the change in time or, more mathematically precise, as the rate of change of speed, and is commonly measured in m/s^2 or ft/s^2.

Strictly speaking, two of the preceding variables should be defined in two ways: in terms of *instantaneous* speed and acceleration, and in terms of *average* speed and acceleration. In this textbook, we will work with constant or "average" accelerations. An average acceleration a is defined as:

$$a = \frac{\Delta v}{\Delta t} \tag{4.9}$$

Example 4.9

A car enters an on-ramp traveling 15 miles per hour. It accelerates for 11 s. At the end of that time interval, it is traveling at 60. miles per hour. What is its average acceleration?

Need: Average acceleration = _____ ft/s^2.
Know: The average acceleration is given by Eq. (4.9) as $a = \Delta v/\Delta t$.
How: Average acceleration = (change in speed)/(change in time).

Initial speed = 15 miles per hour = 22 ft/s

Final speed = 60. miles per hour = 88 ft/s

Time interval, $\Delta t = 11$ s

Solve: In Engineering English units, the average acceleration is:

$$a = \frac{\Delta v}{\Delta t} = \frac{(88 - 22)[\text{ft/s}]}{11[\text{s}]} = \textbf{6.0 ft/s}^2$$

4.7.2 The speed versus time diagram

One could deal with problems of distance, speed, acceleration, and time just using words and equations, but a much more useful tool makes possible an insightful description of motion problems. This versatile tool is the **speed** versus **time diagram** shown in Fig. 4.6. For brevity, we call this diagram the $v-t$ diagram. The horizontal axis represents time, with zero being the instant the situation being considered began. For each instant of time, the speed is plotted in the vertical direction.

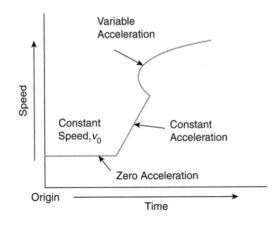

FIGURE 4.6

Acceleration is the slope of the speed–time diagram.

Example 4.10

Plot the v–t diagram in SI units from Example 4.9.

   Need: The v–t diagram describing the situation "a car enters an on-ramp traveling 15 miles per hour. It accelerates constantly for 11 s. At the end of that time interval it is traveling at 60. miles per hour."

   Know: First, convert all speeds to SI: 15 mph = 6.7 m/s and 60. mph = 27 m/s.

   How: Then, plot speed for $t = 0$ and speed when $t = 11$ s and join the two points with a *straight line*. (What tells us the line is straight is the phrase "constant acceleration.")

   Solve.

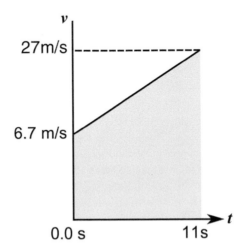

Why is this particular graph so important? There are two reasons.

1. The slope of the v–t graph is the acceleration.
2. The shaded area under the v–t graph is the distance traveled.

No other graph involving these four variables summarizes so much information in so intuitive a manner. The *slope* of a line is its vertical "rise" divided by its horizontal "run." In Cartesian x, y coordinates, the slope is $\Delta y/\Delta x$. That the slope of the v–t graph gives you acceleration is evident from considering dimensions. The "rise" of the graph has units of speed, m/s. The "run" of the graph has units of time, s. So, slope $=$ rise/run $= \Delta v/\Delta t$ in units of [m/s]/[s] or m/s², the dimensions of acceleration.

The second statement relating the area under the curve to the distance traveled is less obvious. In Example 4.10, the *average* speed over the 11-second period of acceleration is $(27 + 6.7)/2 = 17$ m/s, and so the vehicle will have covered 17 [m/s] × 11 [s] $=$ 190 m (to two significant figures). It also can be approached dimensionally. The "height" or ordinate has units of [m/s]. The horizontal axis has units of seconds. So, length × height has units of [s] × [m/s] $=$ [m], a unit of distance.

Example 4.11

What is the distance traveled in steadily accelerating from 15 mph to 60. mph in 11s in SI units?
　　Need: Distance $=$ _____ m.
　　Know–How: The distance is the area (shaded in the diagram for Example 4.10) beneath the v–t graph. Geometry tells us we can break the area of a trapezoid into a rectangle with a length of 11 s and height of 6.7 m/s and a triangle with a base of 11 s and height $(27 - 6.7)$ of m/s.
　　Solve: Distance $=$ Area $=$ Area of rectangle $+$ Area of triangle

$$= 6.7\ [\mathrm{m/s}] \times 11\ [\mathrm{s}] + (1/2) \times 11\ [\mathrm{s}] \times (27 - 6.7)\ [\mathrm{m/s}]$$

$$= 74\ \mathrm{m} + 110\ \mathrm{m} = \mathbf{190\ m}$$

(Note that there are only two significant figures in this answer).

It is recommended that you use the v–t diagram on the problems and exercises of this chapter. The use of this tool develops a visual and intuitive appreciation of motion problems that cannot be obtained merely by manipulating equations. Formulae are useful once you fully understand the process, but the v–t diagram is the best way to achieve that understanding.

The answers to a number of other engineering problems are also contained in the v–t diagram. These range from the design of highway intersections to the design of a cannon that might be used to shoot payloads to the Moon. Example 4.12 shows the power of this method even for a relatively complex application.

Example 4.12

Provide a table and graph of the length of a highway on-ramp as a function of acceleration ranging from 1.0 m/s² to 10. m/s² in increments of 1.0 m/s². Assume that the vehicle enters the on-ramp with a speed of 15 mph (6.7 m/s) and leaves the on-ramp at a speed of 60. mph (27 m/s).
　　Need: A table and a graph with on-ramp entry lengths d versus acceleration, a.
　　Know–How: The v–t diagram gives the relationships among a, d, and Δt.

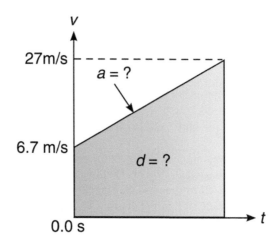

Solve: From the definition of acceleration as the slope of the $v - t$ line

$$a = \frac{\Delta v}{\Delta t} = \frac{(27 - 6.7)}{\Delta t} \text{ or}$$

$$\Delta t = \frac{(27 - 6.7)}{a} [m/s][s^2/m] = \frac{20.}{a} [s]$$

Also, d is the area under the $v-t$ curve, so

$$d = 6.7(\Delta t) + \frac{27 - 6.7}{2}(\Delta t) \, [m/s][s] = 6.7(\Delta t) \times 10.(\Delta t)[m] = 17(\Delta t) \text{ to two significant figures}$$

Now, substituting $\Delta t = 20./a$ gives $d = 340/a$ meters. Take these relationships to a spreadsheet and prepare the table, and from the table, prepare graphs of a versus Δt and d versus Δt as shown below.

a, m/s²	d, m	Δt, s
1.00	340	20.
2.00	170	10.
3.00	114	6.7
4.00	85	5.0
5.00	68	4.0
6.00	56	3.3
7.00	49	2.9
8.00	43	2.5
9.00	37	2.2
10.00	34	2.0

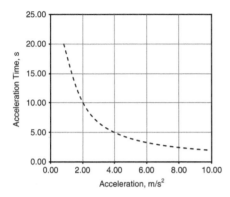

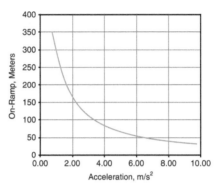

Summary

This chapter introduced the basic concepts of forces and motion. These concepts are necessary for many engineers as part of their fundamental knowledge.

The study of forces and their effects on objects was formulated by Isaac Newton when he developed his three laws of motion. **Newton's first law (statics)** deals with bodies that do not move unless forces act on them. **Newton's second law (dynamics)** relates the acceleration of a body to its mass and the unbalanced forces acting on it. Newton's second law is written as $F = ma$ in SI units of newtons, kg, and s, and as $F = ma/g_c$ in English Engineering units of lbf, lbm, and s, where $g_c = 32.2$ lbm·ft/(lbf·s^2). **Newton's third law** is the action/reaction law for interacting bodies.

Kinematics is the study of motion without regard to the forces that produce the motion. Kinematic relationships among distance, speed, and acceleration were presented through a versatile tool, the speed versus time (v–t) diagram. Newton's laws (statics and dynamics) are used in a wide variety of applications, and the subject of kinematics is used in a number of engineering fields, such as mechanical, civil, aerospace, and biomedical. Together, these concepts lie at the center of much of modern engineering. They allow engineers to design much of our modern world.

Exercises

Remember that when we say that a variable is "**exactly**" some number, that number is to be treated as an integer (or integer fraction) with an infinite number of significant figures. For example, if a mass is given as "exactly 5 kg" or as "exactly 1/3 kg," then the 5 is an integer and the 1/3 is an integer fraction and both have an infinite number of significant figures. This will help you determine the correct number of significant figures to use in your answers.

1. Suppose the mass in Example 4.2 is 50.0 slugs. What would its weight be in lbf (pounds force)?
2. What would the 5.00 slug mass in Example 4.2 weigh on the Moon, where the acceleration of gravity is only one-sixth of that on Earth?
3. What would the force on the body in Example 4.3 be if its mass were 856 g?
4. What would the weight of the body in Example 4.4 be on the Moon, where the acceleration of gravity is just $g_{moon} = 1.64$ m/s^2?
5. What force would be necessary in Example 4.5 if the mass is 735 lbm?
6. What is the value and units of g_c in the Engineering English system on the Moon?
7. Acceleration is sometimes measured in g, where $1.0 \ g = 9.8$ m/s^2. How many g's correspond to the steady acceleration of a car going from exactly 0 to 60. mph in 10. seconds? (**Ans. 0.27 g**)
8. What is your mass in kilograms divided by your weight in pounds? Do you have to step onto a scale to answer this question? How did you answer the question?
9. If power (measured in W, or watts) is defined as work (measured in J, or joules) performed per unit time (measured in s), work is defined as force (measured in N or newtons) × distance (measured in m), and speed is defined as distance per unit time (measured in m/s), what is the power being exerted by a force of 1000. N on a car traveling at 30. m/s. (Assume force and speed are in the same direction and treat all numbers as positive.) (**Ans. 3.0 × 10⁴ W**)
10. A rocket sled exerts 3.00×10^4 N of thrust and has a mass of 2.00×10^3 kg. How much time does it take to go from exactly 0 to 60.0 mph? How many g's (see Exercise 7) does it achieve?
11. A person pushes a crate on a frictionless surface with a force of 100. lbf. The crate accelerates at a rate of 3.00 feet per second^2. What is the mass of the crate in lbm? (**Ans. 1.07 × 10³ lbm.**)
12. The force of gravity on the Moon is exactly one-sixth as strong as the force of gravity on Earth. An apple weighs 1.00 N on Earth. (a) What is the mass of the apple on the Moon in lbm? (b) What is the weight of the apple on the Moon, in lbf? (Conversion factor: 1.00 kg = 2.20 lbm.)
13. How many lbf does it take for a 4.0×10^3 lbm car to go from exactly 0 to 60. mph in 10. seconds? (**Ans. 1.1 × 10³ lbf.**)
14. Suppose a planet exerted a gravitational force at its surface that was 0.6, the gravitational force exerted by Earth. What is g_c on that planet?

   It is strongly suggested you use the **v−t diagram** for most of the following exercises.
15. A car is alone at a red light. When the light turns green, it starts at constant acceleration. After traveling 100. m, it is traveling at 15 m/s. What is its acceleration?

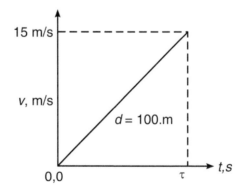

16. A car is alone at a red light. When the light turns green, it starts at constant acceleration. After traveling exactly one-eighth of a mile, it is traveling at 30. mph. What is its acceleration in m/s^2?
17. An electric cart can accelerate from exactly 0 to 60. mph in 15 s. The Olympic champion sprinter Usain Bolt can run 100. m in 9.69 s. Which would win a 100. m race between this cart and the world champion sprinter? (**Ans. The sprinter wins by 16.7 m**)

18. A car starts from a stop at a traffic light and accelerates at a rate of 4.0 m/s^2. Immediately on reaching a speed of 32 m/s, the driver sees that the next light ahead is red and instantly applies the brakes (reaction time = 0.00 s). The car decelerates at a constant rate and comes safely to a stop at the next light. The whole episode takes 15 s. How far does the car travel? (**Partial Ans.** See the following figure.)

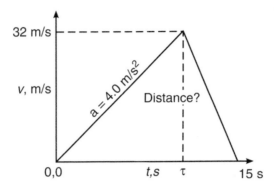

19. For the previous problem, supply a table and draw a graph showing the distance, speed, and acceleration of the car versus time.

20. Based on your experience as a driver or a passenger in a car, estimate the maximum deceleration achieved by putting maximum pressure on the brakes when traveling at 30. mph.

21. A car leaves a parking space from a standing stop to travel to a fast-food restaurant 950. meters away. Along the journey, it has to stop after 325 m at a stop sign. It has a maximum acceleration of 3.0 m/s^2 and a maximum deceleration of $-10. \text{ m/s}^2$. It never exceeds the legal speed limit of 15 m/s. What is the least possible time it can take until the car comes to a full stop in front of the fast-food restaurant? (**Ans. 70. s.**)

22. You are an engineer designing a traffic light. Assume a person can see a traffic light change color from red to yellow, and it takes 1 second to respond to a change in color. Suppose the speed limit is 15 m/s. Your goal is to enable drivers always to stop after seeing and responding to the yellow light with a maximum deceleration of -5.0 m/s^2. How long should the yellow light last?

23. You are a driver responding to the traffic light in the previous exercise. If it was correctly designed according to that problem, at what distance from the light should you be prepared to make your "to stop or not to stop" decision? Assume you are a safe driver who neither speeds up to get through the yellow light nor stops more suddenly than the deceleration rate of

    -5.0 m/s^2.

24. Suppose the deceleration of a car on a level off-ramp is -3.0 m/s^2. How long would the off-ramp have to be to allow a car to decelerate from 60. mph to 15 mph? (**Ans. 110 m**)

25. An early proposal for space travel involved putting astronauts into a large artillery shell and shooting the shell from a large cannon.[2] Assume that the length of the cannon is 30. M and the speed needed by the shell to achieve orbit is 15,000 m/s. If the acceleration of the shell is constant and takes place only within the cannon, what is the acceleration of the shell in g's?

26. Suppose that a human body can withstand an acceleration of 5.0 g's, where 1 g is 9.8 m/s^2. How long would the cannon have to be in the previous exercise to keep the acceleration of the humans within safe limits? (**Ans.** $4.6 \times 10^6 \text{ m}$)

27. You wish to cover a 2.00-mile trip at an average of 30.0 mph. Unfortunately, because of traffic, you cover the first mile at just 15.0 mph. How fast must you cover the second mile to achieve your initial schedule?

28. You are an engineer with the responsibility of choosing the route for a new highway. You have narrowed the choice to two sites that meet all safety standards and economic criteria. (Call them Route A and Route B.) Route B is arguably slightly superior in terms of both safety and economics. However, Route B would pass by the site of an expensive new house recently built by your favorite niece, and the proximity to the highway would severely depress the value of the house. As your niece's last name is different from yours and you have never mentioned her at the office, you would be unlikely to be "caught" if you chose Route A to save the value of her house. What do you do? (Use the Engineering Ethics Matrix.)

    (a) Choose Route A.

    (b) Choose Route B.

    (c) Ask to be relieved from the responsibility because of a conflict of interest.

29. You are an engineer on a team designing a bridge for a state government. Your team submits what you believe to be the best design by all criteria, at a cost that is within the limits originally set. However, some months later the state undergoes a budget crisis. Your supervisor, also a qualified engineer, makes design changes to achieve cost reductions that he believes will not compromise the safety of the bridge. You are not so sure, though you cannot conclusively demonstrate a safety hazard. You request that a new safety analysis be done. Your supervisor denies your request on the grounds of time and limited budget. What do you do? (Use the Engineering Ethics Matrix).

    (a) Go along with the decision. You have expressed your concerns, and they have been considered.

    (b) Appeal the decision to a higher management level.

    (c) Quit your job.

    (d) Write your state representative.

    (e) Call a newspaper reporter and express your safety concerns.

---

[2] A number of military satellites have been delivered to space using specially adapted naval cannons from the Vandenberg Air Force Base in California.

Energy

5.1 Introduction

Without energy, we would not exist. Our bodies store energy from the food we eat and this energy allows us to walk, run, read, write, play games, and think. Electricity is perhaps the most versatile type of energy. It can be converted into mechanical energy to power cars, ships, and factories. It can also be used to create heat, and it is essential for powering the computers modern society depends upon.

Is the world running out of energy? Where does the energy come from that provides the light by which you are reading these words? Does a car possess more energy when it is sitting in your driveway or 15 min later when it is traveling down the highway at 65 miles per hour? In this chapter, you will learn how to answer these

Exploring Engineering. https://doi.org/10.1016/B978-0-443-13541-5.00019-2

questions, as well as how to address the more quantitative ones involving energy that engineers of all types encounter in their work.

Energy comes in many different forms, and with modern technology, we can convert it from one form to another. A fundamental property of energy is that *it cannot be created or destroyed because it is conserved*. Consequently, the total amount of energy in the universe remains constant.

By understanding energy conversion processes, engineers can design technologies with a wide range of applications, such as new fuels for automobiles, methods of protecting buildings from earthquakes or lightning strikes, or understanding the processes required to replace fossil fuels with renewable energy resources such as solar and wind power, biomass, or geothermal. To achieve these wide-ranging technologies, engineers must understand the concept of energy and its conversion from one form to another.

Mechanical work is based on **Newton's Second Law of Motion** because it introduced the notion of a force in motion. In the simplest mathematical terms, work (W) is defined as:

$$\text{Work} = W = Force \times distance \ = \ F \times d \qquad (5.1)$$

in which a force F moves through a distance d. To do 5 J of work, you could apply a force of 5 N of force to move something through a distance of 1 m, or 1 N moving through 5 m, and so on.

Fig. 5.1 represents one of the most important principles you will have to master as an engineer. An engineer analyzes energy flow using the concept of a "**system**" that surrounds the object of interest. The object might be an engine, a sailboat, a lawn mower, an electric toaster, a crane, or anything that uses energy. It is identified by drawing a dotted line, called the **system boundary**, loosely around it. You must use engineering skills and know-how to construct a useful system boundary so that you can monitor the energy that flows across it. When you do this, the **Law of**

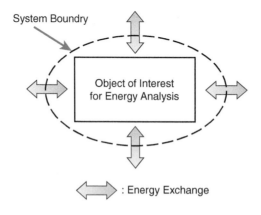

FIGURE 5.1

Energy exchanges.

Conservation of Energy will enable you to determine the energy required for an object contained inside the system boundary to operate properly.

5.2 Energy has the capacity to do work

Understanding the variable "energy" begins by defining variables for the concepts of force and work. Force is a variable that may be thought of as a push or a pull, as measured with a spring balance or by the weight of an object in a gravitational field. As we have already seen, the force has the units of **newtons (N)** in the (SI) system and **pounds force (lbf)** in the Engineering English system of units. For example, at standard gravity, the weight in the SI units required of a book of those mass is 1.0 kg is

$$\mathbf{Weight = mass \times gravity = mg} = 1.0 \,[\text{kg}] \times 9.81 \,[\text{m/s}^2] = 9.81 \,[\text{kg} \cdot \text{m/s}^2]$$
$$= 9.8 \text{ newtons} = \mathbf{9.8 \text{ N}}.$$

In the Engineering English system, the mass of this book is 1.0 [kg] × 2.2 [lbm/kg] = 2.2 lbm, and its weight is

$$\mathbf{Weight} = \frac{\mathbf{mg}}{g_c} = \frac{2.2 \,[\text{lbm}] \times 32.2 \,[\text{ft/s}^2]}{32.2 \,[\text{lbm} \cdot \text{ft}/(\text{lbf} \cdot \text{s}^2)]} = 2.2 \text{ lbf}$$

The units of work are **joules (J)** in the SI system, and **foot pounds force** (ft·lbf) in the Engineering English system. As work is force times distance, a joule is defined to be a newton × meter (**N·m** or equivalently a **kg·m²/s²**). For example, raising a book with a mass of 1.0 kg and weighing 9.8 N from the top of your desk to 1.0 m above your desk requires a work equal to

$$\text{Work} = \text{Force} \times \text{Distance} = \text{Weight} \times \text{Distance} = \text{mg} \times d$$
$$= 9.8 \,[\text{N}] \times 1.0 \,[\text{m}]$$
$$= 9.8 \,[\text{N} \cdot \text{m}] = 9.8 \text{ joules}.$$

In the Engineering English units system, the weight of this book is 2.2 lbf, and the distance is 1.0 [m] × 3.28 [ft/m] = 3.3 feet. Then the work required is

$$\text{Work} = \frac{\text{mg} \times d}{g_c} = \frac{2.2 \,[\text{lbm}] \times 3.3 \,[\text{ft}] \times 32.2 \,[\text{ft/s}^2]}{32.2[\text{lbm} \cdot \text{ft}/(\text{lbf} \cdot \text{s}^2)]} = 7.3 \text{ ft} \cdot \text{lbf}$$

Energy may be stored in objects, such as gasoline, solid uranium, or a speeding train. The stored energy in an object can be released, such as by water running through a turbine, heat from burning gasoline, or electrical current from a generator. More precisely, any object that does an amount of work consisting of some number of joules or ft·lbf will see its energy content decrease by that same amount. Also, any object that has an amount of work done on it will have its energy increased by that work.

Often, an engineer wants to know how fast work was done or how rapidly the amount of energy possessed by an object has changed. The time rate of change of work is called **power**. Power is work divided by the time it took to do the work, or

$$\textbf{Power} = P = \textbf{Work/Time}$$

It is measured in **watts** (W) in SI, where 1 W is equal to 1 J per second. For example, if a person takes 2.0 s to lift a 1.0 kg book at a height of 1.0 m above the surface of their desk, that person uses

$$Power = Work/Time = (Force \times height)/Time$$

$$= ((1.0 \text{ [kg]} \times 9.8 \text{ [m/s}^2\text{]}) \times 1.0 \text{ [m]})/(2.0 \text{ [s]}) = 4.9 \text{ J/s} = 4.9 \text{ watts of power.}$$

In the Engineering English unit system, power is normally expressed in units of ft·lbf/s or in horsepower,[1] where one horsepower is equal to 550 ft·lbf/s. It is roughly the power exerted by one horse and is roughly 5—10 times the power exerted by a person doing physical labor such as shoveling or carrying. Some other examples of quantities of energy are shown in Table 5.1.

In the past three centuries, a vast increase has occurred in the amount of energy available to the ordinary person in the United States. In 1776, when Thomas Jefferson was writing the Declaration of Independence, the average adult in the United States would use less than 1 kilowatt-hour (kW·h) of energy per day.[2]

With less than 5% of the world's population, the United States now consumes nearly 20% of the world's energy supply. The average person in the United States typically uses more than 250 kW·h of energy every day in the form of domestic electrical energy and chemical energy in the fuels used in automobiles, trucks, and airplanes. At the same time, vast populations elsewhere in the world are still at the less-than-one-kilowatt-hour level (or worse, a one-person-energy level). Resolving this global energy discrepancy in an environmentally acceptable manner is one of the great engineering and social challenges of the 21st century.

5.3 Different kinds of energy

Energy comes in various forms. Energy due to motion includes both **translational kinetic energy (TKE)** and **rotational kinetic energy (RKE)**. We also have **thermal energy** and **gravitational potential energy (GPE)**, and, less obviously, **chemical energy** and **electrical energy**. There are other important energy types such as

---

[1] The Scottish engineer James Watt measured the rate of work that a good horse could sustain as a way to advertise the power produced by his early steam engines.
[2] Energy is measured in joules (J) and 1 J = 1 watt second (W·s). However, it is more convenient to measure energy usage in modern technologies in kilowatt hours (kW·h) where 1.0 kW·h = 3. 6 × 10^6 J = 3.6 MJ

Table 5.1 Typical energy magnitudes.

Joules	Approximate energy contained in:
1	A small apple falling 1 meter
10	A person swinging a baseball bat
10^2	A lighted match
10^3	One hamburger
10^4	A speeding bullet
10^5	A small car at 65 mph
10^6	A small meal
10^7	One kg of burning wood
10^8	One gallon of gasoline
10^9	A lighting strike
10^{10}	One ton of oil
10^{11}	One gram of uranium in fission
10^{12}	1000 tons of TNT
10^{13}	An atomic bomb
10^{14}	The annihilation of 1 gram of matter
10^{44}	A super nova explosion

Einstein's discovery that mass is a form of energy, as embodied in his famous equation $E = mc^2$ (see the chapter on Nuclear Energy).

TKE is the energy of a mass moving in a straight line. It is calculated by the equation:

$$\text{TKE} = \frac{mv^2}{2} \text{ (in SI units) or TKE} = \frac{mv^2}{2g_c} \text{ (in English units)} \tag{5.2}$$

where m is mass and v is the velocity (speed). When TKE is assumed to be the only form of energy of motion, it is simply called "kinetic energy" and then abbreviated "**KE**."

Note first that Eq. (5.2) results in the correct SI units of joules, since speed is m/s, and mass is in kilograms, so ½ mv^2 has units $[\text{kg}][\text{m}^2/\text{s}^2] = [\text{N} \cdot \text{m}] = \text{joule [J]}$. In Engineering English units, we have to divide by the conversion constant g_c to get the proper units of $[\text{ft} \cdot \text{lbf}]$ or,

$$\frac{\left[\text{lbm} \left(\frac{\text{ft}}{\text{s}} \right)^2 \right]}{\left[\frac{\text{lbm} \cdot \text{ft}}{\text{lbf} \cdot \text{s}^2} \right]} = [\text{ft} \cdot \text{lbf}]$$

Example 5.1

What is the TKE of an automobile with a mass of 1.0×10^3 kg traveling at a speed of 65 miles per hour (29 m/s)?

Need: TKE of vehicle.
Know: Mass is 1.0×10^3 kg, speed is 29 m/s.
How: Apply Eq. (5.2), TKE $= \frac{1}{2}mv^2$.
Solve: TKE $= \frac{1}{2} \times (1.0 \times 10^3 \ [\text{kg}]) \times (29^2 \ [\text{m/s}]^2) = 423{,}410 \ [\text{kg m}^2/\text{s}^2] = 4.2 \times 10^5$ J
$= \mathbf{420 \ kJ}$.

Anything that has mass and is moving in a straight line has TKE. Prominent examples of TKE in nature include the winds and ocean tides.

Example 5.2

Estimate the total kinetic energy of the wind on Earth. As an introduction to this problem, consider that the wind is a movement of air that is produced by temperature differences in the atmosphere. Since hot air is lighter than cold air, air heated by the Sun at the equator rises until it reaches an altitude of about 6.0 miles (9.0 km) and then it spreads north and south. If the Earth did not rotate, this air would simply travel to the North and South Poles, cool down, and return to the equator along the surface of the Earth as wind. However, because the Earth rotates, the prevailing winds travel in a west-east rather than a south-north direction in the northern hemisphere.

Need: Total TKE of the wind in joules.
Know: The atmosphere is about 9.0 km thick (i.e., the height of Mt. Everest). The radius of the Earth is about 6.4 million meters. The surface area of a sphere is $4\pi R^2$, so the volume of an annular "shell" of thickness T around the Earth[3] is about $4\pi R^2 T$, since the thickness T is very small compared to the radius R. Air has a density of about 0.75 kg per cubic meter (averaging from sea level to the top of Mt. Everest). Also, this air has a speed of about $v = 10$. m/sec.
How: Find the mass of air in that 9.0 km thick shell around the Earth, and apply Eq. (5.2), TKE $= \frac{1}{2} mv^2$.
Solve: Volume or air around the Earth is $4\pi R^2 T$, so

$$V = 4\pi \times (6.4 \times 10^6 \ [\text{m}])^2 \times (9.0 \times 10^3) \ [\text{m}] = 4.6 \times 10^{18} \ \text{m}^3.$$

The mass of air around the Earth is its density $\times$ volume, or

$$m = (0.75 \ [\text{kg/m}^3]) \times (4.6 \times 10^{18} \ [\text{m}^3]) = 3.5 \times 10^{18} \ \text{kg}.$$

So, the TKE $= \frac{1}{2} \times (3.5 \times 10^{18} \ [\text{kg}]) \times 10.^2 \ [\text{m/s}]^2 = 1.7 \ \text{kg} \cdot \text{m}^2/\text{s}^2 = \mathbf{1.7 \times 10^{20} \ J}$.

This is a number sufficiently large as to be meaningless to most people. But to an engineer it should inspire such questions as: Where does this energy come from? Where does it go? In fact, the total Sun's energy reaching the Earth is about 1.4 kW for every m^2 of the Earth's surface. Of this, about 31% is reflected back into space. Thus, we receive about 1 kW/m^2 net solar radiation, and over the Earth this amounts to about 1.7×10^{17} kW or 1.7×10^{20} W.

---

[3] Since the volume of a sphere is $(4/3)\pi R^3$, then the volume of the atmosphere of thickness T is $(4/3)\pi(R + T)^3 - (4/3)\pi R^3$, which is approximately $(4/3)\pi R^2 T$ if R is much greater than T.

Since we receive this for 12 hours every day, our beneficent Sun delivers $(1.7 \times 10^{20}$ [J/s]) $\times$ (12 [h/day] $\times$ (60 [m/h]) $\times$ (60 [s/m]) $= 7.3 \times 10^{24}$ J/day. So our estimate of the wind's TKE accounts for only $(1.7 \times 10^{20}/7.3 \times 10^{24}) \times 100 \approx 0.002\%$ of the daily solar energy we receive.

The role that the winds or other solar energy sources might play in meeting the needs of those billions of people who still live at an eighteenth-century energy standard is still debated. The major problems are that this source of energy is diffuse, often unpredictable, and fluctuates daily and seasonally.

Thermal energy is often referred to as *heat* and is a very special form of kinetic energy because it is the *random* motion of trillions and trillions of atoms and molecules that leads to the perception of temperature. Heat is simply the motion of things too small to see, an insight captured by the nineteenth-century German physicist Rudolf Clausius when he defined thermal energy as "the kind of motion we call heat."

Our analysis of the meaning of temperature will begin with the motion of a single particle of gas (either an atom or a molecule) in a box. The **Ideal Gas Law** for a fixed mass or equivalently, a fixed number of moles of gas is:

$$p\,V = NR\,T \qquad (5.3)$$

where p is the pressure, V is the volume of the gas, T is its absolute temperature, and R is the universal gas constant per **mole** of gas, and N is the number of moles of gas.[4] This law is accurate for many gases over a wide range of conditions.

We can use the Ideal Gas Law to analyze **pressure** (i.e., the force per unit area) from the point of view of the motion of a single gas atom in a box. The pressure on the walls of the box is the result of the atom repeatedly hitting the walls of the box and transferring its momentum to the walls. If there are few collisions, the internal pressure on the box walls is small, but if there are many collisions, the pressure is larger. If the temperature rises, the atom has more speed and thus more momentum and consequently, the pressure on the walls increases. Now we need to calculate the rate of change of momentum (mv) as our single atom hits a wall of the box.

Assume that an atom is moving with speed v perpendicularly toward the box's wall (see Fig. 5.2). On impact with a wall, the atom's momentum will be transferred to the wall of the box, and the resulting force on the wall is proportional to its rate of change of momentum as (the symbol $\propto$ means "proportional to")

$$F \propto mv/t \qquad (5.4)$$

As the average time for an atom or molecule to reach the wall of the box is proportional to L/v, where L is a measure of the size of the box, the *force* on the wall is therefore proportional to mv^2/L. The wall area for the impact of atoms or molecules is L^2 and thus the pressure on this wall by our chosen particle is

---

[4] A *mole* of gas is its mass divided by its molecular mass. For example, helium has a molecular mass of 4 kg per kg-mole so that 1 kg-mole of it has a mass of 4 kg, and 2 kg-moles $= 8$ kg, and so on. For oxygen gas (a molecule with two oxygen atoms each of molecular mass 16 kg/kg-mole) it has a molecular mass of 32 kg per kg-mole meaning 1 mole of it has a mass of 32 kg.

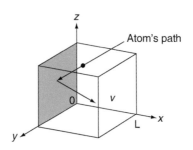

FIGURE 5.2

Cubical box of side *L*, volume *V* containing just one atom.

$$P \propto \frac{Force}{Area} \propto \frac{mv^2/L}{L^2} \propto \frac{mv^2}{L^3} \propto \frac{mv^2}{V} \tag{5.5}$$

where $V = L^3$ is the box's volume. But there are trillions of atoms (or molecules) in just 1 cm³ of gas. Let this number of atoms be **n**. Thus the total pressure on each of the box's walls is

$$p \propto n \times \left(\frac{mv^2}{V} \right) \tag{5.6}$$

It is hard to comprehend just how large *n* really is. The next example will show you why.

Example 5.3

If the number of atoms in 1 kg-mole of a gas is 6.022×10^{26}, how many atoms of this gas occupy just 1.00 cm³ at the standard atmospheric pressure of 1.00×10^5 N/m² and a temperature of 273 K (0°C).

    Need: *n, the number of* atoms in 1.00. cm³ $= 1.00 \times 10^{-6}$ m³ of a gas at $P = 1.00 \times 10^5$ N/m² and $T = 273$ K.

    Know: 1 kg mole of the gas contains 6.022×10^{26} molecules, and the universal gas constant is $R = 8314$ J/(kg mole·K).

    How: Use the ideal gas law $p V = N R T$ to find N (the number of moles), and then convert N, into n, the number of atoms.

    Solve:

$$N = \frac{pV}{RT} = \frac{1.00 \times 10^5 \times 1.00 \times 10^{-6}}{8314 \times 273} \quad \left[\frac{N}{m^2} \times m^3 \times \frac{\text{kgmole K}}{J} \times \frac{1}{K} \right]$$

$$= 4.41 \times 10^{-8} \text{ kg moles in } 1.00 \ cm^3$$

Finally, convert N into the number of atoms in 1 cm³ we have

$$n = 4.41 \times 10^{-8} \times 6.022 \times 10^{26} \text{ [kg moles/cm}^3\text{][molecules/kg mole]} = 2.65 \times 10^{19} \text{ atoms}$$

This is a very large number and there is no way of monitoring the motion of all of these gas atoms or molecules as they collide with each other and the walls in random directions and at random speeds. Thus, we use *averaging* techniques to describe the behavior of large numbers of atoms or molecules.

Returning to our calculation of the pressure on the walls of a container, it is just the force per unit area, (*F*/A), and we can see that:

$$pV \propto n \times mv^2 \qquad (5.7)$$

Note that $n \times m$ is the total *mass* of all of the gas in the box.

Our two expressions for pV, the ideal gas law $pV = NRT$ and our derived equation, $pV \propto n \times mv^2$ are very similar so that we can conclude that the absolute temperature T is:

$$T \propto \tfrac{1}{2} mv^2 \qquad (5.8)$$

So the absolute temperature of a gas is thus proportional to the kinetic energy of the atoms or molecules, $\tfrac{1}{2} mv^2$ so when you feel hot it is because the kinetic energy of the molecules of air striking your body has increased.

Example 5.4

If you feel hot at 35°C and cold at 5.0°C, what the % change in the average speed of molecules contacting your skin? **Hint**: convert the temperatures in degrees C into absolute temperatures in degrees kelvin by adding 273 degrees.

> **Need:** % change in average molecular speed between a cold gas (at 5.0°C = 278 K) and a hot gas (at 35°C = 308 K).
>
> **Know-How:** From Eq. (5.8) we have that $T \propto \tfrac{1}{2} mv^2$
>
> **Solve:** Form the ratio $\frac{T_H}{T_C} = \frac{\frac{1}{2} mv_H}{\frac{1}{2} mv_C} = \left(\frac{v_H}{v_C}\right)^2$ where the subscripts refer to hot and to cold. Then we have $\frac{v_H}{v_C} = \sqrt{\frac{T_H}{T_C}}$
>
> Therefore $\frac{v_H}{v_C} = \sqrt{\frac{T_H}{T_C}} = \sqrt{\frac{308}{278}} = 1.05$
> So the difference in the average air molecule's speed is only 5% between these temperatures.

The key principle here is that temperature is a result of the kinetic energy of atoms and molecules. If we want the temperature of a roast in the oven, or a piece of steel being welded, or a block of ice, we cannot measure the speeds of all the molecules. In practice, we use thermometers and other kinds of devices that automatically average the speed of the atoms or molecules.

In 1848 William Thomson (Lord Kelvin) developed an absolute temperature scale based on the Celsius degree. Later an absolute temperature scale based on the Fahrenheit degree was developed by the Scottish engineer William Rankine. The relationship between these temperature scales is shown below.

$$T(°F) = (9/5) \times T(°C) + 32 = T(R) - 460.$$

$$T(°C) = (5/9) \times [T(°F) - 32] = T(K) - 273$$

$$T(R) = (9/5) \times T(K) = (1.80) \times T(K) = T(°F) + 460.$$

$$T(K) = (5/9) \times T(R) = T(R) \div 1.80 = T(°C) + 273$$

in which $T(R)$ and $T(K)$ (*without* the degree symbol on the R and K) represent the temperature in Rankin and Kelvin.

Gravitational potential energy (GPE) is the energy acquired by an object by virtue of its position in a gravitational field—typically by being raised above the surface of the Earth. In SI units, it has the unit of joule and is calculated by the equation:

$$(\text{GPE})_{\text{SI}} = mgh \tag{5.9}$$

where h is the height above or below some reference level, usually the local ground level. In Engineering English units, this definition is modified by the constant g_c to ensure that it comes out in units of ft·lbf:

$$(\text{GPE})_{\text{English}} = \frac{mgh}{g_c} \tag{5.10}$$

Example 5.5

Wile E. Coyote holds an anvil of mass 100. lbm at the edge of a cliff, directly above the Road Runner who is standing 1000. feet below. Relative to the position of the Road Runner, what is the GPE of the anvil?

Need: GPE.

Know: The anvil has a mass of 100. lbm. The reference level is where the Road Runner is standing, so the height h of the anvil is 1000. feet above the Road Runner. Because we are calculating in Engineering English units, we will also have to use g_c.

How: Eq. (5.10) is in Engineering English units, GPE $= \frac{mgh}{g_c}$.

Solve: GPE $= \frac{100.\ [\text{lbm}]\ 1000.\ [\text{ft}]\ 32.2\ [\text{ft/s}^2]}{32.2\ [\ \text{lbm} \cdot \text{ft} / \text{lbf} \cdot \text{s}^2]} = 1.00 \times 10^5$ ft·lbf

to three significant figures as the mass of the anvil has the least number (3) of significant figures.

Electrical energy (often just called "electricity") is a form of energy that is typically carried by electric charges (negatively charged electrons) moving through wires or electromagnetic waves moving through space. Like all forms of energy, electrical energy is measured in joules, and electrical power is measured in watts. Consider the electrical power used by a device that is connected by wires to a battery to create an electrical circuit as shown in Fig. 5.3.

An **electrical current**, represented by the letter I, flows into and out of the device. Electrical current is measured in units called **amperes** (symbol **A**), which is

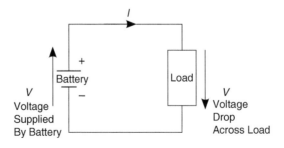

FIGURE 5.3

An electrical circuit.

simply a measure of the number of electrons passing through any cross-section of the wire every second. Electrical potential is measured in units called **volts (V)** and originates in the battery that keeps the current flowing through the wire. The voltage has its most positive value at the terminal on the battery marked plus (+), and drops throughout the circuit, reaching its minimum value at the battery terminal marked minus (−). By measuring the voltages at any two points along the wire and subtracting them, one can determine the voltage drop between those two points. The voltage drop can be thought of as an "electrical pressure" pushing the current from the point of higher voltage to the point of lower voltage.

In mechanical systems, power is the rate of doing work, and is computed as the product of force × speed. In an electrical system, the force on a charge of Q **coulombs (C)** in a voltage drop of V/d is $Q \times V/d$ (d is the distance over which the voltage changes from 0 to V)—hence, the *work* in moving the charge is $d \times QV/d = QV$. If the charge is moved in time t, the power required to move it is VQ/t.

We define the rate of movement of charge as the **electric current**, $I = Q/t$. If Q is in coulombs and t is in seconds, then I is in amperes (1 A = 1 C/s)). Further, if V is in volts, then the power is in watts (1 W = 1 V·A). Simply put, what you need to remember is that the electrical power is expressed as:

$$\textbf{Electric Power} = \textbf{\textit{P}} = \textbf{\textit{Voltage}} \times \textbf{\textit{Current}} = \textbf{\textit{V}} \times \textbf{\textit{I}} \tag{5.11}$$

Example 5.6

A 3.0 V battery produces a current of 0.1 A through a small lightbulb. What is the power of the lightbulb?

Need: Power of lightbulb in watts (W).
Know: Voltage across bulb is 3.0 V and the current through lightbulb is $I = 0.1$ A.
How: Use power = voltage × current, or $P = V \times I$, Eq. (5.11).
Solve: $P = (3.0 \text{ [V]}) \times (0.1 \text{ [A]}) = 0.3 \text{ [V·A]} = 0.3$ W.

Note that the power into the light bulb leaves the electric circuit, but it does not disappear. Instead, some of it is radiated away from the bulb in the form of visible "light," and some of it is in the form of invisible heat that is lost to the light bulb's local surroundings.

Chemical energy is another form of energy that is determined by the relative distribution of electrons in the atoms that make up the structure of chemical molecules. Chemical energy is so important to 21st century engineering that we devote an entire chapter to it. Suffice it here to say that it too is measured in joules.

5.4 **Energy conversion**

Sunlight drives the winds. The potential energy of an anvil on a cliff top is converted into kinetic energy as it falls toward an unwary Road Runner below. Burning fossil fuel (a form of stored solar energy) is converted into the rotational kinetic energy (RKE) of a turbine, which is then converted into electrical energy in a generator to light up a city. All these occurrences suggest a second key fact about energy: its various kinds can be converted from one form to another.

We mentioned earlier that the Sun delivers more than 10^{24} J of energy daily to the Earth in the form of the solar energy that we call sunlight. What happens to that energy? Part of this energy heats the atmosphere and drives the winds as well as driving the water cycle. Part of this solar energy is converted into chemical energy via photosynthesis in trees, grass, and agricultural crops. Part of it simply heats the Earth and the ocean, and some of it is reflected back into space. If all that energy is arriving from the Sun every day, and much of it is converted to thermal energy (heat), then why isn't the Earth accumulating more and more of this energy and continuously getting hotter?

To be in equilibrium, the Earth must reflect or radiate into space the same amount of energy that it receives from the Sun. Should anything happen, either due to natural or human causes, to interfere with this energy balance between absorption and radiation of solar energy, the Earth could either cool (as it did in the various ice ages) or warm — as it appears to be doing right now according to many scientists who believe that global warming is taking place. So understanding energy conversion is crucial to protecting the future of life on our planet.

We can perform a similar energy balance on a technological system, such as an automobile, a house, a toaster, or anything we choose. In the case of an automobile, the initial source of energy is the chemical energy pumped into the vehicle at the gas station. In the automobile's engine, that chemical energy is converted to the TKE of a piston in a cylinder. That TKE is then converted into RKE of the crankshaft, transmission, axles, and wheels. That RKE in turn is converted into the TKE of the automobile as it moves down the road. In this process, part of that energy—a majority of the initial chemical energy—is transferred to the environment in the form of heat through the car's radiator, exhaust pipes, road friction, and air resistance. Energy conversion is often not very efficient and much of it ends up as thermal energy (heat) in the atmosphere.

Example 5.7

A gallon of gasoline can provide about 1.30×10^5 kJ of chemical energy. Based on Example 5.1, if all the chemical energy of a gallon of gasoline could be converted into the TKE of an automobile, how many gallons of gasoline would be equivalent to the vehicle's TKE if the automobile is traveling at 65 miles per hour on a level highway?

Need: Gallons of gasoline to propel the automobile at 65 mph (29 m/s).

Know: TKE of vehicle $= 4.2 \times 10^2$ kJ (to two significant figures at 65 mph as per our previous calculation in Example 5.1). Energy content of gasoline is 1.30×10^5 kJ/gallon.

How: Set TKE of vehicle equal to chemical energy in fuel. Let $X =$ number of gallons needed to accomplish this.

Solve: $(X \text{ [gallons]}) \times (1.30 \times 10^5 \text{ [kJ/gallon]}) = 4.2 \times 10^2 \text{ [kJ]}$.
Solving for X gives $X = (4.2 \times 10^2 \text{ [kJ]})/(1.30 \times 10^5 \text{ [kJ/gallon]}) = \mathbf{3.2 \times 10^{-3}}$ **gallons.**

This is a tiny amount of gasoline. As we will see later, this is a misleadingly small amount of gasoline. In fact, most of the gasoline is *not* being converted into useful KE but is mostly "wasted" elsewhere. The fault is in the statement "*if all* the chemical energy of a gallon of gasoline could be converted into the translational kinetic energy of an automobile." This simply cannot be done. The lesson to be learned here is to beware of your assumptions!

5.5 Conservation of energy

A "conserved" quantity can neither be created nor destroyed. In nature, only energy, momentum, and electric charge are conserved. Engineers have a method of applying the fact that energy is conserved to practical problems. This method is called "**control boundary**" analysis and was schematically illustrated in Fig. 5.1.

Control volume analysis consists of isolating the particular object or system under consideration and observing the ways in which the object exchanges energy with its surroundings. To begin this analysis, you make a sketch of the object or system and draw a dotted line around it that represents the system boundary. This becomes your control boundary for energy analysis. Then draw arrows that cross the control boundary representing specific types of energies that flow between the object or system and its surroundings.

As a simple example, consider a 1.0 kg book on your desk, and that the desk is 1.0 m high. The book has GPE $= mgh = 1.0$ [kg] $\times 9.81$ [m/s^2] $\times 1.0$ [m] $= 9.8$ [kg·m/s^2] $= 9.8$ J with respect to the floor. Suppose that the book now falls to the floor, thus losing all of its GPE. Where did the GPE go? Have we violated the conservation of energy?

The fact is that GPE is still in the room but not in its original form. What physically happens is: (1) the GPE of the falling book is converted to TKE as it falls; (2) when the book hits the floor, it sets up some sound waves in the floor and in the surrounding air; and finally (3) all the energy in (1) and (2) end up heating the room and its contents by exactly 9.8 J of energy (ultimately all forms of energy degrade to heat) see Fig. 5.4.

The principle of conservation of energy says energy is never created or destroyed, it is merely transformed from one form to another form (this is also called the **First Law of Thermodynamics**). Control boundary analysis is a very useful way to account for energy conversions and flows. This process can be highly useful in engineering analysis and design.

Fig. 5.5 shows a control boundary analysis applied to an automobile. It illustrates the "energy accounting" of a typical automobile trip. As indicated by the arrows crossing the system dotted line, an automobile traveling at 65 miles per hour (29 m/s) transfers about 40. kW to the atmosphere in the form of thermal energy from the radiator, about 20. kW to the atmosphere in the form of heat and chemical

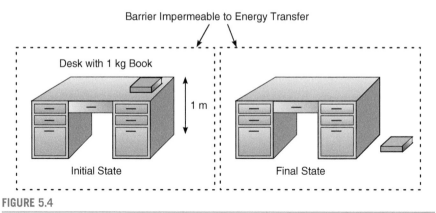

FIGURE 5.4

Where did the book's potential energy go?

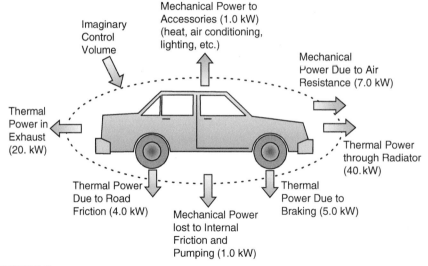

FIGURE 5.5

Control boundary analysis model of an automobile.

energy out of the exhaust, about 7.0 kW of mechanical power to the atmosphere in overcoming air resistance (aerodynamic drag), about 10. kW in frictional work (mainly the result of elastic compression and expansion of the rubber in the tires and applying the brakes), and about 1.0 kW of power for other purposes such as pumping and operating such accessories as heat, lights, and air conditioning. All of these contributions total about 78 kW.

Example 5.8

If gasoline contains 1.30×10^5 kJ/gallon, how many gallons of gasoline must be used per second to provide the power needed to sustain a car traveling at 65 miles per hour (29 m/s) on a level road? From this, estimate the car's fuel economy in miles per gallon (mpg).

Need: Amount of gasoline consumed per second in gallons, and the car's mpg.

Know: An automobile traveling at 65 miles per hour (29 m/s) transfers about 78 kW or 78 kJ/s into various energy forms.

How: Apply the principle of the conservation of energy. The energy lost by the car through the boundary of the control surface must come from somewhere. The only place it can come from within the car is by the decrease in the chemical energy of some of the gasoline. By the principle of conservation of energy, the decrease in chemical energy must equal the transfer out of the dotted lines of all the other types of energy.

Solve: Chemical energy needed per second = 78 kW or 78 kJ/s. Since gasoline contains about 1.30×10^5 kJ/gallon of chemical energy, this means that

(78 [kJ/s])/(1.30×10^5 [kJ/gallon] = 6.0×10^{-4} gallons/s of gasoline are consumed

Is this a reasonable answer? Note that a car going at 65 miles per hour travels

(65 [mph])/(3600 [s/h]) = 0.018 miles per second, so its fuel economy in miles per gallon is: (0.018 [miles/s])/(6.0×10^{-4} [gallon/s]) = 30. mpg

This is a reasonable estimate given the roughness of our calculations.

Summary

The concept of energy is one of the indispensable analytical tools of an engineer. That understanding begins with the following key ideas. **(A)** Energy has the capability to do work (mechanical, electrical, etc.), **(B)** energy comes in many forms that can be converted from one form to another, and **(C)** energy is neither created nor destroyed and is always conserved.

The **principle of the conservation of energy** can help answer the questions that began this chapter. **(1)** Is the world running out of energy? **(2)** Where does the energy come from that provides the light by which you are reading these words? **(3)** Does a car possess more energy when it is sitting in your driveway or 15 min later when it is traveling down the highway at 65 miles per hour?

To answer the first question, we draw a dotted line around a sketch of the Earth and apply the **control boundary analysis**. In doing so we find that the amount of energy on Earth is either constant or increasing slightly (due to global warming). So, the answer is no—the world certainly is not running out of energy, although useful primary energy sources like coal and petroleum are in decline.

For the second question, the light that you are reading this by probably came from a lightbulb powered by a fossil-fuel-burning electric power plant. As you read these words, electrical energy is being converted into visible light which is then converted into heat in your room—a form of energy of very little use for anything beyond keeping you warm.

For the third question, a control boundary analysis of the automobile tells us that the automobile is continually converting the chemical energy from its gas tank into other forms of energy as it travels down the highway. So, it does possess less energy traveling on the highway at 65 miles per hour than it did 15 min earlier sitting in your driveway. (The answer is yes it does - can you explain why?).

Exercises

Pay attention to the number of significant figures in your answers!

Conversion factors: $1.00 \text{ kJ} = 738 \text{ ftl} \cdot \text{bf}$, $1.00 \text{ kg} = 2.20 \text{ lbm}$, and $g_c = 32.2 \text{ lbm} \cdot \text{ft/lbf} \cdot \text{s}^2$.

1. Determine the TKE of the automobile in Example 5.1 if its speed was reduced to 55 miles per hour.

2. Determine the TKE in Engineering English units of the automobile in Example 5.1 if its mass was increased to 4.00×10^3 lbm.

3. Determine the TKE of the atmosphere in Example 5.2 if the average air velocity increased to 15 m/s.

4. Repeat the calculation of Example 5.5 in Engineering English units. Check that your answers agree with the solution in Example 5.5 using the appropriate conversion factors.

5. What would the gravitational potential in SI units of the anvil in Example 5.5 be if its mass was 100. kg and the cliff was 1000. meters high?

6. Determine the GPE of an 8.00×10^3 kg truck 30. m above the ground. (**Ans. 2.4×10^6 J** to two significant figures, since h is known only to two significant figures.)

7. A spring at ground level—that is, at a height of exactly 0 m—shoots a 0.80 kg ball upward with an initial kinetic energy of 245 J. Assume that all of the initial TKE is converted to GPE. How high will the ball rise (neglecting air resistance)?

8. Chunks of Earth's orbital debris can have speeds of 2.3×10^4 miles per hour. Determine the TKE of a 2.0×10^3 lbm chunk of this material in SI units. (**Ans. 4.8×10^{10} J** to two significant figures.)

9. An airplane with a mass of 1.50×10^4 kg is flying at a height of 1.35×10^3 m at a speed of 250.0 m/s. Which is larger—its TKE or its GPE with respect to the earth's surface? (**Ans. TKE = 4.69×10^5 kJ; GPE = 1.99×10^5 kJ**, Therefore, the TKE is **greater than** GPE.)

10. Determine the amount of gasoline required in Example 5.7 if the automobile was traveling at 55 miles per hour.

11. Suppose the 1.0 kg book in Section 5.5 fell from a height of 2.5 m. What would be the energy increase of the classroom?

12. A vehicle of mass 1.50×10^4 kg is traveling on the ground with a TKE of 4.69×10^8 J. By means of a device that interacts with the surrounding air, it is able to convert 50% of the TKE into GPE. This energy conversion enables it to ascend (fly) vertically. To what height above the ground does it rise?

13. Aeronautical engineers have invented a device that achieves the conversion of kinetic to potential energy as described in Exercise 12. The device achieves this conversion with high efficiency. In other words, a high percentage of the TKE of motion is converted into vertical "lift" with little lost to air "drag." What is the device called? (**Hint:** This is not rocket science.)

14. A hypervelocity launcher is an electromagnetic gun capable of shooting a projectile at very high speed. A Sandia National Laboratory hypervelocity launcher shoots a 1.50 g projectile that attains a speed of 14.0 km/s. How much electrical energy must the gun convert into TKE to achieve this speed? Solve in SI. (**Ans. 1.5×10^2 kJ**)

15. Solve Exercise 14 in Engineering English units. (Also check your answer by converting the final answer to Exercise 14 into Engineering English units.)

16. Micrometeoroids could strike the International Space Station with impact velocities of 19 km/s. What is the TKE of a 1.0 gram micrometeoroid traveling at that speed? (**Ans. 1.8×10^5 J**)

17. Suppose a spaceship is designed to withstand a micrometeoroid TKE impact of a million joules. Suppose that the most massive micrometeoroid it is likely to encounter in space has mass of 3 grams. What is the maximum speed relative to the spaceship that the most massive micrometeorite can be traveling at for the spaceship to be able to withstand its impact?
18. A stiff 10.0 gram ball is held directly above and in contact with a 600. gram basketball and both are dropped from a height of 1.00 m. What is the *maximum* theoretical height to which the small ball can bounce?
19. What would be the power required by the lightbulb in Example 5.6 if it had a voltage drop of 120. V?
20. What would be the current in the lightbulb in Example 5.6 if it had a voltage drop of 120. V and required a power of 100. W?
21. An electric oven is heated by a circuit that consists of a heating element connected to a voltage source. The voltage source supplies a voltage of 110. V, which appears as a voltage drop across the heating element. The resulting current through the heating element is 1.0 A. If the heating element is perfectly efficient at converting electric power into thermal power, what is the thermal power produced by the heating element? (**Ans. 1.1×10^2 W** to two significant figures.)
22. A truck starter motor must deliver 15 kW of power for a brief period. If the voltage of the motor is 12 V, what is the current through the starter motor while it is delivering that level of power?
23. A hybrid car is an automobile that achieves high fuel efficiency by using a combination of thermal energy (gasoline) and electrical energy (battery) for propulsion. One of the ways it achieves high fuel efficiency is by regenerative braking. That is, every time the car stops, the regenerative braking system converts part of the car's TKE into electrical energy which is then stored in a battery that can later be used to propel the car. The remaining part of the TKE is lost as heat. Draw a control surface diagram showing the energy conversions that take place when the hybrid car stops. (**Ans.** See diagram.)

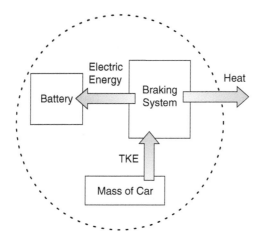

24. Suppose the car in Exercise 23 has a mass of 1000. kg and is traveling at 33.5 miles per hour. As it comes to a stop, the regenerative braking system operates with 75% efficiency. How much energy per stop can the regenerative braking system store in the battery? Illustrate with a control boundary showing the energy flows.

25. Suppose the car in Exercises 23 and 24 has stored 1.00×10^2 MJ of energy in its battery. Suppose the electric propulsion system of the car can convert 90% of that energy into mechanical power. Suppose the car requires 30. kW of mechanical power to travel at 33.5 miles per hour. How many miles can the car travel using the energy in its battery? (**Ans. 28 miles.**)

26. Determine the amount of gasoline consumed per second by the automobile in Example 5.8 if it was traveling at 41 m/s (note that this is 41 m/s not 41 mph).

27. To maintain a speed v on a horizontal road a car must supply enough power to overcome air resistance. That required power goes up with increasing speed according to the formula:

$$P = \text{Power in kW} = K \times V^3$$

where v is the speed measured in miles/hour and K is a constant of proportionality. Suppose it takes a measured 7.7 kW for a car to overcome air resistance alone at 30. mph.

a. What is the value of K in its appropriate units?

b. Using a spreadsheet, prepare a graph of power (kW on the y axis) as a function of speed (mph on the x axis) for speeds from exactly 0 mph to exactly 100. mph.

28. Review Exercises 21−24 in Chapter 3 concerning the dynamics (and consequent fate) of bungee jumpers. Draw a control surface around the jumper and cord. Show the various forms of energy possessed by the jumper and cord, along with arrows showing the directions of energy conversion inside and across the control surface: (a) when the jumper is standing on the cliff top, (b) when the jumper is halfway down, and (c) when the cord brings the jumper to a safe stop.

29. After working for a company for several years, you feel you have discovered a more efficient energy conversion method that would save your company millions of dollars annually. Since you made this discovery as part of your daily job you take your idea to your supervisor, but he/she claims it is impractical and refuses to consider it further. You still feel it has merit and want to proceed. What do you do?

a. You take your idea to another company to see if they will buy it.

b. You contact a patent lawyer to initiate a patent search on your idea.

c. You go over your boss's head and talk to his/her supervisor about your idea.

d. You complain to your company's human resources office about having poor supervision.

30. Your course instructor claims that energy is not really conserved. He/she uses the example of a spring that is compressed and then tied with a nylon string. When the compressed spring is put into a jar of acid, the spring dissolves and the energy it contains is lost. How do you react?

a. Ignore him/her and follow the established theories in the text.

b. Go to the department chairperson and complain that the instructor is incompetent.

c. Say nothing, but make detailed statements about the quality of the instructor on the course evaluation at the end of the term.

d. Respectfully suggest that the energy in his/her spring example really is conserved and show why it is conserved.

Engineering economics

6

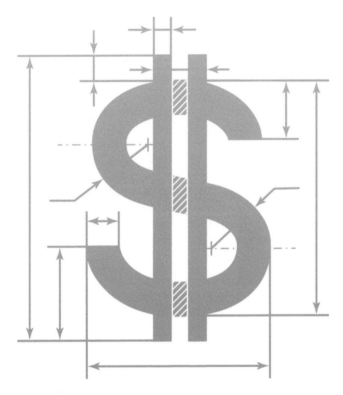

Source: kulykt/Shutterstock.com

6.1 Introduction

This chapter is applicable to virtually *all* the fields of engineering since engineer's design and make things that people buy. Many subdisciplines of engineering use this specialized form of economics, whether to financially embrace new products

Exploring Engineering. https://doi.org/10.1016/B978-0-443-13541-5.00022-2

121

or to evaluate existing ones. All use engineering economics as these disciplines are often involved in the cost management of engineering projects.

6.2 Why is economics important?

We all make economic decisions every day. Can I afford that steak dinner, a new laptop computer, an iPhone, or should I get a hamburger, continue to use my old desktop computer, and keep the old cell phone I have used for years? Can I afford a new car? What about a mortgage for a new house? Engineers make similar decisions all the time to determine if what they are doing makes economic "sense." Unlike personal economic decisions, the criteria used by engineers are multidimensional and have several components.

An engineer should be familiar with several financial terms. These include **principal**, **interest**, **time value of money**, **present value**, **first cost**, and **future value** to name just a few. These terms are needed to assess the true cost of an investment, such as building a new factory to manufacture a part that you currently purchase from another manufacturer. Other criteria used to evaluate the economics of a new investment include **breakeven point (BEP)**, **process improvement (PI)**, and **return on investment (ROI)**. These terms represent a few of the many concepts used before an engineer expends large sums of money on developing new products.

One's answer to almost all economic decisions made for personal use rests on **first cost**. First cost is what you pay for the item when you buy it. Few of us think in terms of the product's life cycle costs such as repairs, depreciation, disposal, and the cost of borrowing money when making a purchase.[1] But, if you are making a major business purchase, you better consider more than just the first cost. In the next section, we will concentrate on how the **cost of borrowing** may be as important as the cost of the item itself.

6.3 The cost of money

Suppose you purchase a used car for $10,000. Is that what it really costs you? Even the most elementary analysis suggests that this is not so. Because, if you had $10,000 to spend, you already have the cash *somewhere*. Suppose it was not kept under your mattress, but in a bank in an interest-bearing account earning 5% **simple** interest. This means that each year you will have a deposit to your account of 5% of $10,000, or $500. Purchasing a car with that cash means that this income stream stops forever. If nothing else changes, your once $10,000 now forgoes the addition of $500 every year.

---

[1] Except possibly for a home mortgage, because the monthly amounts are usually a large fraction of your income.

It is very unusual to make cash purchases in a business (except for small "petty cash" purchases). If you want to make a significant purchase (a $10,000 car for example) a business would probably borrow the money from a bank. Suppose the bank charges 5% simple interest on the loan. Then once a year you owe the bank $500. With simple interest, it is as if once a year you paid back the $10,000 and gave the bank $500 for the use of the $10,000, and then you immediately borrowed another $10,000 at the same interest for the next year. If you pay back the full amount after 5 years, you will pay the bank $10,000 (the "**principal**," P) plus 500 $/year × 5 years = $2500 (the "**interest**," I). In this case, the **time value of money** is $2500 on a principal of $10,000.

But, if you were a banker and knew that you would have to wait 5 years before seeing a penny back of the principal, wouldn't you also think that you have lent the business another $2500 spread over 5 years and on which you received nothing in return? The banker wants interest on the interest as well as on the original principal, so the banker charges **compound interest** for the loan. Compound interest is also called the **future value**.

The **future value**, F of the **present value**, P of money is: $F = P(1 + r)^N$ where r is the interest rate per period and N is the number of compounding periods. The "period" might be years, quarters, months, and so on. It all depends on how frequently interest is to be compounded. So, the future value of your $10,000 at 5% interest ($r = 0.05$) compounded yearly after 5 years ($N = 5$) is: $F = (\$10,000) \times (1 + 0.05)^5 = \$12,763$. You see that money itself has a value, and a proper model of how to account for this value is important in an engineer's decision to buy a new car or anything else.

It works this way. Year 1 of the loan is on $10,000, year 2 is on $10,500, and year 3 is on $11,025 (the interest being 5% of $10,500 in year 3), and so on until in year 5 you repay the banker the sum of

$$P = \$10,000 + I$$

in which

$$I = \$500 + \$525 + \$551 + \$579 + \$608 = \$2,763$$

instead of just $I = \$2500$. Note that each yearly interest is $(1 + 0.05) \times$ the previous year's balance (which, if multiplied by the principle $P = \$10,000$, is the future value, F, of P). Table 6.1 summarizes this situation.

Table 6.1 defines the **future value**, F, of the **present value** of the principal, P. In other words, the future value is what today's principal is worth under some estimate of its future financial behavior. *Virtually all business transactions use the compound interest formula*.

Table 6.1 Generalizations of interest accrual equations (r is the interest rate of the loan).

Type of interest	Period	Beginning of period	End of period	Future value, F
Simple[a]	1	P	$P + rP$	$P(1 + r)$
	2	P	$P + rP$	$P(1 + r) + rP = P(1 + 2r)$
	3	P	$P + rP$	$P(1 + 2r) + rP = P(1 + 3r)$
	...	...	...	...
	N	**P**	**P + rP**	**P(1 + Nr)**
Compound	1	P	$P(1 + r)$	$P(1 + r)$
	2	$P(1 + r)$	$P(1 + r) + rP(1 + r)$	$P(1 + r)^2$
	3	$P(1 + r)^2$	$P(1 + r)^2 + rP(1 + r)^2$	$P(1 + r)^3$
	...	...	...	...
	N	**$P(1 + r)^{N-1}$**	**$P(1 + r)^{N-1} + rP(1 + r)^{N-1}$**	**$P(1 + r)^N$**

[a] *While we have also included a simple interest table, it is for pedagogical reasons only because it is easy to understand and not because it is used in practical engineering economics.*

Note: In the problems involving financing we will often ignore our engineering rules concerning significant figures and conform to the rules of the financial industry. Also, we will normally express calculations in whole dollars.

Example 6.1

You wish to borrow $100,000 for 10 years at a 5.0% annual interest rate. What is the difference in the cost of the loan if it is compounded (1) yearly, $N = 10$; (2) monthly, $N = 10 \times 12 = 120$; or (3) daily, $N = 10 \times 360 = 3650$?

Need: Cost of borrowing $100,000 for 10 years at 5.0% annually under assumptions of (1) 10 annual payment periods, (2) 120 monthly periods, and (3) 3650 daily periods.

Know − How: The formula for compound interest from Table 6.1 is: $F = P(1 + r)^N$.

Solve: $P = \$100,000$.

1) $N = 10$, $r = 5.0\% = 0.050$ annually, therefore

$$F = P(1 + r)^N = \$100,000 \times (1 + 0.050)^{10} = \$\mathbf{162,889}.$$

2) $N = 120$, but since the 5.0% is an annual interest rate, the monthly "period" interest is $r = 0.050/12 = 0.00417$. Then

$$F = P(1 + r)^N = \$100,000 \times (1 + 0.00417)^{120} = \$\mathbf{164,701}.$$

3) $N = 3650$, and the daily "period" interest is $r = 0.050/365 = 1.37 \times 10^{-4}$, then

$$F = P(1 + r)^N = \$100,000 \times (1 + 1.37 \times 10^{-4})^{3650} = \$\mathbf{164,866}.$$

Notice that the cost of the loan in Example 6.1 goes up when the interest is compounded more often. Note too that interest rates are nominally quoted in annual terms but they must be adjusted to reflect the daily or monthly periods of compounding.

Example 6.2

Nuclear power plants cost billions of dollars. This reflects the need for the highest possible confidence in the construction of the nuclear reactor. A spending plan for a large generic nuclear power plant is shown in the figure below.

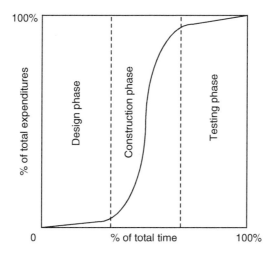

In a simplified case, each phase takes about 1/3 of the total time to build the power plant with the heaviest spending naturally occurring during the construction phase. A manufacturer of nuclear plants would typically get construction loans as needed to minimize the interest expense. Notice that significant loan costs are incurred during the final test phase since the construction money has already been borrowed and spent for activities that have yet to bring in any income.

Suppose we have an outstanding debt of $5.0 billion at the beginning of the test phase of a 1000 MW nuclear power plant. Suppose further that the final test phase takes 4 years. If the interest rate is 12.0% with quarterly compounding, how much does the testing phase cost in finance charges alone?

Need: Testing phase finances charges = $_____.

Know: Outstanding capital, $P = \$5.0$ billion. Period of testing $= 4$ years. Costs are compounded over $N = 4$ quarters/year $\times$ 4 years $= 16$ quarters, and the interest rate $r = 12.0\%/4 = 3.0\% = 0.030$ per quarter.

How: Use the compound interest (future value) law from Table 6.1: $F = P(1 + r)^N$.

Solve: $F = P(1 + r)^N = (\$5.0 \times 10^9) \times (1 + 0.030)^{16} = $ **$8.0 billion**. The finance charges during the testing phase are then $= F - P = 8.0 - 5.0 = $ **$3.0 billion**.

The financial expenditures during the testing phase swamp the actual costs of testing itself by a considerable amount. For a project with a long construction time up front, the viability of the project may depend on financing charges more than any other single factor.

Rather than do the arithmetic on a calculator, any spreadsheet program you use almost certainly contains many functions developed especially for financial analysis. Excel has the functions described in Table 6.2 (plus *many* other financial functions).

Table 6.2 Some Excel financial functions.

Function	Use	Arguments
FV(rate, nper,pmt,pv,type)	Calculates future value FV	• $rate = r$ • $nper = N$ • pmt = periodic payments of P and r • $pv = P$ • fv is optional. It is the future value that you'd like the investment to be after all payments have been made. If this parameter is omitted, the NPER function will assume that fv is 0. • $type = 0$ if the payment is at the start of the period. • $type = 1$ if the payment is at the end of the period.
PV(rate, nper,pmt,fv,type)	Calculates present value PV	
NPER(rate, pmt, pv, fv, type)[a]	Calculates how long it will take to increase PV to FV	

[a] *For example, suppose you deposit $10,000 in a savings account that earns an interest rate of 8%. To calculate how many years it will take to double your investment, use NPER as follows: =NPER(0.08, 0, −10000, 20000,0). This will return an answer of 9.01, which indicates that you can double your money in about 9 years.*

Example 6.3

Repeat Example 6.1 using the Excel (or equivalent) spreadsheet future value function, FV.

Need: Cost of borrowing $100,000 for 10 years at $r = 5.0\% = 0.050$ under assumptions of 1) $N = 10$ payment periods, 2) $N = 120$ payment periods, and 3) $N = 3650$ payment periods.

Know - How: Use the FV function in Excel

Solve: FV(rate, nper, pmt, pv, type)

	A	B	C	D
1	P	r, %		
2	$100,000	5%		
3				
4	Case #	N	r per period	FV
5	1	10	5.00%	$162,889
6	2	120	0.42%	$164,701
7	3	3650	0.014%	$164,866

Below is the same spreadsheet after simultaneously pressing the control and tilde keys (Ctrl ~).

	A	B	C	D
1	P	r, %		
2	100,000	0.05		
3				
4	Case #	N	r per period	FV
5	1	10	=$B$2	=FV(C5,B5,,-$A$2,1)
6	2	120	=$B2/12	=FV(C6,B6,,-$A$2,1)
7	3	3650	=$B2/365	=FV(C7,B7,-$A$2,1)

Notice the convention that your principal (or, "investment") is a negative number (-$A$2), which is logical but confusing. If you enter it as a negative number in $A$2, then you don't have to put a negative sign in front of $A$2 in the FV function. Also, notice that unless you are prepaying the owed principal or interest, you do not need to enter a number for *pmt* in the FV function. Just leave a blank space (or enter 0) for *pmt* so that, ", *pmt*" becomes " , ," as shown in the second spreadsheet.

Example 6.4

You want to reap $1,000,000 from an investment of $500,000. How many years will you have to wait if $r = 7.50\% = 0.0750$?

Solve: Use the Excel spreadsheet function NPER(*rate, pmt, pv, fv, type*).

	A	B	C
1	P	r	FV
2	$500,000	7.50%	1,000,000
3			
4		N, years	r per period
5		9.58	7.50%

Below is the same spreadsheet after simultaneously pressing the control and tilde keys (Ctrl ~).

◢	A	B	C
1	P	r	FV
2	500000	0.075	1000000
3			
4		N, years	r per period
5		=NPER(B2,-A2,C2,1)	=$B$2

Thus, you need to wait just over 9½ years to double your money.

To summarize what it really costs to make an engineering purchase, you need to include the present value (the principal) P and the interest charges on that amount. Clearly, you also must add other costs such as labor costs, material costs, and overhead costs (such as the rent for your factory, the cost of power to keep the lights on, the cost of heat to keep the factory warm, and so on) to get the total costs.

6.4 When is an investment worth it?

But how do you know if your project is worth the effort? Most engineering businesses use one or more criteria to assess the value of a project. A major indicator occurs when you start to make a profit by selling enough of your products. This is called the **Breakeven Point** or **BEP**, and it is a standard measure of a project's value. It has a simple definition:

BEP occurs when the project has earned back what it took to make it.

Example 6.5

Assume the cost of producing a new product is $1,000,000. Then the BEP occurs when your net profit reaches $1,000,000. Let's say the profit is $1.00 per product sold, and you're selling 1000/day.
  Need: BEP = _____ years for a project costing $1,000,000 assuming you are selling 1000/day with a profit of $1.00 per product.
  Know: How: Equate cost to total money stream.
  Solve: 1000 [products/day] × 1.00 [$/product] × D [days] = $1,000,000.
    Solving for D gives: D = 1000 days = **2.74 years**.

In Example 6.5 it will take 2.74 years to reach the BEP. Is this good enough? It depends on the industry. Many companies would prefer a BEP of 18 months or less.

Example 6.6

One "long horizon" industry is the electric power industry. In Example 6.2, our 1000 MW nuclear power plant cost about $8.0 billion, a very large sum of money. Yet its product sells for about 10 cents per kWh. What is its best possible BEP after it starts to produce electricity?

Need: BEP = _____ years for a 1000 MW nuclear power plant that cost $8.0 billion and selling electrical energy for 10.0 cents/kWh = 0.10 $/kWh.

Know: For t hours of operation a 1000 MW nuclear power plant can produce a maximum of $1,000,000 \times t$ kWh of electricity.

How: Equate costs to money produced by the power plant.

Solve: $(1,000,000 \times t \text{ [kWh]}) \times 0.10 \text{ [\$/kWh]} = \$8.0 \times 10^9$, or

$$t = \left(8.0 \times 10^9 \text{ [\$]}\right) / \left(1,000,000 \text{ [kW]} \times 0.10 \text{ [\$ / kWh]}\right)$$
$$= 8.0 \times 10^4 \text{ hours, or } \quad \textbf{BEP} = \textbf{9.2 years.}$$

Is the BEP in Example 6.6 good enough? While nuclear power plants currently cost much more than fossil fuel power plants, the advantages of uranium over fossil fuel ultimately might increase as world oil and gas supplies dwindle and concerns about greenhouse gases grow.

Example 6.7

An engineer proposes an improvement to an existing process. The cost required to make this **process improvement (PI)** is $100,000. Suppose the process makes 100,000 units/day. If the proposed process improvement saves one cent per unit ($0.01/unit), what is its BEP for the PI?

Need: BEP = _____ for a **PI** costing $100,000 assuming you are selling 100,000 items/day with a **PI** savings of $0.01/unit.

Know − How: You will save 100,000 [units/day] × 0.01 [$/unit] = 1000 $/day with the new and improved manufacturing process.

Solve: The **PI BEP** is given by how long it takes to recover your $100,000 investment. This is ($100,000)/(1000 $/day)] = 100 days = **3.3 months**.

Is this PI BEP in Example 6.7 good enough? As a rule of thumb, a nominal BEP should be 18 months or less, and less than 12 months is preferable. In Example 6.5, the new widget is a marginal investment. But in Example 6.7, the process improvement BEP is a definite go.

One other financial term that is often used to determine if an engineering investment is satisfactory is the (annual) **Return on Investment** or "**ROI.**" This is defined as

$$\textbf{ROI (in\%)} = \frac{\textbf{Annual Return}}{\textbf{Cost of the Investment}} \times \textbf{100}$$

In the simplest case, if an investment of $500,000 produces an income of $40,000 per year, its ROI = ($40,000/$500,000) × 100 = 8%. Many companies would not invest for such small returns on their money. Many successful large companies operate with ROI's of 15% or more.

As Table 6.3 shows, there is a big difference among the ROIs of companies. You should not be surprised to see that Exxon Mobil has the highest ROI in this table since its profits, and therefore the numerator of the ROI formula is high while the processing cost for the equipment for converting crude oil to refined products is low relative to the cost of the processed material.

Table 6.3 2007 ROI of several large chemical companies.

Company	ROI, annual %
Dow Chemical	10.5
Exxon Mobil	22.4
DuPont	18.5
PPG Industries	20.2
Air Products	11.0
Eastman Chemical	10.9
W.R. Grace	9.8

Example 6.8

Calculate the annual ROI for the process improvement in Example 6.5. Assume factory is operating 300 days per year.

Need: ROI annual % for a project costing $1,000,000 assuming you are selling 1000 units per day at a profit of $1.00 per unit for 300 days per year.

Know: Profit is 1000 [units/day] $\times$ 1.00 [$/unit] = 1000 $/day

How: Compare annual profits to the investment cost.

Solve: Profit (i.e., "return") = 1000 [$/day] $\times$ 300 [days/year] = $300,000 $/year.

Since the investment is $1,000,000 then the

$$\textbf{ROI} = (\$300,000 \, / \, \$1,000,000) \times 100 = \textbf{30}\%.$$

Is the ROI in Example 6.8 good enough? Almost surely yes! This is a wonderful place to invest $1,000,000 (assuming all the assumptions and numbers are correct).

Summary

This chapter was concerned with engineering economics, a discipline that is important to almost every branch of engineering, and particularly to industrial, manufacturing, or management engineering.

Terms used in this chapter include *principal, compound interest, present value*, and *future value*. All are concerned with the *time value of money* in making economic decisions. In some cases, the cost of financing new products may be the dominant cost in the product rather than the investment in hardware, materials, and labor.

Further guidelines are given on how to assess whether a proposed project will be commercially viable. These are the *breakeven point* (BEP), a measure of how long it takes to recover one's investments, and the *return on investment* (ROI), a ratio in percent of the annual profit divided by the investment that produces it. Successful engineering businesses typically want to see a BEP of less than 18 months and an ROI greater than 15%.

Exercises

(1) Find the cost of borrowing $100,000 in Example 6.1 if the interest is compounded hourly.

(2) Rework Example 6.1 for a $15,000 loan at a 10% annual interest.

(3) You need to borrow $12,000 to pay your tuition plus room and board. One bank offers to loan you the money for 10 years at 5.0% interest compounded annually. Another bank offers you the loan at 4.75% interest compounded monthly. Which bank has the better deal?

(4) You are a big financial success and you want to purchase the Remlab Company for $35 billion. You have $5 billion in cash but need to borrow the remaining $30 billion from your friendly banker. You banker says fine, I'll lend you the money for 10 years, but at 7.5% annual interest compounded quarterly. How much interest will you pay to the bank over the life of this loan?

(5) What will be the testing phase finance charges in Example 6.2 if the nuclear power plant cost $10 billion and everything else stays the same?

(6) It turns out that the testing phase in Example 6.2 takes 6 years instead of 4 years, and the entire project takes 14 years. What will the finance charges be if everything else remains the same?

(7) Repeat Example 6.3 using the Excel spreadsheet (or equivalent) FV function to determine the cost of borrowing $100,000 for 10 years at 5% for 87,600 hourly payment periods.

(8) Repeat Example 6.3 using the Excel spreadsheet (or equivalent) FV function to determine the cost of borrowing $100 billion for 10 years at 8% compounded quarterly.

(9) Repeat Example 6.3 using the Excel spreadsheet (or equivalent) FV function to determine the cost of borrowing $3000 on a credit card for 1 year at 28% compounded monthly.

(10) Use the NPER Excel function (or equivalent) in Example 6.4 to determine how long it will take to double your $500,000 investment if you earn 7.5% interest?

(11) Use the NPER Excel function (or equivalent) in Example 6.4 to determine how long it will take to triple your $500,000 investment if you earn 11.3% interest?

(12) Use the NPER Excel function (or equivalent) in Example 6.4 to determine how long it will take to increase an investment of $3500 to $4200 if you earn 3.5% interest?

(13) How long will it take to double your money at an interest rate that is $<< 100\%$ starting with the expression $F = P(1 + r)^N$? **Hint**[2]: $\ln(1 + r) \approx r$ if $r << 1$. (**Answer**: Years to double the investment is $70 \div r\%$, which is known as the **Rule of Seventy**.[3])

(14) Compare the answer to Exercise 10 with the "Rule of Seventy" described in Exercise 13.

(15) Suppose the profit on each item sold in Example 6.5 was $10.00 instead of $1.00. What would the BEP be if everything else remained the same?

(16) Suppose the new product in Example 6.5 has a profit of $5.00 per item, but the sales projection is off by 50% and the company is only able to sell 500 per day. What is the new breakeven point?

(17) You are a manufacturing engineer and have been asked to project the costs required to produce a new line of products for your company. After contacting the appropriate equipment manufacturers and consulting with management about labor and overhead costs, you determine that the manufacturing line will cost $8.3 million to get into operation. The marketing department tells you that the company can expect to see a profit of $37.50 per item produced and that they can sell 1500 items per day in big chain stores. You run into your boss at the coffee machine and he/she casually asks you what the breakeven point will be for your new line. What do you tell him/her?

(18) The owners of the nuclear power plant in Example 6.6 want a breakeven point of 5 years. To do this, they hope to raise the price of the electricity they sell. How much do they need to sell the electricity for in order to meet the BEP?

(19) In Example 6.6, the owners of the nuclear power plant decide to use substandard materials to reduce the cost of building and testing the power plant from $8.0 billion to $7.35 billion. If everything else remains the same (and no one finds out what they've done), what will be their new BEP?

(20) Another "long horizon" industry is the commercial aircraft industry. A Boeing 747—8 Freighter costs $300 million. A commercial air freight company purchases this plane and keeps it in the air 12 h per day for 280 days per year. If the air freight company makes a profit of $4500 per flight hour, how long before it breaks even with the cost of the plane?

(21) In Example 6.7, it turns out that the new process does not save one cent per unit. It only saves 0.3 cents per unit. What is the new BEP, and is it worth the expense to make this improvement?

(22) You are a manufacturing engineer and discover that you can purchase a machine that will replace a labor-intensive manual operation in your manufacturing line. The machine costs $1.5 million, but the labor savings costs are $7.80 per unit manufactured. Can you get a BEP in less than 18 months if you make 8500 units per day?

(23) An automobile manufacturer would like to add a new feature to their standard production model. The new feature will cost the manufacturer $8.67 per car to produce, and they expect to sell 100,000 cars per years. If the cars normally sell for $25,990 each, how much do they need to increase the price to reach a breakeven point in 18 months?

(24) Calculate the annual ROI for the process improvement in Example 6.5 if the factory operates 300 days per year, but only sells 500 widgets per day.

(25) What is the ROI for the process improvement in Example 6.7 if the factory operates 300 days per year? Would you invest in this company?

Use the Engineering Ethics Matrix to analyze the following situations.

(26) As a junior engineer in a small company you are involved in designing a production line for a new product. The company CEO believes that this new product will make a great deal of money and wants to get it into production as soon as possible. Your immediate supervisor wants to please the CEO and get things moving, so he/she starts ordering equipment and hires contractors to expand the building. While glancing at a copy of the new project your boss gave to the CEO, you discover an error. The breakeven point is improbably low and the return on investment is improbably high both by a factor of about 10. When you ask your immediate supervisor about this you are told that it is none of your concern. What do you do?

(27) You are the CEO of a large corporation and a recent economic downturn requires you to eliminate all upcoming dividends to your company's stockholders. This is a highly confidential decision because it will negatively affect your stock price. Several of your wealthy neighbors own a considerable amount of your company stock. At a recent picnic in your backyard, one of them asks if she should sell some of your company's stock, considering how badly the stock market is doing in general. If you say "no," she could lose a considerable amount of money. What do you do?

---

[2] "ln" means the base of natural logarithms, $e = 2.718$ ....
[3] Sometimes called the Rule of Seven.

Minds-on

Source: Sudowoodo/iStockphoto.com

Aeronautical engineering

VanderWolf/Shutterstock.com

It is better to be on the ground wishing you were flying, rather

than up in the air wishing you were on the ground

—Aviation proverb.

7.1 Introduction

Aeronautical engineers design, test, and help manufacture commercial and military aircraft, missiles, and spacecraft. They are also engaged in defense systems, aviation, and space exploration.

A large modern aircraft is the result of literally thousands of engineers and technicians working in discrete parts of its design and construction. While aeronautical engineers are responsible for the overall system design, including the structural aspects of the aircraft, electronic systems, and structural aspects of the aircraft, this chapter will focus on just one aspect of flight, how aerodynamic "lift" is produced.

An airplane in level flight and at constant altitude balances two major forces, (1) lift versus weight and (2) thrust versus drag. Imagine a Boeing 787 jet liner weighing 250 tons sitting on a runway. In steady level flight, "lift" is the force needed to overcome gravity, so **lift = weight**. The "drag" is the force produced by air resistance as the plane flies. Since the wings of a Boeing 787 have an area of about 325 m^2 (or 3500 square feet), the magnitude of the lift is an amazing 1 ton/m^2 of wing area.

Exploring Engineering. https://doi.org/10.1016/B978-0-443-13541-5.00004-0

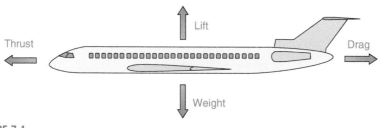

FIGURE 7.1

Forces on an airplane flying straight and level.

In level flight, each of the two engines for a Boeing 787 provides enough thrust (about 267,000 N or 60,000 lbf) to propel the aircraft at more than 500 mph. Unfortunately, the faster an aircraft flies, the more thrust is needed to overcome the air resistance, so in steady level flight **drag = thrust**. Drag is mostly due to the change in momentum of the air flowing over the wings, the fuselage, and engines. As the aircraft's speed increases, so does its drag, and more fuel is used. Lift versus weight and drag versus thrust are illustrated in Fig. 7.1.

The Wright brothers' experimental airplane flew for a short distance, but they lacked a theoretical understanding of aerodynamics. The engineer/scientist who filled that gap was born in Russia about 20 years before the Wright brothers. His name was Nikolay Zhukovsky. He was the first person to understand the basics of **airfoil** (wing) design.

To understand airfoils and how they generate lift, we need to explore the related topics of imaginary numbers, conformal mapping, Joukowski airfoil theory, and the Kutta correction.

7.2 Airfoils and lift

It's the wings that provide the lift by which an airplane defies gravity. To create lift you need to divide the aircraft engine's thrust (which is approximately parallel to the wing) into two discrete forces, one parallel to the wing (drag) and one perpendicular to the wing (lift). The shape and orientation of the wings is crucial to producing adequate lift.

The term "airfoil" is a two-dimensional slice through the wing section. The flow across an infinite airfoil is called two-dimensional or simply 2D flow. This is an approximation to a 3D wing (see Fig. 7.2).

At the leading edge of the wing, the incoming flow is split into two streams: one over the top surface of the wing and the other along the bottom. The flow over the

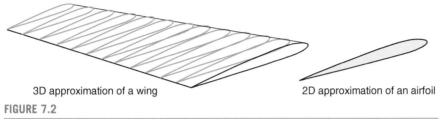

3D approximation of a wing   2D approximation of an airfoil

FIGURE 7.2

3D model of wing approximated by 2D cross sections.

wing's upper surface will experience an increase in its velocity and a corresponding decrease in its pressure. The pressure difference between the bottom and top of the wing thus produce a net vertical force on the wing (lift).

This flow can be accurately calculated by assuming the air is "frictionless". We will use the Joukowski transform to convert a simple 2D shape—a circle—into a complex 2D airfoil shape. But first we need to take a brief diversion into the topic of **imaginary numbers**.

7.3 **The algebra of imaginary numbers**

What is an imaginary number?[1] You were always taught that you cannot take the square root of a negative number, so an equation like $x^2 = -9$ has no solution. Ah, but you can "***imagine***" the solution to be $x = 3\left(\sqrt{-1}\right)$ because when you square x you get $x^2 = -9$.

Today, **imaginary numbers** are an important part of engineering analysis. We let the symbol i represent the imaginary number $\sqrt{-1}$, or $i = \sqrt{-1}$ and $i^2 = -1$. An imaginary number is any real number multiplied by the imaginary unit $i = \sqrt{-1}$. For example, let y be any real number, then iy is an imaginary number and its square is $(iy)^2 = -y^2$

An imaginary number iy can be added to a real number x to form a **complex number** z, as in $z = x + iy$. Here, x is called the **real part** and y is called the **imaginary part** of the **complex number** z (we often write this as $x = \mathrm{Re}(z)$ and $y = \mathrm{Im}(z)$). For example, a complex number could be $z = 3.5 - i7.3$ or $z = 21.7 + i45.8$. You can plot these numbers in a "complex plane" called an **Argand diagram** (Fig. 7.3).[2]

---

[1] Imaginary numbers were introduced in the 17th century and were generally regarded as fictitious or useless "imaginary" mathematical trivia.

[2] Named after the Swiss mathematician Jean Robert Argand, it is a system of rectangular coordinates in which the complex number $x + iy$ is represented by the point whose coordinates are x and iy.

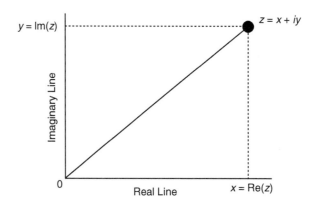

FIGURE 7.3

The complex or Argand plane for the point $z = x + iy$.

Example 7.1

Find x and y in the following complex numbers:

a) $(4 + i)(2 + i3) = x + iy$

b) $\frac{3-i}{2+i7} = x + iy$

c) What are the real and imaginary parts of the complex number $A = \frac{1}{x+iy}$?

Need: Complex number algebraic calculations.

Know-How: Simple algebra and $i = \sqrt{-1}$ so that $i^2 = -1$.

Solve:

a) $(4 + i)(2 + i3) = 8 + i12 + i2 + i^2 3 = = 8 + i14 + 3(-1) = 5 + i14$

Therefore, $x = 5$ and $y = 14$

b) $\frac{3-i}{2+i7} = \left(\frac{3-i}{2+7i}\right)\left(\frac{2-7i}{2-7i}\right) = \frac{6-21i-2i+7i^2}{4-14i+14i-49i^2} = \frac{-1-23i}{53} = -\frac{1}{53} - \frac{23}{53}i$

Therefore, $x = -1/53$ and $y = -23/53$

c) $A = \frac{1}{x+iy} = \left(\frac{1}{x+iy}\right)\left(\frac{x-iy}{x-iy}\right) = \frac{x-iy}{x^2 - ixy + ixy - i^2 y^2}$

$= \frac{x-iy}{x^2 - i^2 y^2} = \frac{x-iy}{x^2 + y^2} = \frac{x}{x^2 + y^2} - \frac{iy}{x^2 + y^2}$

The real part of A is $\mathbf{Re}(A) = \frac{x}{x^2+y^2}$ and the imaginary part of A is $\mathbf{Im}(A) = -\frac{y}{x^2+y^2}$

Example 7.2

Convert the complex number $A = (x + iy) + \frac{K^2}{x+iy}$ (where K^2 is a constant) into the standard complex number format $A = B + iC$.

Need: The complex number A in the format $A = B + iC$.

Know-How: Simple algebra and $i = \sqrt{-1}$ so that $i^2 = -1$.

Solve:

$$A = (x + iy) + \frac{K^2}{x+iy} = (x + iy) + \frac{K^2}{x+iy}\left(\frac{x-iy}{x-iy}\right)$$

$$= (x + iy) + \frac{K^2(x-iy)}{x^2+y^2} = \left(1 + \frac{K^2}{x^2+y^2}\right)x + \left(1 - \frac{K^2}{x^2+y^2}\right)iy$$

$$= B + iC$$

Thus, $B = \mathbf{Re}(A) = \left(1 + \frac{K^2}{x^2 + y^2}\right)x$ and $C = \mathbf{Im}(A) = \left(1 - \frac{K^2}{x^2 + y^2}\right)y$

Note the appearance of $x^2 + y^2$ in the results of Example 7.2. This expression defines a circle of radius R about the origin (e.g., $R^2 = x^2 + y^2$). The arithmetic can be simplified by introducing dimensionless coordinates x/R and y/R which defines a *unit* circle described by the equation $x^2 + y^2 = 1$ (Fig. 7.4).

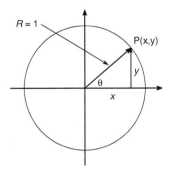

FIGURE 7.4

$x^2 + y^2 = 1$ for any point $\mathbf{P}(x, y)$ on the circumference of the unit circle.

Example 7.3

Simplify the complex number from Example 7.2 on the circumference of a unit circle.

Need: The complex number $A = B + iC = \left(1 + \frac{K^2}{x^2 + y^2}\right)x + i\left(1 - \frac{K^2}{x^2 + y^2}\right)y$ on a unit circle.

Know: The circumference of a unit circle lies on $x^2 + y^2 = 1$

How: Simplify the expression for A using $x^2 + y^2 = 1$

Solve: When $x^2 + y^2 = 1$, then $B = x(1 + K^2)$ and $C = y(1 - K^2)$

Thus, on the circumference of a unit circle, $A = (1 + K^2)x + i(1 - K^2)y$

All this algebra appears to have nothing to do with the theory of flight; however, the complex number $A = z + K^2/z$, where $z = x + iy$, is so important in the theory of flight it has been given a name, the **Joukowski Transform**. We will pursue why it is so important after we have had a look at the topic of **Conformal Mapping**.

7.4 Conformal mapping

Conformal mapping[3] is the process of transforming a relationship from one complex number (or Argand) plane, $z = x + iy$, to another, $A = B + iC$, while preserving all the angles between the lines (Fig. 7.5).

---

[3] The word "conformal" means mapping of one surface onto another surface so that all the angles between intersecting curves remain unchanged.

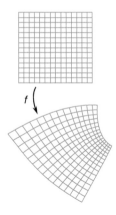

FIGURE 7.5

Example of a conformal mapping.

The heart of what we need to do is to have two Argand planes, one in (x, y) co-ordinates and one in (B, C) coordinates, and then *transform* a circular cylinder in the x, y Argand plane to an airfoil shape (i.e., a wing's cross section) in the B, C Argand plane. This is done using the Joukowski transform. Fig. 7.6 summarizes what we are trying to do.

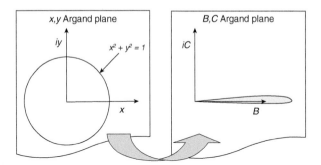

FIGURE 7.6

Conformal mapping.

Example 7.4

Using the Joukowski transform $A = (1+K^2)x + i(1-K^2)y$ on a unit circle for which $y = \pm\sqrt{(1-x^2)}$, plot the resulting graph in the $B = \mathbf{Re}(A)$, $C = \mathbf{Im}(A)$ plane using a spreadsheet and explore the effects of varying the magnitude of K^2.

Need: A graph in the $B = \mathbf{Re}(A)$, $C = \mathbf{Im}(A)$ plane of the transformation.

Know-How: Real and imaginary parts for A are already known from Example 7.3:

$$A = (1+K^2)x + i(1-K^2)y \text{ so } B = (1+K^2)x \text{ and } C = (1+K^2)y$$

where $y = \pm\sqrt{(1-x^2)}$.

Solve: See the spreadsheet solution in Table 7.1.

Table 7.1 Spreadsheet solution for Example 7.4

The Joukowski transform from the x, y plane to the B, C plane

$B = x(1 + K^2)$				$K^2 = 0.980$		
Note range of x is from -1 to $+1$				$K = 0.990$		
$C = y(1 - K^2)$ in which $y = +/- \mathrm{SQRT}(1 - x^2)$						
x	$y = +\mathrm{SQRT}(1 - x^2)$	B Pos root	C Pos root	$y = -\mathrm{SQRT}(1 - x^2)$	B Neg root	C Neg root
−1.000	0.000	−1.98	0.000	0.000	−1.98	0.000
−0.950	0.312	−1.881	0.006	−0.312	−1.881	−0.006
−0.900	0.436	−1.782	0.009	−0.436	−1.782	−0.009
−0.850	0.527	−1.683	0.011	−0.527	−1.683	−0.011
−0.800	0.600	−1.584	0.012	−0.600	−1.584	−0.012
−0.750	0.661	−1.485	0.013	−0.661	−1.485	−0.013
−0.700	0.714	−1.386	0.014	−0.714	−1.386	−0.014
−0.650	0.760	−1.287	0.015	−0.760	−1.287	−0.015
−0.550	0.835	−1.089	0.017	−0.835	−1.089	−0.017

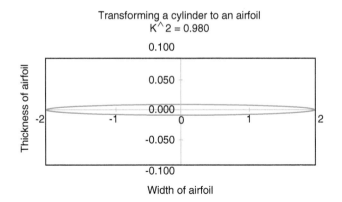

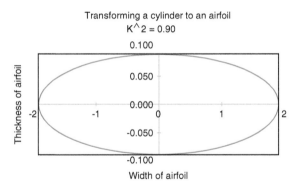

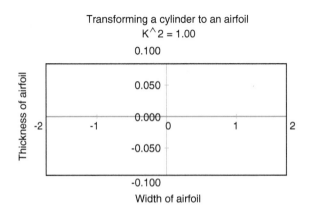

At $K^2 = 0.98$ we have generated a cross section of a symmetrical airfoil. Because this airfoil is symmetrical and, if the air flow impinges parallel to the airfoil, there is no lift because the lift on the topside of the airfoil is exactly balanced by the negative lift on the underside of the airfoil. This simplified calculation is very useful since it shows the purpose of transforming from one Argand plane to another.

Notice that the airfoil in Example 7.4 gets thicker with smaller K^2. At $K^2 = 0.90$, there would be heavy drag on the airfoil. On the other hand, the maximum K is 1.00 and the resulting airfoil is totally flat and there is no lift.

7.5 The Joukowski airfoil theory

So far we have shown how to transform a circle into an airfoil shape in Example 7.4, but you have yet to see how a simple flow over a circle transforms into the actual airflow. The calculated flow over an airfoil is shown in Fig. 7.7.

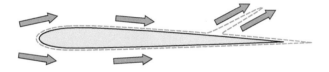

FIGURE 7.7

Airflow over a Joukowski airfoil.

Note that the flow has an obvious discontinuity at trailing edge of the airfoil. We will deal with this in the next section. So, ignoring it for a moment, see how the oncoming flow has split at the leading edge and is diverted around the airfoil, which therefore shows different flow patterns on its top and undersides. The airfoil's lift depends on these differences in flow patterns on the two sides of the airfoil.

7.6 The Kutta correction

The flow separation as shown in Fig. 7.7 should feel intuitively unreasonable because you should expect the flow to move smoothly off the trailing edge and not to back up against the local flow in that region. The discontinuity in Fig. 7.7 can be dealt with by using the **Kutta** correction; it is named after a German applied mathematician who made important early contributions to the theory of flight. The effect is related to the **circulation** of the flow which is denoted by an upper case Greek letter gamma (**Γ**). Circulation is, as the name implies, about the rotation of the flow. The Kutta correction fixes the apparent defect in airfoil flow by applying circulation as follows:

A body with a sharp trailing edge which is moving through a fluid will create about itself a circulation of sufficient strength to hold the rear stagnation point at the trailing edge.

What this is saying is that flow in the close vicinity of an airfoil can be rotated until it is tangential to the trailing edge of an airfoil—see Fig. 7.8.

FIGURE 7.8

The Kutta correction.

The corresponding mathematics says that the amount of circulation you add by this rotation provides the lift required of the airfoil. The equation that is used to calculate the lift L' per *unit length of wingspan* from the **circulation Γ** is as follows:

$$L' = \rho V \Gamma \text{ in SI units} \qquad (7.1)$$

$$L' = \frac{\rho V \Gamma}{g_C} \text{ in English engineering units} \qquad (7.2)$$

in which the lift L' (lbf/ft of *wingspan* or N/m of *wingspan*) is equal to the product of the air density ρ (lbm/ft^3 or kg/m^3), the airspeed V (ft per second or m per second), and the circulation Γ (ft^2 per second or m^2 per second). Eqs. (7.1) and (7.2) are known as the **Kutta–Joukowski theorem**. It is very worthwhile noting that the lift term L' is applied to the entire contour of the wing.[4]

Example 7.5

A newly designed small airplane has a total mass of 2000. kg and a wingspan of 10.0 m. It is designed to fly at low altitude at a fixed speed of 150. km per hour. At this speed and altitude, the circulation is $\Gamma = 50.0$ m^2 per second and the air density is $\rho = 1.05$ kg/m^3. Is the calculated total lift, $L = L' \times$ *wingspan*, sufficient given the plane's weight?

Need: Total lift on a small airplane of mass $m = 2000.$ kg.

Know: Air density is 1.05 kg/m^3. Circulation $\Gamma = 50.0$ m^2 per second and the acceleration of gravity is 9.81 m per second^2. Convert to SI units: 150. km per hour = 41.7 m per second, and 2000. kg = 2000. × 9.81 [kg] [m per second^2] = 19,620 N.

How: Use the Kutta–Joukowski theorem in SI units.

Solve: $L' = \rho V \Gamma = 1.05 \times 41.7 \times 50.0 [\text{kg/m}^3][\text{m per second}][\text{m}^2 \text{ per second}] = 2190 \text{ N/m}$

If the lift is constant over a wingspan of 10.0 m, the total lift is

$$L = L' \times \text{wingspan} = 2190 \times 10.0 \text{ [N / m][m]} = 21,900 \text{ N}$$

which is larger than 19,620 N, the weight of the airplane.

---

[4] This is important because it is obvious that the total lift is dependent on the *area* of the wings; the more wing area, the more lift. The area is implicit in L' being per unit of span but, less obviously, the circulation Γ implicitly contains the width of the wing.

The airplane in Example 7.5 will not need a wingspan quite as large as 10.0 m to maintain level flight. The designer might reduce or otherwise change the wingspan to save manufacturing costs or engine fuel, or just keep it "in reserve"

Example 7.6

A jumbo jet with a take-off weight of 292 tons and a wingspan of 200. ft has a wing area of 3.50×10^3 ft². At its cruising altitude where the air density is 0.022 lbm/ft³ and its cruising speed is 567 mph, what is

a) the lift per ft² of wing area and

b) the circulation Γ needed to enable this design?

Need: The lift per ft² of wing area and the circulation Γ (ft²/s) to keep a 292 ton jumbo jet flying at 567 mph where the air density is 0.0.022 lbm/ft³.

Know: Weight is 292 [tons] × 2000 [lbf][1/ton] = 584×10^3 lbf. Speed is 832 ft per second, wingspan = 200. ft, and the wing area = 3.50×10^3 ft². Also, $g_c = 32.2$ lbm·ft/(lbf·s²) in the English engineering system.

How: Lift/ft² comes directly from the geometry and circulation via the Kutta–Joukowski equation. Lift per foot of wingspan is as follows:

$$L' = \left(584 \times 10^3\right) / 200. \ [\text{lbf}] / [\text{ft}] = 2920 \ \text{lbf/ft}$$

Solve: a) Lift/ft² of wing area = airplane's weight/wing area

$$= \left(584 \times 10^3\right)/\left(3.50 \times 10^3\right)[\text{lbf}] / \left[\text{ft}^2\right] = \mathbf{167 \ lbf/ft^2}.$$

b) The circulation can be calculated from the Kutta–Joukowski theorem.

$$\Gamma = \frac{L' g_C}{\rho V} = \frac{2920 \times 32.2[\text{lbf/ft}]\left[\text{lbm·ft}/\left(\text{lbf·second}^2\right)\right]}{0.022 \times 832\left[\text{lbm/ft}^3\right] \times [\text{ft per second}]} = 5140 \ \text{ft}^2 \ \text{per second}$$

7.7 Symmetric airfoils

By now you have realized that the circulation around an airfoil relates to its lift, the lift itself being generated by the flow differences between the flow on top of and the flow below an airfoil. What then happens in a symmetric airfoil with the same geometry top and bottom? Even more extreme is the airfoil shape designed in Example 7.3—it's elliptical. How can you get any lift?

7.7.1 The angle of attack

Fig. 7.9 shows how a symmetric airfoil can achieve lift even in the absence of circulation.

Basically the lift arises due to the resolution of the components of forces on the wing. Have you ever put your hand out of the window of a moving car? If your hand is perpendicular to the direction of motion, you feel a force pushing your hand back. If you rotate your hand parallel to the direction of motion, you feel very little force. But at a modest angle of attack your hand feels both the resistance and the lift.

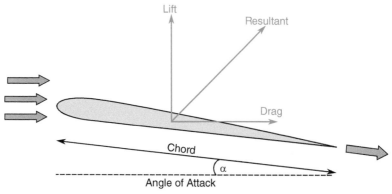

FIGURE 7.9

Lift on a symmetric airfoil.

7.8 Major factors in aircraft economy

For an airline to buy an airplane, there are several factors it must consider. A major item is the cost of aviation fuel. The fuel is burnt in a turbine and converted to thrust. The thrust propels the aircraft and has to be sufficient to provide the desired speed. A slower aircraft might be more economical in fuel consumption but the savings in fuel are offset by the reduced number of trips (and therefore a reduced number of tickets sold).

An engine's thrust can be increased by supplying more fuel to it. A variable in analyzing aircraft engine performance is its **specific fuel consumption**, abbreviated as **SFC**. It is defined as follows:

$$\text{SFC (in English units)} = (\text{lbm/hour of fuel consumed}) / (\text{lbf of thrust}) \quad (7.3)$$

$$\text{SFC (in SI units)} = (\text{kg/second of fuel consumed}) / (\text{N of thrust}) \quad (7.4)$$

There are oddly mixed units in both SFC definitions. Note that aviation fuel flow is measured in mass units and not in volume units, since its volume will vary with temperature.

Example 7.7

At a cruising speed of 567 mph, a Boeing 787 consumed 2.73×10^4 US gallons of jet fuel (equivalent to 1.82×10^5 lbm) while flying 8000. miles. Its two engines delivered a total thrust of 2.16×10^4 lbf. What is the SFC of the engines?

Need: SFC in English engineering units for a Boeing 787 at cruising conditions. It uses 1.82×10^5 lbm of fuel for an 8000. mile flight. During the flight, its engines deliver 2.16×10^4 lbf of thrust.

Know-How: SFC = (lbm per hour of fuel)/(lbf of thrust)

Solve: Time of flight is 8000./567 [miles]/[miles per hour] = 14.1 hours.
Fuel consumption rate = 1.82×10^5/14.1 [lbm]/[hour] = 1.29×10^4 lbm per hour.
Therefore the **SFC** = 1.29×10^4/2.16×10^4

$$= \textbf{0.598}\textbf{(lbm per hour of fuel)} / \textbf{(lbf of thrust)}.$$

Example 7.8

During a flight, the SFC of a Boeing 787 is 0.587 (lbm per hour of fuel)/(lbf of thrust) and its engines produce a steady thrust of 2.16×10^4 lbf. Its cruising lift-to-drag force ratio is 15.0. What is
a) the fuel consumption rate in lbm per hour, and
b) the weight and mass of the aircraft (including fuel and passengers)?

Need: Fuel consumption rate when the thrust = 2.16×10^4 lbf, and the weight and mass of the aircraft when the total lift-to-drag force ratio is 15.0.

Know: SFC of this aircraft is 0.587 (lbm per hour)/(lbf of thrust). In steady level flight, the drag must be equal to the thrust.

How: Fuel consumption rate can be calculated from the SFC, and lift = weight.

Solve:

a) The fuel consumption rate is

$$\text{SFC} \times \text{Thrust} = 0.587\ [\text{lbm per hour}]\ /\ [\text{lbf}] \times 2.16 \times 10^4\ [\text{lbf}] = \mathbf{1.27 \times 10^4\ lbm\ per\ hour}$$

b) The total lift-to-drag force ratio is 15.0 and, since the drag and thrust must be equal, then the lift = (lift-to-drag force ratio) (drag) = (lift-to-drag force ratio) (thrust) = $15.0 \times 2.16 \times 10^4$ lbf = 3.24×10^5 lbf.

In steady cruising, the lift = the weight = 3.24×10^5 lbf

and the mass is mass = Weight $\times (g/g_c) = 3.24 \times 10^5$ lbm

Summary

An aircraft flies because of lift generated by the flow of air across its wings. It is the difference in flow between the upper and lower surfaces of the wing that generates the lift. The wings are airfoils directing the flow. In the simplest model of lift, an airfoil is mapped conformally between two complex planes or Argand diagrams. The mapping of $A = z + K^2/z$ predicts the shape of the Joukowski airfoil and the air flow around; it is the key step in understanding the origins of lift.

A difficulty that soon appears is that flow over the Joukowski airfoil's trailing edge detaches in a retrograde direction; it must be corrected by inserting circulation around the wing to force it to detach smoothly from the trailing edge (known as the Kutta correction). This correction leads to an equation for lift per unit of wingspan, $L' = \rho V\Gamma$ in which ρ is the air density, V is the aircraft's speed, and Γ is the circulation to correct the Kutta correction.

Total lift is equal to weight in a cruising aircraft in level flight. The engine's thrust equates to the air drag, again while cruising in level flight.

Various ratios are used in understanding an aircraft's attributes. Important ones include the total lift to drag ratio, the specific fuel consumption, SFC, and others related to fuel consumption and costs.

Exercises

1. Complete the following table where $i = \sqrt{-1}$.

i^{-3}	i^{-2}	i^{-1}	i^{0}	i^{1}	i^{2}	i^{3}	i^{4}	i^{-5}	i^{6}
i			1	i	-1	$-i$			-1

2. Expand $(1 + i)^2 = ?$ (**Answer:** $2i$)
3. Carry out the multiplication $(3 + 2i)(1 + 7i) = ?$ (**Answer:** $-11 + 23i$)
4. What is $(a + bi)(a - bi) = ?$ (**Answer:** $a^2 + b^2$)
5. The absolute value of a complex number is defined to be the distance from the origin to the number in the Argand plane using the Pythagorean theorem. What is the absolute value of $z = x + iy$? (**Answer:** the absolute value of z is $\sqrt{x^2 + y^2}$)
6. Simplify: $\frac{4-i}{7+i}$. **Answer:** $\frac{27}{50} - \frac{11i}{50}$
7. Simplify: $\frac{(3+2i)^2}{(-1+3i)}$.
8. What are the real and imaginary parts of $A = 1/z^2$, where $z = x + iy$. Write your answer in the form $A = B + iC = \text{Re}(A) + i\text{Im}(A)$.
9. What is the total lift in steady level flight on an airfoil 21.0 m wingspan traveling at 150. m per second if the air density is 0.525 kg/m^3 and the circulation is 55.0 m^2 per second?
10. In Example 7.8, the SFC for a Boeing 787 is 0.587 lbm per hour/lbf leading to a fuel consumption rate of 12,680 lbm per hour. If a standard aviation jet fuel (JP- 4) costs \$3.50 per US gallon and has a density of 6.84 lbm per US gallon, how much does fuel cost for each hour of level flight?
11. An aircraft has a specific fuel consumption SFC of 0.450 lbm per hour per lbf of thrust. Assume this is a constant during normal steady cruise conditions. The thrust over this period is a constant 5120 lbf. On a flight lasting 2.00 hours, how much fuel is consumed by this aircraft? How much is the cost of fuel over this 2 hours flight? Assume 1.00 lbm of fuel = 0.150 US gallons and that it costs \$3.50 per US gallon. Ignore the weight changes due to fuel consumption en-route.
12. Most large passenger planes fly at a cruising speed of about Mach 0.85.[5] The Mach number is defined as the ratio of actual speed V to the local speed of sound, c, as $M = V/c$. The Mach number is an important variable in most speeding objects moving at or exceeding the local speed of sound. The speed of sound varies directly with the square root of the absolute temperature ($c \sim \sqrt{T}$). At 20°C, the speed of sound is 343 m per second or 767 mph. At an altitude of 10,000 m, the atmosphere has cooled to about -45°C = 228 K. What is the speed of sound at 10,000 m in m per second and mph?
13. Refer to Example 7.8: How much *extra* fuel does it take if another passenger (including luggage) is added to the plane for a total mass gain of 175 lbm? Assume the flight time is 5.25 hours and the SFC is constant during level cruising at 0.587 (lbm per hour)/(lbf of thrust). The original thrust was 21,600 lbf.
14. A new development in airfoil design is the so-called "scimitar winglet". They work by redirecting the airflow around the wingtip, which reduces the amount of drag that is created by the wing. What is the net improvement in lift-over-drag ratio given the unmodified plane had a lift-to-drag ratio of 17.0 while the drag has been reduced by 4.00%? Your solution must also compensate for the combined 220.0 lbm of two winglets. Assume the unmodified weight of the plane was 176,000 lbf.
15. Using the results of Exercise 14 what is the reduction in the fuel rate of consumption in steady level flight? Assume the fuel rate is proportional to the thrust when in steady level flight.
16. Nevil Shute was a British/Australian author and aeronautical engineer highly successful in both careers. He wrote an influential book entitled *No Highway* about a post-World War II British airliner he called the "Reindeer".[6]

The narrator is a hard-driving government-employed aeronautical engineer, Dr. Scott, who takes over a laboratory staffed by over-the-hill specialists, most of whom he quickly retires. But one metallurgist, Mr. Honey, is a quiet and unassuming scientist (and iron willed). He is working on stress fracture[7] of the tail plane of the Reindeer. Mr. Honey says the tail unit will catastrophically fail in a certain number of flight hours and Dr. Scott wonders if Mr. Honey is just another over-the-hill specialist. In fact, a full scale experiment of Mr. Honey's is already underway to test for stress fractures on the tail section. Mr. Honey's projection for a catastrophic failure is already overdue.

Then a Reindeer crashes killing all on board in remote Canada (while unfortunately carrying a Soviet ambassador and causing an international incident).

Could Mr. Honey have been right all along? Should Dr. Scott have grounded the Reindeer when he *first* heard of Mr. Honey's theory? Or was his impression of Mr. Honey as ineffectual correct?

Is there an ethical dilemma here? If Dr. Scott thinks Mr. Honey is wrong, why has Mr. Honey been allowed to waste the government's money on a large and useless experiment? On the other hand, if Dr. Scott believes Mr. Honey, why did he not ground the Reindeer immediately?

Use an Engineering Ethics Matrix with these top-line choices for Dr. Scott.

NSPE Cannons of ethics	Fire Mr. Honey	Encourage Mr. Honey's experiment	Ground all reindeers until fracture possibility rules out	Effects on aeronautical engineering	Contact a newspaper reporter
Hold paramount the safety, health, and welfare of the public					
Perform services only in the area of your competence					
Issue public statements only in an objective and truthful manner					
Act for each employer or client as faithful agents or trustees					
Avoid deceptive acts					
Conduct themselves honorably …					

[5] The Mach number is named after Ernst Mach who was an Austrian physicist known especially for his work on shock waves in air. You have heard a shock wave when lightning strikes nearby.

[6] The world's first passenger jet was the Comet. (Reindeer and Comet seem to be related by a well-known Santa poem!) The Comet *did* suffer from stress fatigue cracking and several aircraft went down with 100% fatalities by the time the problem was fixed.

[7] Take a piece of a tin can and bend it back and forth. It will eventually crack. This is a stress fracture.

Chemical engineering

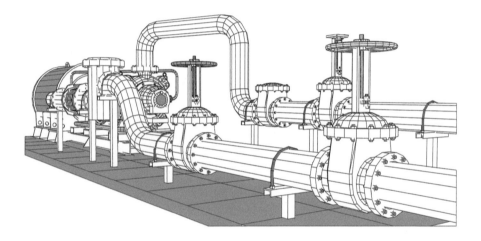

From Mirexon/Shutterstock.com

8.1 Introduction

Chemical engineering is a branch of engineering that applies physical (and increasingly biological) sciences as well as mathematics in the design, development, and maintenance of large-scale chemical processes that convert raw materials into useful or valuable products. Chemical engineers aid in the manufacture of a wide variety of products such as fuels, fertilizers, insecticides, plastics, explosives, detergents, fragrances, flavors, electronic chip manufacture, biotechnology, and pharmaceuticals.

The three primary physical laws underlying chemical engineering design are (1) **the conservation of mass**, (2) **the conservation of momentum**, and (3) **the conservation of energy**. The movement of mass and energy around a chemical process is evaluated using mass and energy balances, as illustrated in the following section.

Exploring Engineering. https://doi.org/10.1016/B978-0-443-13541-5.00030-1

151

8.2 Chemical energy conversion

How is the chemical energy of gasoline converted into the kinetic energy of an automobile moving down the road or the chemical energy of natural gas converted into electrical energy at a power plant? The conversion in both cases is part of a multistep process, and in this chapter, we consider the first step in that process, the conversion of chemical energy into thermal energy.

Combustion is a chemical reaction of a fuel with oxygen to generate heat. It can be slow (burning wood) or rapid (an explosion) depending on the circumstances. While a combustion process requires keeping tabs on all the atoms in the reaction, other processes, such as those found in oil refineries, can be analyzed just by keeping tabs on molecules. A prime example of the latter is distillation, a major process that is used in oil refineries to separate useful chemical compounds from crude petroleum.

The key points to understanding combustion and the field of chemical engineering are the concepts of:

- **Atoms, molecules**, and **chemical reactions.**
- **mol**, a name for a very large number of atoms or molecules.[1]
- **Stoichiometry**, a "chemical algebra" method.
- **Air-to-fuel ratio**, a measure of the amount of air present when combustion occurs.
- **Heating value**, the amount of heat produced by the combustion of a fuel.

One useful principle that is very helpful is the discovery by the Italian scientist Amedeo Avogadro that *equal volumes of gases at the same temperature and pressure contain the same number of molecules*.

8.3 Atoms, molecules, and chemical reactions

Atoms are the basic building blocks of matter. In chemical reactions, the number and type of participating atoms must remain constant. When two or more **atoms** chemically bond with each other, the resultant chemical structure is called a **molecule**. For example, two **A** atoms and one **B** atom can chemically bond to form a hypothetical molecule A_2B as shown in Fig. 8.1.

---

[1] In chemistry the word mol usually means gram-mole (gmol) and kmol means kilogram-mol (kgmol). In this chapter we will use gmol and kgmol instead of mol and kmol.

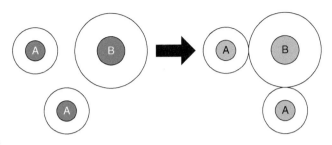

FIGURE 8.1

Two atoms **A** bond with atom **B** to form a molecule **A₂B**.

Chemical reactions occur when two or more atoms bond together to form molecules or when molecules are broken apart. The substances at the beginning of a chemical reaction are called the **reactants**, and the substances produced at the end of the reaction are called the **products**. An arrow drawn between the reactants and products indicates the direction of the chemical reaction. This produces a chemical reaction "equation" that looks like this:

$$\textbf{Reactants} \rightarrow \textbf{Products} \qquad (8.1)$$

For example, the chemical reaction shown in Fig. 8.1 has the following chemical reaction equation $\mathbf{2A + B \rightarrow A_2B}$. This equation means that two atoms of **A** combine with one atom of **B** to form one molecule of **A₂B**. Note that there are two **A** and one **B** atoms to the left of the arrow and two **A** and one **B** atoms to the right of the arrow. The conservation of mass requires that in a chemical reaction equation, the number of *atoms* remains constant, but the number of *molecules* is usually not constant.

8.4 The mol

As molecules are extremely small entities, it takes enormous numbers of them to provide useful amounts of energy for powering automobiles or performing an engineering task. So rather than counting individual molecules, they are combined into very large groups of molecules called moles. The term "mol" is an abbreviation for *mole*, which in turn is an abbreviation for the word *molecule*. Without a prefix, mol or mole always means a *gram*-mole.

Effective May 20, 2019, a **mole** is defined as the amount of a chemical substance that contains exactly $6.02,214,076 \times 10^{23}$ (Avogadro's number) constitutive particles such as atoms, molecules, ions, and so on.

Constitutive particles may be such things as atoms, molecules, ions, electrons, or other well-defined particles or groups of particles. Just as a dozen eggs is a way of referring to exactly 12 eggs, a gram mole (gmol) of hydrogen atoms is a way of referring to exactly $6.02,214,076 \times 10^{23}$ hydrogen atoms (and a kilomol of hydrogen atoms contains $6.02,214,076 \times 10^{26}$ hydrogen atoms). Similarly, a mole of people would contain $6.02,214,076 \times 10^{23}$ people (quite a crowd!).

A mole unit is an SI unit of measure used to indicate the number of individual items. Continuing the dozen-egg analogy, the constitutive particles might consist of five individual chicken eggs and seven individual turkey eggs. If so, notice that not every egg has the same mass. The atomic masses of some common elements are given in Table 8.1. The unit of atomic or molecular mass in this table can be either g/gmol, kg/kgmol, lbm/lbmol, and so on.[2] In addition, the number of moles n of a substance with mass m and molecular (or atomic) mass M is given by Eq. (8.2).

$$n = \frac{m}{M} \qquad (8.2)$$

Table 8.1 Atomic and molecular mass (M) of some common chemicals.

Hydrogen, H	1.00	Nitrogen, N	14.0
Oxygen, O	16.0	Helium, He	4.00
Carbon, C	12.0	Argon, Ar	40.0
Sulfur, S	32.1	Chlorine, Cl	35.5
Water, H_2O	18.0	Methane, CH_4	16.0
Carbon dioxide, CO_2	44.0	Ethanol, C_2H_5	46.1

Example 8.1

a. How many gram moles (gmol) of water are in 10.0 kg of water?
b. How many kilogram moles (kgmol) of water are in 10.0 kg of water?
   Need: Number of gram moles and kilogram moles in 10.0 kg of H_2O.
   Know: Atomic masses of O and H are 16.0 and 1.00, respectively.
   How: The number of moles n of a substance with a mass m that has a molecular mass M is given by $n = m/M$.
   Solve: The molecular mass of water (H_2O) is

$$M = 2 \times (1.00) + 1 \times (16.0) = 18.0 \text{ g/gmol} = 18.0 \text{ kg/kgmol} = 18.0 \text{ lbm/lbmol}$$

   Then, for 10.0 kg of water,
   a. $10.0 \text{ kg} = 100. \times 10^2$ g, then

$$n = m/M = 100. \times 10^2 \text{ [g]}/[18.0 \text{ g} / \text{gmol}] = \textbf{556 gram moles}$$

   b. $n = m/M = 10.0 \text{ [kg]}/[18.0 \text{ kg} /\text{kgmol}] = \textbf{0.556 kilogram moles}$

---

[2] Therefore, every kgmol of water has a mass of 18.0 kg, as the atomic mass of every hydrogen atom is 1.00 kg/kgmol, and the atomic mass of every oxygen atom is 16.0 kg/kgmol. Similarly, every pound mole (lbmol) of water has a mass of 18.0 lbm.

Example 8.2

Determine the effective molecular mass of air, assuming it is composed of 79.0% nitrogen and 21.0% oxygen.

Need: Molar mass of air with 21.0% O_2 and 79.0% N_2. Therefore, air is a mixture of "constitutive particles." A kgmol of a mixture of constitutive particles must still have Avogadro's Number of particles, be they oxygen or nitrogen molecules. So, we need the combined mass of these two kinds of entities in the correct ratio, remembering that each of them has a different mass.

Know: Molar mass of O_2 is 32.0 kg/kgmol, and the molar mass of N_2 is 28.0 kg/kgmol.

How: Proportion the masses of each constituent according to their concentration.

Solve : $M_{air} = \%N_2 \times M_{N2} + \%O_2 \times M_{O2} = 0.790 \times 28.0 + 0.210 \times 32.0 = $ **28.8 kg/kgmol.**
Note that we have defined a kgmol of air even though "air" molecules per se do not exist, and in so doing we preserved the notion that every mole of air should have Avogadro's Number of entities.

8.5 Stoichiometry

The goal of many chemical engineering studies is to determine the energy produced by the combustion of a fuel. For example, how does an automobile engine convert 80. kJ per s of chemical energy into the same quantity of mechanical and heat energy. The first step in modeling a combustion process is writing the chemical reaction Eq. (8.1) in a form where the "reactants" are fuel and oxygen and the "products" are the combustion products:

$$\textbf{Fuel} + \textbf{Oxygen} \rightarrow \textbf{Combustion Products} \tag{8.3}$$

For example, in the combustion equation for hydrogen, the "reactants" are molecular hydrogen, H_2, and molecular oxygen, O_2, and the "product" is water, H_2O.

$$2H_2 + O_2 \rightarrow 2H_2O$$

For most of the common combustion processes, the fuel is a **hydrocarbon**—a compound containing only two chemical elements: hydrogen and carbon. For example, "natural gas" is methane, CH_4, which has four hydrogen atoms bonded to each carbon atom. Gasoline is a hydrocarbon that can be modeled as isooctane, C_8H_{18}, a molecule containing eight carbon atoms and 18 hydrogen atoms as shown schematically in Fig. 8.2.

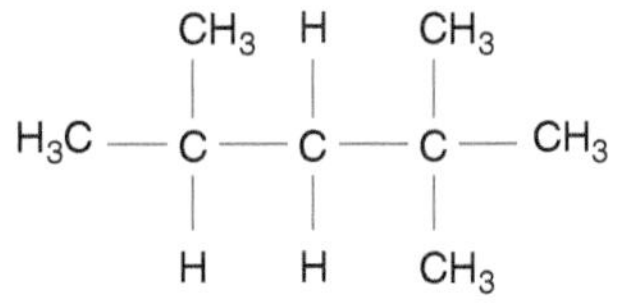

FIGURE 8.2

Chemical structure of isooctane (gasoline).

The energy history of the modern world might be summed up as an increase in the hydrogen–carbon ratio of the dominant fuel. Coal, the dominant fuel in the 19th century, has about one atom of hydrogen per atom of carbon. The 20th century saw increasing use of petroleum-based fuels with about two hydrogen atoms per carbon atom. In the early 21st century, societies relied on natural gas with about four hydrogen atoms per carbon atom. A goal for the 21st century is a "hydrogen economy," with an infinite hydrogen-to-carbon ratio—that is, no carbon at all in the fuel. Advocates of this continuing reduction of the carbon content of fuels point out that the combustion of carbon produces carbon dioxide, a major "greenhouse gas" implicated in global warming. Solar, nuclear, wind, and hydroelectric energy are contenders for future noncarbon energy sources. Solar and nuclear are also the first two entries in the National Academy of Engineering list of Engineering Challenges for the 21st Century.

In an engineering analysis, we know the fuel we want to "burn" and the combustion products that result. However, we do not initially know the number of kilogram moles of oxygen needed to combine with the fuel, and we do not know the number of kilogram moles of each kind of combustion product that will result. We can express this situation symbolically by assuming that we have just 1-kg mole of fuel and put undetermined coefficients in front of the other chemicals present to represent those unknown amounts. For example, for the molecular hydrogen–oxygen combustion reaction, this process would look like this:

$$\textbf{1 kgmol } H_2 + a \textbf{ kgmol } O_2 \rightarrow b \textbf{ kgmol } H_2O \qquad (8.4)$$

Determining the numerical value of the coefficients a and b is done by a "chemical algebra" called **stoichiometry**. It relies on a key fact mentioned earlier: the number of each kind of atom in a chemical reaction remains constant. So, we simply write an equation for each type of atom that expresses this equality. Then, we algebraically solve for the unknown coefficients. In the reaction shown in Eq. (8.4) we have.

- **Hydrogen atomic balance equation**: $2 = 2b$ (because there are 2 kgmol of hydrogen atoms on the left, and $2b$ kgmol of hydrogen atoms in the water molecules on the right).
- **Oxygen atomic balance equation**: $2a = b$ (because there are $2a$ kgmol of oxygen atoms on the left and b kgmol of oxygen atoms in the water molecules on the right).

Solving those two equations with two unknowns yields $b = 1$ and $a = \frac{1}{2}$. Substituting this back into our original reaction results in the **stoichiometric equation**:

$$H_2 + \tfrac{1}{2}O_2 \rightarrow H_2O \qquad (8.5)$$

You will note that this equation looks slightly different from the one that was presented earlier ($2H_2 + O_2 \rightarrow 2H_2O$), but you can reassure yourself that it has the same meaning as the earlier equation by dividing each of the coefficients in that equation by 2.

The process of finding the stoichiometric coefficients can be more difficult as the number of atoms in the participating molecules increases and yet more complicated when there are repeat molecules on both sides of the chemical reaction. Consider some common fossil fuels: natural gas (methane, CH_4), coal (which is approximately $C_{137}H_{97}O_9NS$ for bituminous), oil (C_nH_{2n+2}, where n varies from 5 to 40 depending on the type of oil), and many others. Finding the stoichiometric coefficients in the combustion equation is not always a simple task. The easiest way is to use a systematic method using tabular entries in tables as shown in the following example.

Example 8.3

What are the stoichiometric coefficients when diethyl ether, $C_4H_{10}O$ is burned in an oxygen atmosphere? The combustion equation has the following form:

$$C_4H_{10}O + aO_2 \rightarrow bCO_2 + cH_2O$$

Need: Stoichiometric coefficients, a, b, and c.

Know: You could certainly find a, b, and c by trial and error.

How: The calculation can be systematized into a simple tabular method of finding three equations in the three unknowns a, b, and c, as per Table 8.2.

Table 8.2 Tabular method of solving stoichiometric combustion problems.

Atoms	LHS	RHS	Solution
C	4	b	$b = 4$
H	10	$2c$	$c = 5$
O	$1 + 2a$	$2b + c$	$a = b + c/2 - \frac{1}{2} = 6$

LHS, *lefthand side of the equation;* RHS, *righthand side.*

Solve: $a = 6, b = 4.$ and $c = 5.$ Therefore, $C_4H_{10}O + 6O_2 \rightarrow 4CO_2 + 5H_2O$.

Notice that you have three unknowns (a, b, and c) and three equations: one for C, one for H, and one for O. Hence, you can solve it uniquely by equating the LHS with the RHS for each element.

It is very important that you make sure that simple balances, such as for C and for H in this example, each involving only one unknown, are solved **before** those that involve more than one variable, such as for O in this instance. Also, make a quick scan of your solution to confirm that you have a "mass balance," that is, that you have neither created nor destroyed atoms. In this case, you can quickly spot that there are 4 C atoms, 10 H atoms, and 13 O atoms on both sides of the equation. If other elements are present, you can add rows in the table to balance each additional element.

8.5.1 The air-to-fuel ratio

Most combustion processes react fuel with the oxygen in the air. Engineers usually treat air as a mixture that has 3.76 kgmol of nitrogen molecules (N_2) for every kgmol of oxygen molecules (O_2), and ignore any other molecules, such as carbon dioxide, water vapor, and argon, that are present in much smaller quantities. This approximate composition of air works out to be 79.0% N_2 and 21.0% O_2 on a molar basis. The N_2 in air is considered totally unreactive in our treatment of combustion, so the N_2 in the fuel-air mixture is the same as the N_2 as the combustion products.

To design efficient combustion systems, it is important to know the amount of air needed to completely burn a specific amount of fuel. This is called the **air-to-fuel ratio**. To determine it, we begin by including the 3.76 kgmol of nitrogen that accompany each kgmol of oxygen in the combustion equation. This is done by recognizing that the stoichiometric coefficient for nitrogen is always 3.76 times the stoichiometric coefficient of oxygen. As, in our model, the nitrogen does not take part in the combustion reaction, it usually has the same coefficient on both sides of the reaction equation. Therefore, the combustion of hydrogen in air is written as:

$$H_2 + \tfrac{1}{2} O_2 + \tfrac{1}{2}(3.76N_2) \rightarrow H_2O + \tfrac{1}{2}(3.76N_2)$$

or

$$H_2 + \tfrac{1}{2}(O_2 + 3.76N_2) \rightarrow H_2O + 1.88N_2 \tag{8.6}$$

The term $O_2 + 3.76N_2$ is *not* 1 kgmol of air. It is 4.76 kgmol of an oxygen plus nitrogen mixture and can be thought of as being 4.76 kgmol of air.

The previous stoichiometric equation makes it possible to determine the air-to-fuel ratio in two ways: as a ratio of numbers of molecules or moles—called the molecular or molar air-to-fuel ratio, and as a ratio of their masses—called the mass air-to-fuel ratio. We use the nomenclature $(A/F)_{molar}$ and $(A/F)_{mass}$ to distinguish between these two ratios. As these ratios have no units, they are best written explicitly as

$$(A/F)_{molar} = [\text{kgmol air / kgmol fuel}]$$

and

$$(A/F)_{mass} = [\text{mass air / mass fuel}]\,(\text{e.g., [kg air / kg fuel]or [lbm air / lbm fuel], etc.})$$

Example 8.4

Determine the molar and mass stoichiometric air-to-fuel ratios for the combustion of hydrogen in air.

Need: $(A/F)_{molar} = $ ___ moles of air per mol of hydrogen and $(A/F)_{mass} = $ ___ mass of air per mass of hydrogen.

Know: Stoichiometric equation: $H_2 + \tfrac{1}{2}[O_2 + 3.76N_2] \rightarrow H_2O + 1.88N_2$.

How: Use the stoichiometric equation to ratio moles; then, multiply by the relative molecular masses from Table 8.1.

Solve: $(A/F)_{molar} = \frac{0.5(1+3.76)}{1} = 2.38$ kgmol of air/kgmol H_2

To express the *mass* ratio, simply use the molecular masses of each component. In Example 8.2 we found that the molar mass of air is $M_{air} = $ **28.8 kg air/kgmol air**. Then,

$$(A/F)_{mass} = \left(2.38 \frac{\text{kgmol of air}}{\text{kgmol H}_2}\right)\left(\frac{28.8 \text{ kg air/kgmol air}}{2 \text{ kg H}_2/\text{kgmol H}_2}\right) = 34.3 \text{ kg air / kg H}_2$$

The fuel-to-air ratios are simply the inverses of the air-to-fuel ratios:

$$(F/A)_{molar} = 1/(A/F)_{molar} = 1/2.38 = 0.420 \text{ kgmol H}_2/\text{kgmol air}$$

$$(F/A)_{mass} = 1/(A/F)_{mass} = 1/34.3 = 0.029 \text{ kg H}_2/\text{kg air}$$

The stoichiometric fuel-to-air ratio is a constant for a given fuel. However, actual fuel-to-air ratios can be varied by deciding just how much air or fuel to add. The term **equivalence ratio**, symbol φ (Greek phi), is defined as

$$\text{Fuel to Air Equivalence Ratio } \varphi = (F/A_{actual})/(F/A_{stoichiometric}) \qquad (8.7)$$

Stoichiometric combustion means that $\varphi = 1.0$. Excess fuel produces a **rich** mixture ($\varphi > 1$) and excess air is a **lean** mixture ($\varphi < 1$), the lean mixture is preferred to ensure none of the fuel escapes in an unburnt state.

8.6 Heating value of hydrocarbon fuels

Beyond the stoichiometric fuel ratios, chemistry can also answer the question, "How much energy can be obtained by the combustion of 1 kg of fuel under stoichiometric conditions?" That, in turn, is a key part of answering a question such as "How many miles per gallon of fuel is it possible for an automobile to travel?"

To determine the amount of energy that can be obtained by the combustion of 1 kg of fuel under stoichiometric conditions, the type of fuel being used must be specified. For simplicity assume that fuel is a generalized hydrocarbon C_xH_y with x atoms of carbon and y atoms of hydrogen, and its combustion is described by the following stoichiometric equation:

$$C_xH_y + [x + (y/4)]O_2 \rightarrow xCO_2 + (y/2) H_2O + \textbf{Heat of Combustion} \qquad (8.8)$$

Table 8.3 lists the heating values for several common hydrocarbon fuels.

Table 8.3 Heating values of hydrocarbon fuels.

Fuel	Heating value, kJ/kg
Methane, CH_4	55,650
Propane, C_3H_8	46,390
Isobutane, C_4H_{10}	45,660
Gasoline, "C_8H_{18}"	45,560
Diesel fuel, "$C_{16}H_{34}$"	43,980
Benzene, "C_6H_6"	42,350
Toluene, "$C_6H_5CH_3$"	42,960

Gasoline is typically modeled as isooctane, where $x = 8$ and $y = 18$. Using Eq. (8.8) you can show that 1 kg of gasoline will react with 3.5 kg of oxygen and produce 3.1 kg of carbon dioxide and about 1.4 kg of water while releasing 45,560 kJ of energy—see Table 8.3.

Example 8.5

An engine burns gasoline with a composition C_8H_{18} at the rate of 25.0 kg/h. How many kilowatts of thermal energy are produced?

Need: kW of thermal energy produced by burning gasoline, C_8H_{18}.

Know: Heat of combustion of gasoline is 45,560 kJ/kg (from Table 8.3).

How: Power = fuel consumption rate × fuel energy

$$\text{Fuel consumption rate} = (25.0 \text{ kg} / \text{h}) \times (1 \text{ h} / 3600 \text{ s}) = 6.94 \times 10^{-3} \text{ kg/s}$$

Solve: Power $= \left(6.94 \times 10^{-3} \text{ kg} / \text{s}\right) \times (45,560 \text{ kJ/kg}) = 316 \text{ [kJ/s]} = \textbf{316 kW}$

8.7 Chemical engineering: How do you make chemical fuels?

Why don't you pull your car up to an oil well and fill it up? Could the oil that comes directly out of the ground run your car? The answer is no! Chemical engineers design oil refineries that convert crude oil from the well into gasoline.

We have seen that hydrocarbons, molecules made of hydrogen and carbon atoms, react with oxygen to produce carbon dioxide, water, and energy in the form of heat. The most convenient sources of hydrocarbons lie concentrated in deposits under the ground, the remains of plant matter that lived hundreds of millions of years ago. The name for these hydrocarbon deposits, *crude oil*, encompasses a wide range of substances. At one extreme, hydrocarbon deposits contain just light gases containing only a few carbon atoms per molecule of hydrocarbon. Sometimes, just one carbon atom joins with four hydrogen atoms forming a hydrocarbon molecule called *methane* (or *natural gas*). At the other extreme, crude oil contains a thick viscous sludge with molecules containing dozens of carbon and hydrogen atoms. Some typical hydrocarbon molecules are shown schematically in Fig. 8.3.

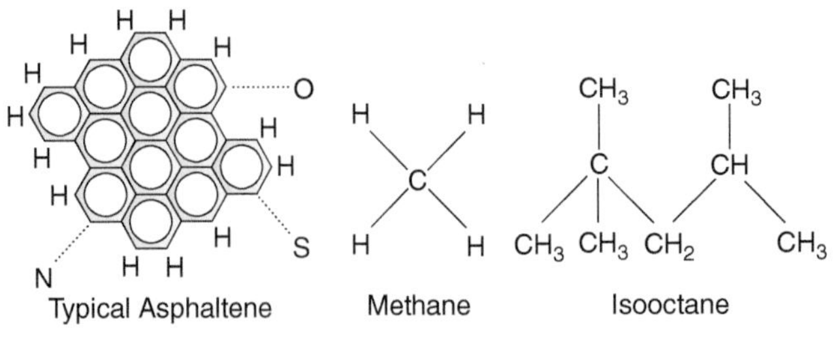

Typical Asphaltene   Methane   Isooctane

FIGURE 8.3

Some of the molecules present in crude oil.

Crude oil from an oil well would fill your tank with an impractical mixture of molecules. The heavy viscous ones would clog the fuel system, and they would not ignite in the combustion chamber. Crude oil also contains significant amounts of foul-smelling, smog-producing, sulfur and nitrogen compounds. However, the lightest hydrocarbons in crude oil evaporate easily and explode readily. A better choice is the Goldilocks' solution: fill the tank with molecules that are not too big, not too small. For automotive engines, this means molecules that have about 5−10 carbon atoms and 10−25 hydrogen atoms. This class of molecules includes the compounds collectively called *gasoline* (isooctane, C_8H_{18}).

Example 8.6

An oil deposit contains crude oil whose molar composition is 7.0% C_5H_{12}, 33% C_8H_{18}, 52% $C_{16}H_{34}$, and 8.0% $C_{38}H_{16}$. What is the percentage of each substance by mass?

Need: By mass $C_5H_{12} = $ ____%, $C_8H_{18} = $ ____ %, $C_{16}H_{34} = $ ____%, and = ____%.

Know−How: Eq. (8.1) tells us how to convert moles to mass, $m = nM$ where m is mass in kg, n is the number of moles in kgmol, and M is the molecular mass of the constituent in kg/kgmol. Now imagine that we have 100 kgmol of this crude oil. Divide it among the various species that we are given:

Species	n in kgmol	Molecular mass, M in kg/kgmol	Mass, $m = n \times M$ in kg	% Mass
C_5H_{12}	7.0	$5(12) + 12(1) = 72$	$7.0 \times 72 = 500$	$(500/20{,}000) \times 100 = \mathbf{2.5\%}$
C_8H_{18}	33	$8(12) + 18(1) = 114$	$33 \times 114 = 3800$	$(3800/20{,}000) \times 100 = \mathbf{19\%}$
$C_{16}H_{34}$	52	$16(12) + 34(1) = 226$	$52 \times 226 = 12{,}000$	$(12{,}000/20{,}000) \times 100 = \mathbf{60\%}$
$C_{38}H_{16}$	8.0	$38(12) + 16(1) = 472$	$8.0 \times 472 = 3800$	$(3800/20{,}000 \times 100 = \mathbf{19\%}$
Totals	100		20,000 kg (to 2 sig. figs.)	100% (to 2 sig. figs.)

Using a tabular format (or spreadsheet) to solve this problem makes the solution much easier to understand.

8.7.1 Process engineering

In the early decades of the 20th century, a new idea emerged. Chemical engineers, such as Arthur D. Little, an independent consultant, and W. W. Lewis, a professor at MIT, looked at chemical manufacturing plants in a new way. They viewed the plants as arrangements of simple elements. These elements, which served as a sort of alphabet for chemical plant construction, came to be known as **process steps** and were analyzed using the concept of ***unit operations***. It turned out that many process steps were quite different but were closely related in their basics. Therefore, several chemical engineering processing steps, such as *distillation, absorption, stripping*, and *liquid extraction*, are all described by essentially the same mathematics. Here, we use distillation as an example.

8.7.1.1 Distillation

In this section we conduct a simplified analysis of a basic chemical engineering process, **distillation**. The first major process in most refineries is a set of distillation columns (see Fig. 8.4). This is a purely *physical* separation process that relies on the different relative volatilities among the various components of crude oil. Generally, they are arranged in sequence with the most volatile components removed first (such as methane, pentane, and isooctane), and in the last distillation columns the least volatile, such as asphaltenes, are removed.

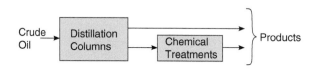

FIGURE 8.4

Process diagram for a refinery.

Distillation is conceptually a simple process operation. The key unit operation is the fact that condensed liquid (condensate) flows down the distillation column while vapor of a different composition flows up as shown in Fig. 8.5. A distillation column

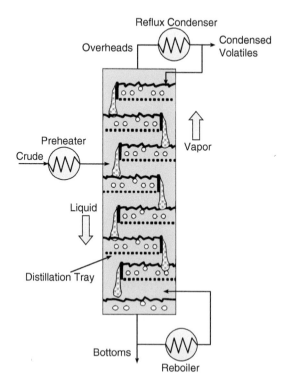

FIGURE 8.5

Schematic of a typical distillation column.

is made up of pipes, pumps, and heat exchangers, plus a large vertical column containing trays. The trays temporarily hold back the liquid condensate to expose it to the up-flowing vapor. The liquid condensate is produced by condensing the vapor at the top of the column in a "reflux condenser." The vapor is produced by heating the "bottoms" in a "reboiler" and as it goes up the column it is washed by liquid condensate flowing down to extract any remaining less-volatile components from the vapor.

Two process diagram symbols for a distillation unit are shown in Fig. 8.6. The more detailed one is the more literal, as it includes the reflux condenser and the reboiler; however, often we do not care about the internal flows and just need the inlet and outlet flow streams. Again, you should imagine the distillation column symbol as surrounded by a control surface "dotted line," through which the arrows denoting the flows penetrate. That control surface reminds us that over enough time, the mass into a distillation unit must equal the mass out. For these two control surfaces, the internal flows are irrelevant, and the two diagrams are completely equivalent.

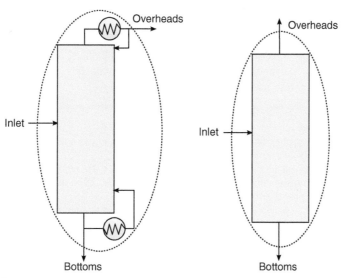

FIGURE 8.6

Process diagrams of a distillation column.

8.8 Modern chemical engineering

Chemical engineering today is much more than just process engineering. Chemical engineers are engaged in the development and production of a diverse range of products that include high-performance materials used in aerospace, automotive, biomedical, electronic, environmental, and space applications. Some examples include ultra-strong fibers; new fabrics, adhesives, and composites for vehicles;

biocompatible materials for implants and prosthetics; gels for medical applications; and pharmaceuticals. Chemical engineers today also work on biological and bioengineering projects, such as understanding biopolymers (proteins) and developing artificial organs.

Summary

Determining useful engineering parameters, such as the miles per gallon achievable by an automobile, requires calculating the amount of thermal energy (heat) available from the chemical energy contained in a fuel. This calculation requires the mastery of several concepts from the field of chemistry. Those concepts include **atoms, molecules**, and **chemical reactions**. A chemical algebra method called **stoichiometry** is used to keep track of the number of atoms and molecules that occur in those reactions. A measure of the amount of air needed for combustion called the **air-to-fuel ratio** and finally a way to understand the **heating value** of a fuel was discussed.

The term **equivalence ratio**, φ, is defined as $(F/A_{\text{actual}})/(F/A_{\text{stoichiometric}})$, and is used to quickly assess combustion conditions: a "fuel-rich" process operates at $\varphi > 1$ and a "fuel-lean" one at $\varphi < 1$. The principles of stoichiometry and stoichiometric balances have further utility when applied to physical, as opposed to chemical, processes. Many oil refinery processes fall into this category, including distillation to separate useful products from crude oil.

Exercises

1. How many kgmol are contained in 3.0 kg of ammonia NH_3? (**Ans. 0.18 kgmol** to two significant figures.)
2. How many kgmol are contained in 1.0 kg of nitroglycerine $C_3H_5(NO_3)_3$?
3. What is the mass of 5.0 kgmol of carbon dioxide CO_2? (**Ans. 2.2 × 10^2 kg**)
4. What is the mass of 1.00 kgmol vitamin B1 disulfide $C_{24}H_{34}N_8O_4S_2$?
5. The effective molecular mass of air in kg/kgmol is defined as the mass of a kgmol of elementary particles of which 78.09% are nitrogen molecules, 20.95% are oxygen molecules, 0.933% are argon atoms, and 0.027% are carbon dioxide molecules. What is the effective molecular mass of air? (Watch your significant figures!) What other factor could affect the effective molecular mass of air?
6. A gallon of gasoline has a mass of about 3.0 kg. Further, a kg of gasoline has an energy content of about 45,500 kJ/kg. If an experimental automobile requires just 10. kW of power to overcome air resistance at a steady speed of 30 miles an hour, and *if* there are no other losses, what would the gas mileage of the car be in miles per gallon? (**Ans. 110 mpg** to two significant figures.)
7. Determine the value of the stoichiometric coefficients for the combustion of an oil (assumed molecular formula CH_2) in oxygen: $CH_2 + aO_2 = bCO_2 + cH_2O$. Confirm your answer is correct! (**Ans. $a = 1.5, b = 1, c = 1$.**)
8. Determine the value of the stoichiometric coefficients for the combustion of coal in oxygen given by the stoichiometric equation

$$CHN_{0.01}O_{0.1}S_{0.05} + aO_2 = bCO_2 + cH_2O + dN_2 + eSO_2$$

9. Determine the value of the stoichiometric coefficients for the combustion of natural gas in air:
   $CH_4 + a(O_2 + 3.76N_2) = bCO_2 + cH_2O + dN_2$

10. Using the stoichiometric coefficients you found in Exercise 9, determine the molar air-to-fuel ratio $(A/F)_{molar}$ for the combustion of natural gas in air. (**Ans. $(A/F)_{molar}$ = 9.52 kgmol of air/ kgmol of fuel.**)

11. Using the results of Exercises 9 and 10, determine the *mass* air-to-fuel ratio $(A/F)_{mass}$ for the combustion of natural gas in air. (**Ans. 17.2 kg of air/kg of fuel.**)

12. Determine the mass air-to-fuel ratio $(A/F)_{mass}$ for the combustion of an oil (represented by CH_2) in air.

13. Determine the mass air-to-fuel ratio $(A/F)_{mass}$ for the combustion of coal in air represented by the equation

$$CHN_{0.01}O_{0.1}S_{0.05} + a(O_2 + 3.76N_2) = bCO_2 + cH_2O + dN_2 + eSO_2.$$

14. Determine the mass air-to-fuel ratio $(A/F)_{mass}$ for the combustion of isooctane C_8H_{18} in air. Its stoichiometric equation is

$$C_8H_{18} + 12.5(O_2 + 3.76N_2) = 8CO_2 + 9H_2O + 47.0N_2.$$

Exercises 15−21 use engineering considerations to give insight on the "global warming" issue.

15. Assume 1 kg of isooctane produces 45,500 kJ of energy. How much carbon dioxide is released in obtaining a kJ of energy from the combustion of isooctane in air? (**Ans. 6.79×10^{-5} kg CO_2/kJ.**)

16. Assume a kg of hydrogen produces 1.20×10^5 kJ of thermal energy. How much carbon dioxide is released in obtaining a kJ of energy from the combustion of hydrogen in air?

17. The amount of carbon dioxide levels in the atmosphere has been increasing for several decades. Many scientists believe that the increase in carbon dioxide levels will lead to "global warming" and trigger abrupt climate changes. You are approached by a future US President for suggestions on what to do about this problem. Based on your answers to Exercises 15 and 16, what might be the *simplistic* approach to reducing the rate of increase of the amount of carbon dioxide in the atmosphere due to combustion in automobiles and trucks?

18. Give three reasons why your suggested solution in Exercise 17 to global warming has *not* already been adopted.

19. A monomeric formula for wood cellulose is $C_6H_{12}O_6$ (it repeats a hexagonal structure based on this formula). The energy produced by burning wood is approximately 1.0×10^4 kJ/kg. Determine the amount of carbon dioxide released in obtaining a kJ of energy by the combustion of wood in air. (**Ans. 1.5×10^{-4} kg of CO_2 released/kJ of energy produced.**)

20. Repeat Exercise 19 for the combustion of coal. For simplicity, assume coal is just carbon, C, with an HV of 32,800 kJ/kg

21. In 1800, the main fuel used in the United States was wood. (Assume wood is cellulose whose representative repeating formula is $C_6H_{12}O_6$.) In 1900, the main fuel used was coal (assume coal in this example can be approximated by pure carbon, C). In 2024, the main fuel used is oil (assume it can be represented by isooctane). Assume that, in 2100, the main fuel will be hydrogen (H_2). Use a spreadsheet to prepare a graph of carbon dioxide released per kJ of energy produced (y-axis) as a function of year (x-axis) from 1850 to 2050.

22. A pipe carries a mixture of, by kgmol (which are proportional to the number of molecules), 15% C_5H_{12}, 25% C_8H_{18}, 50% $C_{16}H_{34}$, and 10% $C_{32}H_{40}$. What is the percentage of each substance by mass?

23. The glass pipe system shown in the following figure has a liquid flowing in it at an unknown rate. The liquid is very corrosive and flow meters are correspondingly very expensive. Instead, you propose to your boss to add a fluorescent dye via a small pump and to deliver 1.00 g/s of

this "tracer" and activate it by UV light. By monitoring the dye fluorescence downstream, you determine its concentration is 0.050 wt.%. What is the unknown flow rate in kg/h? (This is a useful method of indirectly measuring flows.)

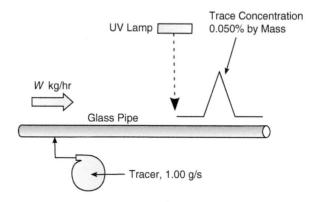

24. You have worked for a petroleum company producing automotive fuels for several years, and in your spare time at home, you developed a new fuel composition that has a higher heating value than ordinary gasoline. When you began work at this company, you signed a "confidentiality agreement" that gave them all your intellectual property. What are your obligations to your employer? (Give your solution using the Engineering Ethics Matrix.)
    a. It is your work on your time, so you have no obligations to your current employer.
    b. Everything you used to develop your new fuel you learned on the job at your company, so your work really belongs to them.
    c. Your company's confidentiality agreement requires you to provide them with all your work.
    d. Give your idea to a friend and let him pursue it while keeping you as a silent partner.
25. As a production engineer for a large chemical company, you need to find a new supplier for a specific commodity. As this contract is substantial, the salespeople you meet with are naturally trying to influence your purchasing decision. Which of the following items are ethical in your opinion?[3] (Give your solution using the Engineering Ethics Matrix.)
    a. Your meeting with a salesperson extends over lunch, and she pays for the lunch.
    b. In casual conversation at the sales meeting, you express an interest in baseball. After the meeting, a salesperson sends you free tickets to your favorite team's game.
    c. After your meeting, a salesperson sends a case of wine to your home with a note thanking you for the "useful" meeting.
    d. As a result of the sales meeting, you are invited on an all-expenses-paid trip to China to visit the salesperson's manufacturing facility.
26. You are attending a national engineering conference as a representative of your company. A supplier to your company has a booth at the conference and is passing out small electronic calculators to everyone who comes to their booth. The calculators are valued at about $25, and you are offered one. What do you do? (Give your solution using the Engineering Ethics Matrix.)
    a. Accept the gift; as it has a small value, there is no need to report it.
    b. Accept the gift and report it to your supervisor.
    c. Decline the gift, explaining that you work for one of their customers.
    d. Ask someone else to get one of the calculators for you.

27. It is December 1928 and your name is Thomas Midgley, Jr.[4] You have just invented a new miracle refrigerant composed of chlorinated fluorocarbons (CFCs) that will make your company a lot of money. At midnight on Christmas Eve, you are visited by three spirits who show you your past, present, and future. In your future, you discover that the chlorine in your miracle refrigerant will eventually destroy the Earth's ozone layer and put the entire human race at risk. What do you do? (Give your solution using the Engineering Ethics Matrix.)
    a. Claim the entire vision was due to a bit of undigested meat and forget it?
    b. Run to the window and ask a passerby to go buy a large goose.
    c. Destroy all records of your CFC work before anyone can use it.
    d. Start work developing nonchlorinated hydrocarbon refrigerants.

28. It is 2021, and you are a process engineer at a large oil refining company. The world is rapidly moving toward a hydrogen energy economy, and your company has been trying to develop an efficient and cost-effective way to extract hydrogen from crude oil. You found a low-cost process, but it produces a considerable amount of undesirable chemical by-product pollutants. However, during your work on this project, you also discovered an effective and inexpensive way to extract hydrogen directly from seawater. You realize that revealing this process would effectively eliminate the world demand for petroleum and would probably cause serious financial damage to your company. What do you do? (Give your solution using the Engineering Ethics Matrix.)
    a. Quit your job and start your own hydrogen-producing company.
    b. Talk to your supervisors and reveal your process to them to see if they wish to pursue implementing it as part of their company.
    c. Contact a patent lawyer not associated with your current employer and try to patent this potentially lucrative new process.
    d. Without your employer's permission, publish an article in a well-read chemical or energy magazine revealing your process and giving it to the world free of charge.

---

[3] How do you distinguish between a token gift and a bribe? Where do you draw the line?
[4] See an abbreviated story of his life: http://www.uh.edu/engines/epi684.htm

Civil engineering

9.1 Introduction

Civil engineering is one of the oldest and broadest engineering professions. It focuses on the infrastructure necessary to support a civilized society. The Roman aqueducts, the great European cathedrals, and the earliest steel bridges were built by highly skilled forerunners of the modern civil engineer. These craftsmen of old relied on their intuition, trade skills and experience-based design rules, or heuristics,

Exploring Engineering. https://doi.org/10.1016/B978-0-443-13541-5.00010-6

derived from years of trial and error experiments but rarely passed on to the next generation.

In contrast, today's civil engineers bring to bear on these problems, a knowledge of the physical and natural sciences, mathematics, computational methods, economics, and project management. Civil engineers design and construct buildings, transportation systems (such as roads, tunnels, bridges, railroads and airports), and facilities to manage and maintain the quality of water resources. Society relies on civil engineers to maintain and advance human health, safety and our standard of living. Those projects that are vital to a community's survival are often publicly funded to ensure that they get done even where there is no clear or immediate profit motive.

9.2 What do civil engineers do?

Civil engineers may find themselves just as easily on the computer optimizing a design or in the field monitoring progress and interacting with customers, contractors, and construction team members. They are involved throughout the life cycle of the project—exploration and measurement of on-site environmental conditions, design and computer modeling, management of construction, system management of operation (e.g., of traffic or water resources), demolition, and recycling of materials.

An important challenge confronted by all civil engineers is the need to deal with the uncertainty of highly variable environmental conditions. Informed design decisions by civil engineers based on a historical knowledge of environmental extremes are our first line of defense against hurricane winds, earthquakes, droughts, flooding, periods of extreme heat or cold, and nonuniform or time-dependent soil properties. When actual conditions exceed predictions of those extremes, disasters happen.

In this chapter, you will gain a flavor of the fundamentals underpinning four traditional areas of civil engineering:

- **Structural engineering** deals with the analysis and design of structures that support and resist loads. Structural engineers are most commonly involved in the design of buildings, bridges, tunnels, heavy machinery, and vehicles or any item where structural integrity affects the item's function and safety. For example, buildings must be designed to endure heavy wind loads as well as changing climate and natural disasters.
- **Geotechnical engineering** deals with the behavior of earth materials. It uses principles of soil mechanics and rock mechanics to investigate underground conditions and materials. Using this information, they design foundations and earthworks structures for buildings, roads, and many other types of projects.
- **Water resources engineering** deals with designing the infrastructure to manage and utilize the nation's water supply. Applications range from irrigation, navigation, and hydro-power generation to water treatment plants and the network

of pipes that convey clean safe drinking water to your tap. It also involves the treatment of domestic and industrial wastewaters before discharging them back into the environment.

- **Transportation engineering** deals with the safe and efficient movement of people and goods. This involves the design of many types of transportation facilities, including highways and streets, mass transit systems, railroads, airports, and shipping ports and harbors. Transportation engineers apply technological knowledge as well as an understanding of the economic, political, and social factors in their projects.

New civil engineering fields and applications are constantly emerging as new technologies replace the old and priorities shift—e.g., environmental engineering, alternative energy systems including new energy extraction methods, earthquake engineering, and green engineering, to name but a few.

9.3 Structural engineering

Civil engineers strive to create efficient structures by minimizing cost or weight subject to constraints on **stiffness** (resistance to deformation) and **strength** (resistance to breaking). Today, the stiffness and strength of a prospective structure can be rapidly evaluated using computational methods. Many of the fundamentals underlying these computational models, such as free-body diagrams (FBDs) and the equations of static equilibrium, have been around for a long time. For example, the American civil engineer Squire Whipple was analyzing truss bridges in the 1830s. But the complex hand calculations required placed severe restrictions on the range of bridge designs that could be safely built.[1] With the development of the computer and its ability to solve huge systems of equations quickly, the variety of structures that can be accurately modeled now seems unlimited.

9.3.1 Truss structures and the method of joints

A **truss** is a special type of structure renowned for its high strength-to-weight and stiffness-to-weight ratios. This structural form has been employed for centuries by designers in a myriad of applications ranging from bridges and race car frames to the International Space Station.

Trusses are easy to recognize—lots of straight slender struts joined end-to-end to form a lattice of triangles, such as the bridge in Fig. 9.1. In large structures, the joints

---

[1] For a perspective on early American bridge failures see *To Engineer is Human*, Henry Petroski, Vintage Books, 1992.

FIGURE 9.1

Truss bridge in Interlaken, Switzerland.

FIGURE 9.2

Joint formed by riveting a gusset plate to converging members.

are often created by riveting the strut ends to a gusset[2] plate as shown in Fig. 9.2. A structure will behave like a truss only in those regions where the structure is fully triangulated; locations where the struts form other polygonal shapes (e.g., a rectangle) may be subject to a loss of stiffness and strength.

The special properties of a truss can be explained in terms of the loads being applied to the individual struts. Consider the three general types of end loadings shown in Fig. 9.3—tension, compression, and bending. If you were holding the

---

[2] A gusset plate is a thick steel plate used for joining structural components. It may be triangular, square, or rectangular. A recent failure of one in Minneapolis caused a bridge to collapse with multiple deaths and injuries.

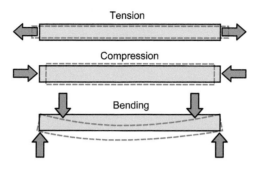

FIGURE 9.3

Different end loading possibilities. The dashes represent the deformed shape produced by the applied forces.

ends of a long thin steel rod in your hands and wanted to break it or at least visibly deform it, bending would be the way to go. Thus, if we could eliminate bending of the struts as a potential failure mode, the overall strength and stiffness of the truss would be enhanced. This is precisely the effect of the truss geometry on the structure, as the stiff triangular lattice serves to keep any bending induced in the struts to a minimum.

If one accepts that all the members of a truss will be in either tension or compression, each joint can be idealized, without loss of accuracy, by a **single pin** connecting the incoming members. The pin implies that the members are free to rotate with respect to the pin, thereby avoiding introduction of any bending as the structure deforms. Both the **live load**[3] (due to traffic) and the **dead load** (due to the weight of the members) must be applied at the pins to preserve the 'no bending' assumption in our mathematical model. Often, the weights of the members are assumed to be negligible compared to the live load.

The forces acting at the ends of each strut in a truss can be found from static equilibrium using a specialized approach known as the **Method of Joints**. This method derives its name from the fact that the static analysis is based on the freebody diagrams (FBDs) of the pins. You may recall from physics that an FBD is a drawing in which the component is shown isolated from its environment with the effects of adjoining members and other external forces (e.g., gravity) replaced by equivalent forces. The feasibility of basing the analysis on the FBDs of just the pins is demonstrated in Fig. 9.4, which shows the FBDs of the pin and struts comprising a typical model of a joint. All the struts are assumed to be in tension, in accordance with the recommended sign convention, while subscripts uniquely associate the forces with the members. To minimize the number of unknowns, forces exerted by adjoining

[3] In 1987, on the 50th anniversary of the opening of the Golden Gate Bridge, the bridge was closed to traffic and opened to pedestrians. So many people (hundreds of thousands) celebrated on the bridge that its live load was the weight of these people. This justifiably scared the bridge operators into rapid stress calculations as the bridge's arch flattened under the load!

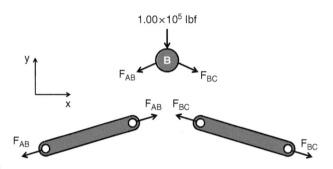

FIGURE 9.4

A mathematical model of a typical truss joint. The gusset plate has been modeled as a pin to indicate that no bending is present. The pin is held in equilibrium by forces exerted by the connecting members and the known live load. Since we already know the end forces of each strut are equal, their free-body diagrams need not be included in the final static analysis.

parts are shown as equal in magnitude and opposite in direction as required by Newton's third law. Equilibrium of the struts yields the trivial result that the end forces are equal, and so may be dropped from the analysis.

9.3.2 Example using the method of joints

Our goal in applying the method of joints will be to determine the forces on the members comprising the truss structure of Fig. 9.5. The first step is to draw the FBDs. The FBDs of the pins are shown in Fig. 9.6. We assume all members are in tension and use subscripts to associate the forces with the members. Forces exerted by the ground supports on the pins are labeled as A_x, A_y, and B_y. There are two force components at pin A since this support prevents the movement of the pin in two directions. The rollers under the support at B mean this support does not limit the horizontal movement in the x-direction at B (i.e., $B_x = 0$). The assumed directions of these support reactions are guesses; later, negative signs in the final calculated results will tell us where those guesses were incorrect.

The next step is to check that the total number of unknowns associated with the pins is equal to the total number of available equilibrium equations. Only when they match will it be possible to solve the equations. Such a problem is referred to as being **statically determinate**. For each pin, there are two equations governing *static equilibrium,* representing the summation of forces in two coordinate directions:

$$\sum_{i=1}^{N} F_{xi} = 0 \quad \sum_{i=1}^{N} F_{yi} = 0 \tag{9.1}$$

where N is the total number of forces acting on the joint; F_{xi} is the x-component of the i-th force; F_{yi} is the y-component of the i-th force; the symbol $\sum$ indicates a

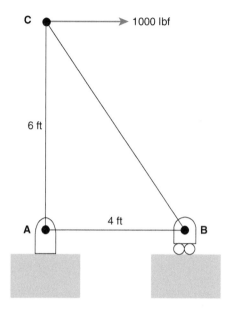

FIGURE 9.5

Truss structure.

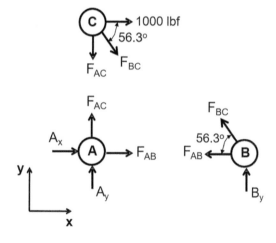

FIGURE 9.6

Pin FBDs.

summation over the N force *components*. The x- and y-components of a force are found by projecting it onto orthogonal axes located at the tail of the force vector, as illustrated in Fig. 9.7. Checking determinacy, we find that there are six unknowns (F_{AB}, F_{BC}, F_{AC}, A_x, A_y, B_y) and six equations (three pin FBDs with two equilibrium equations per FBD). This tells us in advance that the equations will be solvable.

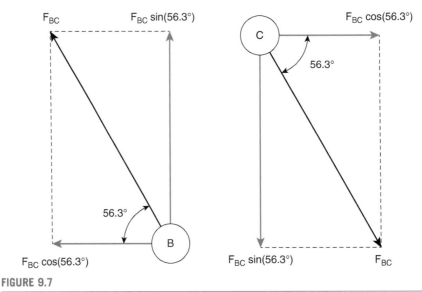

FIGURE 9.7

Resolution of the force F_{BC} into components at pins B and C.

The determinacy check clears the way to write the equilibrium equations. The FBD of joint A in Fig. 9.6 yields the following two equations in four unknowns describing equilibrium.

$$\text{Pin A:} \quad \sum F_x = 0 \quad F_{AB} + A_x = 0$$
$$\sum F_y = 0 \quad F_{AC} + A_y = 0$$

where the signs are determined by the directions of the force components relative to the coordinate axes. Thus, x-components that point to the right are positive and those that point to the left are negative, and y-components that point up are positive and those that point down are negative. The equilibrium equations for the remaining pins are as follows:

$$\textbf{Pin B:} \quad \sum F_x = 0 \quad -F_{AB} - F_{BC}\cos(56.3°) = 0$$
$$\sum F_y = 0 \quad B_y + F_{BC}\sin(56.3°) = 0$$

$$\textbf{Pin C:} \quad \sum F_x = 0 \quad F_{BC}\cos(56.3°) + 1000 = 0$$
$$\sum F_y = 0 \quad -F_{BC}\sin(56.3°) - F_{AC} = 0$$

9.3.3 Solution of the equations using excel

The solution of these equations by hand is both cumbersome and subject to error. We therefore turn to a computational approach based on first translating the equations into a **matrix** form that can be read by a computer, and then solving the equations with the matrix operations available in a spreadsheet.

Two intermediate steps are required to set up the matrix equations. First, any constants in the equations need to be shifted to the right-hand side of the equal (=) signs. Of course, this will change the sign on those constants. Second, the coefficients multiplying the unknown forces need to be identified. The resulting right-hand side constants (*RHS*) and the coefficients of the forces are listed in Table 9.1 for each equation (*EQ*). The order of the forces across the top of the table is totally arbitrary.

Table 9.1 Matrix set up of equations.

EQ	F_{AB}	F_{BC}	F_{AC}	A_x	A_y	B_y	RHS
1	1	0	0	1	0	0	0
2	0	0	1	0	1	0	0
3	−1	−cos(56.3 degrees)	0	0	0	0	0
4	0	sin(56.3 degrees)	0	0	0	1	0
5	0	cos(56.3 degrees)	0	0	0	0	−1000
6	0	−sin(56.3 degrees)	−1	0	0	0	0

The matrix form of the equilibrium equations, given below, was extracted directly from this table:

$$
\begin{bmatrix}
1 & 0 & 0 & 1 & 0 & 0 \\
0 & 0 & 1 & 0 & 1 & 0 \\
-1 & -\cos(56.3^\circ) & 0 & 0 & 0 & 0 \\
0 & \sin(56.3^\circ) & 0 & 0 & 0 & 1 \\
0 & \cos(56.3^\circ) & 0 & 0 & 0 & 0 \\
0 & -\sin(56.3^\circ) & -1 & 0 & 0 & 0
\end{bmatrix}
\begin{Bmatrix}
F_{AB} \\
F_{BC} \\
F_{AC} \\
A_x \\
A_y \\
B_y
\end{Bmatrix}
=
\begin{Bmatrix}
0 \\
0 \\
0 \\
0 \\
-1000 \\
0
\end{Bmatrix}
$$

where the brackets [ ] denote a matrix with multiple rows and columns; the braces { } denote a single column matrix commonly referred to as a vector. For ease of reference, we can express this equation using shorthand matrix notation as follows:

$$[A]\{x\} = \{b\} \tag{9.2}$$

where the coefficient matrix [A] and the right-hand side vector {b} come straight from the table. The order of the unknown force magnitudes in the vector {x} must correspond to the force labels across the top of the table so that the same multiplications are implied. Because the coefficients of [A] premultiply the variables in {x}, this operation is known as a **matrix multiplication**.

The unknowns can be isolated by premultiplying both sides of Eq. (9.2) by the **matrix inverse** of [A]. Denoted by $[A]^{-1}$, and having the same dimension as [A], the matrix inverse can be found using one of the special functions available in Excel. Premultiplying a matrix by its inverse is equivalent to premultiplying a number by its reciprocal. Therefore, Eq. (9.2) simplifies to:

$$\{x\} = [A]^{-1}\{b\} \tag{9.3}$$

Thus, having found $[A]^{-1}$ using Excel, we can then apply another special function in Excel to perform this matrix multiplication and, in so doing, determine the values of the unknown force magnitudes. The details of how to apply Excel to this example are summarized in the following procedure. Results are shown in the worksheet of Fig. 9.8.

1. Open an Excel worksheet.
2. Enter the values of [A] into a block of cells with six rows and six columns.
3. Enter the values of {b} into another nearby block of cells with six rows and one column.
4. Highlight an empty block of cells with six rows and six columns. Then type the following:

$$= \textbf{MINVERSE(B3: G8)}$$

and, instead of pressing Enter, press the following three keys simultaneously:

Ctrl − Shift − Enter

to compute the inverse of [A] and have the results appear in the highlighted block of cells. Note that B3:G8 defines the range of cells belonging to [A] and MIN-VERSE is a function that computes the inverse of the matrix defined by the range. This range could also have been entered by highlighting the cells of [A].

5. Highlight another empty block of cells; this one with six rows and one column. Then type the following:

$$= \textbf{MMULT(B11: G16, J3: J8)}$$

and press the following three keys simultaneously:

Ctrl − Shift − Enter

to multiply $[A]^{-1}$ and {b} in accordance with Eq. (9.3) and have the results appear in the highlighted column of cells. Note that the function MMULT re-quires us to define two ranges of cells, one for each of the matrices being multiplied.

	A	B	C	D	E	F	G	H	I	J	K
1	Solution of the Equations for the Truss Example										
2											
3		1	0	0	1	0	0			0	
4		0	0	1	0	1	0			0	
5	[A]=	-1	-0.5547	0	0	0	0		{b}=	0	
6		0	0.8321	0	0	0	1			0	
7		0	0.5547	0	0	0	0			-1000	
8		0	-0.8321	-1	0	0	0			0	
9											
10											
11		0	0	-1	0	-1	0			1000	FAB
12		0	0	0	0	1.803	0			-1803	FBC
13	inv[A]=	0	0	0	0	-1.500	-1		{x}=	1500	FAC
14		1	0	1	0	1	0			-1000	Ax
15		0	1	0	0	1.500	1			-1500	Ay
16		0	0	0	1	-1.500	0			1500	By
17											

FIGURE 9.8

Excel worksheet for solving the equations.

The results from Excel are displayed along with the truss structure in Fig. 9.9. The original Excel results for the support reactions A_x and A_y have minus signs, indicating that our original guesses for the directions of those forces in the FBDs of Fig. 9.6 were incorrect. These forces are shown without minus signs in Fig. 9.9 because the forces have been redrawn in their correct directions. The minus signs

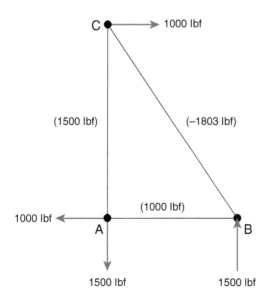

FIGURE 9.9

Results for the truss structure (member forces are indicated in parentheses).

on the member forces (shown in parentheses) have been retained to indicate if members are under compression (member BC) or tension (members AB and AC).

These steps just carried out for the truss—(1) draw the FBD, (2) check determinacy, (3) write the equilibrium equations, and (4) solve the equations—define a general methodology for equilibrium analysis that is applicable to any statically determinate structure.

9.4 Geotechnical engineering

In the previous example, we assumed the truss structure was secured to rigid supports at A and B. This is a big assumption if we consider that the ground beneath those supports may be subject to consolidation (i.e., compaction), the influences of a rising water table, or the possibility of the soil failing. The same concerns are present when designing the foundation for a building, highway, or dam.

To design and construct a proper foundation, a geotechnical engineer must have a working knowledge of both **soil mechanics** and **foundation engineering**. The aim of soil mechanics is to predict how the soil will respond (e.g., consolidate or fail) to water seepage and forces at the construction site. With knowledge of the forces exerted on the proposed foundation by the surrounding soil well in hand, a strong and stable foundation can be designed and constructed using the principles of foundation engineering.

9.4.1 Properties of soils

Unlike man-made materials like steel or concrete, whose properties can be tailored to the application, the geotechnical engineer has limited control of the properties of the soil at the proposed site. Rather, it is more a matter of determining the soil properties through exploration and testing and then reacting to those properties when designing the foundation.

The Tower of Pisa is a classic example of the role soil properties can play in the design and stability of a foundation. The Tower was built over a period of 197 years, after starting construction in 1173 AD. If the work had progressed more quickly, the tower would have fallen over during construction due to settling of the soil below. Instead, the builders had time to compensate for the observed tilting and built the upper floors along a more vertical axis. Tilting of the tower continued through the last century, though at a much slower rate than during construction. The history of tilt can be understood in terms of progressive consolidation of the soil layers (shown in Fig. 9.10) beneath the tower. Initial tilting has been traced to the highly compressible upper clay layer. With time, consolidation of the upper clay stabilized, enabling construction to continue. More recent tilting movements have been attributed to Layer A, whose soil properties are believed to be sensitive to seasonal rises and falls of the water table.

The compressibility of soils and their sensitivity to the presence of water are the results of a highly porous and mobile microstructure, as depicted in Fig. 9.11. A sponge, in the dry and saturated states, provides a good but imperfect analogy. The grains of a soil are free to associate and disassociate as it plastically and permanently deforms under load. In contrast, the skeleton of a sponge, while highly flexible, will return to its original shape when the load is removed.

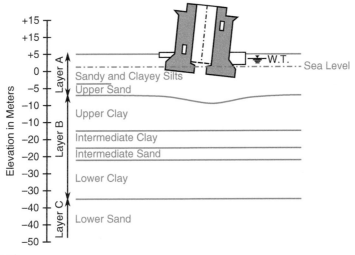

FIGURE 9.10

Soil layers below the Tower of Pisa.

While the structure of a soil typically varies from point to point, average properties can be defined to characterize the porous microstructure. We begin by introducing the general definition of the **weight density**[4] (γ) as follows:

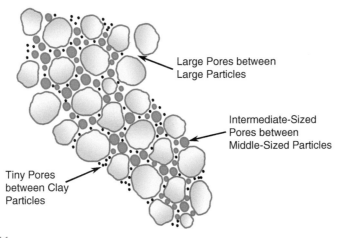

FIGURE 9.11

Soil structure consisting of large particles (sand) and small particles (clay) interspersed with air- or water-filled pores.

---

[4] The weight density γ has units of N/m^3 or lbf/ft^3 (not kg/m^3 or lbm/ft^3 which is mass density).

$$\textbf{Weight Density} = \gamma = \frac{W}{V} \qquad (9.4)$$

which defines the weight (W) of a substance per unit volume (V). In the case of soil, the total volume (V) consists of two parts—the volume of the solid particles (V_s) and the volume of the pores (V_p), with corresponding weights W_s and W_p, respectively. Another property, **porosity** (n), relates pore volume to the total volume:

$$\textbf{Porosity} = n = \frac{V_p}{V} \qquad (9.5)$$

With these two definitions in hand, we can now proceed to define the dry and saturated densities of soil. For **dry soil**, since the pores are air-filled, $W_p = 0$ and all of the weight is due to the solid particles:

$$\gamma_{\text{dry}} = \frac{W_s}{V} \qquad (9.6)$$

By contrast, the pores in **saturated soil** are completely filled with water. The weight of the water in the pores is significant and so must be accounted for when defining the weight density of saturated soil:

$$\gamma_{\text{sat}} = \frac{W_s + W_p}{V} = \gamma_{\text{dry}} + n\gamma_w \qquad (9.7)$$

where γ_w is the weight density of water (lbf/ft^3 or N/m^3). Use was made of Eq. (9.6), the equality $W_p = \gamma_w V_p$ and the definition of porosity in arriving at Eq. (9.7).

Example 9.1

Typical dry weight densities and porosities of sand, clay, and silt are given in Table 9.2. Calculate the weight density of each soil type when fully saturated with water. The weight density of water is 62.4 lbf/ft^3.

Table 9.2 Dry weight densities of different soils.

Soil type	Dry density (lbf per ft^3)	Porosity
Sand	94.8	0.375
Clay	74.9	0.550
Silt	79.9	0.425

Need: Saturated soil weight densities (γ_{sat}) of sand, clay, and silt
Know: The dry weight densities and porosities of sand, clay, and silt (from Table 9.1) and the weight density of water.
How: Apply Eq. (9.7).
Solve:

$$\text{SAND: } \gamma_{\text{sat}} = \gamma_{\text{dry}} + n\gamma_w = 94.8 + (0.375)(62.4) = \textbf{118. lbf/ft}^3$$

$$\text{CLAY: } \gamma_{\text{sat}} = \gamma_{\text{dry}} + n\gamma_w = 74.9 + (0.550)(62.4) = \textbf{109. lbf/ft}^3$$

$$\text{SILT: } \gamma_{\text{sat}} = \gamma_{\text{dry}} + n\gamma_w = 79.9 + (0.425)(62.4) = \textbf{106. lbf/ft}^3$$

9.4.2 **Effective stress principle**

The goal of this section is to introduce the **effective stress principle**, considered by many to be the most important concept in soil mechanics. Here, we will apply it to investigate how pressure naturally varies with depth in soil.

In general, **pressure** (p) is caused by forces pressing against a surface, as in Fig. 9.12, and is defined as follows:

$$p = \lim_{A \to 0} \frac{F_n}{A} \tag{9.8}$$

in which F_n is the resultant force acting normal to a surface of area A, and p is the associated pressure; the symbol $\lim_{A \to 0}$ is read as "in the limit as A approaches zero" and means that pressure can vary from point to point on the surface. The magnitude of pressure is indicative of the tendency of F_n to compact, or consolidate, the volume of material under A. When pressure acts on an internal surface, it is called **stress** though it has the same definition and units as pressure.

Before proceeding, we need to make a few assumptions. For the sake of simplicity, we will assume that the soil properties are uniform over the range of depths under consideration; in Example 9.2, we will generalize the equations to two layers, each with its own set of soil properties. We will also assume that the water table has reached ground level, so that all of the soil is fully saturated. Finally, we will assume that there is nothing man-made on top of the soil bearing down on it, such as a building or road. In other words, the source of all pressures below ground is the soil itself.

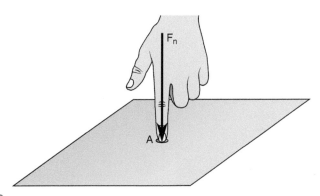

FIGURE 9.12

Pressure is what you feel when your finger presses with force F_n on the area A (the size of the finger tip).

To determine pressure, we must first find force, which is best done using an FBD. Our system is the large chunk of soil shown in Fig. 9.13 which extends from the surface down to a depth, z, and has the lateral dimensions L by T. The depth is kept general so that our final result will be applicable to all depths; the exact lateral dimensions will be of no consequence, as will soon become evident. There are just two vertical forces acting on our system—the downward acting weight of the soil as given by weight density times volume, and a resisting force, F_n, exerted on the lower face by the soil just below. Static equilibrium in the vertical direction requires that these opposing forces be equal in magnitude, as expressed by the following:

$$F_n = \gamma_{sat} z L T \tag{9.9}$$

Substituting this force into Eq. (9.8) and noting that $A = LT$ leads to an expression for pressure on the bottom face that varies linearly with depth:

$$\sigma = \gamma_{sat} z \tag{9.10}$$

where the symbol, σ, for stress has been used to denote the pressure since it acts on a hypothetical surface "inside" the soil. Eq. (9.10) defines a "total stress" in the sense that it accounts for forces exerted by both soil grains and water-filled pores.

Experiments have shown that the water pressure in the pores "does not" in fact contribute to the compaction of soil such as occurred beneath the Tower of Pisa. We therefore remove its contribution to stress by first assuming water pressure in the pores varies with depth as it would if the saturated soil were replaced by an equal volume of water:

$$p_p = \gamma_w z \tag{9.11}$$

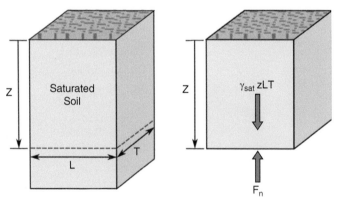

FIGURE 9.13

Soil section (*left*) being analyzed for stress on the horizontal plane (*dashed line*) at depth z. As the free-body diagram (*right*) demonstrates, the weight of the saturated soil induces a reaction force, F_n, of equal magnitude on the bottom face of the block. This force is distributed uniformly over the bottom face creating a constant pressure (or stress).

In this context, water pressure is referred to as pore pressure (p_p). Subtracting the pore pressure from the total stress then yields the **effective stress** (σ'):

$$\sigma' = \sigma - p_p = (\gamma_{\text{sat}} - \gamma_w)z \qquad (9.12)$$

This is the component of stress exerted by the soil skeleton. According to the **effective stress principle**, it is the effective stress, not the total stress, which determines if the soil will consolidate or fail.

Even though we subtracted pore pressure to define effective stress, Eq. (9.12) still models the buoyancy force exerted on the soil skeleton by the water.[5] In other words, the effective stress at given depth is smaller for saturated soil than for dry soil because of the soil skeleton's tendency to float in water.

Example 9.2

The two top layers of a soil sample taken from a potential building site are shown in the figure. If the thicknesses of the layers are $H_A = 2.50$ ft and $H_B = 2.00$ ft, determine (a) the effective stress at a depth of $Z_A = 2.00$ ft in Layer A and (b) the effective stress at a depth of $Z_B = 3.00$ ft in Layer B. Assume both layers are fully saturated with weight densities of $\gamma_A = 78.6$ lbf/ft^3 and $\gamma_B = 91.2$ lbf/ft^3. The weight density of water is $\gamma_W = 62.4$ lbf/ft^3.

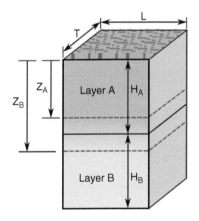

Need: the values of effective stress at depths of 2.00 and 3.00 ft.
Know: depths of interest are $Z_A = 2.00$ ft and $Z_B = 3.00$ ft; layer thicknesses are $H_A = 2.50$ ft and $H_B = 2.00$ ft; weight densities of saturated soil and water are $\gamma_A = 78.6$ lbf/ft^3, $\gamma_B = 91.2$ lbf/ft^3 and $\gamma_w = 62.4$ lbf/ft^3; both soil layers are fully saturated.
How: (a) Since all the soil above $z = Z_A$ is of a uniform composition and fully saturated, Eq. (9.12) applies.

---

[5] According to Archimedes' Principle, the upward acting buoyancy force is equal to the weight of the water displaced by the soil skeleton.

(b) With two layers of soil above $z = Z_B$ to consider, Eq. (9.12) cannot be used directly. Therefore, rederive Eq. (9.12), this time including the weights of each layer above $z = Z_B$ in the FBD. Follow the same procedure that was used to develop Eq. (9.12):
   (1) draw the FBD of the soil section,
   (2) find F_n from summation of forces in the vertical direction,
   (3) substitute F_n into the definition of pressure to get total stress, and
   (4) subtract pore pressure from total stress to get effective stress.

Solve: (a) From Eq. (9.12):

$$\sigma' = (\gamma_A - \gamma_w)z_A = (78.6 - 62.4) \times (2.00) \left[\text{lbf}/\text{ft}^3\right][\text{ft}] = \mathbf{32.4\ lbf/ft^2}$$

(b) The FBD now includes two weight vectors counteracted by the normal force on the bottom face:

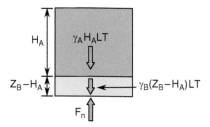

Static equilibrium in the vertical direction yields the normal force:

$$F_n = \gamma_A H_A LT + \gamma_B (Z_B - H_A)LT$$

Divide by the area of the bottom face to get total stress:

$$\sigma = \gamma_A H_A + \gamma_B (Z_B - H_A)$$

Subtract pore pressure to get effective stress, and substitute in values:

$$\sigma' = \gamma_A H_A + \gamma_B (Z_B - H_A) - \gamma_w Z_B$$

$$\sigma' = (78.6) \times (2.50) + (91.2) \times (3.00 - 2.50) - (62.4) \times (3.00)\ \left[\text{lbf}/\text{ft}^2\right]$$

$$\sigma' = \mathbf{54.9\ lbf/ft^2}$$

Note that the effective stress in this example is much smaller than the pore pressure, yet it is the effective stress that will determine if the soil skeleton fails or consolidates.

9.5 Water resources engineering

Availability of clean water is taken for granted by most of us who live in the more temperate regions of the northern hemisphere. But that common perception is changing rapidly as connections to the global economy strengthen and multiply. With modernization often associated with increased water usage, e.g., due to rising standards of cleanliness and the demands of industry, there are now concerns that

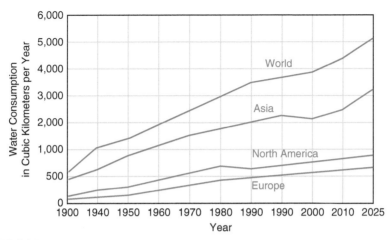

FIGURE 9.14

Forecast of water scarcity in 2025 (No data available for Greenland, Iceland, and Mongolia).

water is becoming an increasingly scarce commodity in some regions of the world (see Fig. 9.14). The goal of **sustainability** of the earth's water supplies, i.e., long-term balancing of need with availability, is one that can only be achieved through international cooperation and well-planned water resources engineering.

Two questions are central to every water resources project: (1) how much water is needed and (2) how much water can be delivered? The answer to the first question is deeply rooted in the social and economic needs of the region. The answer to the second question is achieved through the application of the fundamentals of water resources engineering.

9.5.1 Reservoir capacity

One method for achieving a sustainable water supply is to build a water reservoir. When properly designed, a reservoir can provide a steady supply of water that reliably meets the needs of regional consumers, even under adverse environmental conditions such as drought.

To explore how this can be achieved, we will consider the hypothetical reservoir site shown in Fig. 9.15. A stream feeds into the reservoir site and serves as the main source of water. On the downstream end of the reservoir, there are plans to construct a triangular-shaped dam of height H and width W to contain the water within. The reservoir itself will be triangular in cross-section, linearly increasing in depth along the length L from near zero at the stream inlet to H at the dam, consistent with the native topography.

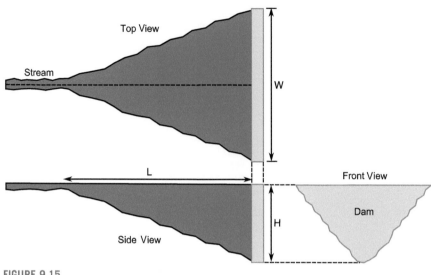

FIGURE 9.15

Reservoir geometry used to determine capacity.

The maximum volume of water that can be stored in the reservoir, and retrieved, is known as "**capacity**", which is typically expressed in units of volume (e.g., acre-ft). The capacity of the reservoir in Fig. 9.16 is given by the following formula:

$$V = \frac{1}{6}LWH \tag{9.13}$$

or by the following:

$$V = \frac{C_1 C_2}{6}H^3 \tag{9.14}$$

if we express the proportions of the reservoir in terms of H as follows: $L = C_1 H$ and $W = C_2 H$, where C_1 and C_2 are constants.

Example 9.3

Develop an expression that relates the instantaneous height of the water level in the reservoir of Fig. 9.15 to the volume of water stored within. Assume reservoir dimensions of $H = 100.$ ft, $W = 300.$ ft, and $L = 3000.$ ft.

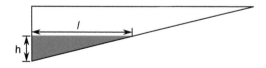

Need: An expression for stored volume (V_s) in terms of instantaneous depth (h).

Know: The proportions of the reservoir ($H = 100.$ ft, $W = 300.$ ft, $L = 3000.$ ft) and a formula for volume (Eq. 9.13).

How: Derive an equation for the volume of water in the partially filled reservoir, recognizing that the dimensions of this smaller volume will have the same relative proportions as the completely filled reservoir.

Solve: First, let the dimensions l, w, and h of the partially filled volume correspond to L, W, and H of the filled reservoir, respectively. Since both volumes have the same relative dimensions, we know that:

$$l/h = L/H = 3000./100. = 30. \quad so\ l = 30.0 \times h$$

$$w/h = W/H = 300./100. = 3.00 \quad so\ w = 3.00 \times h$$

Substituting both results into the equation for volume leads to the desired expression in terms of h:

$$V_s = \frac{1}{6} l\, w\, h = \frac{1}{6}(30.0h)(3.00h)\, h = 15.0\, h^3$$

Similar relationships are used to monitor the amount of water stored in a reservoir in real time. In actual practice, the local terrain is usually too complex to be approximated by so simple a geometric shape. Under such circumstances, computational methods are used in conjunction with topographical surveys to generate curves of reservoir volume versus h.

9.5.2 Conservation of mass

The water level in the reservoir will continuously rise and fall as environmental conditions and consumer demand change over time. If this were not the case, there would not be any need for a reservoir. Any changes in water level will depend on the amount of water coming in (mass inflow, Δm_{in}) and the amount of water going out (mass outflow, Δm_{out}) during a given time interval (Δt), as related through **conservation of mass**:

$$\Delta m_{in} - \Delta m_{out} = \Delta m_{stored} \tag{9.15}$$

in which Δm_{stored} is the net mass of water added to the reservoir during Δt; m_{in} is mainly attributable to the incoming stream water; m_{out} goes toward meeting various consumer demands (e.g., drinking water) and diverting water downstream; and Δt is typically defined to be a month. Thus, the total amount of water in the reservoir at a particular instant in time may be found by summing up the changes in stored mass starting from some initial time:

$$m_{stored}^n = \Delta m_{stored}^1 + \Delta m_{stored}^2 + \Delta m_{stored}^3 + \dots + \Delta m_{stored}^n = \sum_{i=1}^{n} \Delta m_{stored}^i \tag{9.16}$$

where Δm_{stored}^i is the change in mass associated with the ith time interval; m_{stored}^n is the total (or cumulative) mass stored in the reservoir after n number of time intervals.

Example 9.4

Monthly data for inflow from a feeder stream at a potential reservoir site are available in the form of the table below. Average monthly water use by regional consumers has been estimated and appears in the table. Plot cumulative water storage over the time span of the historical record, allowing for negative storage.

Need: Water storage plotted as a function of time.

Know: Monthly values of inflow and consumer demand for a 12-month period.

How: Calculate net inflow for each month using Eq. (9.15). Then, for each month, add up storage increments, up to and including that month's contribution, to define cumulative storage values as per Eq. (9.16). Finally, plot results using Excel.

Solve: Results, shown in the table, can be obtained using Excel without the use of special functions.

Month	Inflow $(\times 10^9 \text{ ft}^3)$	Demand $(\times 10^9 \text{ ft}^3)$	Net inflow $(\times 10^9 \text{ ft}^3)$	Cumulative storage $(\times 10^9 \text{ ft}^3)$
1	2.36	3.11	−0.75	−0.75
2	2.49	2.85	−0.36	−1.11
3	1.66	3.37	−1.71	−2.82
4	2.62	3.63	−1.01	−3.83
5	13.58	3.89	9.69	5.86
6	3.45	5.18	−1.73	4.13
7	4.48	5.96	−1.48	2.65
8	4.20	6.48	−2.28	0.37
9	4.25	4.15	0.10	0.47
10	4.92	3.63	1.29	1.76
11	3.16	3.37	−0.21	1.55
12	2.70	3.11	−0.41	1.14

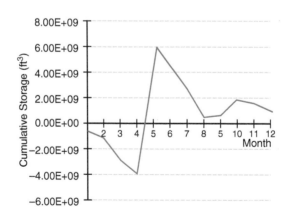

9.5.3 Estimation of required capacity and yield

An estimate of required reservoir capacity may be found from monthly data for mass inflow and consumer demand using a graphical method known as **sequent-peak analysis**. The sequent peaks are the succession of next highest peaks, or maxima, that occur as you move from left to right (i.e., chronologically) along the cumulative storage versus time curve, as illustrated in Fig. 9.16. Between each pair of sequent peaks, the required storage is calculated as the difference between the first sequent peak and the lowest valley in the interval. Required **capacity** is then the largest of these differences over the span of the historical record being examined. If the calculated capacity exceeds the economic or physical constraints of the project, expectations of demand may have to be adjusted down.

Another important design parameter, **yield**, defines how fast the water can be drawn out of the reservoir without compromising its ability to meet future demands. It therefore will typically have units of volume per unit time (e.g., m^3 per second). Yield is calculated as the average inflow over a span of time which is often taken to be the period of lowest inflow from the stream on record:

$$Y = \frac{\frac{1}{M}\sum_{i=1}^{M}\Delta m_{in}^{i}}{\Delta t} \tag{9.17}$$

where Y is the yield, M is the number of time intervals spanning the period in question, and Δt is the time associated with each equal-sized time interval.

Example 9.5

Assuming Fig. 9.16 represents cumulative storage (i.e., inflow minus demand summed over time) at a proposed reservoir site, estimate the required reservoir capacity.

Need: Reservoir capacity based on Fig. 9.16.

Know: 4 years' worth of historical data for cumulative storage.

How: First identify sequent peaks and associated minima by reference to Fig. 9.16. Compute the difference between each sequent peak and its minimum. Largest difference is required capacity.

Solve: Values corresponding to sequent peaks and their minima are listed in the table along with computed differences. The largest difference defines required capacity:

No.	Peak, P_i ($\times 10^9$ ft^3)	Min, M_i ($\times 10^9$ ft^3)	$P_i - M_i$ ($\times 10^9$ ft^3)
1	2.5	−4.0	6.5
2	6.7	0.6	6.1
3	9.4	−2.3	11.7

$$C = \text{largest value of } (P_i - M_i) = \mathbf{11.7 \times 10^9 \ ft^3}$$

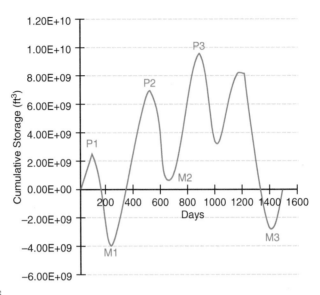

FIGURE 9.16

Cumulative storage versus time plotted over a period of 4 years. Sequent peaks (P1, P2, P3) and their corresponding minima (M1, M2, M3) have been indicated. The sequent peaks are defined such that P1 < P2 < P3. This graph was adapted from actual historical data available online from the U.S. Department of the Interior.

9.6 Transportation engineering

Whether designing a new transportation facility or proposing that a traffic light be added to a new intersection, a civil engineer working in the field of transportation has the additional challenge of having to deal with the unpredictability of the human factor. For example, the traffic slowdown caused by a roadside accident or a new billboard design is a human reaction that cannot be predicted by computer alone. For this reason, a necessary skill for a transportation engineer is the ability to go into the field and collect good data, e.g., by counting cars at a highway location or by measuring delays at an intersection.

In this section, we will introduce the concept of highway capacity. It is analogous to "reservoir capacity" in some respects. For example, the inflows and outflows of traffic that occur via entrance and exit ramps, respectively, must satisfy conservation of mass. Also, a highway should be designed to handle the fluctuations in traffic that are a fact of life. However, the determination of highway capacity for a given roadway is not just a matter of available space. It is also dependent on driver tendencies and reaction times, vehicle performance, and state laws.

9.6.1 Highway capacity

How many cars can a superhighway deliver to a city at rush hour? Consider a single lane of such a highway. Imagine, at rush hour, you stood beside the highway and counted the number of cars in any one lane that passed you in the space of 1 hour (assuming you have sufficient patience). What sort of answer do you think you might get? It's unlikely that your answer was less than 100 cars per hour, or more than 10,000 cars per hour. But how can one narrow down further and justify an answer?

The goal is no single answer, but it suggests a basis for a trade-off between the key variables. The key (and closely interrelated) variables in the case of highway capacity are:

- **Capacity (cars per hour).** The number of cars that pass a certain point during an hour.
- **Car speed (miles per hour, mph).** In our simple model, we will assume all cars are traveling at the same speed.
- **Density (cars per mile).** Suppose you took a snapshot of the highway from a helicopter and had previously marked two lines on the highway a mile apart. The number of cars you would count between the lines is the number of cars per mile.

You can easily write down the interrelationship among these variables by using dimensionally consistent units: [cars per hour] = [miles per hour] × [cars per mile], or

$$\textbf{Capacity} = \textbf{Speed} \times \text{Density} \tag{9.18}$$

For simplicity, we will consider these variables for the case of only a single lane of highway. Once again, our initial goal is a tool that provides insight into the problem, and that helps us solve it intuitively and visually. As a first step toward that tool, study Example 9.6.

Example 9.6

Three of you decide to measure the capacity of a certain highway. Suppose you are flying in a helicopter, take the snapshot mentioned above, and count that there are 160. cars per mile. Suppose further that your first partner determines that the cars are crawling along at only 2.50 miles per hour. Suppose your second partner is standing by the road with a watch counting cars as described earlier. For one lane of traffic, how many cars per hour will your second partner count?

Need: Capacity in cars per hour.

Know–How: You already know a relationship between mph and cars per mile from Eq. (8.18),
Capacity = Speed × Density or [cars per hour] = [miles per hour] × [cars per mile]

Solve: Capacity = [2.50 miles per hour] × [160. cars per mile] = **400. cars per hour.**

9.6.2 Follow rule for estimating highway capacity

Now let us use the result of Example 9.6 to plot a new graph. It's one of two related graphs, each of which can be called a **highway capacity diagram** (Fig. 9.17). It is made by plotting car density on the vertical axis and speed, in mph, on the horizontal axis.

How might Fig. 9.17 be made even more useful? If we knew a *rule* relating car density and speed, we could plot a whole curve instead of just one point. The **Follow Rule** is a first pass at highway design. This rule states that when following another car, leave one car length between vehicles for every 10 miles per hour. In equation form, this becomes the following:

$$\textbf{Number of car lengths between cars} = v[\text{mph}]/\textbf{10}. \, [\text{mph}] \qquad (9.19)$$

where v is the speed (or velocity) of the cars. The effect of this rule is graphically illustrated in Fig. 9.18. The separation distance per car is more as the car speed increases. At a dead stop, the cars are bumper to bumper. Thus, the car density falls with increasing speed.

Each car effectively occupies its own footprint plus the separation distance between them. Thus, the length of road, L_{road}, occupied by one car is as follows:

$$L_{\textbf{road}} = L_{\textbf{car}}\left(1 + \frac{v}{10.}\right) \quad [\text{miles / car}] \qquad (9.20)$$

where L_{car} is the average length of a car, about 4.0 m (or 0.0025 mile). As the units would suggest, density is simply the reciprocal of L_{road}, with L_{car} defined in miles and v in mph:

$$\text{Density} = \frac{1}{L_{\textbf{road}}} = \frac{1}{0.00025\left(1 + \dfrac{v}{10.}\right)} \, [\text{cars per mile}] \qquad (9.21)$$

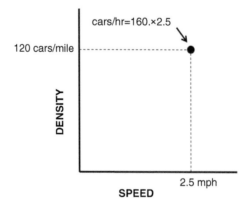

FIGURE 9.17

The highway capacity diagram.

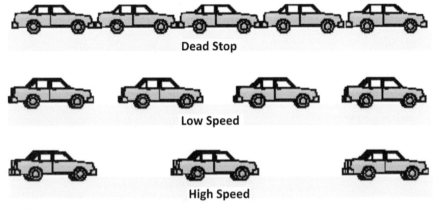

Dead Stop

Low Speed

High Speed

FIGURE 9.18

The follow rule in terms of car length separation.

The product of density and speed then yields the highway capacity for the **Follow Rule**:

$$\text{Capacity} = \frac{v}{0.00025\left(1 + \dfrac{v}{10.}\right)} \quad \text{[cars per hour]} \qquad (9.22)$$

Example 9.7

Using the Follow Rule, draw the full highway capacity diagram.

Need: Table and chart.

Know–How: Put the follow rule into a spreadsheet and graph.

Solve:

Speed, mph	Density, cars per mile	Capacity, cars per hour
0	400	0
10	200	2000
20	133	2667
30	100	3000
40	80	3200
50	67	3333
60	57	3429
70	50	3500
80	44	3556
90	40	3600
100	36	3636

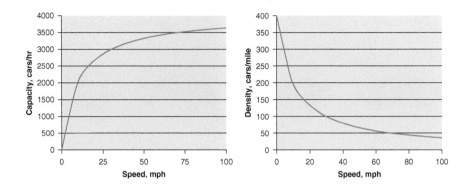

Either of these two curves could be appropriately called a "highway capacity diagram."

Notice what happens to the capacity as the car gains speed. It begins to level off and approach a constant value of 4000 cars per hour. There is a gain of only 300 cars per hour as we increase the speed from 50 to 100 mph. Eq. (9.22) for capacity developed above shows why this is occurring: The denominator of the equation is dominated by the spacing between vehicles and not by the footprint of the car. The former was mathematically expressed as v [mph]/10 [mph] » 1 so that at high speeds:

$$\textbf{Capacity} = \frac{v}{0.0025\left(\frac{v}{10}\right)} = 4000 \textbf{ cars per hour}$$

Notice too that you can always deduce the third member of the triumvirate of speed, density, and capacity by manipulating the other two from Eq. (9.18). For example, multiplying the density by the speed gives the capacity at every point (as per Fig. 9.17), or you can get the density by dividing the capacity by the speed.

The implications are that for any superhighway, if the drivers observe the follow rule, the capacity of each lane for delivering cars is a little less than 4000 cars per hour, *if the cars are moving at high speeds*. This insight enables engineers to solve practical problems such as the following.

Example 9.8

You are a highway engineer hired by the New York State Thruway Authority. About 7000 citizens of Saratoga, New York, work in the nearby state capital, Albany, New York. They all have to arrive in Albany during the hour between 8 and 9:00 a.m. These workers all travel one to a car, and all drivers obey the follow rule. How many highway lanes are needed between Saratoga and Albany?

Need: Number of highway lanes.

Know—How: We have just discovered a universal law of highways. A highway lane can deliver 4000 safe drivers per hour to their destination. So, number of highway lanes = (number of drivers per hour)/(4000 drivers per hour per lane).

Solve: Assume the number drivers per hour is uniform at 7000. Then the number of highway lanes is 7000/4000 = 2 lanes (rounded to the proper number of significant figures).

Not all drivers obey the follow rule (surprise!). Highways such as Interstate 87 that links Saratoga and Albany do not merely link two cities but many cities. I-87 runs from the Canadian border to New York City. Even within a small city, a highway typically has more than one on-ramp and off-ramp. While some highways run between cities, others are beltways that ring a city.

To meet these more realistic conditions, more complicated versions of the traffic rules, as well as many additional considerations, are needed. As an example, engineers have devised criteria for rules called "level-of-service" from *A* (major interstate highways) to *F* (rutted, damaged, under repair, etc., where one cannot exceed 10 mph).

Summary

In this chapter, we examined representative applications of four traditional areas of civil engineering. First, you learned a general methodology that a **structural engineer** uses to determine the forces in a truss. The same methodology can be applied to find the forces in a structure composed of hundreds of triangulated members, provided of course that a computer and an equation solver, such as the one in Excel, are available to solve the resulting system of equilibrium equations.

Second, you learned that the nonuniform composition and highly porous structure of soils leads to complex behaviors that can challenge **geotechnical engineers** tasked with the job of designing a stable foundation. From the effective stress principle, we know that it is the properties of the soil skeleton, and not the pressurized water in the pores, that will determine if the soil compacts or fails. This observation greatly simplifies the design of experiments aimed at measuring the properties of soils under varying conditions.

Third, you learned that one method used by **water resources engineers** to achieve a sustainable water supply in the face of variable demand and environmental conditions is to build a reservoir. One of the keys to accurate estimates of required storage (capacity) and delivery rates (yield) is an accurate historical record of the outflows (i.e., demand, seepage and evaporation) and inflows (from incoming streams).

Finally, you learned that the **transportation engineer** also has to maintain a balance of inflows and outflows, of cars in this case, but that the amount of highway space required to achieve this balance is dependent on driver tendencies, i.e., the human factor.

Exercises

Problems for structural engineering

1. When estimating the live load for a new bridge design, you want that estimate to be **conservative**. In other words, you want to err on the safe side by basing the estimate on the worst

possible scenario, e.g., bumper-to-bumper traffic, the heaviest vehicles, the worst environmental conditions, etc. With this in mind, estimate the live load for the following proposed bridge designs:

a) A foot bridge with a 174 ft span and two separate 4.0 ft wide lanes to allow for pedestrian traffic in both directions.
b) A 4-lane highway bridge with a 300. ft span.

For each of the trusses shown in Exercises 2–5, determine the forces on all members by the method of joints. Use a spreadsheet to solve the equations.

2. 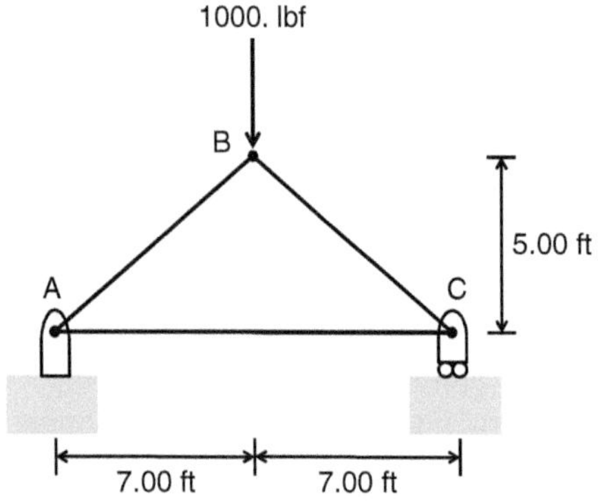

3. 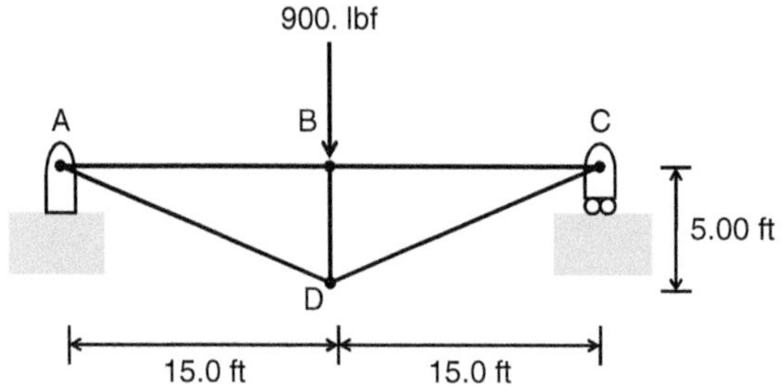

4. 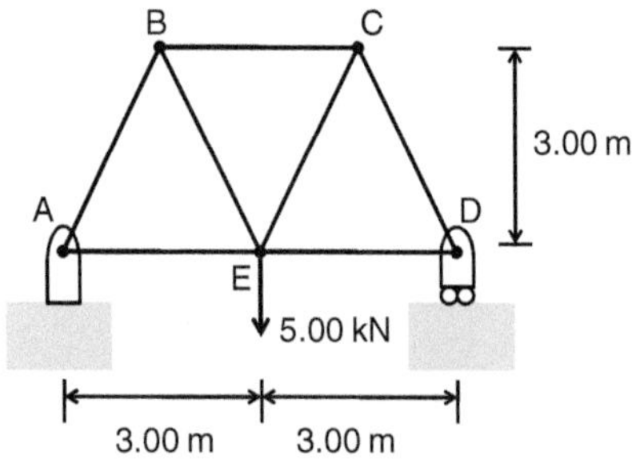

5. 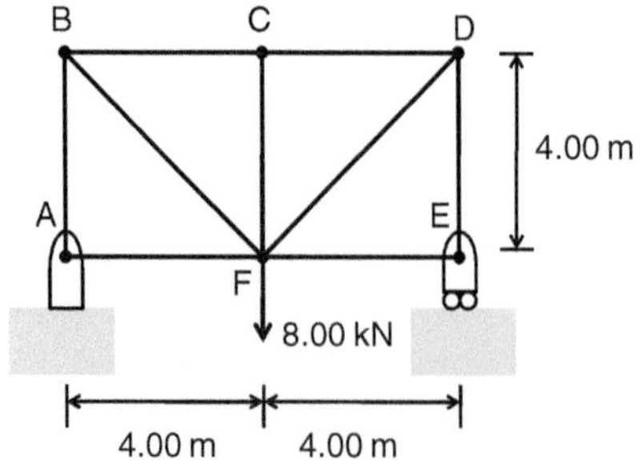

6. A truss that is pinned at one end and on rollers at the other will be statically determinate if it satisfies the following equation:

$$0 = 2j - m - 3$$

where j is the number of joints and m is the number of members. This is a quick way to check for static determinacy without drawing the FBDs. Apply this equation to:
   a) Verify that the trusses in Exercises 2–7 are statically determinate.
   b) Sketch 3 statically determinate truss bridges with the following numbers of members: (i) 11 members, (ii) 15 members, and (iii) 19 members.
7. Download the West Point Bridge Designer software from https://www.cesdb.com/west-point-bridge-designer.html and use it to design a truss.

Problems for geotechnical engineering

Refer to this table for Exercises 8 and 9.

Dry weight densities of different soils		
Soil type	Dry density (lbf/ft^3)	Porosity
Sand	103	0.423
Clay	87.2	0.615
Water	62.4	—

8. A recently extracted soil sample is 1.50 ft^3 in volume and composed of saturated sand. Calculate (a) the weight density of the saturated sand, (b) the weight of the solid particles in the soil sample, and (c) the weight of the water in the soil sample. (Refer to table of dry densities above.)
9. The soil at a proposed construction site is composed of saturated clay. Calculate (a) the weight density of the saturated clay, (b) the total stress at a depth of 20.0 ft, (c) the effective stress at a depth of 20.0 ft, and (d) the buoyancy force on the soil skeleton. (Refer to table of dry densities above.)
10. In arriving at Eq. (9.12), we assumed that the water table was even with ground level. Now assume the water table is located a distance, d, below ground level and develop a new expression for effective stress valid for $z \geq d$. (**Hint:** Follow a procedure similar to the one used in Example 9.2b, i.e., (1) draw the FBD of the soil section, this time including the weights of both the dry and saturated layers of soil, (2) find F_n from summation of forces in the vertical direction, (3) substitute F_n into the definition of pressure to get total stress, (4) redefine pore pressure to be $p_p = \gamma_w(z - d)$, (5) subtract pore pressure from total stress to obtain effective stress in terms of γ_{sat}, γ_{dry}, γ_w, z and d.
11. Referring to Example 9.2 in the section on geotechnical engineering, use Excel to plot depth, z, versus effective stress for values of z ranging from 0.00 to 4.50 ft. Feel free to use any results obtained in Example 9.2 to assist you with creating the graph.
12. The top three layers of soil at the proposed site of a new building are shown in the figure along with each layer's thickness and saturated weight densities. Assuming the water table is even with ground level, calculate the effective stress at the following depths:
    (a) $z = 4.00$ ft (**b**) $z = 8.00$ ft (**c**) $z = 12.0$ ft

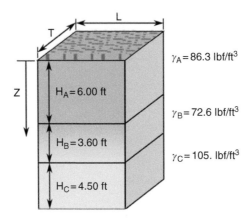

Problems for water resources engineering

13. A proposed reservoir has the geometry described by Fig. 9.15 and Eq. (9.13). The local topography constrains the dimensions H, W, and L of this geometry to have the relative proportions 1 to 5 to 20, respectively. Assuming the required capacity has been estimated at 5.00×10^8 ft^3, use Excel to plot water level height versus stored volume for water level heights ranging from 0 to H.

14. The average flow rates into and out of a small reservoir are given in the table for each hour of a 5-hour time period. If, at the start of this time period, 90,000. ft^3 of water is being stored in the reservoir, how much is left after 5 hours?

Hour	Flow in (ft^3 per second)	Flow out (ft^3 per second)
1	0.436	0.645
2	0.567	0.598
3	0.734	0.553
4	0.810	0.541
5	0.832	0.583

15. Use sequent-peak analysis, with the cumulative storage versus time graph provided, to estimate reservoir capacity.

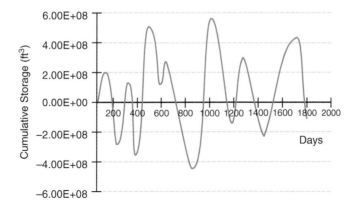

16. The monthly inflow of water to a proposed reservoir site is listed in the table for a critical 3-year period associated with low rainfall. Average consumer demand for water over the same 3-year period was estimated to 2.86×10^8 ft^3 per month. Based on these data, determine:
    (a) An estimate of yield in ft^3 per month.
    (b) An estimate for reservoir capacity in ft^3.
    (c) The required dimensions of the reservoir if it has the geometry described by Fig. 9.15 and Eq. (9.13), and H, W, and L have the relative proportions 1 to 4 to 40, respectively.

Year 1 inflow ($\times 10^7$ ft^3)	Year 2 inflow ($\times 10^7$ ft^3)	Year 3 inflow ($\times 10^7$ ft^3)
1.26	1.42	1.72
0.985	1.27	1.68
1.01	3.06	2.01
1.78	6.85	4.87
4.86	4.71	7.93
3.50	4.98	3.97
5.91	6.21	4.97
4.33	4.79	3.38
3.01	4.03	8.34
2.37	4.21	3.76
1.89	2.23	2.89
1.35	1.90	2.26

Problems for Transportation Engineering:
17. You are a TV reporter who rides in a helicopter to advise commuters about their travel time from their bedroom community to a city 20. miles away. You notice from your helicopter there are 160 cars per mile. Assume that all the drivers are observing the rule-of-thumb "follow rule" (see Fig. 9.18). What should you tell your audience that the travel time between the cities will be? (**Ans. 1 hour, 11 minutes**)
18. This chapter has shown that for cars adhering to the follow rule at high speeds, the number of cars per hour completing a trip is independent of speed. Consider now a slightly more realistic follow rule: mph $= 1800/(\text{cars per mile})^{0.9} - 4$. Find the speed for which the number of cars per hour is a maximum. (**Ans. 40 mph.**)

19. If you want to achieve a flow of 13,000 cars per hour in each direction on a highway, and the follow rule applies, how many lanes must the highway have in each direction?

20. Imagine a smart car of the future with radar and computer control capabilities that make it safe to operate on today's highways with twice the speed *and* half the spacing of today's human-driven cars. Using the follow rule, compare the capacity in terms of cars per hour for smart cars with cars per hour for human-driven cars. (**Ans. Four times as much as for human driven cars.**)

Engineering Ethics Problems:

21. In 1985, a judge found the structural engineers for the Hyatt Regency Hotel guilty of "gross negligence" in the July 17, 1981, collapse of two suspended walkways in the hotel lobby that killed 114 and injured 200 people. Many of those killed were dancing on the 32-ton walkways when an arrangement of rods and box beams suspending them from the ceiling failed.

    The judge found the project manager guilty of "a conscious indifference to his professional duties as the Hyatt project engineer who was primarily responsible for the preparation of design drawings and review of shop drawings for that project." He also concluded that the chief engineer's failure to closely monitor the project manager's work betrayed "a conscious indifference to his professional duties as an engineer of record." Responsibility for the collapse, it was decided, lay in the engineering design for the suspended walkways. Expert testimony claimed that even the original beam design fell short of minimum safety standards. Substantially less safe, however, was the design that actually was used.[6]

    Use the engineering Ethics matrix to analyze the ethical issues that occurred in this case.

22. Sara, a recent graduate, accepts a position at a small engineering design firm. Her new colleagues form a tightly knit, congenial group, and she often joins them for get-togethers after work.

    Several months after she joins the firm, the firm's president advises her that his wife has objected to her presence on the staff, feeling that it is "inappropriate" for a young, single female to work and socialize with a group of male engineers, many of whom are married. The president's wife encouraged him to terminate Sara's employment, and although the president himself has no issues with Sara or her work, he suggests to her that she should look for another employer.

    Everyone in the firm become aware of the wife's objections, and Sara begins to notice a difference in her work environment. Although her colleagues are openly supportive of her, she nevertheless feels that the wife's comments have altered their perception of her. She stops receiving invitations to her company's parties and is excluded from after-hours gatherings. Even worse, while she had previously found her work both interesting and challenging, she no longer receives assignments from the firm's president and she begins to sense that her colleagues are treating her as someone who will not be a long-term member of the staff. Believing that she is no longer taken seriously as an engineer and that she will have little opportunity to advance within the firm, she begins searching for a new position. However, before she can do so, her supervisor announces a downsizing of the firm, and she is the first engineer to be laid off.[7]

    Use the Engineering Ethics Matrix to examine potential ethics violations for an engineering employer to exclude and ultimately discharge an employee based on sex, age, or marital status.

---

[6] Modified from http://wadsworth.com/philosophy_d/templates/student_resources/0534605796_harris/cases/Cases/case68.htm

[7] Modified from http://pubs.asce.org/magazines/ascenews/2008/Issue_01-08/article6.htm

Computer engineering

10

Source : Fourleaflover/iStock.com

10.1 Introduction

Why do cars, trucks, planes, ships, and almost everything else today contain computers? After all, until the 1960s, we got along perfectly well without computers. As late as 1965, the question "Should a car or truck or plane or ship or anything

Exploring Engineering. https://doi.org/10.1016/B978-0-443-13541-5.00014-3

else contain a computer?" would have seemed ridiculous. In those days, a computer was not only much more expensive than a car, but also bigger.

10.2 Moore's law

In the late 1960s, the integrated circuit was invented, and computers began to shrink in both size and cost. In the 1990s, it was also asserted that "if a Cadillac had shrunk in size and cost as fast as a computer did since 1960, you could now buy one with your lunch money and hold it in the palm of your hand." The basic message, however was, and remains, valid. From the 1960s until the present, integrated circuits, the building blocks of computers, have doubled their computing power every year or two. This explosive progress is the result of a technological trend called *Moore's Law* that states that the number of transistors on an integrated circuit "chip" will double every year or two. The name honors electronics pioneer Gordon Moore, who proposed it in 1965. Moore's law is not a law of nature, but an empirical rule of thumb, and it has held true for four and a half decades.

Today's computers are so small that in most applications size is no longer an issue. For example, a computer for controlling the air/fuel mixture in an automobile is typically housed along with its power supply and communication circuitry in an assembly about the size of a small book. As a result, people reliably control not just the hundreds of horsepower of an automobile, but the delivery of energy routinely available to them in modern industrial societies while minimizing pollution emitted in the course of energy conversion.

The subjects of control and binary logic have taken on a life of their own in the form of computers and other information systems. Indeed, such systems have become so pervasive as to lead many people to assert that the industrialized world changed in the late 20th century from an industrial society to an information society.

The computers that carry out these logical and arithmetical tasks are essentially collections of electrical/electronic circuits. In more descriptive terms, they are boxes of switches that are turning each other on and off. So the question for this chapter might be rephrased as follows: How can switches turning each other on and off respond to inputs and perform useful calculations? We will give a modest answer to this question by the end of this chapter.

Answering that question introduces the topics of digital logic and computation, which are central parts of computer engineering. The term **digital** can apply to any number system, such as the base 10 system used in ordinary arithmetic. However, today's computers are based on a simpler number system: the base two or **binary** system. Therefore, this chapter focuses on binary logic and computation.

Computation was implemented first by mechanical devices, then by **analog computers**, and today by **digital computers**. A particularly convenient way to use computation to accomplish control is through the use of **binary logic**. A means of summarizing the results of binary logic operations is through **truth** tables that can be expressed as **electrical logic circuits**. This technique is useful not only in control, but also in other areas of computation, such as **binary arithmetic** and **binary codes**. Binary arithmetic and information are the basis of computer software. Binary logic also enables us to define the engineering variable *information.* In this chapter, you will learn how to perform the above operations. These topics will be mostly illustrated by reference to a familiar automotive application: deciding when a seat belt warning light should be turned on.

As an afterword, the process by which actual computers do these things, the **hardware-software connection**, will be qualitatively summarized.

The topics in this chapter extend far beyond single computers. Millions of *embedded* computers (that is, computers placed within other systems and dedicated to serving those systems) now can be found in thousands of applications. Electronic and computer engineers have the jobs of reproducing this computer population, which brings forth a new generation every 2 or 3 years. The jobs these engineers do range widely. Some develop the electronic circuits that are the basis of electronic computation. Others devise new computer architectures for using those circuits. Yet others develop the "software", which are the instructions that make computers perform as intended. Many other engineers apply these tools to everything from appliances to space vehicles.

10.3 Analog computers

For hundreds of years, engineers proved ingenious and resourceful at using mechanical devices to control everything from the rotation of water wheels and windmills to the speed of steam engines and the aiming of guns on battleships. However, such mechanical systems had major disadvantages. They were limited in their speed and responsiveness by the mechanical properties of the components and they had to be custom designed for every control challenge.

By the 20th century, this had led engineers to seek more flexible, responsive, and general types of controls. As a first step, engineers put together standardized packages of springs, wheels, gears, and other mechanisms. These systems also proved capable of solving some important classes of mathematical problems defined by differential equations. Because these collections of standard mechanisms solved the problems by creating a mechanical analogy for the equations, they were called analog computers. Bulky and inflexible, they often filled an entire room and were "programmed" by a slow and complex process of manually reconnecting wires and components in order to do a new computation. Despite these drawbacks, in such applications as predicting the tides, determining the performance of electrical transmission lines, or designing automobile suspensions, analog computers marked a great advance over previous equation-solving methods.

10.4 **From analog to digital computing**

Meanwhile, a second effort, underway since about 1800, sought to calculate solutions numerically using arithmetic done by people. Until about 1950, the word *computer* referred to a person willing to calculate for wages. Typically, these were selected members of the workforce whose economic status forced them to settle for relatively low pay. It was not until about 1900 that these human computers were given access to mechanical calculators and not until the 1920s that practical attempts at fully automated mechanical calculations were begun.

Meanwhile, as early as the 1820s, beginning with the ideas of the British scientist Charles Babbage, attempts had been made to do arithmetic accurately using machinery. Because these proposed machines operated on digits as a human would (rather than forming analogies), they were called *digital* computers. Digital techniques were also used to control machinery. For example, the French inventor Joseph Marie Jacquard (1752−1834) used cards with holes punched in them as a digital method to control the intricate manipulations needed to weave large complex silk embroideries.

Humans learn digital computing using their 10 fingers.[1] Human arithmetic adapted this ten-digit method into the decimal system. Digital computers can be built on a decimal basis, and some of the pioneers, such as Babbage in the 1840s and the team at the University of Pennsylvania who, in the 1940s built a very early electronic digital computer, the ENIAC, adopted this decimal system.

10.5 **Binary logic**

However, computer engineers quickly found that a simpler system less intuitive to humans proved much easier to implement with electronics. This is the binary system, based on only two digits: one and zero. The Jacquard loom was such a binary system, with a hole in a card representing a 1 (one) and the lack of a hole representing 0 (zero). In other applications, turning on a switch might represent a 1, and turning it off might represent a 0. The logic and mathematics of this system were developed mainly by the British mathematician George Boole (1815−64), whose contributions were so important that the concept is often referred to as **Boolean algebra**.

Binary logic begins with statements containing variables symbolized by letters such as X or Y. The statement can assert anything whatsoever, whether it is "The switch is open" or "There is intelligent life on a planet circling the star Procyon." The variable representing the statement can be assigned either of two values, 1 or

---

[1] The word *digitus* is literally "finger" in Latin, indicating the discrete nature of such counting schemes.

0, depending on whether the statement is true or false. (We will adopt the convention of using the value 1 for true statements and the value 0 for false ones.)

Binary logic permits three, and only three, operations to be performed, **AND**, **OR**, and **NOT**:

> **AND** (sometimes called "intersection" and indicated by the symbol • or *) means that, given two statements X and Y, if both are true, then $X • Y = 1$. If either one is false, then $X • Y = 0$. For example, the statement "It is raining (X) and the sun is out (Y)" is true only if both it is raining *and* the sun is out.
>
> **OR** (sometimes called "union" and indicated by the symbol +) the inclusive "or" means that, given two statements X and Y, if either one or both are true, then $X + Y = 1$, while only if both are false is $X + Y = 0$. For example, the statement "It is raining (X) *or* the sun is out (Y)" is true in all cases except when *both* are not true.
>
> **NOT** (sometimes called "negation" and indicated here by the postsymbol ′ as in X', but sometimes indicated in other texts with an overbar as in $\overline{X}$) is an operation performed on a single statement. If X is the variable representing that single statement, then $X' = 0$ if $X = 1$, and $X' = 1$ if $X = 0$. Therefore, if X is the variable representing the statement "It is raining," then X' is the variable representing the statement "It is not raining."

These three operations provide a remarkably compact and powerful tool kit for expressing any logical conditions imaginable. A particularly important type of such a logical condition is an "if–then" relationship, which tells us that **if** a certain set of statements has some particular set of values, **then** another related statement, often called the **target statement**, has some particular value.

Consider, for example, the statement "If a cold front comes in from the south or the air pressure in the north remains constant, but not if the temperature is above 50°F, then it will rain tomorrow." The target statement and each of the other statements can be represented by a variable. In our example, X can be the target statement "It will rain tomorrow," and the three other statements can be expressed by $A = $ "a cold front comes in from the south," $B = $ "the air pressure in the north is constant," and $C = $ "the temperature is above 50°F". The connecting words can be expressed using their symbols. Thus, this long and complicated sentence can be expressed by the short and simple **assignment** statement:

$$X = A + B • C'$$

This is not a direct *equivalence* in which information flows both ways across the equation. Specifically, in an assignment,[2] the information on the right-hand side is

---

[2] In some programming languages such logic statements are written with an *assignment* command, ": = " and not just an " = " command so that our Boolean statement could be written as $X:$ $= A + B · C'$ to reinforce the fact that these are not reversible equalities.

assigned to X but not vice versa. This distinction is required because of the way that computers actually manipulate information.

However, as written above, the statement still presents a problem. In which order do you evaluate the operations? Does this make a difference? A simple example will show that it *does* make a difference! Consider the case $A = 1$, $B = 0$, $C = 1$. Let the symbol $+$ be used first, the symbol $\bullet$ next, and the symbol $'$ last. Then in the preceding statement, $A + B = 1 + 0 = 1$ and $(A + B) \bullet C = 1 \bullet C = 1 \bullet 1 = 1$. But since $1' = 0$, then $X = 0$. But if the symbols are applied in the reverse order ($'$ first, $\bullet$ next, and $+$ last), then $C' = 1' = 0$, $B \bullet C' = 0 \bullet 0 = 0$, $A + B \bullet C' = 1 + 0 = 1$, and $X = 1$.

So, to get consistent results when evaluating logic statements, a proper order must be defined. This is similar to standard precedence rules used in arithmetic. That order is defined as follows:

1) All NOT operators must be evaluated **first**, then
2) all AND operators (starting from the right if there is more than one) **second**, and
3) all OR operators (starting from the right if there is more than one) are evaluated **third**.

If a different order of operation is desired, that order must be enforced with parentheses, with the operation within the innermost remaining parentheses being evaluated first, after which the parentheses is removed. In our example above, the proper answer with explicit parentheses would have been written $X = A + (B \bullet C')$ and evaluated as $X = 1 + (0 \bullet 1') = 1$. However, the value of the expression $X = (A + B) \bullet C'$ is $X = (1 + 0) \bullet 1' = 1 \bullet 1' = 1 \bullet 0 = 0$.

Example 10.1

Consider the following statement about a car: "The seat belt warning light is on." Define the logic variable needed to express that statement in binary logic.

Need: Logic variable (letter) = "..." where the material within the quotes expresses the condition under which the logic variable has the value 1 = true and 0 = false.

Know–How: Choose a letter to go on the left side of an equivalence. Express the statement in the form it would take if the content it referred to is true. Put the statement on the right side in quotes.

Solve: W = "the seat belt warning light is on."

This example is trivially simple, but more challenging examples can arise quite naturally. For example, if there are two or more connected constraints on a given action, then the methods of Boolean algebra are surefire ways of fully understanding the system in a compact way.

Example 10.2

Consider the statement involving an automobile cruise control set at a certain speed (called the "set speed"): "Open the throttle if the speed is below the set speed and the set speed is not above the speed limit." Express this as a logic formula, and evaluate the logic formula to answer the question: "*If the speed is below the set speed and the set speed is above the speed limit, then will the throttle be opened?*" Answer using these variables: the car's speed is 50. miles per hour, the set speed is 60. miles per hour, and the speed limit is 45 miles per hour.

Need: A binary logic formula expressing the statement "Open the throttle if the speed is below the set speed and the set speed is not above the speed limit," and an evaluation of the formula for the situation when the speed is 50. miles per hour, the set speed is 60. miles per hour, and the speed limit is 45 miles per hour.

Know: Any statement capable of being true or false can be represented by a variable having values 1 = true and 0 = false, and these variables can be connected by AND ($\bullet$), OR ($+$), and NOT ($'$).

How: Define variables corresponding to each of the statements, and use the three connectors to write a logic formula.

Solve: Let X = "throttle is open."

Let A = "speed is below set speed."

Let B = "set speed is above speed limit."

Then the general logic formula expressing the target statement is $X = A \bullet B'$

If the speed is 50. miles per hour, and the set speed is 60. miles per hour, then $A = 1$. If the set speed is 60. miles per hour, and the speed limit is 45 miles per hour, then $B = 1$. Substituting these values, the general logic formula gives $X = 1 \bullet 1'$. Evaluating this in the proper order gives $X = 1 \bullet 0 = 0$.

In plain English, if the speed is below the set speed, and the set speed is above the speed limit, the throttle will not open.

Example 10.3

Suppose we want to find a Boolean expression for the truth of the statement "W = the seat belt warning light should be on in my car," using all of the following Boolean variables:

Case 1: D is true if the driver seat belt is fastened.

Pb is true if the passenger seat belt is fastened.

Ps is true if there is a passenger in the passenger seat.

Case 2: For actuation of the warning light, include the additional Boolean variable:

M is true if the motor is running.

Need: $W = ?$

Know−How: Put the W variable on the left side of an assignment sign $W =$ and then array the variables on the other side of the assignment sign.

Case 1

a) Put D, Ps, and Pb on the right side of the assignment sign. Thus, the temporary (for now, incorrect) assignment statement is $W = D\ Ps\ Pb$.

b) Connect the variables on the right side with the three logic symbols $\bullet$, $+$, and $'$ so that the relationship among the variables on the right side correctly represents the given statement.

A good way to do this is to simply put the symbols the way they should appear in the if statements. For example, since part of the if statement Pb contains the words "the passenger's seat belt is fastened," it should also contain a "not". The corresponding logic variable will appear as Pb'. Also, you only care about the passenger's seat belt if the passenger is sitting in the seat, that is, the intersection between these two variables.

Solve: A solution in English is: "if the driver's seat belt is not fastened" or "if there is a passenger in the passenger seat" and "the passenger's seat belt in not fastened", then "the seat belt warning light should be on in my car", or $W = D' + Ps \cdot Pb'$.

Once written, a logic equation can now be solved for any particular combination of variables. This is done by first plugging in the variable and then carrying out the indicated operation.

Case 2

For activation of the warning light, the light will go on if the motor is running, and if either the driver's seat belt is not fastened or if there is a passenger in the passenger seat and that seatbelt is not fastened. This is written as follows:

$$W = M \bullet \left(D' + Ps \bullet Pb' \right)$$

10.6 Truth tables

It is often convenient to summarize the results of a logic analysis for all possible combinations of the values of the input variables of an "if … then" statement. This can be done with a truth table. It is simply a table with columns representing variables and rows representing combinations of variable values. The variables for the "if" conditions start from the left, and their rows can be filled in systematically to include *all possible combinations* of inputs. The column at the far right represents "then." Its value can be computed for each possible input combination.

Example 10.4

Consider the condition in Example 10.2: open the throttle if the speed is below the set speed and the set speed is not above the speed limit. The "**if**" conditions are A = "speed is below the set speed" and B = "set speed is above the speed limit." The "**then**" condition is X = "open the throttle." The truth table is set up as shown here.

A	B	B'	X = A · B'

Need: All 16 entries to the truth table.
Know: Negation operator, ' and the AND operator •.
How: Fill in all possible binary combinations of statements A and of B.
Solve: One convenient way of making sure you insert all the possible input values is to "count" in binary from all zeroes at the top to all ones at the bottom. (If you are not already able to count in binary, this is explained on the next page or so.) In this example, the top line on the input side represents the binary number 00 (equal to decimal 0), the second line is 01 (decimal 1), the third is 10 (decimal 2), and the fourth is 11 (decimal 3).[3]

A	B	B'	X = A · B'
0	0		
0	1		
1	0		
1	1		

---

[3] If the input had three columns, A, B, and C, you would count from 0 to 7 in binary (000−111). If four columns, the 16 entries would count from 0 to 15 in binary (000−1111), and so on.

Next the value of the "then" ("open the throttle") is computed for each row of inputs (this is done here in two steps, first computing B′, then computing $X = A \cdot B'$).

A	B	B′	X = A · B′
0	0	1	0
0	1	0	0
1	0	1	1
1	1	0	0

In English, this truth table is telling us that the only condition under which the control will open the throttle ($X = 1$) is when both the speed is below the set speed and the set speed is below the speed limit. This is the way we *should* want our cruise control to operate!

Truth tables can also be conveniently expressed as electric circuits. Indeed, this capability is the essence of computing. This capability is further explored in the exercises.

10.7 Decimal and binary numbers

In the decimal or base 10 number system, digits are written to the left or right of a dot called the **decimal point** to indicate values greater than one or less than one. Each digit is a **placeholder** for the next power of 10. The digits to the left of the decimal point are whole numbers, and as you move to the left, every number placeholder increases by a factor of 10. On the right of the decimal point the first digit is **tenths** (1/10), and as you move further right, every number placeholder is 10 times smaller (see Fig. 10.1).

The number **6357**. has four digits to the left of the decimal point, with "7" in the **units** place, "5" in the **tens** place, "3" in the **hundreds** place, and 6 in the **thousands** place. You can also express a decimal number as a whole number plus tenths, hundredths, thousandths, and so forth. For example, in the number 3.76, the 3 to the left of the decimal point is the "whole" number, the 7 on the right side of the decimal point is in the "tenths" position, meaning "7 tenths", or 7/10, and the 6 is in the hundredths position. So, 3.76 can be read as "3 and 7 tenths and 6 hundredths".

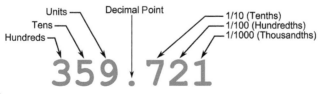

FIGURE 10.1

The structure of a decimal number.

There is nothing that requires us to have 10 different digits in a number system. The **base-10** number system probably developed because we have 10 fingers, but if we happened to have eight fingers instead, we would probably have a base-8 number system. In fact, you can have a **base-anything** number system. There are often reasons to use different number bases in different situations.

The binary[4] (base 2) number system is similar to the decimal system in that digits are placed to the left or right of a "point" to indicate values greater than one or less than one (see Fig. 10.2). For binary numbers, the first digit to the left of the "binary point" is called the **units**. As you move further to the left of the binary point, every placeholder increases by a factor of **2**. To the right of the binary point, the first digit is **half** (1/2), and as you move further to the right, every placeholder number becomes **half again smaller**. Table 10.1 below shows a few equivalent decimal and binary whole numbers.

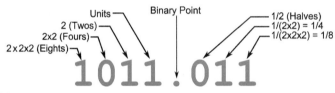

FIGURE 10.2

The structure of a binary number.

Table 10.1 Decimal and binary numbers.

Decimal:	0	1	2	3	4	5	6	7	8	9	10	11	12	13	14	15
Binary:	0	1	10	11	100	101	110	111	1000	1001	1010	1011	1100	1101	1110	1111

So, how do you convert binary numbers to decimal numbers? Several examples are given below to show the steps used to convert several binary numbers to decimal numbers.

1. **What is 1111 in decimal?**
   - The "1" in the leftmost position is in the "$2 \times 2 \times 2$" (2^3) position, so that means $1 \times 2 \times 2 \times 2 = 8$
   - The next "1" is in the "2×2" (2^2) position, so that means $1 \times 2 \times 2 = 4$
   - The next "1" is in the "2" (2^1) position, so that means $1 \times 2 = 2$

---

[4] The word **binary** comes from "bi-" meaning two. We use it in words such as "bicycle" (two wheels) and "binocular" (two eyes).

- The last "1" is in the units ($2^0 = 1$) position, so that means $1 \times 1 = 1$
- Answer: $1111 = 8 + 4 + 2 + 1 = 15$ in decimal.

2. What is 1001 in decimal?
- The "1" in the leftmost position is in the "$2 \times 2 \times 2$" (2^3) position, so that means $1 \times 2 \times 2 \times 2 = 8$
- The next "0" is in the "2×2" (2^2) position, so that means $0 \times 2 \times 2 = 0$
- The next "0" is in the "2" (2^1) position, so that means $0 \times 2 = 0$
- The last "1" is in the units (2^0) position, so that means $1 \times 1 = 1$
- Answer: $1001 = 8 + 0 + 0 + 1 = 9$ in decimal

3. What is 1.1 in decimal?
- The "1" on the left side of the binary point is in the units (2^0) position, so that means $1 \times 1 = 1$
- The 1 on the right side is in the "halves" (2^{-1}) position, so that means $1 \times (1/2) = 0.5$
- So, 1.1 is "1 and 1 half" $= 1.5$ in decimal

4. What is 10.11 in decimal?
- The "1" in the leftmost position is in the "2" (2^1) position, so that means $1 \times 2 = 2$
- The "0" is in the units (2^0) position, so that means $0 \times 1 = 0$
- The first "1" on the right of the point is in the "halves" (2^{-1}) position, so that means $1 \times (1/2) = 0.50$
- The last "1" on the right side is in the "quarters" (2^{-2}) position, so that means $1 \times (1/4) = 0.25$
- So, 10.11 is $2 + 0 + 1/2 + 1/4 = 2.75$ in decimal

A single binary digit (0 or 1) is called a "**bit**". For example, the binary number 11,010 has 5 bits. The word **bit** is made from the words "**bi**nary dig**it**". Bits are usually combined into 8-bit collections called **bytes**. With an 8 bit byte, you can represent 256 values ranging from 0 to 255, as shown below:

$$0 = 00000000$$
$$1 = 00000001$$
$$2 = 00000010$$
$$\cdots\cdots\cdots\cdots$$
$$254 = 11111110$$
$$255 = 11111111$$

Bytes often come with **prefixes** like kilo, mega, and giga, as in kilobyte, megabyte, and gigabyte. Table 10.2 lists the actual sizes of these binary numbers.

Table 10.2 Binary number byte prefixes.

Name	Size
Kilo (K)	$2^{10} = 1024$
Mega (M)	$2^{20} = 1,048,576$
Giga (G)	$2^{30} = 1,073,741,824$
Tera (T)	$2^{40} = 1,099,511,627,776$
Peta (P)	$2^{50} = 1,125,899,906,842,624$
Exa (E)	$2^{60} = 1,152,921,504,606,846,976$
Zetta (Z)	$2^{70} = 1,180,591,620,717,411,303,424$
Yotta (Y)	$2^{80} = 1,208,925,819,614,629,174,706,176$

A terabyte hard drive actually stores 10^{12} bytes.[5] How could you possibly need a terabyte of disk space? When you consider all the digital media available today (music, games, and video), it is not difficult to fill a terabyte of storage space. Terabyte storage devices are fairly common, and indeed there are some petabyte storage devices available.

10.8 Binary arithmetic

The value of a **bit** depends on its position relative to the "binary point." For example, the binary number 11010.101 has a decimal value computed from the second and third rows of Table 10.3 as $16 \times 1 + 8 \times 1 + 4 \times 0 + 2 \times 1 + 1 \times 0 + 0.500 \times 1 + 0.250 \times 0 + 0.125 \times 1 = 26.625$.

Table 10.3 Converting the binary number 11,010.101 to a decimal number.

Placeholder	2^4	2^3	2^2	2^1	2^0	.	2^{-1}	2^{-2}	2^{-3}
Bit	1	1	0	1	0	.	1	0	1
Decimal value	16	8	4	2	1	.	1/2 = 0.500	1/4 = 0.250	1/8 = 0.125

---

[5] The capacities of computer storage devices are typically advertised using their SI standard values, but the capacities reported by software operating systems use the binary values. The standard SI terabyte (TB) contains 1,000,000,000,000 bytes = 1000^4 or 10^{12} bytes. However, in binary arithmetic, a terabyte contains 1,099,511,627,776 bytes = 1024^4 or 2^{40} bytes.

Let us begin with the **Rules of Binary Addition**.

$$0 + 0 = 0$$
$$0 + 1 = 1$$
$$1 + 0 = 1$$
$$1 + 1 = 10, \text{ so carry the 1 to the next bit and save the 0}$$

For example, adding 010 (digital 2) to 111 (digital 7) gives the following:

$$
\begin{array}{r}
010 \\
+111 \\
\hline
1001 \text{ (digital 9)}
\end{array}
$$

Binary addition is conceptually identical to decimal addition. However, instead of carrying powers of 10, one carries powers of two. Here are the formalized steps to follow for the carry digits:

1. Starting at the right, $0 + 1 = 1$ for the first digit (No carry needed)
2. The second digit is $1 + 1 = 10$ for the second digit, so save the 0 and carry the 1 to the next column
3. For the third digit, $0 + 1 + 1 = 10$, so save the zero and carry the 1
4. The last digit is $0 + 0 + 1 = 1$
5. So the answer is 1001 (digital 9—you can see it is correct since decimal $2 + 7 = 9$)

Binary subtraction is conceptually identical to decimal subtraction. However, instead of borrowing powers of 10, one borrows powers of two.

10.8.1 Rules of binary subtraction

$$0 - 0 = 0$$
$$0 - 1 = -1$$
$$1 - 0 = 1$$
$$1 - 1 = 0$$

The following examples illustrate "borrowing" in binary subtraction.

$$
\begin{array}{rrr}
10 & 100 & 1010 \\
-1 & -10 & -110 \\
\hline
1 & 10 & 100
\end{array}
$$

Can you complete Table 10.4? You can self-check against their decimal equivalents.

Table 10.4 Simple binary arithmetic examples.

Example 1		Example 2		Practice 1		Practice 2	
Binary	Decimal	Binary	Decimal	Binary	Decimal	Binary	Decimal
1001	9	1001	9	1011		1011	
+101	+5	−101	−5	+110		−110	
1110	14	100	4				

The process of binary subtraction may be viewed as the addition of a negative number. For example, 3−2 may be viewed as 3 + (−2). To do this, you must determine the negative representation of a binary number. One way of doing this is with the **one's complement**.

The one's complement of binary number is found by changing all the ones to zeroes and all the zeroes to ones as shown below:

Number	One's complement
10011	01100
101010	010101

To subtract a smaller number from a larger number using the one's complement method, you:

1. Determine the one's complement of the smaller number,
2. Add the one's complement to the larger number,
3. Remove the final carry and add it to the result (this step is called the 'end-around carry').

Example 10.5

Do the following subtraction: 11001 (decimal 25) − 10011 (decimal 19).

  Need: $11001 - 10011 = $ _____? (a binary number)

  Know−How:

  Step 1: The one's complement of 10011 is 01100

  Step 2: Adding the one's complement to the larger number gives

$$01100 + 11001 = 100101$$

  Step 3: Removing the final carry and adding it to the result gives

$$00101 + 1 = 00110$$

  Solve: $11001 - 10011 = 110$. To verify that this is correct, convert each base 2 number to decimal and repeat the subtraction, or **$25 - 19 = 6$.**

To subtract a larger number from a smaller number, the one's complement method is as follows:

1. Determine the one's complement of the larger number,
2. Add the one's complement to the smaller number (the result is the one's complement of the answer),
3. Take the one's complement of the result to get the final answer. Don't forget to add the minus sign since the result is negative.

Example 10.6

Do the following subtraction: 1001 (decimal 9) − 1101 (decimal 13)

    Need: $1001 - 1101 = ?$ (a binary number)

    Know−How:

    Step 1: The one's complement of the larger number 1101 is 0010

    Step 2: Adding the one's complement to 1001 gives

$$0010 + 1001 = 1011$$

    Step 3: Add a minus sign to the one's complement of 1011 to get

$$1001 - 1101 = -0100$$

    Solve: $1001 - 1101 = -100$. To verify that this is correct, convert each base 2 number to decimal and repeat the subtraction, or **$9 - 13 = -4$**.

The rest of the familiar arithmetic functions can also be carried out in binary. **Fractions** can be expressed in binary by means of digits to the right of a binary point. Once again, powers of 2 take the role that powers of 10 play in digital arithmetic. Thus, the decimal fraction 0.5 (i.e., ½) is the binary fraction 0.1 and the decimal fraction ¼ is the binary fraction 0.01, and so on. A fraction that is not an even power of 1/2 can be expressed as a sum of binary numbers. Thus, the decimal $^3/_8 = 0.011$ in binary, which is the sum of decimal ¼ + ⅛.

Multiplication and **division** can be carried out using the same procedures as in decimal multiplication and "long division." The only complication is, again, systematically "carrying" and "borrowing" in powers of two, rather than powers of 10. Here, we only give the rules for binary multiplication.

10.8.2 Rules for binary multiplication

$$0 \times 0 = 0$$
$$0 \times 1 = 0$$
$$1 \times 0 = 0$$
$$1 \times 1 = 1, \text{ and no carry or borrow bits}$$

10.8.3 Rules for binary division

The good news is that binary division is a little easier than decimal division because instead of having to guess how many times the divisor fits into the dividend, in binary division, the answer will either be 0 or 1. But because division involves multiplication and subtraction operations which may also involve "borrowing" from the next digit, binary division is somewhat more complicated than binary multiplication. For example, suppose we want to divide 15 by 5 in binary. Since 5 in binary is 101 and 15 in binary is 1111, their binary division looks like this:

$$
\begin{array}{r}
11 \\
\hline
101\ )\overline{1111} \\
-1010 \\
\hline
0101 \\
-101 \\
\hline
0
\end{array}
$$

So, the answer is 11 (or 3 in decimal).

Example 10.7

If a powerful race car has an air to fuel ratio (A/F) of 15 [kg air]/[kg fuel] and the air intake draws in 1.5 kg of air per second, how much fuel must be injected every second? Solve in binary to five significant binary digits.

Need: Fuel rate = _____ kg fuel/s in binary?

Know–How: Fuel flow rate = (F/A) × 1.5 kg air/s = (1/15) [kg fuel]/[kg air] × 1.5 [kg air/s] = 0.10 [kg fuel/s].

To illustrate binary arithmetic, let us break down this problem into two separate ones. While unnecessary to do this, one will illustrate binary division and binary multiplication.

The division problem will be 1/15 (decimal) = 1/1111 (binary), and the multiplication problem will take the solution of that problem and then multiply it by 1.5 (decimal) = 1.1 (binary).

Solve: Start with 1/1111, then binary multiply that answer by 1.1.

$$
\begin{array}{r}
.00010001 \\
\hline
1111\ \big|\ 1.00000000 \\
1111 \\
\hline
10000 \\
1111 \\
\hline
1
\end{array}
\qquad
\begin{array}{r}
0.00010001 \\
\times 1.1 \\
\hline
0.00010001 \\
0.00001000 \\
\hline
0.00011001
\end{array}
$$

Therefore, the **fuel flow rate = 0.00011001 kg/s (in binary)**.

Checking in decimal: (1/15) × 1.5 = **0.10** and 0.00011001 = 0/2 + 0/4 + 0/8 + 1/16 + 1/32 + 0/64 + 0/128 + 1/256 = **0.098** (to get the closer answer of 0.10, we would need to use more significant (binary) figures).

Many times each second, a computer under the hood of an automobile receives a signal from an air flow sensor, carries out a binary computation such as the one shown in this example, and sends a signal to an actuator that causes the right amount of fuel to be injected into the air stream in order to maintain the desired air—fuel ratio. The result is much more precise and reliable control of fuel injection than was possible before computers were applied to automobiles.

10.9 Binary codes

We can now see how 0 and 1 can be used to represent "false" and "true" as logic values, while also of course 0 and 1 are numeric values. It is also possible to use *groups* of 0s and 1s as "codes." The now outdated Morse code is an example of a binary code, while the genetic code is based on just the pairings of four, rather than two, chemical entities known as bases.

Suppose we want to develop unique codes for the following nine basic colors: red, blue, yellow, green, black, brown, white, orange, and purple. Can we do this with a three-bit code (three 0s or 1s [bits])? No, since there are only eight combinations of 3 bits (note: $2^3 = 8$): namely 000, 001, 010, 011, 100, 101, 110, 111. Table 10.5 contains a possible four-bit code that would work, but of course it is only one of several, since there are $2^4 = 16$ possible combinations of a four-bit code. If we have N bits, we can code 2^N different colors.

Table 10.5 Possible set of binary numbers for colors.

Color	Binary equivalent	Color	Binary equivalent
Red	0000	Brown	0101
Blue	0001	White	0110
Yellow	0010	Orange	0111
Green	0011	Purple	1000
Black	0100		

10.10 How does a computer work?

How can these abstract ideas of Boolean algebra, binary logic, and binary numbers be used to perform computations using electrical circuits, particularly switches (which we will study in a later chapter)? The modern computer is a complex device, and any answer we give you here is necessarily oversimplified. But the principles are sufficient to give you some insight. For this discussion, we will need to know what a **central processing unit,** or **CPU,** does and what a computer **memory** is (which can take such forms as **read-only memory,** or **ROM,** and **random-access memory,** or **RAM**).

The "smart" part of the computer is the CPU. It is just a series of **registers,** which is nothing but a string of switches. These switches can change their voltage states

Table 10.6 Eight-bit register.

Bit #	7	6	5	4	3	2	1	0
Voltage	0 or 5 (0,5)	0 or 5 (0,5)	0 or 5 (0,5)	0 or 5 (0,5)	0 or 5 (0,5)	0 or 5 (0,5)	0 or 5 (0,5)	0 or 5 (0,5)
Bits	0 or 1 (0,1)	0 or 1 (0,1)	0 or 1 (0,1)	0 or 1 (0,1)	0 or 1 (0,1)	0 or 1 (0,1)	0 or 1 (0,1)	0 or 1 (0,1)

from "off" (nominally a no voltage state) to "on" or $+5$ V above ground. The early personal computers (or PCs) used only 8-bit registers and modern ones use 64 or 128, but we can think in terms of the 8-bit registers. (The notation x, y in Table 10.6 means either state x or state y.)

This register can contain 2^8, or 256, discrete numbers or addresses. What the CPU addresses is the memory in the computer. You can think of memory as a pigeonhole bookcase with the addresses of each pigeonhole preassigned.[6]

If we have just 256 of these pigeonholes in our memory, our CPU can address each of them. If we want to calculate something, we write a computer code in a suitable language that basically says something like: Add the number in pigeonhole 37 to that in pigeonhole 64, and then put the contents in pigeonhole 134. The binary bits can represent numbers, logic statements, and so on. All we then need to do is to be able to send out the results of our binary calculations in a form we can read. For example, we can decide that the contents of pigeonhole 134 are an equivalent decimal number.

If, in the above example, we identify the program input as what is in pigeonholes 37 and 64, and the program output is pigeonhole 134, the computer is constructed essentially as in Fig. 10.3 to accomplish its mission.

A computer is made up of **hardware** (electrical circuits containing such components as transistors). How the computer is instructed to execute its functions is managed by **software** (the instructions that tell the components what to do).

Part of the hardware of the computer that we have so far ignored is an internal clock. One might view the software as a list of instructions telling the computer what to do every time the clock ticks. This list of instructions includes the normal housekeeping functions that the computer carries out regularly (such as checking whether a user has entered in any keystrokes on the keyboard in the very short time interval since the last time this was checked). But it can also include downloading into a portion of the computer's memory called **program memory**, a special list of instructions called a **stored program**, and then executing that stored program. Examples of stored programs are a word processor and a spreadsheet program.

Once this stored program is downloaded into the memory, the CPU carries out this stored program step by step. Some of those steps involve carrying out

---

[6] M. Sargent III and R. L. Shoemaker, *The IBM PC from the Inside Out* (Reading, MA: Addison–Wesley Publishing Co. Inc., revised edition, 1986), p. 21.

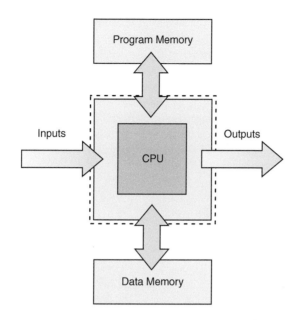

Program Memory

Inputs

Outputs

CPU

Data Memory

FIGURE 10.3

The central processing unit (CPU).

computations, which are executed in binary arithmetic by the CPU using the methods described in this chapter. In some cases, the result of the computation determines which of the stored program's instructions is the next one that should be carried out. This flexibility regarding the order in which instructions are carried out is the basis of the computer's versatility as an information processing system.

Software in this binary form is called **machine language**. It is the only language that the computer understands. It is, however, a difficult language for a human to write or read. So computer engineers have developed programs that can be stored in the computer that translate software from a language that humans understand into machine language.

The types of language that humans understand are called **higher-level languages.** These language consist of a list of statements somewhat resembling ordinary English. For example, a higher-level language might contain a statement such as "if $x < 0$, then $y = 36$." Examples of higher-level languages are C++, BASIC, and Java. Most computer programs are originally written in a high-level language.

The computer then **translates** this high-level language into machine language. This translation is typically carried out in two steps. First, a computer program called a **compiler** translates the statements of the high level language into statements in a language called **assembly language** that is closer to the language that the computer understands. Then another computer program called an **assembler** translates the assembly language program into a **machine language program**. It is the machine language program that is actually executed by the computer.

The personal computers that we all use, typically by entering information through such devices as a keyboard or mouse and producing outputs on a screen or via a printer, are called **general-purpose computers**. As the name suggests, they can be used for a wide range of purposes, from game playing to accounting to word processing to monitoring scientific apparatus.

Not all computers need this wide range of versatility. So there is another important class of computers directed at narrower ranges of tasks. These computers are called **embedded computers** (Fig. 10.4).

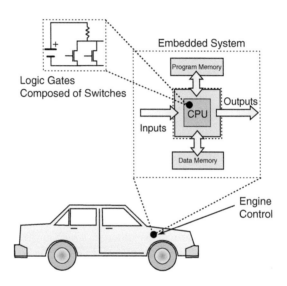

FIGURE 10.4

Schematic for an embedded automotive computer.

As their name suggests, these computers are embedded within a larger system. They are not accessible by keyboard or mouse, but rather receive their inputs from sensors within that larger system. In an automobile, there may be dozens of embedded computers. There might be, for example, an embedded computer for controlling a car's stereo system, another recording data for automated service diagnostics, another for the operation of the brakes, another for the steering conditions, and yet another for fuel control.

10.11 Computer security

Computer security (or cyber security) is the protection of computer systems from theft or damage to their hardware, software, or electronic data, as well as from disruption or of the services they provide. The field is growing in importance because of our increasing use of the Internet and of wireless networks such as Bluetooth and Wi-Fi, and of "smart" devices such as cellphones.

Malware is short for **mal**icious soft**ware** that can be used to compromise computer functions, steal data, bypass access controls, or otherwise cause harm to your computer. Malware is a broad term that refers to a variety of malicious programs. The most common types of malware are shown in Table 10.7.

Table 10.7 Common types of computer malware.

Trojan horse	A Trojan Horse is a program that is packaged with an application, usually free, such as a screen saver or computer game. Once the program initiates, a virus is released that creates problems for your computer without your knowledge.
Virus	Computer viruses infect computers to gain control and steal data. They are spread by visiting an infected website, clicking on an executable email file, viewing an infected advertisement, or connecting to infected removable storage devices such as USB drives.
Worm	A worm is software that copies itself repeatedly into a computer's memory using up all available RAM. When you open an email attachment containing a worm it looks through your address books choosing names at random and sends them copies of itself.
Adware	Pop-up advertisements. It is not uncommon for adware to come bundled with spyware that is capable of tracking user activity and stealing information.
Bot	A bot (short for "robot") is an automated internet program that performs functions such as capturing email addresses from website contact forms, address books, and email programs, then adds them to a spam mailing list.
Spyware	Spyware is software that monitors a user's activity without their knowledge, such as collecting keystrokes and data harvesting (account information, internet logins, financial data, etc.).
Phishing	Phishing is a term used to describe individuals who try to scam users. They send emails or create web pages that look like legitimate companies designed to collect an individual's online bank, credit card, or other login information.

These steps will help to safeguard your computer:

- Use antivirus protection and a firewall
- Get antispyware software
- Increase your browser security settings
- Avoid questionable Web sites
- Only download software from sites you trust.
- Don't open messages from unknown senders
- Immediately delete messages you suspect to be spam

Summary

Although analog computers have been of some historical importance, digital computers do the most important work in today's advanced technologies. An engineer today must understand the principles of digital computation that rest on the immensely powerful concepts of **Boolean algebra**, **binary logic**, **truth tables**, **binary arithmetic**, and **binary codes**. These concepts enable an engineer to make a first effort at defining the concept of information. Consequently, **computer security** is one of the major computer challenges of the 21st century.

Exercises

1. A popular musical ditty of the late 19th and early 20th-century railroad era had the following words.

   Passengers will please refrain

   From flushing toilets

   While the train

   Is standing in the station

   I love you.

   It is sung to the tune of *Humoresque* by the nineteenth-century Czech composer Antonin Dvorak.[7] For this little ditty, define a variable S expressing whether or not the train is in the station, a variable M expressing whether or not the train is moving ("standing" meaning "not moving"), and a variable F expressing the fact that the toilet may be flushed. (**Ans.: S** = "the train is the station," M = "the train is moving" (or you could use its negation, M' **meaning the train is stationary**), and F = "the toilet may be flushed")

2. For the ditty in Exercise 1, **(a)** express as a logic formula the conditions under which one may flush the toilet, **(b)** evaluate the formula you wrote in (a) for $M = 1$ and $S = 1$, and **(c)** express in words the meaning of your answer to (b). (Assume that any behavior not explicitly forbidden is allowed.)

3. Rework Example 10.2 to express and evaluate the logic formula to answer the question "If the speed is below the set speed and the set speed is above the speed limit, then will the throttle be opened?"

4. In Example 10.3, include the additional Boolean variable that for actuation of the warning light the driver's door must be closed (D_{door} is true if the driver's door is closed).

5. Consider the following logic variables for a car:

   Db = "the driver's seat belt is fastened"

   Pb = "the passenger seat belt is fastened"

   W = "the seatbelt warning should be on in my car"

   Write a sentence in English that expresses the logic equation $W = Db' + Pb'$. (**Ans: W = "If either the driver's seat belt is not fastened or the passenger's seat belt is not fastened, the seatbelt warning light should be on."**)

6. Consider the following logic variables for a car:

   W = "the seatbelt warning light is on"

---

[7] You can find the music at: http://www.youtube.com/watch?v=WmAZoexenx8

D = "a door of the car is open"

Ps = "there is a passenger in the passenger seat"

K = "the key is in the ignition"

M = "the motor is running"

Db = "the driver's seat belt is fastened"

Pb = "the passenger seat belt is fastened"

Write a logic equation for W that expresses the following sentence: "If all the doors of the car are closed, and the key is in the ignition, and either the driver's seat belt is not fastened or there is a passenger in the passenger seat and the passenger's seat belt is not fastened, then the seatbelt warning light should be on."

7. Consider the following logic variables for a car:

M = "the motor is running"

Db = "the driver's seat belt is fastened"

In the early 1970s, the government ordered all seat belt warnings to be tied to the motor in a manner expressed by the following sentence: "If the driver's seat belt is not fastened, then the motor cannot be running."

a. Write a logic equation for M in terms of Db that expresses this sentence.

b. Write a truth table for that logic equation.

In practice, this seat belt light logic caused problems. Think of trying to open a manual garage door or pick up the mail from a driveway mailbox. The government soon retreated from an aroused public. **Ans.: a. $M = Db$; b. see table below.**

Db	M
0	0
1	1

8. Consider the sentence "If a customer at a restaurant is over 21 and shows proper identification, then she can order an alcoholic beverage." Using the following logic variables:

A = "a customer at a restaurant orders an alcoholic beverage"

I = "the customer shows proper identification"

M = "a customer at a restaurant is over 21"

a. Express this sentence as a logic equation, and

b. write a truth table for the logic equation.

9. Consider the following variables expressing a football team's strategy.

T	L	P
0	0	0
0	1	1
1	0	1
1	1	1

T = "it is third down"

L = "we must gain more than 8 yards to get a first down"

P = "we will throw a pass"

The team's strategy is expressed by this truth table. Write a logic equation for P in terms of T and L. **(Ans.: $P = T + L$)**

Exercises 10 and 11 use electrical circuits to effect logical statements. If you are uncertain about electrical circuits, you should review ahead to the chapter on Electrical Engineering. In particular, an electrical switch (shown as an inclined line when open) means there is no current flowing from the battery. The circuit is off and is then said to be in a "0" or a "false" state; contrarily, when the switch is closed, current flows from the battery and a voltage appears across the lamp L. The circuit is now in an "on" position and is said to be "1" or a "true" state.

10. Consider the following electric circuit and the variables L = "the light is on," A = "switch A is closed," and B = "switch B is closed." Express the relationship depicted by the electric circuit as a logic equation for L in terms of A and B. (**Ans.: $L = A + B$.**)

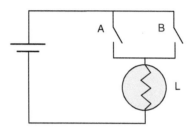

11. Consider the following circuit diagram and the variables L = "the light is on," A = "switch A is closed," B = "switch B is closed," and C = "switch C is closed." Write a logic equation for L in terms of A, B, and C.

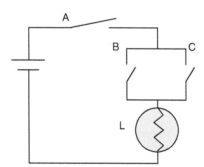

12. Consider the following logic variables for a car:
    1. W = "the seat belt warning light is on"
    2. Ps = "there is a passenger in the passenger seat"
    3. Db = "the driver's seat belt is fastened"
    4. Pb = "the passenger seat belt is fastened"
    Draw a circuit diagram for the logic equation $W = Db' + (Ps \cdot Pb')$.

13. Explain the following sentences.
    a) There are 10 kinds of people in the world, those who understand binary numbers, and those who don't."
    b) Binary is as easy as 1, 10, 11."
14. Convert the following numbers from binary to decimal: (a) 110, (b) 1110, and (c) 101,011. **Partial Ans.: (a) in the table below.**

—	2^5	2^4	2^3	2^2	2^1	2^0	Decimal equivalent
—	32	16	8	4	2	1	—
(a) 110	0×32	0×16	0×8	1×4	1×2	0×1	6
(b) 1110							
(c) 101011							

15. Convert the following numbers from decimal to binary: (a) 53, (b) 446, and (c) 1492. **Partial Ans.: (a) in the table below.**

Decimal	1024	512	256	128	64	32	16	8	4	2	1
Binary place	2^{10}	2^9	2^8	2^7	2^6	2^5	2^4	2^3	2^2	2^1	2^0
(a) 53	0	0	0	0	0	1	1	0	1	0	1
(b) 446											
(c) 1492											

16. Do the binary additions in the table below. Check your answer by converting each binary number into decimal. **Partial Ans.: the first addition in the table.**

Binary	Decimal	Binary	Decimal	Binary	Decimal
1010 +110	10 6	11101 +10011		10111 +10	
10000	16				

17. Do the binary subtractions in the table below. Check your answer by converting each binary number into decimal. **Partial Ans.: The first subtraction in the table.**

Binary	Decimal	Binary	Decimal	Binary	Decimal
1010 −1101	10 −6	11101 −10011		10000 −1	
0100	4				

18. If a powerful race car has an $(A/F)_{Mass}$ of 12.0 (kg air)/(kg fuel), and the air intake draws in 1.000 kg of air per second, how much fuel must be injected every second? Solve in binary to three significant binary digits. (**Ans. 0.000101 kg**)

19. Suppose we want to devise a binary code to represent the fuel levels in a car:
    a. If we need only to describe the possible levels (empty, 1/4 full, 1/2 full, 3/4 full, and full), how many bits are needed?
    b. Give one possible binary code that describes the levels in (a).

c. If we need to describe the levels (empty, 1/8 full, 1/4 full, 3/8 full, 1/2 full, 5/8 full, 3/4 full, 7/8 full, and full), how many bits would be needed?

d. If we used an 8-bit code, how many levels could we represent?

20. Construct a spreadsheet[8] that converts binary numbers from 0 to 111 to decimal numbers, print as formulas using the "control tilde" command. Check your spreadsheet against Exercise 14. **(Ans. e.g., binary 110 $\equiv$ 1 $\times$ 2^2 + 1 $\times$ 2^1 + 0 $\times$ 2° = decimal 6)**

21. Construct a spreadsheet that converts decimal number 53 to binary. Print as formulas using the "control tilde" command. Check your spreadsheet against Exercise 15. **Hint:** 5 (decimal) can be divided by 2^2 to yield an integer "1" and remainder 1; 1 cannot be divided by 2^1 [therefore, integer "0"]; and "1" can be divided by 2° for the last integer "1." Check for a spreadsheet function that will divide two numbers and display their result with no remainder.

22. Construct a spreadsheet that does binary subtraction with one's complement. Test it on Exercise 17.

23. In the game "Rock, Scissors, Paper" we have the following rules:

| Rock breaks Scissors |
| Scissors cuts Paper |
| Paper covers Rock |

If we were to create a code to represent the three entities, Rock, Scissors, and Paper, we would need 2 bits. Suppose we have the following code, where we call the first bit X and the second bit Y:

	X	Y
Rock:	0	0
Scissors:	0	1
Paper:	1	0

Now if we have two players who can each choose one of these codes, we can play the game. For example,

Player 1	Rock (0,0)	**Player 2**	Scissors (0,1)	**Player 1 wins**
Player 1	Scissors (0,1)	**Player 2**	Scissors (0,1)	**Tie**
Player 1	Rock (0,0)	**Player 2**	Paper (1,0)	**Player 2 wins**

We see that there are three possible outcomes of the game: Player 1 wins, Tie, Player 2 wins. Complete the table below that describes all the possible outcomes of the game.

---

[8] Spreadsheets have some built-in functions for decimal conversions to and from binary. It is recommended that you try first to use the actual mathematical functions described in this chapter and then *check* your answers using these functions to confirm those answers.

Player 1 X,Y	Player 2 X,Y	Player 1 wins	Tie	Player 2 wins
0,0	0,0	0	1	0
0,0	0,1	1	0	0
0,0	1,0			
0,1	0,0			
0,1	0,1			
0,1	1,0			
1,0	0,0			
1,0	0,1			
1,0	1,0			
Totals				

24. A company purchased a computer program for your part-time job with them. The license agreement states that you can make a backup copy, but you can only use the program on one computer at a time. Since you have permission to make a backup copy, why not make copies for friends? What do you do? (Use the Engineering Ethics Matrix.)
    a. Go ahead, since your friends only use one computer at a time, and these are backup copies.
    b. Make the backup copy but sharing it with anyone clearly violates the license agreement.
    c. Ask your supervisor if you can use the backup copy at home, and then make as many copies as you wish.
    d. Use the program discretely, since software license agreements cannot be enforced anyway.
25. You are a software engineer at a small company. You have written a software program that will be used by a major manufacturer in a popular product line. Your supervisor asks you to install a "back door" into the program that no one will know about so that he can monitor its use by the public. What do you do? (Use the Engineering Ethics Matrix.)
    a. Install the back door, since it sounds like a fun experiment.
    b. Tell your supervisor that you cannot do it without authorization from the end user.
    c. Install the back door but then deactivate it before the software is implemented.
    d. Stall your supervisor while you look for another job.

Electrical engineering

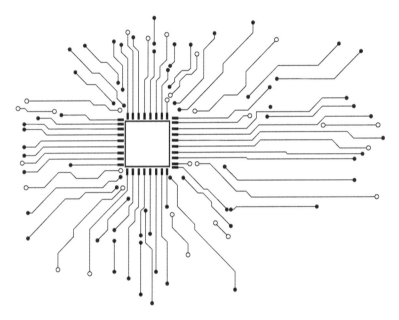

Source : Green Tech/Shutterstock.com

11.1 Introduction

Electrical engineering deals with the study and application of electricity, electronics, and electromagnetism. One of the most convenient ways to make energy useful is by converting it to electromagnetic energy, called **electricity**.

The easiest way of getting electricity to go where it is needed is by means of an **electrical circuit**. In this chapter, we learn what electrical circuits are, and apply a simple model for analyzing their operation. The model has as its basic variables charge, current, voltage, and **resistance**, and consists of **Ohm's Law**, the **Power**

Exploring Engineering. https://doi.org/10.1016/B978-0-443-13541-5.00024-6

231

Law, and **Kirchhoff's Voltage** and **Current Laws**. Using these laws, we can analyze the operation of two classes of direct current circuits called series circuits and parallel circuits.

The field of electrical engineering is devoted to the design and analysis of electrical circuits to meet a wide range of uses. Electrical engineers work in such varied areas as melting metals, keeping food frozen, projecting visual images, powering cell phones, taking elevators to the top of skyscrapers, and moving submarines through the ocean depths. Rather than attempt to describe the entire range of electrical engineering applications, we focus on one case study: the development of ever-faster **switches**. Different kinds of switches can be implemented in many ways, using the properties of electricity, such as the ability of an electrical current to produce **magnetism** and the use of electric resistance to control the flow of current.

11.2 Electrical circuits

An electric circuit is a *closed* loop of wire connecting various electrical components, such as batteries, lightbulbs, switches, and motors. The word *circuit* should suggest the idea of a closed loop. We first recognize the concept of electric charge, measured in an SI unit called **coulombs**, that can flow through the electrically conductive wires just as water flows through pipes. This is the essence of the model we use for understanding electrical circuits. And, just as water is made up of very small particles that we call molecules, where each molecule has the same tiny mass of 3×10^{-26} kg, the electrical charge in a wire is made up of very small particles called **electrons**, each of which has the same tiny charge of -1.60×10^{-19} coulombs (notice the minus sign). Because it is inconvenient to measure the mass of water by counting the number of molecules present, instead we use gallons, liters, or whatever, which contain a huge number of molecules. It is similarly inconvenient to measure charge by counting the number of electrons present; we measure it in coulombs, where a coulomb is the total charge produced by a huge number of electrons (about 10^{20}). A car battery might hold one million or more coulombs of charge, while the mid-sized C cell used in a flashlight might hold 10,000 coulombs.

Water would not be useful to us if it just sat in a reservoir, and electric charge is not useful to us if it just sits in a battery. Think of a Roman aqueduct system, in which water from a high reservoir can flow down to a city. It progressively loses its original potential energy as it flows. Electrons similarly flow from a high potential to a lower one. We call this potential **voltage**. Electric charges are "pumped up" by a battery or generator to a high voltage (potential) for use in electrical components such as lights, motors, and air conditioning, and then they return to the battery or generator by means of wires for their reuse.

Just as the flow of water from a hose can be separated into two or more streams who's combined mass equals the mass of entering water (the **conservation of mass**), the total number of electrons leaving an electrical junction is exactly equal to the number of arriving electrons. This is called the **conservation of charge**. The flow of charge is called **current**, and is measured in the SI unit of **amperes**, where 1 ampere is equal to a flow of 1 coulomb per second.

The hollow interior of a water pipe permits the water flow. Wires are not hollow, but they do allow the flow of electricity in the same way that a hollow pipe allows the flow of water. We call materials used in wires **conductors**, which include metals. Materials that do not allow the flow of electricity through them are called **insulators**. They include such nonmetals as ceramics and plastics. Materials called **semiconductors** have properties somewhere between those of conductors and insulators. We distinguish among conductors, semiconductors, and insulators by a physical property called **electrical resistance**. Because electrons are negatively charged they are attracted (flow) from a negative to a positive voltage (potential). Unfortunately, by early convention, electrical current is defined as flowing in the opposite direction, from positive to negative.[1] It was discovered after this convention had been adopted that electrical current (electrons) actually flows in the opposite direction (see Fig. 11.1). This figure also introduces you to the visual symbol for a battery, with the longer line indicating the "plus" side of the battery and the shorter line representing the "minus" side of the battery.

The motion of water through a pipe may be caused by a pump, which exerts a pressure on the water forcing it to move through the pipe. In the same way, a battery,

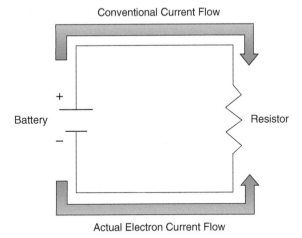

FIGURE 11.1

Blame Ben Franklin!

---

[1] This convention was suggested by Benjamin Franklin in the mid-18th century and is still used today. Franklin used the word "battery" for his charged electrical cells because it reminded him of a military artillery battery (a row of guns firing simultaneously). He considered the source of the electrical energy in his battery to be "positive" just as the source of the artillery battery energy would be positive.

in addition to holding electrical charge, serves as a sort of electrical pump to force electrons to move through a wire. This electrical analog of "pressure" is "voltage", measured in SI unit of **volts**. Just as a hand pump might be used to apply a pressure of 12 pounds per square inch to move water through a pipe, a car battery might be used to apply 12 volts to move current through a the car's electrical wires.

Resistive devices are called **resistors** and may be made from coils of wire, pieces of carbon, or other materials that conduct electricity. In fact, all materials except electrical **superconductors** exhibit some amount of resistance to current flow. Some metals, such as pure silver and pure copper, have a very low electrical resistance. Other metals, such as aluminum and gold, have somewhat higher resistance but are still used as conductors in some applications. We usually use copper for circuit connections, since we do not want much resistance there. Semiconductors offer appreciable resistance to electron flow, and we use those where we want to exploit that property.

To summarize, in our model of an electric circuit, the voltage (i.e., the electrical "pressure") pushes a moving current of charge though a conductor. This charge (i.e., electrons) never disappears but simply circles around the circuit again and again, doing its various functions each time around the circuit. As the electrons travel around the circuit, they are pumped to high potential energy by the voltage source then lose this energy as they perform their functions.

Symbolically, the simplest possible electrical circuit consists of a battery, a wire, and an electrical resistive component located somewhere along the wire, as shown in Fig. 11.2.

The zigzag line represents a resistor, and the straight lines are the wires connecting it to the battery. The current is considered to flow from the positive pole of the battery to the negative pole. Since a circuit goes in a circle and ends up where it started, the voltage must also return to its initial value after a trip around the circuit. Since the voltage is raised across the battery, it must be lowered back somewhere else in the circuit. In our model, we assume that none of this voltage "drop" occurs in the wires, but rather all of it occurs in the component or components in the circuit.

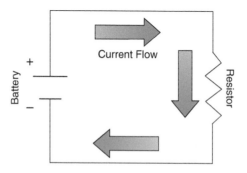

FIGURE 11.2

Current flow in an electrical circuit.

11.3 **Resistance, Ohm's law, and the "power law"**

As water flows through a pipe, it experiences energy loss through viscous resistance in the water. Pressure is needed to keep water moving through the pipe, and the magnitude of the pressure is determined by the amount of viscous resistance in the water.

In the same way, as charge flows through a wire, it experiences interactions that tend to reduce the voltage "pushing" the charge. We initially assume that these interactions do not occur in the connecting wires but are confined to the components hooked together by the wires. The interactions that reduce the voltage are described by resistance. It is measured in SI units called **ohms** and abbreviated with the upper-case Greek letter omega, Ω. The voltage drop (that is, an electrical "pressure drop") needed to keep the charges moving through the wire is determined by the resistance. This relation between current, voltage, and resistance constitutes the first basic law of our model of electrical circuits, **Ohm's law**. It states that the resistance, which is defined as the ratio of voltage drop to current, remains *constant* for all applied voltage drops. Mathematically stated,

$$R = \frac{V}{I} \tag{11.1}$$

If V is in volts and I is in amperes, R is defined in the units of ohms $[\Omega] = [V]/[A]$. For example, consider a 2.00-ohm resistor made of a material that obeys Ohm's law. If a voltage drop of 1.000 V is applied across the resistor, its current is exactly 0.500 A. If a voltage drop of 1000. V is applied, the current is exactly 500. A.

Few if any real materials obey Ohm's law in this exact manner. The law is, however, a very useful approximation for most practical electric circuits. We now have a way of using numbers in our model of a circuit.

Example 11.1

Find the current flowing in amperes through the wire in a circuit consisting of a 12-volt battery, a wire, and a 10. ohm resistor.

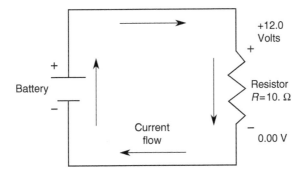

Need: Current flowing through the wire in amperes.
Know: Voltage provided by battery is 12 volts, and the resistance is 10. ohms.
How: Sketch the circuit and apply Ohm's law. We are using Franklin's convention for the current direction, but you can easily use the alternative that follows the actual electron flow.
Solve: The voltage drop across the resistor is 12 volts and, as given by Ohm's law,

$$V = I \times R \text{ or } I = V/R = 12/10. = \textbf{1.2 A.}$$

Ohm's law implies a second important property about electric circuits. Recall that, for the circuit as a whole, the total drop in voltage in the components other than the battery must be equal in magnitude but opposite in sign to the increase in voltage caused by the battery. Ohm's law then enables us to find this voltage drop, including, if necessary, its implied sign.

In continuing our water analogy, hydraulic resistance to water flow may be thought of as flow restrictions in the water pipe—think of a water valve quarter open. However, opening the valve say from one quarter open to one half open reduces its hydraulic resistance. Since water is forced through pipes under pressure, it provides energy that might be used, for example, to power a water wheel to run your washing machine. To attempt this, it would be useful to know the relation between water pressure, amount of water flow, and power output from the water wheel. In the same way, electric charge can provide energy to run a wide variety of electrical appliances, from light bulbs to computers. To attempt this, it is useful to know the relation between voltage, current, and power.

That relation is the second leg of our model—the "**Power Law.**" We have seen that electrons with a charge of Q coulombs "fall" through an electric potential ΔV in a resistor. This is analogous to a mass m falling through a gravitational potential Δh to produce a gravitational energy change $mg\Delta h$. The equivalent electric work done is $Q\Delta V$. The power produced is therefore $Q/t \times \Delta V$, in which Q/t is the rate of charge flow, which we call *current*. The "Power Law" simply states that electric power is given by: **Power = Current × Voltage**, or

$$P = I \times V \tag{11.2}$$

Note that power is not a new variable, but it is the same old one that we used earlier in talking about energy flows. We inserted quotes around "Power Law" to emphasize that it is not an independent law but perfectly derivable from the definition of power as the rate of working. Power is still measured in watts, where a watt is a joule per second. This fact allows us to relate our model of electrical circuits to our earlier energy models. The analogy between water flow and electricity is thus reasonably complete (see Table 11.1).

Evidence of the basic "Power Law" relation is provided by a second important property of a resistor in addition to its ability to limit the flow of electricity. That property becomes evident if we force enough charge through the resistor at a high enough voltage and observe that the resistor gets hot enough to "glow". This is the basis of a very useful invention, the incandescent lightbulb.

Table 11.1 Analogy between water flow in a pipe and electricity flow in a wire.

Water flow in a pipe	Electricity flow in a wire
Pressure	Voltage
Flow rate	Current
Hydraulic resistance	Electrical resistance
Power in water stream = pressure drop × flow rate	Power in electrical circuit = voltage drop × electric current

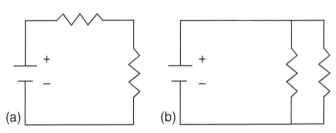

FIGURE 11.3

Series circuit (a) and parallel circuit (b).

11.4 Series and parallel circuits

There are two basic types of electric circuits. In a **series circuit**, shown in Fig. 11.3a, the current takes a single path. In a **parallel circuit**, shown in Fig. 11.3b, the wire branches out into two or more paths and these paths subsequently join up again. In a parallel circuit, we should envision the current as being divided at the point where the wires branch out then joining together again where the wires come together.

As previously stated, by the conservation of charge principle, the current in each of the two branches has to add up to the current that went in; and conversely, the total current that goes out of the two branches has to equal the total current that went in.

The mathematics for determining the voltages and currents in series and parallel circuits rests on two very simple principles: (1) two resistors in a series have the same current passing through each of them, and (2) two resistors in a parallel have the same voltage drop across each of them.

Example 11.2

Consider the series circuit in Fig. 11.3a. Assume that the battery voltage equals 12 volts and that each resistor has a resistance of 100. ohms. What is the current in the circuit and the voltage drop across each resistor?

Need: The current in the circuit I and the voltage drop across each resistor.

Know: Battery voltage = 12 volts; resistance of each resistor = 100. ohms.

How: Call the unknown current I and the resistance in each resistor R. By the first principle just stated, the current I must be the same in each series resistor. So, by Ohm's law, the voltage drop across each resistor must be $I \times R$. Since the total voltage drop around the circuit must be zero,

$$V(\text{battery}) + V(\text{resistor}_1) + V(\text{resistor}_2) = 0, \text{ or } V(\text{battery}) - I \times R_1 - I \times R_2 = 0.$$

Solve: $12 \text{ [volts]} - I \times 100. \text{ [ohms]} - I \times 100. \text{ [ohms]}$

$$= 12 \text{ V} - I \times 200. \text{ [ohms]} = 0, \text{ or } I = \textbf{0.06 A.}$$

Then **voltage drop** across each resistor is $V = I \times R = 0.06 \times 100. = \textbf{6.0 V}.$

Notice that in this example we could have simply added the two 100. ohm resistances together and used their sum of 200. ohms directly to get the current by Ohm's law in one step: $I = 12/200. = 0.06$ A. In a series circuit, the equivalent series resistance R is always the sum of the individual resistances: $R_{eq} = R_1 + R_2 + R_3 + \ldots.$

Example 11.3

Consider the parallel circuit in Fig. 11.3b with the battery voltage = 12 volts and each of the resistors in parallel have a resistance of 100. ohms. What is the current drawn from the battery?

Need: The current drawn from the battery.

Know: Battery voltage = 12 volts; resistance of each parallel resistor = 100. ohms.

How: Call that current the unknown, I. By the second principle stated, the voltage drop must be the same across each parallel resistor. Since the total voltage drop around the circuit must be zero, $V(\text{battery}) = V(\text{resistor}_1) = V(\text{resistor}_2)$.

Now assume that the current divides between the two resistors with I_1 going through one of the resistors and I_2 going through the second. By Ohm's law,

$$V(\text{battery}) = V(\text{resistor}_1) = I_1 \times R_1 = V(\text{resistor}_2) = I_2 \times R_2$$

Solve for I_1 and I_2. Since the current coming out of the branches is the sum of the current in each branch, $I = I_1 + I_2$.

Solve: $12 \text{ volts} = I_1 \times 100. \text{ ohms} = I_2 \times 100. \text{ ohms}$. Therefore, $I_1 = I_2 = 0.12$ A (since here both resistors have the same value). Then, $I = I_1 + I_2 = \textbf{0.24 A}.$

Note that, in this case, the parallel circuit draws more current than the series circuit. This is true for parallel circuits. It's the same thing as adding a second hose to a garden faucet; provided the supply pressure remains constant, we spray more water (analogous to an electric current) with two hoses than with one.

One form of series circuit enables us to add more realism to our simple model. Earlier, wires were modeled as conductors with zero resistance. However, a real wire has a small resistance, which is typically proportional to the length of the wire and inversely proportional to its cross-sectional area. In more realistic circuit models, this resistance of the wire is modeled by a resistance element (our now familiar wiggly-line symbol) inserted at an arbitrary location into the wire. This resistance henceforth is called the *wire* resistance, and the resistance of a component such as a lightbulb is called the *load* resistance.

Example 11.4

For the circuit shown, assume (1) the wire resistance (distributed along the whole length of the wire) totals 1.0 ohm, (2) the load resistance is 10. ohms, and (3) the battery voltage is 6.0 volts. Compute the efficiency of the circuit, where the efficiency is defined as the power dissipated in the load divided by the power produced by the battery.

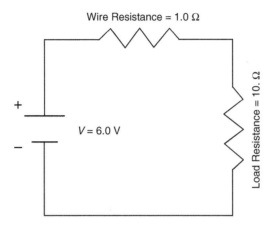

Wire Resistance = 1.0 Ω

$V = 6.0$ V

Load Resistance = 10. Ω

Need: Efficiency of circuit = (power dissipated in load)/(power produced by battery).

Know: Battery voltage = 6.0 volts; wire resistance = 1.0 ohms; load resistance = 10. ohms.

How: Computing efficiency requires two computations of power. Computation of power, using the "Power Law" $P = I \times V$ requires knowing the voltage across and the current through each circuit element. So, first, find the voltage across and current through each element using Ohm's law. Then, compute the power. Then, compute the efficiency.

Solve: In a series circuit, the total resistance is the sum of the individual resistors.

So, $R = R(\text{wire}) + R(\text{load}) = 10. + 1.0 = 11$ ohms. The battery voltage is 6.0 volts, so, by Ohm's law, $I = V/R = 6.0/11 = 0.55$ A.

Now, applying the "Power Law," $P(\text{battery}) = V(\text{battery}) \times I = 6.0 \times 0.55$ [V][A] = 3.3 W. This is the *total* power drained from the battery.

Applying Ohm's law to the wire's resistance, its overall voltage drop is $V(\text{wire}) = 1.0 \times 0.55$ [Ω][A] = 0.55 V.

Applying the "Power Law" to wire losses, $P(\text{wire}) = 0.55 \times 0.55$ [V][A] = 0.30 W. Therefore, the "load" power = 3.3−0.30 = 3.0 W and the **efficiency** is 3.0/3.3 = **0.91 or 91%**.

One task of an engineer might be to maximize the efficiency of application of power, since higher efficiency generally results in lower cost, and lower cost is generally desired by customers. The previous exercise gave one version of that challenge. The next example gives another.

Edison, in 1879, applied the "Power Law" and as a result made the first useful incandescent lightbulb. He did so by recognizing that a lightbulb is simply a resistor that converts electricity first into heat then renders part of that heat (only about 4%) into visible light. In the example that follows, treat the resistor (indicated by the zigzag line in the figure) inside the bulb (indicated by a circle) as an ordinary, if very high-temperature, resistor that obeys our two electrical laws. The lightbulb's resistor is called its *filament*.

Example 11.5

In the diagram shown, assume the battery delivers a constant 10.0 A of electric current whatever the voltage across the filament may be. If one wishes to design the lightbulb to have a power of 100. W, what should the resistance of the lightbulb be? What is the voltage drop across the filament?

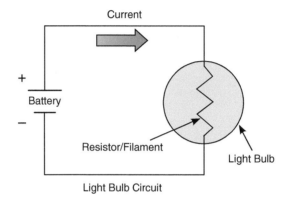

Current

Battery

Resistor/Filament

Light Bulb

Light Bulb Circuit

Need: $R =$ ____ Ω and $V =$ _____ V.
Know: $P = 100.$ W and $I = 10.0$ A.
How: $P = I \times V$ and $V = R\,I$.
Solve: $P = I \times V$ or $V = P/I = 100./10.0$ [W]/[A] = **10.0 volts**.
 Therefore, $\boldsymbol{R} = V/I = 10.0/10.0$ [V/A] = **1.00 ohms**.

If the conductor wire in this example also had a 1.00 ohm resistance, as much power would be dissipated in it as in the bulb's filament. It would also have "consumed" another 10.0 volts of voltage, so the battery would have had to supply 20.0 volts.

Having resistance in the wire comparable to that of the filament complicates the problem. The key to Edison's problem of maximizing the bulb's efficiency was in maximizing the efficiency of the *entire* circuit, rather than concentrating on the bulb alone. Higher efficiency meant lower costs, eventually low enough for electricity and the lightbulb to become mass consumer products.

In all the circuits we discussed so far, the current always flows in the same direction (shown, according to convention, flowing from positive to negative around a circuit). These are called **direct current** circuits. However, another class of electrical circuits allows the current to reverse its direction many times a second. These **alternating current** circuits are the basis for the vast number of electrical technologies beyond the lightbulb. They range from the 1000-megawatt generators that supply electricity, to the motors that put electricity to work, to the radio wave generators that enable us to send out radio signals or to thaw frozen food in minutes in microwave ovens.

11.5 Kirchhoff's laws

Gustav Robert Kirchhoff (1824–87) was a German physicist who contributed to the fundamental understanding of electrical circuits, spectroscopy, and the emission of thermal radiation by heated objects. Kirchhoff formulated his circuit laws through experimentation in 1845, while still a university student.

11.5.1 Kirchhoff's voltage law

Kirchhoff's Voltage Law is a form of the Conservation of Energy Law. Kirchhoff's voltage law can be stated as follows:

The algebraic sum of the voltage drops in a closed electrical circuit is equal to the algebraic sum of the voltage sources (i.e., increases) in the circuit.

Mathematically, Kirchhoff's voltage law is written as

$$\sum V_{(closed\ loop)} = \sum IR_{(closed\ loop)} = 0 \qquad (11.3)$$

where $\sum$ is the summation symbol (e.g., $\sum V = V_1 + V_2 + V_3 + ...$).

A *closed loop* can be defined as any path in which the originating point in the loop is also the ending point for the loop. No matter how the loop is defined or drawn, the sum of the voltages in the loop must be zero. Both loop 1 and loop 2 in Fig. 11.4 are closed loops within the circuit. The sum of all voltage drops and rises around loop 1 equals zero, and the sum of all voltage drops and rises in loop 2 must also equal zero. There is also a third closed loop in this circuit, the one that goes around the outside from A to B to C to D and back to A again. But this loop is not independent of the other two, so it provides no additional information.

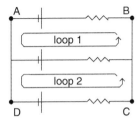

FIGURE 11.4

Closed loops in a simple electrical circuit.

11.5.1.1 Voltage dividers

If two resistors are in series, there is a voltage drop across each resistor, but the current through both resistors must be the same. A simple circuit with two resistors in series with a source (voltage supply) is called a **voltage divider**.

In Fig. 11.5, the voltage V_{in} is progressively dropped across resistors R_1 and R_2. If a current I flows through the two series resistors, by Ohm's law, $I = V_{in}/(R_1 + R_2)$. Then, $V_{out} = IR_2$ or

$$V_{out} = V_{in} \times \left[\frac{R_2}{(R_1 + R_2)}\right] \qquad (11.4)$$

Hence, V_{in} is "divided," or reduced, by the ratio $R_2/(R_1 + R_2)$.

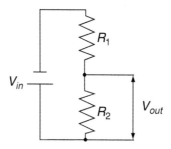

FIGURE 11.5

Voltage divider circuit.

Example 11.6

Suppose you want to build a voltage divider that reduces the voltage by a factor of 15. You visit your local electronics store and find that it has 5.0, 10., 15., and 20. ohm resistors in stock. Which ones do you choose to make your voltage divider in the simplest and most economical way?

Need: An inexpensive voltage divider such that $V_{out}/V_{in} = 1.0/15 = 0.067$.

Know: The voltage divider Eq. (10.4), and the supply of resistors.

How: Since we know that $V_{out}/V_{in} = R_2/(R_1 + R_2) = 0.067$, we could just try the values of the resistors in stock to see if we can get close to the required ratio.

Solve: By inspection, we see that R_1 must be much bigger than R_2 to make V_{out}/V_{in} a small number. Therefore, we choose the smallest available resistor and make $R_2 = 5.0$ ohms. Now solve the voltage divider equation for R_1. This gives $R_1 = 66$ ohms. Using the available resistors, we can come close to this value by connecting six 10. ohm resistors in series with a 5.0 ohm resistor. This makes $R_1 = 65$ ohms and $V_{out}/V_{in} = 0.067$ (to 2 significant figures).

11.5.2 Kirchhoff's Current Law

Kirchhoff's Current Law is a form of the Conservation of Charge Law. Kirchhoff's Current law can be stated as

The algebraic sum of all the currents at a node must be zero,

where a "node" is any electrical junction in a circuit. Kirchhoff's Current Law can be written mathematically as

$$\sum I \,(\text{at a node}) = 0 \qquad\qquad (11.5a)$$

where $\sum I = I_1 + I_2 + I_3 + \ldots$ is the summation symbol. Stated differently,

The sum of all the currents entering a node is equal to the sum of all the currents leaving the node, or

$$\sum I \,(\textbf{entering a node}) = \sum I \,(\textbf{leaving the node}) \qquad\qquad (11.5b)$$

Current flows through wires much like the analog of water flows through pipes. The amount of water that enters a branching junction in a pipe system must be the same amount of water that leaves the junction. The number of pipes in the branching junction does not change the net amount of water (or current in our case) flowing (see Fig. 11.6).

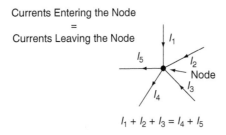

FIGURE 11.6

Currents entering and leaving a circuit node.

11.5.2.1 Current dividers

If two resistors are in parallel, the voltage across them must be the same, but the current divides according to the values of the resistances. A simple circuit with two resistors in parallel with a voltage source is called a **current divider**.

The equivalent resistance of two resistors R_1 and R_2 in parallel in Fig. 11.7 is

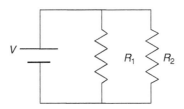

FIGURE 11.7

Parallel resistor current divider.

$$\frac{1}{R_{eq}} = \frac{1}{R_1} + \frac{1}{R_2}$$

which can be algebraically rearranged as

$$R_{eq} = \frac{(R_1 \times R_2)}{(R_1 + R_2)} \qquad (11.6)$$

Now, if the voltage across these resistors is V, then the current I in the circuit, before the division, is, according to Ohm's law,

$$I = \frac{V}{R_{eq}} = V \times \frac{(R_1 + R_2)}{(R_1 \times R_2)}$$

The current through R_1 according to Ohm's law is

$$I_1 = \frac{V}{R_1}$$

Dividing this equation by the previous one and solving for I_1 gives

$$I_1 = I \times \frac{R_2}{(R_1 + R_2)} \qquad (11.7)$$

We can also show that

$$I_2 = I \times \frac{R_1}{(R_1 + R_2)} \qquad (11.8)$$

Thus, the current I has been "divided," or reduced, by the resistance ratio.

Example 11.7

Find the current through and voltage across the 40. ohm resistor, R_3, in the circuit that follows.

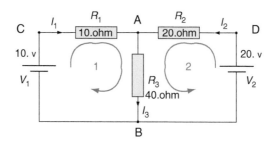

Need: $I_3 = $ _____ A and $V_3 = $ _____ V.
Know: Kirchhoff's current and voltage laws, Eqs. (11.3) and (11.5a and b).

How: This circuit has two nodes (A and B) and two independent loops. Write the current law equations for the two nodes and the voltage law equations for the two loops. Then, solve the resulting set of algebraic equations for the currents.

Using Kirchhoff's current law, at node A the equation is: $I_1 + I_2 = I_3$

Using Kirchhoff's voltage law around loops 1 and 2, the equations are

Loop 1: $10. = R_1 \times I_1 + R_3 \times I_3 = 10. \times I_1 + 40. \times I_3$

Loop 2: $20. = R_2 \times I_2 + R_3 \times I_3 = 20. \times I_2 + 40. \times I_3$

Since I_3 is the sum of $I_1 + I_2$, we can substitute this into the loop equations and rewrite them as

Loop 1: $10. = 10. \times I_1 + 40. \times (I_1 + I_2) = 50. \times I_1 + 40. \times I_2$

Loop 2: $20. = 20. \times I_2 + 40. \times (I_1 + I_2) = 40. \times I_1 + 60. \times I_2$

Solve: Now, we have two simultaneous equations that can be solved algebraically to give the values of I_1 and I_2. Algebraically solving the loop one equation for I_2 and substituting this expression into the loop two equation allows us to solve for I_1 as: $\mathbf{I_1 = -0.14\ A}$. Now, substituting this value of I_1 into either loop equation allows us to solve for I_2 as $I_2 = +\ 0.43$ A. Since $I_3 = I_1 + I_2$, the current flowing in resistor R_3 is $\mathbf{I_3 =} -0.14 + 0.43 = \mathbf{0.29\ A}$, and from Ohm's law, the voltage drop across the resistor $\mathbf{R_3}$ is $0.29 \times 40. = \mathbf{11\ V}$.

The negative sign we got for I_1 in this example simply means that we initially assumed the wrong direction for the current flow, but the numerical value is still correct. In fact, the 20-volt battery is actually "charging" the 10-volt battery (i.e., putting electrical energy back into the battery by forcing an electric current through it).

11.6 Switches

Here, we concentrate on a humble circuit element whose importance was appreciated by Edison: the switch. In our water flow model, a switch is equivalent to a faucet. It turns the flow on or off. Surprisingly, this simplest of components is the basis of the most sophisticated of our electrical technologies, the computer. In its essence, a computer is nothing more than a box filled with billions of switches, all turning each other on and off. The faster the switches turn each other on and off, the faster the computer can do computation. The smaller the switches, the smaller the computer can be. The more reliable the switches, the more reliable is the computer. The more efficient the switches, the less power the computer uses. So the criteria engineers have used in developing better switches are higher speed, smaller size, greater reliability, and increased efficiency.

Unlike the elements we discussed so far, a switch only has two **states**. As indicated in Fig. 11.8a and b a switch can be open, in which case it presents an infinite resistance to the flow of electric current (that is, no current flows at all), or a switch can be closed, in which case ideally it presents zero resistance to electric current.

A **relay** is a switch operated by an electromagnet, Fig. 11.8b. The relay simply closes the switch against the spring when the external power source is turned on. When the external power source is turned off, the spring returns the switch to its

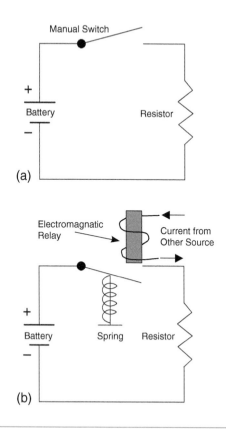

FIGURE 11.8

(a and b) Simple switches.

open position. It got its name from its original use, relaying telegraph messages over longer distances than could be accomplished by a single circuit. The relay consists of two parts. The first is an electromagnet powered by one circuit, which we call the **driving circuit**. The second part is another circuit with a metal switch in it, called the **driven circuit**. The driven circuit is placed so that, when the driving circuit is on, its electromagnet attracts the moving metal part of the driven circuit's switch. This results in closing the gap in the driven circuit and turning it on.

Relays continue to be used in control systems because they are highly reliable, but they are high in power demands, slow, and bulky. In the 20th century, the basis of a faster switch emerged, the transistor, a solid-state switch. Fig. 11.9 shows one important type of transistor, the metal oxide semiconductor field effect transistor (MOSFET).

A transistor is made of semiconductors—materials that have properties midway between those of a conductor and those of an insulator. Most of today's transistors

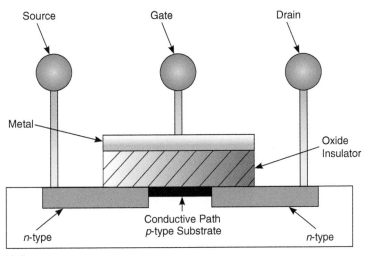

FIGURE 11.9

Principles of MOSFET construction.

use semiconductors made out of the element silicon. By "doping" silicon by adding trace amounts of other elements, its electrical conductivity can be controlled. As a simple model of a particularly useful type of transistor called the *field effect transistor*, we can imagine two such pieces of doped silicon, a "source" that serves as a source of electrical charge and a "drain" that serves as a sink into which the charge flows. The source and drain are connected by a semiconductor channel, which is separated by an insulator from a metal electrode called a *gate*. Some of the semiconductor is doped with material that sustains an excess of electrons and is called *n-type* (*n* for "negative"). Other semiconductors are doped with materials that sustain a deficiency of electrons called *p-type* (*p* for "positive"). When the gate is given a positive voltage in the *npn* MOSFET shown in Fig. 11.9, it promotes the flow of charge from source to drain, effectively making a thin layer of continuous *n*-type conductive channel between the source and the drain. When the gate is given a negative voltage, current cannot flow from source to drain, and the transistor acts as an open switch.

Transistors (switches) are small, fast, efficient, and reliable. But, to get to the modern computer, one more step was needed. This was the **integrated circuit**, independently coinvented by Nobel Prize—winning engineer Jack Kilby and Intel cofounder Robert Noyce in the early 1960s.

A modern integrated circuit contains billions of transistors connected together to form a logic circuit on a single silicon chip the size of a fingernail. This is done by a process called **photolithography**, which literally means "using light to write on a stone." The circuit designer first makes a large drawing of the circuit. Optical lenses are then used to project a very small image of the drawing onto a fingernail-sized chip of silicon, which is coated with special chemicals.

Integrated circuits are the hardware that made personal computers possible. But integrated circuits by themselves do not constitute a computer. What enables a collection of switches to carry out computations? That is the job for a sequence of instructions, called **software**. A crucial distinction needed to understand computers is **hardware** versus software. Hardware is the collection of mechanical, electrical, or electronic devices that make up a computer. Among the hardware in a modern computer, two elements are crucial: **(1)** a system called a **central processing unit** or **CPU**, which carries out binary logic and arithmetic; and **(2)** memory, often called **RAM** (random access memory) or **ROM** (read-only memory), which store the ones and zeroes corresponding to inputs and outputs of the CPU. Both CPU and memory are integrated circuits composed of many millions of switches (see Chapter 10).

Summary

The flow of electricity though wires can be modeled in the same way one might treat the flow of water through pipes. The model has as its basic variables **charge, current, voltage,** and **resistance**. The simplest model consists of **Ohm's Law**, the "Power Law," and **Kirchhoff's Voltage** and **Current** Laws. Using this model, one can analyze the operation of a particular class of electric circuits called **direct current** circuits and, in particular, two classes of direct current circuits called **series circuits** and **parallel circuits**.

An important type of electric circuit component is the **switch**. On the hardware side, the development of the computer can be viewed as the search for ever-faster switches, based first on **relays**, then on **vacuum tubes**, then on **transistors**, and finally on the form of electronics called **solid state integrated circuits**. This form of electronics provides the hardware, which in conjunction with a sequence of instructions, called **software**, makes modern computation and control possible.

Exercises

If no circuit sketch is already given, it is highly recommended you *first draw each circuit* to answer these problems. It is generally useful to add numbers to your diagram with what you think you know, such as voltages and currents at each point.

Note the equivalence among these electrical units:

$$[\Omega] = [V]/[A], [W] = [V][A] = [\Omega][A]^2 = [V]^2/[\Omega]$$

The circuits for Exercises 1–8 are given in the figure below.

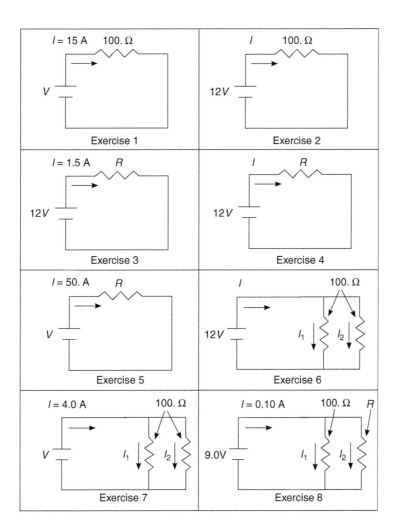

1. For the circuit shown, find the voltage V. (**Ans. 15×10^2 V.**)
2. For the circuit shown, find the current I.
3. For the circuit shown, find the resistance R.
4. For the circuit shown, a power of 100. watts are dissipated in the resistor. Find the current I.
   (**Ans. 8.3 A.**)
5. For the circuit shown, a power of 100. watts are dissipated in the resistor. Find the resistance, R.
6. For the circuit shown, find the current I. (**Ans. 0.24 A.**)
7. For the circuit shown, find the voltage V.
8. For the circuit shown, find the resistance R.

9. For the circuit that follows, a power of 100. watts are dissipated in each resistor. Find the current I and the voltage V

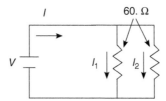

For Exercises 10—19, draw the circuit and solve for the unknown quantity requested. It is suggested that you include a small solid circle to identify each point where you can specify the local voltage.

10. A circuit consists of a 3.0-volt battery and two resistors connected in series with it. The first resistor has a resistance of 10. ohms. The second has a resistance of 15. ohms. Find the current in the circuit. (**Ans. 0.12 A.**)

11. A circuit consists of a 12.0-volt battery and a resistor connected in series with it. The current is 105. A. Find the resistance.

12. A circuit consists of a 9.0-volt battery and two parallel branches, one containing a 1500 ohm resistor and the other containing a 1.0×10^3 ohm resistor. Find the current drawn from the battery. (**Ans. 0.015 A.**)

13. A circuit consists of a 12-volt battery attached to a 1.0×10^2-ohm resistor, which is in turn connected to two parallel branches, each containing a 1.0×10^3 ohm resistor. Find the current drawn from the battery.

14. An automobile's 12-volt battery is used to drive a starter motor, which for several seconds draws a power of 3.0 kW from the battery. If the motor can be modeled by a single resistor, what is the current in the motor circuit while the motor is operating? (**Ans. 250 A.**)

15. An automobile's 12-volt battery is used to light the automobile's two headlights. Each headlight can be modeled as a 1.00-ohm resistor. If the two headlights are hooked up to the battery in series to form a circuit, what is the power produced in each headlight? Why would you *not* wire car lights in series?

16. An automobile's 12-volt battery is used to light the automobile's two headlights. Each headlight can be modeled as a 1.00-ohm resistor. If the two headlights are hooked up to the battery in parallel to form a circuit, what is the power produced in each headlight? Why is a parallel circuit preferred for this application?

17. An automobile's 12-volt battery is used to light the automobile's two headlights. Each headlight can be modeled as a single resistor. If the two headlights are hooked up to the battery in parallel to form a circuit and each headlight is to produce a power of 100. W, what should the resistance of each headlight be? (**Ans. 1.4 ohms per bulb**.)

18. Suppose one of the headlights in Exercise 17 suddenly burns out. Will be the power produced by the other headlight increase or decrease?

19. Suppose a car has headlights operating in parallel but with different resistances—one of 2.0 ohms and the other of 3.0 ohms. Suppose the headlight parallel circuit is connected in series with a circuit for a car stereo that can be modeled by a 1.0 resistor. What is the current being drawn from a 12-volt battery when both the lights and the stereo are on? (**Ans. $I = 5.5$ A.**)

20. Consider the circuit that follows. What are the currents I_0 through I_6? (**Hint:** What are the voltages V_a and V_b? Note the symmetry between $R_1/R_2 = 5.0/5.0$ and $R_3/R_4 = 10.0/10.0$.)

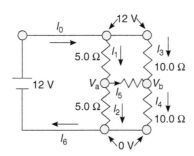

21. Determine the output voltage, V_{out}, in Fig. 11.5 if $V_{in} = 90$. volts, $R_1 = 10$. ohms, and $R_2 = 50$. ohms.
22. Use the same stock resistors in Example 11.6 to design the construction of a voltage divider that reduces the voltage by a factor of 30.
23. Use Kirchhoff's voltage and current laws to determine the voltage drop and current in each of the three resistors that follow if $R_1 = 10$. ohms, $R_2 = 20$. ohms, and $R_3 = 30$. ohms.

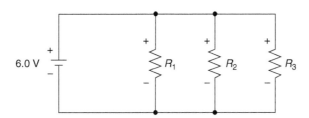

24. If the resistors in Exercise 23 are all equal to 10. ohms ($R_1 = R_2 = R_3 = 10$. ohms), what is the current supplied by the 6.0-volt battery?
25. Determine the currents I_1 and I_2 in the current divider illustrated in Fig. 11.7 if $V = 50.3$ volts, $R_1 = 125$. ohms, and $R_2 = 375$. ohms.
26. Repeat Example 11.7 with the resistor R_1 changed to 100. ohms and all other values remaining unchanged.
27. Repeat Example 11.7 with V_1 increased to 100. volts and V_2 reduced to 5.0 volts.

28. Determine the currents I_1, I_2, I_3, and I_4 in the figure that follows.

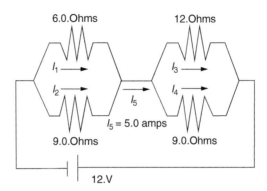

29. It is the last semester of your senior year, and you are anxious to get an exciting electrical engineering position in a major company. You accept a position from company A early in the recruiting process, but you continue to interview, hoping for a better offer. Then, your dream job offer comes along from company B. More salary, better company, more options for advancement. It is just what you have been looking for. What do you do?
    a) Accept the offer from company B without telling company A (just don't show up for work).
    b) Accept the offer from company B and advise company A that you have changed your mind.
    c) Write company A and ask them to release you from your agreement.
    d) Write company B thanking them for their offer and explain that you have already accepted an offer.

    (Summarize using the Engineering Ethics Matrix.)

30. A female student in your class mentions to you that she is being sexually harassed by another student. What do you do?
    a) Do nothing; it is none of your business.
    b) Ask her to report the harassment to the course instructor.
    c) Confront the student accused of harassment and get his or her side of the issue.
    d) Talk to the course instructor or the college human resource director privately.

    (Summarize using the Engineering Ethics Matrix.)

Industrial engineering

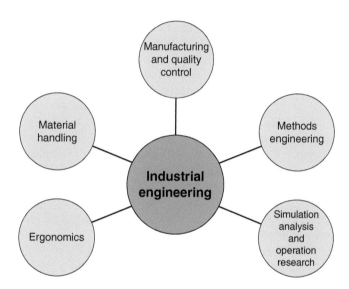

12.1 Introduction

Industrial engineering[1] is one of the most diverse fields of engineering because the word "Industry" has a very broad meaning. It refers to the production of any economic goods within an economy. Because there are so many different kinds of "industries," they are divided into the following three different categories:

---

[1] Originally the name "industrial engineering" applied only to manufacturing. Today it has grown to encompass any procedure that includes how any process, system, or organization operates. Some universities have changed the name "industrial engineering" to a broader term such as "production engineering" or "systems engineering".

Exploring Engineering. https://doi.org/10.1016/B978-0-443-13541-5.00037-4

- **Primary industries:** These industries deal with the *extraction of resources* directly from the Earth such as farming, mining, petroleum, and logging.
- **Secondary industries:** These industries are involved in *manufacturing products* from the resources provided by the primary industries. They include the manufacturing of steel, automobiles, furniture, food, electronics, and so forth.
- **Tertiary industries:** These industries compose the *service industry*. They include education, health care, package delivery, software development, financial institutions, government organizations, and so forth.[2]

In general, industrial engineers develop integrated systems consisting of people, knowledge, equipment, energy, and materials. Examples of activities where industrial engineering is used include designing assembly lines and workstations, process efficiency analysis, developing a new financial algorithm for a bank, streamlining emergency room use in a hospital, planning complex distribution schemes for materials or products, and shortening lines at a bank, hospital, or a theme park.

Modern industrial engineering began in the 20th century. The first half of the century was characterized by mass production with an emphasis on increasing efficiency and reducing costs. In 1911, Frederick Taylor published his now famous book on *The Principles of Scientific Management*, which included an analysis of human labor, a systematic definition of methods, and tools and training for employees. Taylor's time and motion studies set standard times for workers to complete a task that subsequently increased productivity, reduced labor costs, and increased the wages of the employees. In 1913, the moving assembly line developed by Henry Ford produced major increases in manufacturing productivity by reducing the assembly time of a car from 12.5 to 1.5 h. In addition, he was a pioneer in providing financial incentives for his employees to increase productivity.

In the second half of the 20th century, the Japanese management theories highlighted the issues of product quality, rapid delivery time, and manufacturing flexibility to improve customer satisfaction.[3] In the 1990s, the industrial globalization process emphasized supply chain management, and customer-oriented business process design.

Today industrial engineering can be broken down into the following categories:

1) Manufacturing and quality control
2) Methods engineering
3) Simulation analysis and operations research
4) Ergonomics
5) Material handling

---

[2] When industrial engineers work in one of the secondary industries, they are often called Manufacturing Engineers. When they work in one of the primary or tertiary industries, they are called Industrial Engineers.

[3] The Japanese also came up with the term *Kaizen*, which means "good change" to represent continuous improvement.

12.2 Manufacturing and quality control

Once a product is designed, industrial engineers guide it through the manufacturing process. Since this textbook contains a separate chapter on manufacturing, students are encouraged to review this chapter before proceeding. Manufacturing information topics will not be repeated here.

Quality control can be broken down into the areas of (a) Statistical Analysis, (b) Probability Theory, (c) Reliability Analysis, and (d) Design of Experiments (DOE).

12.2.1 Statistical analysis

Statistics is a branch of mathematics in which groups of measurements or observations are studied. Some of the more common descriptive terms used in statistics are defined below:

Population—a group of items that is being studied.
Sample—a small group of items selected from a population.
Random samples—occur when items have an equal chance of being selected.
Data[4]—numbers or measurements that are collected in a population.
Variables—characteristics that allow us to distinguish one item from another.
Constants—items whose value never changes

Example 12.1

Which of the following is a variable?
a) Scores obtained on a math exam.
b) The number of days in the year.
c) The time it takes you to get to this class.
  Need: To decide which of these three items is a variable.
  Know: A "variable" has characteristics that allow us to distinguish one item from another.
  How: Apply the definition of "variable" to these three items.
  Solve: Items (a and c) are variables because they are not constants. Item (b) is a constant.

Some of the common mathematical terms used in statistics are defined in Table 12.1.

---

[4] The word "data" is plural (therefore we should say "data are" "not data is"). Data represents a set of measurements. A single element of that set is called a "datum" as in "there is only one datum point".

Table 12.1 Some of the common mathematical terms used in statistics.[5]

Term	Definition	Formula
Mean	The **mean** $\bar{x}$, or **average**, is 1 divided by the number of samples N multiplied by the sum of all the data points.	$\bar{x} = \frac{1}{N}\sum_{i=0}^{i=N} x_i$ (12.1)
Median	The **median** M is the middle value of a set of data containing an odd number of values, or the average of the two middle values of a set of data with an even number of values.	Median of a set $x_1, x_2, ..., x_M, ..., x_N$ M = Middle value x_M if N is an odd number M = the average of the two middle values if N is and even number (12.2)
Variance	The **variance** σ^2 of a set of data is 1 divided by the number of samples (data) minus 1, multiplied by the sum of the data points subtracted from the mean and then squared.	$\sigma^2 = \frac{1}{N-1}\sum_{i=0}^{i=N}(x_i - \bar{x})^2$ (12.3)
Standard deviation	The **standard deviation** σ is the square root of the variance.	$\sigma = \left[\frac{1}{N-1}\sum_{i=0}^{i=N}(x_i - \bar{x})^2\right]^{1/2}$ (12.4)
Standard error	The **standard error** of the mean is the standard deviation divided by the square root of the number of samples.	$SE = \frac{\sigma}{\sqrt{N}}$ (12.5)
Z-score	The **Z-score** is the number of standard deviations an observation or datum is *above* the mean.	$Z = \frac{x_i - \bar{x}}{\sigma}$ (12.6)

Example 12.2

Ten employees of a department store earn the following weekly wages: $400., $650., $660., $625, $660., $650., $680., $630., $670., and $650.
a) Find the median weekly wage.
b) What is the mean weekly wage?
c) Find the standard deviation of the set of wages.
   Need: The median, mean, and standard deviation of the given set of wages.
   Know: The set of wages in question.

[5] Various computer programs such as Microsoft Excel contain routines that perform these calculations.

How: Use Eqs. (12.1), (12.2) and (12.4)

Solve:

a) The wages can be listed in order as: $400., $625, $630., $650., $650., $660., $660., $665, $670., and $680. Since this set has an even number of items, in this case, Eq. (12.2) gives the **median** as the average of the two middle values, or ($650. + $660.)/2 = **$655**.

b) The **mean** (or **average**) weekly wage is computed from Eq. (12.1) as

$$\bar{x} = \frac{1}{N} \sum_{i=0}^{i=N} x_i$$

$$\text{Mean weekly wage} = \frac{1}{10} \sum_{i=0}^{i=10} (\$400. + \$625 + ... + \$680.) = \$589$$

c) **The standard deviation** of the set of wages is computed from Eq. (12.4) as

$$\sigma = \left[\frac{1}{N-1} \sum_{i=o}^{i=N} (x_i - \bar{x})^2 \right]^{1/2}$$

$$= \frac{1}{(10-1)} \sum_{i=0}^{i=10} \left[(\$400. - \$589)^2 + (\$625 - \$589)^2 + ... + (\$680. - \$589)^2 \right]^{1/2}$$

$$= \$30.8$$

12.2.2 Probability theory

Probability is a number that represents the likelihood that a specific event will occur. It is the ratio of the number of actual occurrences to the number of possible occurrences.

$$P(A) = \text{Probability of event A happening} = \frac{\text{Number of ways A can happen}}{\text{Total number of possible outcomes}} \quad (12.7)$$

When $P(A) = 0$, then event A cannot occur (e.g., the probability that you will grow younger rather than older is zero), and when $P(A) = 1$, then event A is certain to occur (e.g., the probability that you will grow older and not younger). There are a number of mathematical rules that govern probability theory. The rules of probability are easily understood by using dice and card games.

12.2.2.1 Probability addition

The probability addition rule is used when we have independent "*or*" events. Suppose that you throw a single die.[6] There are six possible outcomes: **1, 2, 3, 4, 5, and 6**. The probability of getting a 6 is

$$P(6) = \frac{\text{The number of ways of getting a 6}}{\text{The total number of possible outcomes from the die}} = \frac{1}{6}$$

[6] Dice is plural, die is singular.

and the probability of getting any other value is $P(5) = P(4) = P(3) = P(2) = P(1) = 1/6$.

Since the values in the face of a die are all independent, the probability of getting either a 5 *or* a 6 is

$$P(5 \textit{ or } 6) = P(5) + P(6) = 1/6 + 1/6 = 2/6 = 1/3$$

If you choose a single card from a pack of 52 cards, the probability of getting an ace of spades is $P(\text{ace of spades}) = 1/52$. The probability of getting any of the four aces is $P(\text{any ace}) = 4/52$. The probability of getting any of the 12 picture cards (jack, queen or king) is $P(\text{any picture card}) = 12/52$.

Since all the cards in the deck are unique, the probability of getting any of the four aces *or* any of the 12 picture cards is

$$P(\text{ace } \textit{or } \text{picture}) = 4/52 + 12/52 = 16/52$$

Notice that the "*or*" event probabilities can be computed by **adding** their individual probabilities if the events are "mutually exclusive", meaning that none of the "*or*" events can occur at the same time (i.e., you cannot have an ace that is also a picture card). So, before adding "*or*" probabilities you need to make sure that the events cannot simultaneous occur and are mutually exclusive (i.e., they are independent, and each excludes the other).

The complete addition rule when A, B, C, ... are mutually exclusive is:

$$P(\text{A } \textit{or } \text{B } \textit{or } \text{C } \textit{or}...) = P(\text{A}) + P(\text{B}) + P(\text{C}) + ... \tag{12.8}$$

12.2.2.2 Probability multiplication

The multiplication rule is used when we have independent "*and*" events. For example, the probability of getting a 6 *and* a spade is:

$$P(6 \textit{ and } \text{spade}) = P(6) \times P(\text{spade}) = 1/6 \times 1/4 = 1/24$$

Notice that the "*and*" event probabilities can be computed by **multiplying** their individual probabilities if the events are "mutually exclusive", meaning that none of the "*and*" events can occur at the same time. For example, the probabilities $P(6)$ *and* $P(\text{spade})$ are independent of each other and therefore are mutually exclusive. Knowing whether or not one has occurred has no effect on the probability of the other happening. In this example, knowing that the thrown die produced a 6 has no effect on what happens when we draw a card from the deck, these two events are statistically independent. The complete probability multiplication rule when A, B, C, ... are mutually exclusive is:

$$P(\text{A } \textit{and } \text{B } \textit{and } \text{C } \textit{and } ...) = P(\text{A}) \times P(\text{B}) \times P(\text{C}) \times .... \tag{12.9}$$

Example 12.3

There are 5 marbles in a bag of which 4 are blue and 1 is red. What is the probability that you can pick a blue marble from the bag?

Need: Probability of picking a blue marble.

Know: The number of blue and red marbles in the bag.

How: Use Eq. (12.7) to compute the probability.

Solve: The number of ways that a blue marble will be chosen is 4, and the total number of possible outcomes is 5 (there are 5 marbles in the bag). Then, Eq. (12.7) gives

$$\text{Probability of getting a blue marble} = \frac{4}{5} = 0.8$$

12.2.3 Reliability analysis

The reliability of a product (or a system) is defined as the probability that a product will perform its required function under specified conditions for a certain period of time. If we have a large number of a certain product that we can test over time, then the reliability of that product at time t is given by

$$R(t) = \frac{\text{Number of products that have \textbf{NOT} failed by time } t}{\text{Total number of products being tested at time } t = 0} \qquad (12.10)$$

At time $t = 0$, the number of products that have not failed is equal to number of products being tested. Therefore, the reliability at $t = 0$ is: $R(0) = 1 = 100\%$. Note that the reliability, $R(t)$, will continuously decline as t increases and products fail.

The **failure rate** of a product at time t is defined as the probability of a product failing at time t. It is represented by the symbol $F(t)$. Let

$N_F =$ number of products that failed during the time interval $t + \Delta t$

and $N_S =$ number of products that have NOT failed by time t.

If Δt is small and the failure rate does not change during this time interval, then the failure rate at time t, can be calculated as follows:

$$F(t) = \frac{\text{Number of product failures between } t \text{ and } t + \Delta t}{(\text{Number of products that have NOT failed by time } t) \times \Delta t} = \frac{N_F}{N_S \times \Delta t} \qquad (12.11)$$

For example, if 250. components have not failed by $t = 8.0$ h, and 3.0 components fail during the next 1.0 h, then the failure at 8 h is as follows:

$$F(8) = \frac{3.0}{250. \times 1.0} = 0.012 = 1.2\% \text{ per hour}$$

While it is relatively easy to define failure rates for individual product components such as springs or gears, it becomes more difficult for complex products like automobiles. For example, a failure in a complex product can be complete breakdown (a "catastrophic" failure) to relatively minor easily repaired failure. While failures in complex products may be repairable, they may not be repairable for individual product components.

The **bath tub curve** shown in Fig. 12.1 represents the failure pattern seen in many products. The vertical axis is the failure rate at each point in time and the horizontal axis is the test time (or the time since the product was released for sale to the public). The bath tub curve is divided into three regions: early failure, useful life, and wear out.

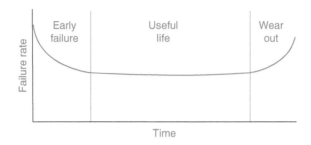

FIGURE 12.1

Failure rate bath tub curve.

1) **Early Failure**: This is the early failure (or break-in) stage. During this period, failures typically occur because products were not designed properly or manufacturing flaws occurred. Failure rates at the beginning of the early failure stage are often high, but then they decrease with time as the early failures are removed.

2) **Useful life**: This is the center stage of the bath-tub curve and is characterized by a constant failure rate. This stage is the most significant stage for reliability prediction and evaluation activities.

3) **Wear out stage**: This is the final stage where the failure rate increases as the products begin to wear out and break down. When the failure rate becomes high after a long period of use, repair or replacement of parts should be performed.

Example 12.4

Would you expect the bath tub curve to apply to:
a) new automobiles,
b) human beings?

   Need: Interpretation of the bath tub curve for new automobiles and human beings.
   Know: How the bath tub curve is defined.
   How: Intuition.
   Solve:

   a) No, because new automobiles are not likely to have a catastrophic early failure. The failure rate of a new car's components may be high as faults from the design or manufacturing process are found and fixed via recall. However, if the manufacturer tested for early faults, then the early failure stage will not occur.

   b) The bath tub pattern does apply to human beings because failure rates (i.e., death rates) are higher in both infancy and in old age.

Many organizations keep failure and repair records for the products they produce. This information can then be used to determine failure rates for those products. Existing failure rate data are also useful when designing similar new products.

Under certain assumptions, the failure rate for a complex system is simply the sum of the individual failure rates of its components. This permits testing of individual components or subsystems, whose failure rates are then added to obtain the total system failure rate.

Example 12.5

One thousand transistors are placed on life test, and the number of failures in each time interval is recorded in the table below. Assume the number of failures and the time intervals are exact values.

Time interval (hours)	0– 100	100– 200	200– 300	300– 400	400– 500	500– 600	600– 700	700– 800	800– 900	900– 1000
Number of failures	160	86	78	70	64	58	52	43	42	36

a) Find the reliability and the failure rate at 0, 100, 200, ... and 1000 h.
b) Plot the failure rate $F(t)$ as the transistors get older. Does this show the bath tub pattern of failure?
c) Plot the reliability $R(t)$ versus time.

   Need: The reliability and failure rate of the transistors plus a plot of their values versus time.
   Know: The test results given in the table.
   How: Use Eqs. (12.10) and (12.11).
   Solve:
   a) At time $t = 0.00$, $N_F = 160$ failures, $N_S = 1000$ nonfailures, and $\Delta t = 100$ h. Using Eq. (12.10) we get

$$R(t=0) = \frac{\text{Number of products that have NOT failed by time } t = 0}{\text{Total number of products being tested at time } t = 0} = \frac{1,000}{1,000} = 100\%$$

and Eq. (12.11) gives the failure rate as

$$F(t=0) = \frac{160}{(1000)(100)} = 0.0016 = 0.16\% \text{ per hour.}$$

The table below shows the remaining calculation results.

Time t	Nonfailures N_S	Failures in next 100 h N_F	Reliability $R(t)$ (%)	Failure rate $F(t)$ (Failures/h)
0	1000	160	100	0.16
100	840	86	84.0	0.10
200	754	78	75.4	0.10
300	676	70	67.6	0.10
400	606	64	60.6	0.11
500	542	58	54.2	0.11
600	484	52	48.4	0.11
700	432	43	43.2	0.10
800	389	42	38.9	0.11
900	347	36	34.7	0.10
1000	311		31.1	

b) The failure rate information from part (a) is shown in the plot below.

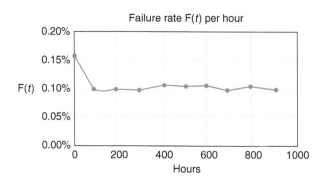

c) The reliability information from part (a) is shown in the plot below.

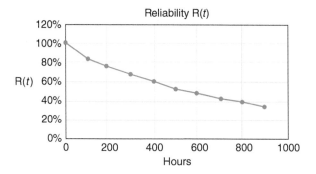

This result shows that the initial failure rate is 0.16% per hour, and then quickly drops to around 0.10% per hour. This is the early failure phase. From then on the failure rate is fairly constant, and represents the useful life phase. There is no wear out phase data, but there are still 311 components left after the 1000 h test. The wear out phase could become apparent if the tests were to be continued.

When the failure rate is constant, the reliability graph is a straight horizontal line and represents the middle phase of a bath tub curve. When there is little break-in failure (early failure), a constant failure rate can be effectively used to predict the reliability of a product to a particular time. If the failure rate F is a constant (i.e., independent of t), then the reliability $R(t)$ at time t is given by

$$R(t) = e^{-Ft} \quad \text{where } F = \text{constant failure rate} \tag{12.12}$$

For example, if there are initially 1000 components and their failure rate is constant at $F = 10\%$ per hour, then using Eq. (12.12), the reliability after 3 h, with $F = 0.1/\text{hour}$, is

$$R(t) = e^{-3F} = e^{-0.3} = 0.741 = 74.1\%$$

The **Mean Time to Failure** (MTTF) is the average time an item may be expected to function before failure and applies to **nonrepairable items**. The MTTF is simply the average of all the times to failure.

$$\text{Mean Time To Failure} = \text{MTTF} = \frac{\text{Total time of operation, } T}{\text{Total number of failures during time } T} \qquad (12.13)$$

For example, if four items are tested and have lasted for exactly 3900 h, 4100 h, 4300 h and 3700 h, respectively, then the MTTF is

$$\text{MTTF} = \frac{3900 + 4100 + 4300 + 3700 \text{ h}}{4 \text{ failures}} = \frac{16000 \text{ h}}{4 \text{ failures}} = 4000 \text{ hours per failure}$$

The **Mean Time Between Failures** (MTBF) applies to **repairable items**. Since this refers to the mean time "between" failures, it is defined as

$$\text{Mean Time Between Failures} = \text{MTBF} = \frac{\sum(\text{Time in operation, } T_i)}{\text{Total number of failures during time } T = \sum T_i} \qquad (12.14)$$

Remember, MTBF is for repairable items and MTTF is for nonrepairable items. For example, if an item has failed 4 times over a period of exactly 8000 h, then its MTBF is 8000/4 = 2000 h. Notice that the MTTF is the same as the MTBF. Thus, for a constant failure rate,

$$F(t) = \frac{1}{\text{MTBF}} = \frac{1}{\text{MTTF}} \qquad (12.15)$$

For example, the item above fails, on average, once every 2000 h, so the probability of failure for each hour is obviously 1/2000. This result depends on the failure rate being constant. Eq. (12.15) can also be written the other way around as MTBF (or MTTF) = $1/F(t)$. For example, if the failure rate is 0.0002 per hour, then MTBF (or MTTF) = 1/0.0002 = 5000 h (to one significant figure).

Example 12.6

The equipment in a furniture manufacturing plant has an MTBF of 1000 h. What is the probability that the equipment will operate for a period of 500 h without failure?

Need: The probability that the equipment will operate for a period of 500 h without failure.

Know: The mean time between failures (MTBF) is 1000 h.

How: Use Eq. (12.15) and Eq. (12.12).

Solve: Assuming the constant failure rate (useful life) model, the failure rate

$$F(t) = 1/\text{MTBF} = 1/1000 = 0.001. \text{ Then, from Eq. (12.12),}$$

$R(500 \text{ h}) = e^{-500 \times 0.001} = e^{-0.5} = 0.6$, and the probability of the equipment operating for 500 h without failure is 60% (to one significant figure).

12.2.4 Calculating system MTTFs

The MTTF of any system with many independent components can be calculated if you have the MTTF of each component in the system. For example, you can calculate the MTTF of a computer if you have the MTTF of the CPU board, disk drives, memory chips, etc. However, computer MTTF analysis typically focuses on disk drive MTTFs because components with moving parts (such as disk drive actuators and motors) typically have significantly smaller MTTFs than nonmoving components (such as CPU boards and memory chips). Large server configurations containing many disk drives obviously have lower MTTFs.

A server's MTTF is calculated from the MTTFs of the independent components that make up the server as follows:

$$MTTF = \frac{1}{\frac{1}{N_1} + \frac{1}{N_2} + \frac{1}{N_3} + \frac{1}{N_4} + ... + \frac{1}{N_x}} \tag{12.16}$$

where N_1 is the MTTF of component #1, N_2 is the MTTF of component #2, and so forth, and x is the total number of components in the system.

If one component, such as a disk drive, has a significantly smaller MTTF than the other components in the system, its MTTF dominates the overall server MTTF. For example, if a computer has a CPU MTTF of 50,000 h and a single drive with an MTTF of 10,000 h then, using Eq. (12.16) and neglecting all of the other components, the computer has an MTTF of

$$Computer\ MTTF = \frac{1}{\frac{1}{CPU\ MTTF} + \frac{1}{Disk\ MTTF}} = \frac{1}{\frac{1}{50,000} + \frac{1}{10,000}} = 8300\ h$$

which is close to the disk drive's MTTF.

Even if all the system components have high MTTFs, the system's overall MTTF is reduced in direct proportion to the number of components in the system. For example, the MTTF of a computer subsystem consisting of two disk drives with identical 10,000 h MTTFs is

$$Disk\ Drive\ Subsystem\ MTTF = \frac{1}{\frac{1}{10,000} + \frac{1}{10,000}} = 5000\ h$$

which is exactly half the MTTF of each disk drive. Similarly, a 10-drive configuration MTTF is one-10th the MTTF of a single drive, or 10,000 h, and a 100-drive configuration is reduced to just 100 h. This is why component redundancy in a complex system is important.[7]

12.2.5 Design of experiments

DOE sounds like an odd topic for an industrial engineer, but it is a very important concept. Manufacturing processes are not carried out in a laboratory under controlled conditions. They are carried out under chaotic factory conditions where

[7] A system with "redundancy" has extra components that provide backup in case of a component failure. A common example of computer redundancy is extra (redundant) independent disk drives.

many uncontrolled variables can affect the quality of the manufactured part. So it is imperative that an industrial engineer understands the relationships between all the input process variables and all of the required design factors in the finished product. For example, in a metal cutting process, the cutting speed, feed rate, depth of cut, and coolant flow are all input variables. The surface finish and dimensional tolerances of the finished part specified by the design engineer are the required output factors.

How do industrial engineers test for the influence of input process variables on the results of a manufacturing process? How do they carry out meaningful experiments in a complex manufacturing environment to determine which of the many input variables has the most influence on the desired output or product quality? Manufacturing equipment is expensive and cannot be taken out of service for long periods of time for extensive experimentation. A common approach employed by many industrial engineers today is the "One-Variable-At-a-Time" (OVAT) approach, where just one input variable is changed at a time keeping all other input variables fixed. This approach depends upon guesswork and luck to be successful. "One Variable-At-a-Time" experiments often are unreliable, inefficient, time consuming (expensive) and may yield false operating conditions for the process.

DOE deals with designing, conducting, and statistically analyzing data to determine how the input variables influence the desired outcome. It allows changing more than one input variable at a time to speed up the experimentation, and it will indicate when variables interact with each other.

A very early DOE occurred in 1747. While serving as surgeon on HM Bark Salisbury, James Lind carried out a controlled experiment to develop a cure for scurvy.[8] He selected 12 men from the ship, all suffering from scurvy, and divided them into six pairs, giving each group different additions to their basic diet for a period of 2 weeks. The treatments were all remedies that had been proposed at one time or another. They were:

1) A quart of cider every day
2) 25 drops of sulfuric acid three times a day upon an empty stomach,
3) One half-pint of seawater every day
4) A mixture of garlic, mustard, and horseradish in a lump the size of a nutmeg
5) Two spoonfuls of vinegar three times a day
6) Two oranges and one lemon every day.

The men who had been given citrus fruits[9] (item 6 above) recovered dramatically within a week. The others experienced some improvement, but nothing comparable to the citrus fruits, which proved to be substantially superior to the other treatments.

---

[8] Patients with scurvy develop anemia, exhaustion, swelling in some parts of the body, and sometimes ulceration of the gums and loss of teeth. The name scurvy comes from the Latin *scorbutus*. This disease existed in ancient Greek and Egypt.

[9] Hence, British sailors were "limeys", a term that has expanded to mean ordinary Brits.

When analyzing a manufacturing process, experiments (or "tests") are often used to evaluate which process input variables have a significant effect on the process output. A well-designed experiment (or "test") can provide answers to the following:

- What are the key input variables that influence the process output?
- What are the input variable settings that would produce the best output results?
- What are the main variable interactions?
- What variable settings would produce the most stable process output?

The four steps of a DOE are as follows:

1) Design the experiment
2) Collection of data
3) Statistical analysis of the data, and
4) Conclusions and recommendations made as a result of the experiment

DOE is a powerful technique for understanding a process, studying the impact of variables or factors affecting the process, and for providing continuous quality improvement possibilities. DOE has proved effective in improving the process yield, process capability, process performance, and reducing process variability.

Example 12.7

An engineer is interested in increasing the yield of a plastic injection molding process. The process seems to be governed by two key process variables, injection pressure and temperature. The engineer decides to perform an OVAT experiment to study the effects of these two variables on the process yield. The first experiment was to keep the temperature T_1 constant and vary the pressure from p_1 to p_2. The results of these experiments are shown in Table 12.2 below. The next experiment was to keep the pressure p_1 constant and vary the temperature from T_1 to T_2. The results of these experiments are shown in Table 12.3 (assume all these values are exact).

The engineer concludes from these experiments that the maximum yield of the process will be 88% when $T_1 = 500°F$ and $p_2 = 2500$ psi. Is this a reasonable conclusion?

Table 12.2 The effect of varying pressure on process yield.

Test	Temperature	Pressure	% yield
1	$T_1 = 500°F$	$p_1 = 2000$ psi	76
2	$T_1 = 500°F$	$p_2 = 2500$ psi	88

Table 12.3 The effect of varying temperature on process yield.

Test	Temperature	Pressure	% yield
3	$T_1 = 500°F$	$p_1 = 2000$ psi	76
4	$T_2 = 800°F$	$p_1 = 2000$ psi	86

Need: An assessment of the engineer's conclusion.
Know: The data in Tables 12.2 and 12.3.

How: Using the principles of DOE.

Solve: The problem with this analysis is that the engineer does not know what the yield will be when $T_2 = 800°F$ and $p_2 = 2500$ psi. This condition did not occur in the tests and it could have had a higher yield than that at T_1 and p_2. The engineer was unable to study this combination or determine whether or not any interaction occurred between temperature and pressure.

Interaction between input variables exists when the effect of one variable on the output is different at different values of another variable. The difference in yields between the Test 1 and Test 2 provides an estimate of the effect of pressure. The effect of pressure on the yield was 12% (i.e., 76%−88%) when temperature was kept constant at $T_1 = 500°F$. But there is no guarantee that the effect of pressure will be the same when the temperature changes.

Similarly, the difference in yields between tests 3 and 4 provides an estimate of the effect of temperature. The effect of temperature was 10% (i.e., 76%−86%) on the yield when pressure was kept constant at $p_1 = 2000$ psi. It is reasonable to expect that we would not get the same effect of temperature when the pressure changes. The OVAT approach to experimentation can be misleading and may lead to unsatisfactory conclusions in real-life situations.

In order to obtain a reliable and predictable estimate of the effects of input variables, it is important that input variables are simultaneously varied over their levels. In the previous example, the engineer should have varied the levels of temperature and pressure simultaneously to obtain reliable estimates of the effects of temperature and pressure. To carry out a complete DOE analysis of the plastic molding process in the previous example, the engineer would still perform four experiments as shown in Table 12.4.

Table 12.4 Complete DOE analysis for Example 12.7.

Test	Temperature (°F)	Pressure (psi)	Yield (%)
1	500	2000	76
2	800	2000	86
3	500	2500	88
4	800	2500	??

In general, the number of experiments (tests) required for a DOE with N variables tested at M levels of each variable is

$$\textbf{Number of DOE Experiments} = N^M \tag{12.17}$$

So, if you have 2 variables and want to test them over 2 levels of each variable you will need $2^2 = 4$ experiments. Similarly, if you have three variables to be tested over two levels you need $3^2 = 9$ experiments.

12.3 Methods engineering

Methods engineering is concerned with the selection, development, and documentation of the methods by which work is to be done. It includes the analysis of input and output conditions, choosing the processes to be used, operations and work-flow analyses, workplace design, tool and equipment selection, human motion analysis and standardization, and the establishment of work time standards.

Methods engineering is the design of the productive process in which a person is involved. The task of the methods engineer is to decide where humans will be utilized in the process of converting raw materials to finished products, and how workers can most effectively perform their tasks. The terms "operation analysis", "work design and simplification", and "methods engineering" are frequently used interchangeably.

A methods engineering procedure utilizes five separate stages to ensure that an existing process is fully analyzed before it is changed or a new process is introduced. The five stages are as follows:

1) Project selection
2) Data acquisition
3) Analysis
4) Development
5) Implementation

Stage one (project selection) is the point at which a methods engineer will be tasked with improving a process. In this stage, a project is identified that either requires an efficiency enhancement or is a new production process where reliability, accuracy, and efficiency are of particular importance.

Stages two and three (data acquisition and analysis) are concerned with collecting and analyzing data from an existing activity or a production line. These data can include output records, detailed design drawings, and marketplace demand or performance records. Data that are collected during the second stage are analyzed in stage three to establish the optimum man-to-machine ratio and production line outputs. This third stage can also suggest improvements in working conditions, production floor layout, and materials handling.

Developing the key improvements identified during the data analysis exercise is carried out in **stage four**. It is in this stage that man-to-machine ratios and the number of operators allocated to a machine or the number of machines to a single operator are established and developed into a workable process. Once final improvements have been established, the economic benefit to the company must be detailed before being presented to management for **stage five** implementation of the method.

12.4 Simulation analysis and operation research

Simulation analysis is an analytical tool that can be used to evaluate the performance of an existing or proposed system under different operating conditions of interest. Simulation analysis can be used before an existing system is modified or a new system is constructed to reduce the chances of failure, eliminate unforeseen problems, and optimize performance. It can be used to answer questions like: What is the best design for a new communications network? What are the material resource requirements? How will a highway perform when the traffic increases by 50%? What will be the impact of a failure?

The United States Department of Homeland Security manages the National Infrastructure Simulation and Analysis Center[10] (NISAC). NISAC conducts computer modeling, simulation, and analysis of the nation's 18 critical infrastructure areas and focuses on the challenges posed by interdependencies and the consequences of disruption.

Simulation analysis is the imitation of the operation of a real-world process or system over time. The act of simulating a system requires that a model of the system be developed. This model represents the key characteristics or behaviors/functions of the selected physical or abstract system or process. The model represents the system itself, whereas simulation analysis represents the operation of the system over time.

Simulation is used in many contexts, such as simulation of technology for performance optimization, safety engineering, testing, training, education, and video games. Often, computer experiments are used to study simulation models. Simulation is also used with scientific modeling of natural systems or human systems to gain insight into their functioning. Simulation can be used to show the eventual real effects of alternative conditions and courses of action. Simulation is also used when the real system cannot be engaged, because it may not be accessible, or it may be dangerous or unacceptable to engage, or it is being designed but not yet built, or it may simply not exist.

Physical simulation refers to simulation in which physical objects are used to model the real system. These physical objects are often chosen because they are smaller or cheaper than the objects in the actual system. These models are instrumented and tested under variety of conditions typical of those to which the actual system may be exposed.

Interactive simulation is a simulation in which a physical simulation includes human operators, such as in a flight or driving simulator and in computer games.

Computer simulation is an attempt to model a real or hypothetical situation on a computer. By changing variables in the simulation, the computer program forecasts results about the behavior of the system. Computer simulation has become a common engineering modeling tool. It is now part of most engineering computer-aided drafting and design (CAD) programs (e.g., ProEngineer, SolidWorks, ANSYS, etc.) and produces stress and deformation information for various loading conditions.

---

[10] NISAC was incorporated by the USA Patriot Act of 2001 into the Department of Homeland Security upon its inception in March 2003. For more information, see http://www.dhs.gov/about-national-infrastructure-simulation-and-analysis-center

As a simple simulation analysis illustration, the cable clamp shown in Fig. 12.2 is first drawn in a CAD program, then a finite element mesh is developed by the program, and finally the stress and deformation of the clamp is computed for various loads. The deformation of the clamp can also be animated to show how the stresses develop for various loading conditions.

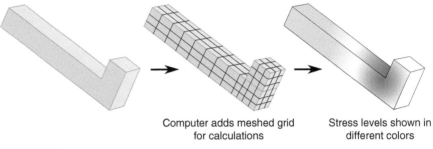

Computer adds meshed grid   Stress levels shown in
for calculations   different colors

FIGURE 12.2

Simulation analysis of a cable clamp.

Operations Research (OR) is a strange title for an industrial engineering topic that does not involve any actual research. The name comes from the military during the Second World War. It was used in England to describe the process they used to research (military) operations. After the war, the process moved into mainstream operational planning and the name stuck. Industrial engineers today use OR methods to produce optimal solutions to complex human–technology interaction problems. These problems often require dealing with a large number of variables and constraints.

OR can be applied to areas such as

- Scheduling planes and crews, pricing tickets, passenger reservations
- Automobile traffic routing and planning
- Financial credit scoring, marketing, banking operations
- Deployment of emergency medical services
- Material transportation networks
- Energy and management

OR also deals with decisions involved in optimizing the allocation of resources such as material, skilled workers, machines, money, and time. Rather than using trial and error methods on the system itself, a mathematical model of the system is developed and then manipulated to find optimum operating conditions. Many mathematical methods and tools have evolved to model and solve complex problems in this field, including linear programming, game theory, queuing theory, network analysis, decision analysis, multifunction analysis, and so forth.

Operational research mathematical models can be either ***deterministic*** or ***stochastic***. A deterministic model contains no random variables. The output is "determined" once the set of input quantities and relationships in the model have been specified. Stochastic models, on the other hand, have one or more random input variables. For this kind of model, probability and statistics are important to measure the prediction uncertainty.

Linear programming is a *deterministic* method to achieve the optimum outcome (such as maximum profit or lowest cost) in a given mathematical model for a set of constraints. It is one of the most widely used tools in OR today. It has been used successfully as a decision-making aid in almost all industries, including financial and service organizations. The "linear" part of the name refers to the fact that the optimization output (i.e., maximization or minimization) is assumed to be a linear function of the input variables. That is, it must involve only the first powers of the input variables with no cross products.

If an OR system has only two input variables, a simple graphical method can be used to solve its linear programming model. If there are more than two input variables, the problem needs to be solved using more advanced analytical methods. The graphical method is illustrated in the following example.

Example 12.8

You are going to manufacture a new product. The product is a Widget, and to make it you need to use a lathe, a milling machine, and a drill press (See Chapter 13 for examples of these machines). The Widget contains only two parts, A and B. Since your machine shop is already busy, you survey the current work load and determine that the machine tools you need are available only for the times shown below.

Machine tool	Availability
Lathe	440 min per day
Milling machine	336 min per day
Drill press	280 min per day

You also determine the time needed on each machine tool to produce parts A and B. This is shown below.

Part	Lathe	Milling	Drilling
A	25 min	8 min	10 min
B	8 min	12 min	8 min

Finally, the profit that you can make on a Widget is based on the estimated unit profit of each of its parts. The profit you can make on each part A is $25 per unit, and the profit on each part B is $10 per unit, so the total profit for your Widget is $25 \times A + $10 \times B$. What is the optimum daily production for parts A and B that will maximize your profit on the Widget? Assume all these values are exact.

Need: The optimum daily production of parts A and B that will maximize profit.

Know: The machine tool availability and machining times required for each part.

How: Using linear programming methods.

Solve: We can write the following set of operational equations for making the Widget.

$$25A_{lathe} + 8B_{lathe} = 440 \text{ min of turning on the lathe}$$

$$8A_{milling} + 12B_{milling} = 336 \text{ min of milling}$$

$$10A_{drilling} + 8B_{drilling} = 280 \text{ min of drilling}$$

Assuming a linear relation between machine tool operations, we can graph these equations with the number of A parts made per day on the horizontal axis and the number of B parts made per day on the vertical axis. The A and B axis intercepts can be computed as follows. When B = 0, then

$A_{lathe} = 440/25 = 17.6$

$A_{milling} = 336/8 = 42$

$A_{drilling} = 280/10 = 28$

and when A = 0,

$B_{lathe} = 440/8 = 55$

$B_{milling} = 336/12 = 28$

$B_{drilling} = 280/8 = 35$

These intercepts are used to plot the straight lines on the A–B graph shown below using an Excel spreadsheet.

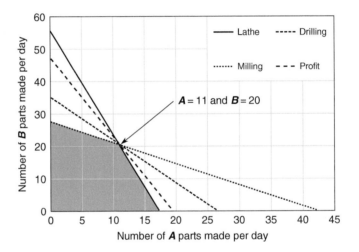

The shaded area in this figure is called the "feasible region" (or solution space) of the problem. Any point inside this region is called a feasible solution and provides values of A and B that satisfy all the requirements. In this case, the "maximum" feasible solution occurs at the intersection of the machining operation lines on the graph, or when 11 units of A and 20 units of B are produced per day. Then, since the total profit for your Widget is $25 × A + $10 × B, the maximum profit is

Maximum Profit = $25 ×11 + $10 ×20 = $475 per day.

12.5 Ergonomics

Industrial engineers also design the work area in which humans interact with machines. This is known as ergonomic engineering. There are two general divisions of ergonomics:

- **Occupational ergonomics** (human factors) is concerned with the strength capabilities of the human body in performing manual work (such as lifting, turning, and stretching), and with the environmental effects of temperature, humidity, vibration, and so forth on the human worker. This is done to reduce occupational injuries and increase work productivity.
- **Cognitive ergonomics** (task analysis) is concerned with understanding the behavior of humans as they interact with machines. This information is used to design machine display interfaces and controls to support operator needs, to limit their workload, and to promote awareness of the operation.

12.5.1 Occupational ergonomics

Occupational ergonomics uses information about human ability in the design of tools, machines, jobs, and work environment for productive, safe, and effective human use. Occupational ergonomics provides methods for optimizing tasks in the workplace. The design of workplaces requires the understanding of the ergonomic principles of posture and movement that produce a safe, healthy, and comfortable work environment. Worker posture and movement is dictated by the task and the body's muscles, ligaments, and joints needed in carrying out the task. Poor posture and movement can produce stress that results in damage to the neck, back, shoulder, wrist, and other work-related injuries.

One of the most common work-related injuries is musculoskeletal disorder. These disorders result in persistent pain, loss of functional capacity, and work disability. For example, a poorly designed tool can adversely impact overall worker performance, create injuries, and produce human task error. Industrial engineers evaluate these tools to determine potential sources of injury and attempt to improve them to fit the needs and workflow of workers.

For example, a worker should be able to operate the electric drill shown in Fig. 12.3 with one hand. To do this safely and comfortably, the weight of the tool should not exceed about two pounds (1 kg) and the center of gravity should be near the center of the gripping hand. The tools should be easy to hold in any position needed. Drills that are front-heavy can produce stress in the wrist and forearm.

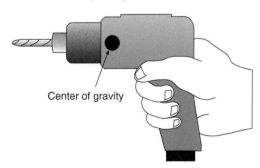

Center of gravity

FIGURE 12.3

The ergonomics of a hand drill.

Example 12.9

For each of the consumer items below identify three ways in which they could be ergonomically improved?
a) Cell phone
b) Alarm clock
c) The electric drill in Fig. 12.3

> **Need:** Representative ergonomic improvements for a cell phone, alarm clock, and electric drill.
> **Know:** The basics of ergonomics.
> **How:** Apply the basics or ergonomics.
> **Solve:** a) Simplify the programming to make it easier to use all the phone's functions.
>> b) Prevent the user from turning off the alarm when he or she tries to activate the snooze option.
>> c) Move the heavy motor back over the center of the hand.
>> (Note: these are not the only improvements you can think of.)

12.5.2 Cognitive ergonomics

A **time and motion study** for repetitive tasks is a method for establishing worker productivity standards in which

- a complex task is broken into small, simple steps
- the sequence of movements taken by the worker in performing those steps is carefully observed to eliminate wasteful motions
- the precise time taken for each movement is measured

A time and motion study is a method to determine the best way to complete a certain task. The simplest way to conduct it is to break the task into a series of individual steps and record the motion and time required for each step. This should be done repeatedly in order to refine the process of completing a task. You can then determine the "normal" or average time for a task by recording how much time is devoted to each part of the task. This can be done as follows:

1) List the steps needed to perform the task
2) Discuss them with the worker
3) Measure each step with a stopwatch as the worker performs the task
4) Repeat the complete process at least 10 times
5) Compute the mean and standard deviation of each step and of the complete task
6) Be aware of worker disruptions and learning curves

From these measurements, production and delivery times and prices can be computed and worker incentive schemes can be devised.

12.6 Material handling

Material Handling is the field concerned with solving the problems of movement, storage, control, and protection of materials throughout the processes of cleaning, preparation, manufacturing, distribution, consumption and disposal of all related goods and their packaging. The focus of material handling is on the methods, mechanical equipment, systems, and controls used to achieve these functions.

There are many ways by which material handling has been defined, but the simplest definition is that material handling means providing the *right* amount of the *right* material at the *right* place at the *right* time for the *right* cost. The **primary** objective of a material handling system is to reduce the unit cost of production. Other objectives are to:

1) Reduce manufacturing cycle time
2) Reduce delays and damage
3) Promote safety and improve working conditions
4) Maintain or improve product quality
5) Promote productivity by having material move short distances, in a straight line, and use the force of gravity whenever possible
6) Encourage increased use of facilities by purchasing adaptable equipment and developing a preventive maintenance program
7) Control inventory

A material handling system often involves warehouses, conveyors, part sorters, feeders, and manipulators. Material handling systems range from simple pallet rack and shelving projects to complex conveyors that use computer controlled storage and retrieval systems with automatic guided vehicles.

A material handling system is a fundamental part of a **flexible manufacturing system** because it connects the different manufacturing processes used to create a finished product. Due to the automated nature of the whole production process, a material handling system must respond in concert with all the requirements of the manufacturing processes. It is something that goes on in every manufacturing plant all the time. It applies to the movement of raw materials, parts in process, finished goods, packing materials, and disposal of scraps. In general, hundreds and thousands of tons of materials are handled daily requiring a large amount power.

The cost of material handling contributes significantly to the total cost of manufacturing. The importance of reducing material handling is greater in those industries where the ratio of material handling cost to processing cost is large. Today material handling is rightly considered as one of the most profitable areas for reducing manufacturing costs. A properly designed and integrated material handling system provides tremendous cost saving and customer service improvement.

A properly designed material handling system will improve customer service, reduce inventory, shorten delivery time, and lower overall handling costs in manufacturing, distribution, and transportation. Some items to consider in designing a material handling system are as follows:

✔ **Performance Objectives**—Define the performance objectives and functional requirements of the system at the beginning of the design process.

✔ **Standardization**—All material handling methods, equipment, controls, and software should be standardized and able to perform a range of tasks in a variety of operating conditions.

✔ **Ergonomics**—The environment should support the abilities of a human worker, reduce repetitive and strenuous manual labor, and promote worker safety.

✔ **Space**—Develop the efficient use of space within a facility by keeping work areas organized and maximizing storage density.

✔ **Automation**—Automated material handling technologies should be used when possible to improve material handling efficiency.

The total cost of material handling per unit is the sum of the following:

1) Cost of material handling equipment—both fixed cost and operating cost calculated as the cost of equipment divided by the number of units of material handled over the working life of the equipment.
2) Cost of labor—both direct and indirect.
3) Cost of maintenance of equipment, damages, lost orders, and expediting expenses.

Example 12.10

Aluminum castings at a small engine manufacturing plant are processed through several different machining operations. Initially, these operations took 3 h per casting. An engineer subsequently installed a unified machining cell that carried out the required drilling, milling, boring, and polishing operations on a casting in just minutes. The plant manager was pleased that the cycle time was reduced from 3 h to a few minutes.

Then the plant manager noticed that the engineer specified the batch size for the new machining cell to be 10,000 castings. Since the castings were heavy, the material handling system would need to deliver 10,000 casting batches to the cell using a fork lift. The cell would then automatically machine multiple castings at the rate of 250 per hour. Once the batch of 10,000 was completed, a fork lift would be called back to deliver a new batch of 10,000 castings and transfer the finished batch of 10,000 machined castings to the assembly line. What problems did this new machining cell create? Assume all these values are exact.

Need: Problems produced by the new machining center.

Know: Cell batch size = 10,000 castings with 250 castings machined per hour.

How: Material handling basics.

Solve: Because the process batch size of 10,000 was incorrectly assumed to be equal to the overall material handling batch size, this created a large backup of finished castings at the assembly line. This required extra space at the assembly line and produced material handling problems at that point. The parts backup at the assembly line would be reduced if the machining cell material handling batch size was reduced from 10,000 castings to a value compatible with the assembly line capacity.

Materials delivered in the wrong sequence will directly result in increased backup (called "work in progress"). Product design and process planning should be done before the layout of the factory and its material handling system. In assembly lines, the sequence in which the parts arrive at an assembly station should match the sequence in which they can be assembled. This is very important when pallets are loaded with a particular part and moved from one station to another with a fork lift.

A **conveyor** is a common material handling system that moves materials from one location to another. Conveyors are able to safely transport materials from one processing station to another and they can be installed almost anywhere. They are also much safer than using a forklift or other machine to move materials, and they can move loads of all shapes, sizes, and weights.

The layout of a conveyor system is determined by the relative positions of processing stations and existing floor space. The width of the conveyor is determined by the size of the part carriers. Usually, parts are put in bins or boxes (called "part carriers") that travel on the conveyor. These carriers are usually rectangular and their size is determined by the size of each part and the number of parts put in each carrier. Given the size of the carrier, it is easy to determine the width of the conveyor. The speed of a conveyer is dictated by its loading and unloading speeds as illustrated in the following example.

Example 12.11

A conveyor belt has boxes attached to it to carry parts to processing stations. The conveyor moves at speed V. The required loading speed of the conveyer is P parts per minute, and the conveyor length used by each part box carrier is L. There are N unloading stations that have an average unloading time of M parts per minute per station.

a) How should the conveyer speed be set relative to the part arrival rate?

b) What is the relation between the rate at which parts are loaded on the conveyer and the rate at which they are unloaded?

  Need: The conveyer speed and the relation between parts loaded and parts unloaded.

  Know: Basic mechanics of motion (physics).

  How: Use the concepts of material handling.

  Solve:

  a) Parts should arrive at the loading station somewhat slower than they can be loaded (to prevent material back-up), so the conveyor speed should be set so that the carrier arrival rate at the loading station (V/L) is greater than the part arrival rate (P).

  b) In order for the unloading stations to properly unload the conveyer, the number of parts unloaded per minute should be somewhat greater than the parts loaded per minute on the conveyer, or $N \times M > P$.

  Note that if every third carrier is empty ($PL/V = 2/3$), then an unloading station with unload times between L/V and $1.5L/V$ will have sufficient capacity. However, it will not be able to unload every second consecutive part carrier that arrives so there will be filled carriers continuously circulating on the conveyor.

Summary

Industrial engineering is the field of engineering concerned with the design, analysis, and operation of systems that range from a single piece of equipment to large businesses. Industrial engineers serve in areas that range from the production of raw materials to manufacturing to the service industry. Industrial engineers determine the most effective way to utilize people, machines, materials, information, and energy to make a product or provide a service.

An important area of industrial engineering is manufacturing. This topic is covered in a separate chapter in this textbook. In this chapter we covered five other important areas of industrial engineering: **Quality Control; Methods Engineering; Simulation Analysis** and **Operation Research; Ergonomics;** and **Material Handling**.

Exercises

1) The median household income in 2012 for the following six nations is shown in the table below.

Nation	Luxembourg	United States	Singapore	Bahrain	Venezuela	Liberia
Median income	$52,493	$43,585	$32,360	$24,633	$11,239	$781

   a) Find the median household income of these six nations.
   b) Find the mean household income of these six nations.
   c) Find the standard deviation of the household income of these six nations

2) An engineer collects data that consists of the following four observations: 1, 3, 5, and 7. What is the (a) variance, (b) standard deviation, and (c) standard error of this data?

3) Every year, 1000 engineers compete in a sports car road race. The mean (average) finishing time is 55 min, with a standard deviation of 10 min. Justin and Cindy completed the race in 61 and 51 min, respectively. Barry and Lisa had finishing times with Z-scores of −0.3 and 0.7, respectively. Since the Z-score is the number of standard deviations *above* the mean, then the lowest Z-score corresponds to the fastest car. List the drivers, fastest to the slowest, based on their Z-scores.

4) A final examination in an engineering class has a mean (average) score of 75 and a standard deviation of 10 points. If Laura's Z-score is 1.20, what was her score on the test?

5) We want to know the chances of choosing a particular student type based on four demographic characteristics. The demographics of the students are illustrated in the table below.

	Caucasian (c)	Minority (m)	Total
Male	55	40	95
Female	65	40	105
Total	120	80	200

   a) What is the probability of choosing a Caucasian student?
   b) What is the probability of choosing a minority student?
   c) What is the probability of choosing a Caucasian male?
   d) What is the probability of choosing a Caucasian female?

6) Suppose the number of parts that can be machined on a lathe in 1 day ranges from 0 (the lathe is not operating) to 250 (the best lathe operator). By collecting data, the following probabilities for machining parts in the first hour were determined. Assume all these values are exact.

Number of parts machined in 1 day	Probability
0	0.10
1–50	0.05
51–100	0.10
101–150	0.30
151–200	0.43
201–250	0.02

This table tells us that the probability of machining between 1 and 50 parts in 1 day is 5.0%, and the probability of machining between 50 and 100 parts in 1 day is 10%. What is the probability of machining "at least" 101 parts in 1 day?
a) What is the probability of machining at least 101 parts in 1 day?
b) What is the probability of not machining at least 101 parts in 1 day?

7) A manufacturer wishes to order material from a wholesale supplier. The supplier has three phone lines with three different numbers that operate *independently*. The probability of phone one being busy is 0.80, the probability of phone 2 being busy is 0.80, and the probability of phone 3 being busy is 0.80. The manufacturer calls to order a large supply of material. What is the probability that the phone lines will be busy?

8) Suppose 10 devices are tested for 500 h. During this test two devices fail.
a) What is the MTBF?
b) What is the MTTF?
c) What is the failure rate?
d) What is the probability that any one device will be operational when $t = $ MTBF?

9) A product has a constant failure rate of 0.2% per 1000 h of operation.
a) What is its MTTF?
b) What is the probability of it successfully operating for 10,000 h?

10) Engineers carry out a reliability test to develop a warranty policy. The test finds that the failure rate is approximately constant with $F(t) = 1/8750 = 1.14 \times 10^{-4}$ failures per hour.
a) Determine the reliability function $R(t)$.
b) Determine the mean time to failure (MTTF).

11) A new automotive valve spring design is tested in continuous use and is found to have a constant failure rate of $F = 0.002$ failures per hour.
a) Determine the probability of failure $(1 - R(t))$ within the first hour of use.
b) Determine the probability of failure within the first 100 h.
c) Determine the probability of failure within the first 1000 h.

12) An electronic circuit board has the failure rate described by: $F(t) = 0.05 \times (t^{-1/4})$ per year (with "t" in years).
a) Determine the reliability function $R(t)$ and the failure probability distribution function
b) Determine the reliability for 1 year of use.
c) Determine the fraction of failed circuits in 3 years.

13) A cell phone manufacturer tested 10,000 phones in a reliability evaluation program. Each phone was turned on-and-off 160 times each day to mimic extreme phone usage for a week. Based on a failure-to-perform criterion, the following failure data were obtained for the first 10 days of test:

Day	1	2	3	4	5	6	7	8	9	10
Failures	180	120	100	70	60	50	40	30	0	10

For the first day, the failure rate is $F(1) = 180/10{,}000 = 0.018$ per day;

For the second day, the failure rate is $F(2) = 120/(10{,}000 - 180) = 120/9820 = 0.0122$ per day.

For the third day, the failure rate is $F(3) = 100/(10{,}000 - 180 - 120) = 100/9700 = 0.010$ per day, and so on. Assume all these values are exact.
   a) Plot $F(t)$ from $t = 0$ to $t = 10$ days.
   b) Fit these data with a curve of the form $F(t) = at^{-b}$ and determine the values of a and b.
   c) Determine the reliability function $R(t)$
   d) Determine the mean-time-to-failure, MTTF.

14) If you have an MTBF of 160 days, it does not mean you can expect an individual device to operate for 160 days before failing. MTBF is a statistical measure, and cannot predict anything for a single unit. However, if you have 1000 devices with this same MTBF operating continuously in a factory, (a) how often would you expect one to fail, and (b) how long would it take for 20 failures to occur? Assume all these values are exact.

15) Using a Design of Experiments process, how many experiments do you need to perform if you have:
   a) two variables and want to test them over two levels of each variable
   b) two variables and want to test them over three levels of each variable
   c) three variables and want to test them over three levels of each variable
   d) five variables and want to test them over four levels of each variable

16) List the five stages in methods engineering.

17) Determine the maximum feasible profit in Example 12.8 if the drilling time for part B was 10 min instead of 8 min. Assume all the remaining conditions are unchanged.

18) Determine the maximum feasible profit in Example 12.8 if the lathe time for part A was reduced from 25 to 10 min. Assume all the remaining conditions are unchanged.

19) Freddy works on an assembly line and uses a handheld pneumatic impact wrench. The assembly line makes up to 1400 products a day and it takes about 3 s for Freddy to tighten each of seven different components. Freddy often has to use poor posture to attach some of the parts. After a few weeks, Freddy found that he was leaving work with shoulder and neck pain. Using ergonomic principles assess Freddy's working conditions and suggest at least four workplace modifications that will reduce Freddy's problems.

20) Janet is a sales engineer at your company. Much of her work involves using a telephone and a computer to make appointments, call customers, and respond to emails. Janet would often hold the telephone between her shoulder and ear while talking on the phone and typing on the computer. Also, Janet's computer screen was difficult to read because of glare and reflections from light through the window in her office. After working for 8 months, Janet found she was leaving work with an aching shoulder, sore eyes, and a headache. Using ergonomic principles, assess Janet's working conditions and suggest at least three workplace modifications that will reduce Janet's problems.

21) A worker's hand was seriously injured while trying to clear a reoccurring blockage in a computer-controlled machining station. Access to the machining station is through a door in its enclosure and all machining operations are stopped when this door is opened. The worker had obtained an override key so he could open the door and enter the enclosure without stopping production. Using ergonomic principles, identify at least three steps that can be taken to help prevent a recurrence.

22) A company uses reusable wooden boxes 36″ square and 48″ long to ship their product. Recently, the cost of purchasing a new shipping box has increased from $32.50 to $81.00, and the maintenance cost for existing boxes has increased from $15 per year to $21 per year. Also, extra truck runs and outside trucking services are required to return the wooden boxes. As a material handling engineer, how would you improve this system by suggestion new methods for shipping the product?

23) The storage area at your company is presently filled to capacity with 20,000 cases of products. Your company recently increased its manufacturing capability by 100% and the finished goods inventory storage area is expected to increase by the same amount. The present storeroom is 250 feet by 375 feet with a 30 foot ceiling height. The product is packaged in cases 15 inches long, 24 inches high, and 32 inches wide and you cannot stack these cases more than 6 feet high without damage. As a material handling engineer, suggest methods for increasing the storage area to accommodate the contemplated increase in finished goods inventory.

24) One of your first assignments when you graduate with an engineering degree and accept a well-paying position at a prominent company is working in the product testing laboratory. Your manager wants you to do a one-variable-at-a-time (OVAT) experiment on the products reliability. You know that a design of experiments (DOE) testing method would save the company time and money and produce more accurate results. Your manager has never heard of DOE and insists you proceed with a long and expensive OVAT experiment. What do you do? (Use the Engineering Ethics Matrix.)
    a) Go over your manager's head and talk to his boss.
    b) Ignore your manager's orders and carry out a DOE without his knowledge.
    c) Ask a colleague for help in convincing your manager to your point of view.
    d) Confront your manager in writing (with a cc to his/her boss) saying that he/she does not understand modern engineering testing methods and should listen to you.

25) You now have an entry level position as an engineer at a hand tool manufacturing company. You are put on a team of engineers that is about to release the design of a new battery powered screw driver to manufacturing. By handling the prototype, you notice that the screw driver has a heavy battery and that the direction that the shaft turns is activated by the user twisting his/her wrist at a severe angle. You point this out at a team meeting and suggest that the design may have serious ergonomic problems that could result in user wrist injury during long term use. The team quietly listens to your comment and replies that profit, not user ergonomics, is their goal. What do you do? (Use the Engineering Ethics Matrix.)
    a) Reassert your position by suggesting that customer satisfaction will also lead to profits.
    b) Contact the company lawyer as a whistle blower about a possible dangerous product.
    c) Talk to the team leader in private and suggest an ergonomics expert be added to the team.
    d) Send an anonymous email to the company president warning him/her about possible law suits.

Manufacturing engineering

13

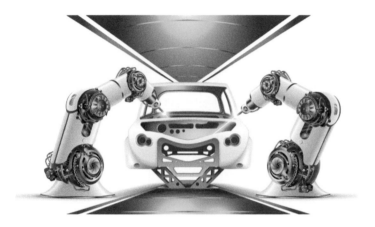

Source: Andrey Suslov/iStockphoto.com

13.1 Introduction

Virtually everything we use at home, at work, and at play was manufactured. Manufactured goods are everywhere: aircraft, bicycles, electronics, coat hangers, automobiles, refrigerators, toys, clothing, cans, bottles, cell phones, and so on. Even lead pencils and paper clips are successes of engineering manufacturing processes.

13.2 What is manufacturing?

You probably have a general idea about manufacturing processes, but let us look at the word itself. The word *manufacture* derives from two Latin words: *manu* (meaning "by hand") and *factum* (meaning "made"). We generally think of manufacturing taking place in a factory.

Exploring Engineering. https://doi.org/10.1016/B978-0-443-13541-5.00023-4

Manufacturing is the process of converting (either by manual labor or by machines) raw materials into finished products, especially in large quantities.

Manufacturing covers a wide variety of processes and products. It involves the production of many types of goods that range from food to microcircuits to airplanes to healthcare equipment. The number and complexity of the processes involved in the production of these items varies. Some products are basic, such as flour, cheese, leather, and iron. Some are merely altered, such as metal ingots, Portland cement, and chemicals like gasoline. Others are moderately changed, like wire rods, metal pipes, glass bottles, soap, cloth, and paper, while things like vehicles, computers, medicines, and cell phones are elaborately transformed.

Manufacturing began more than 10,000 years ago when nomadic people settled in one geographic area and began to plant and grow food and domesticate animals. Skilled artisans did the manufacturing and later the guild system (the predecessors of modern trade unions) evolved to protected their privileges and trade secrets.

Today's manufacturing engineers apply scientific principles to the production of goods. Manufacturing engineers design the processes and systems to make products with the required functionality, high quality, at the lowest price, and in ways that are environmentally friendly. Fig. 13.1 illustrates a typical manufacturing process, from artisans to their market.

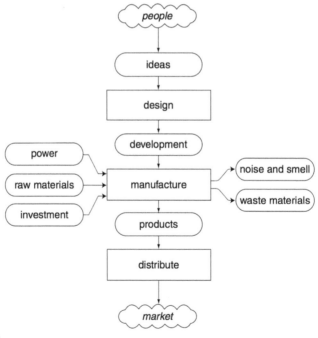

FIGURE 13.1

A typical manufacturing process.

In this chapter, we discuss several machining processes, including **drilling**, **lathe work**, **milling**, **welding**, **extrusion**, **pultrusion**, **blow molding**, and **thermoforming**. Modern manufacturing methods include **Just-In-Time** (JIT) inventory control, **flexible**, **lean**, and **life cycle** manufacturing. A key concept in manufacturing is the recognition of **variability** and **tolerance**, which we describe by introducing statistical methods that include a relatively recent paradigm called **Six Sigma**. An important variable in all of this is the **standard deviation** which is central to **statistical** methods.

13.3 Early manufacturing

Before the Industrial Revolution, the production of goods for sale occurred in homes on farms and in villages to provide additional income. This "cottage" industry provided extra income to a farming family that would take in weaving, sewing, dying, pins, and so forth that was then sold to a retailer.

In 1776, the Scotsman Adam Smith[1] (1723−90) published his influential book *An Inquiry into the Nature and Causes of the Wealth of Nations*. This was one of the first studies on the production techniques in the colonial era. In his book, he discussed how the division of labor could significantly increase production. One example he used was the making of pins. One worker could probably make only 20 pins per day. But if 10 people divided up the 18 steps required to make a pin, they could make 4800 pins per day (240 times as many).

Eli Whitney (1765−825) is most famous for his invention of the cotton gin.[2] However, this was a minor accomplishment compared to his concept of interchangeable parts. He developed the concept of interchangeable parts about 1799 when he promised the U.S. Army he could manufacture 10,000 muskets for only $13.40 each. Whitney is credited for inventing the *American system of manufacturing*—the combination of power machinery, interchangeable parts, and the division of labor that would become one of the foundations of the Industrial Revolution.

---

[1] Adam Smith is known for understanding how self-interest and competition can lead to economic well-being and prosperity without government intervention. The French expression *laissez-faire* (roughly translatable as "to leave alone") became closely associated with his name.
[2] The term *gin* is a contraction of the word "engine," since Whitney's device was a creative machine (engine) for removing seeds from hand-picked cotton.

13.4 **Industrial revolution**

The Industrial Revolution[3] began in Great Britain during the second half of the 18th century and spread through Europe and the United States in the 19th century. In the 20th century, industrialization extended to Asia and to the Pacific Rim. While modern manufacturing processes and economic growth continue to spread throughout the world, many people have yet to experience the benefits of improved economic and social conditions typical of an industrial revolution.

The key technology that brought about the Industrial Revolution was the invention and subsequent improvement of the steam engine. First invented by the English blacksmith Thomas Newcomen in the late 17th century, it was used to pump water out of flooded coal mines. Increased coal production subsequently led to the smelting of iron, which in turn produced a vast array of new metals and associated technologies.

The Industrial Revolution produced sweeping social changes, including the movement of people from farms to cities, the availability of a variety of material goods, and new ways of doing business. Goods that had been made in the home began to be manufactured at lower cost in a factory in a city. The resulting economic development combined with superior military technology made the nations of Europe and the United States the most powerful in the 18th and 19th century world.

However, as products moved from one process to the next in factories, several unresolved questions arose: What should happen between the processes? How should sequential processes be carried out? What was the best way for workers to carry out their tasks?

In the 1890s, Frederick W. Taylor (1856−1915) studied the process of how factory work was carried out. Taylor felt that industrial management was poor and that the best results would come from a partnership between management and workers. He felt that, by working together, there would be no need for trade unions. Taylor's "scientific" management process consisted of four principles:

1. Replace rule-of-thumb methods with techniques based on a scientific study of the tasks.
2. Train workers rather than have them train themselves.
3. Provide detailed instruction and supervision for each worker.
4. Managers apply scientific principles to planning the work and how the workers perform the tasks.

---

[3] The Industrial Revolution is called a *revolution* because it significantly and rapidly changed society. There has been only one other group of changes as significant as the Industrial Revolution, the stone age *Neolithic Revolution*, in which people moved from hunting and gathering to agriculture and the domestication of animals. This led to the rise of permanent settlements and, eventually, urban civilizations.

Taylor's ideas became known as **scientific management**. The concept of applying science to management was useful, but Taylor unfortunately ignored the science of human behavior. It seems that he had a peculiar attitude toward workers. He felt that they were not very bright and generally incapable of understanding what they were doing, even for simple tasks.

In the early 1900s, Frank Gilbreth (1868–1924) developed the concept of a **time and motion study** and invented **process charting** to focus attention on the entire work process, including nonvalue-added steps that occur between production processes. He later collaborated with his future wife, Lillian Moller Gilbreth[4] (1878–1972), to study the work habits of factory and clerical workers in several industries to find ways to increase output and make their jobs easier. She brought psychology into the mix by studying the motivations of workers and how their attitudes affect their work. She and her husband Frank were genuine pioneers in the field of industrial engineering.

Around 1910, Henry Ford developed the first efficient manufacturing plan. As a farm boy in Michigan, he saw how animals were "disassembled" by a line of workers in a slaughterhouse. By simply reversing this process, he created an "assembly" line process for manufacturing products. Ford took all the elements in his manufacturing system—people, machines, tooling, and supplies—and arranged them in a long continuous line to efficiently assemble (manufacture) his early Model T automobiles. By the 1920s, you could buy a new Ford car for only $300 (this is about $5000 in 2024 inflation-adjusted dollars). Ford was so successful that he quickly became one of the world's richest men.

W. Edwards Deming (1900–1993), an American quality control expert, was a consultant to the occupying forces in Japan after World War II. He introduced statistical quality control methods that allowed Japanese industries to manufacture very high-quality products. Deming's methods at first were not understood or appreciated by the rest of the world until Japan became a world leader in exporting superior goods and technology. Motorola expanded Deming's statistical quality approach to reduce its defective products to an amazing 1 defect in 3.4 million products. They call their process **Six Sigma**, which has now been adopted by most European and American manufacturing industries.

13.5 Manufacturing processes

Modern manufacturing can be divided into the following four main categories: subtractive, additive, continuous, and net shape.

---

[4] She was the first woman elected into the National Academy of Engineering. She served as an advisor to Presidents Hoover, Roosevelt, Eisenhower, Kennedy, and Johnson on matters of civil defense, war production, and rehabilitation of the physically handicapped.

13.5.1 Subtractive processes

Subtractive processes involve material removal via machining (e.g., turning, milling, boring, grinding, cutting, and etching). The three principal machining processes are classified as **turning**, **drilling**, and **milling**.

13.5.1.1 Turning

Turning is a process that uses a lathe (see Fig. 13.2) to produce "solids of revolution." The lathe can be operated by a person or a computer. A lathe that is controlled by a computer is known as a CNC (**computer numerical control**) machine. CNC is commonly used with many other types of machine tools in addition to the lathe.

In a turning operation, the part is rotated (i.e., turned) while it is being machined. Turning produces straight, conical, curved, or grooved workpieces such as shafts, spindles, and pins. The majority of turning operations involve the use of a single point cutting tool, such as that shown in Fig. 13.3.

The *material removal rate* (MRR) is defined as the amount of material removed per unit time, and has units of mm^3/min or in^3/min. In turning, each revolution of the workpiece removes a nearly ring-shaped layer of material. The material removed has a cross-sectional area equal to the axial distance the tool travels in 1 revolution (called the **feed**, f) multiplied by the depth of the cut, d. Then the volume of material removed per revolution is approximately equal to the product of the circumference and the feed multiplied by the depth of cut, or

$$\text{Material removed per revolution of the workpiece} = \pi D_{\text{avg}} f d$$

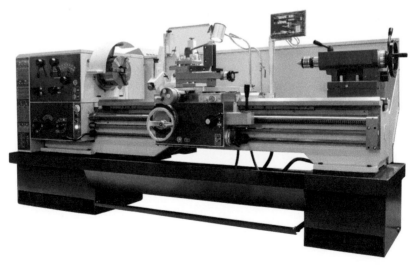

FIGURE 13.2

A typical lathe.

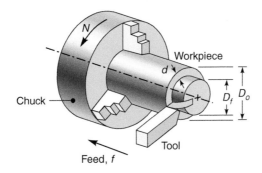

FIGURE 13.3

The basic elements of a turning operation.

where D_{avg} is the average diameter of the workpiece during the cutting operation, or (see Fig. 13.3)

$$D_{\text{avg}} = \frac{1}{2}\left(D_o + D_f\right) \tag{13.1}$$

If the workpiece is rotating at N RPM, then the MRR in turning is

$$MRR_{\text{turning}} = \pi D_{\text{avg}} f d N \tag{13.2}$$

The **cutting speed**, $V_{\textbf{cutting}}$, is the speed at which material is removed, and in turning it is

$$V_{\text{cutting}} = \pi D_{\text{avg}} N \tag{13.3}$$

Then, we can write the MRR as

$$MRR_{\text{turning}} = f d V_{\text{cutting}} \tag{13.4}$$

In the design of manufacturing operations, the time required to machine a part directly affects the manufacturing cost of the part. In turning, the cutting tool has a **feed rate** (FR) in mm/min or in/min, determined by the tool's feed f times the rotational speed N of the workpiece, or

$$FR = fN \tag{13.5}$$

If the axial cutting distance along the workpiece is L, then the time required to make the cut is

$$t_{\text{machining}} = L/FR \tag{13.6}$$

Example 13.1

A shaft 6.0 inches long with an initial diameter of 0.50 inches is to be turned to a diameter of 0.48 inches on a lathe. The shaft is rotating at 500. RPM on the lathe and the cutting tool feed rate is 4.0 inches/min. Determine

1) The cutting speed.
2) The material removal rate.
3) The time required to machine the shaft.

Need: $V_{cutting}, MRR_{turning}$, and $t_{machining}$.

Know: $L = 6.0$ inches, $D_o = 0.50$ inches, $D_f = 0.48$ inches, $N = 500.$ RPM, and the feed rate, $FR = 4.0$ in/min.

How: The cutting speed, MRR, and cutting time are given by Eqs. (13.3), (13.4), and (13.6).

Solve: 1. Eq. (13.3) gives the cutting speed as

$$V_{cutting} = \pi D_{avg} N, \text{ where } D_{avg} = (D_o + D_f)\,/\,2 = (0.50 + 0.48)\,/\,2 = 0.49 \text{ inches}$$

Then, $V_{cutting} = \pi(0.49)(500.) = $ **770 inches/min**.

2. The depth of the cut is $d = (D_o - D_f)/2 = (0.50-0.48)/2 = 0.010$ inches, and the feed is $f = FR/N = 4.0/500.$ [in/min]/[rev/min] $= 0.008$ in/rev. Then, Eq. (13.4) gives the MRR in turning as

$$MRR_{turning} = \pi(0.49)(0.008)(0.010)(500.) \text{ [in]} \times \text{[in]} \times \text{[in / rev]} \times \text{[rev / min]}$$

$$= \textbf{0.062 in}^3\textbf{/min}.$$

3. The time required to machine the shaft is given by Eq. (13.6) as

$$t_{machining} = L/FR = 6.0/4.0 \text{ [in]}/\text{[in / min]} = \textbf{1.5 min}.$$

The power required by a machining operation, $P_{machining}$, is equal to the material's average machining energy per unit volume removed (see Table 13.1) multiplied by the MRR, or

$$P_{machining} = (\text{Average Machining Energy per Unit Volume}) \times MRR \qquad (13.7)$$

Table 13.1 Average machining energy per unit volume of material removed.[a]

Material	Average machining energy per unit volume.[b]	
	$W \cdot s/mm^3$	$hp \cdot min/in^3$
Cast iron	3.3	1.2
Steels	5.5	2.1
Stainless steel	3.5	1.4
Aluminum alloys	0.70	0.28
Magnesium alloys	0.45	0.15
Nickel alloys	5.8	2.2
Titanium alloys	3.5	1.4

[a] Condensed from Kalpakjian and Schmid, *Manufacturing Engineering and Technology, 5th ed.* Prentice Hall, 2006.
[b] hp, *horsepower*.

and, since the machining power is the product of the cutting torque (in $N \cdot m$ or $ft \cdot lbf$) and the rotational speed of the workpiece (in radians/min[5]), then the cutting or machining torque is given by

$$T_{\text{cutting}} \text{ or } T_{\text{machining}} = P_{\text{machining}} \, / \, (2\pi N) \qquad (13.8)$$

Example 13.2

Determine the machining power and cutting torque required to machine the shaft in Example 13.1 if it is made from stainless steel.

Need: T_{cutting} and P_{cutting} for the shaft in Example 13.1.

Know: The material is stainless steel, and from Example 13.1, we know that

$$MRR_{\text{turning}} = 0.062 \text{ in}^3/\text{min, and } N = 500. \text{ RPM}$$

How: Using Table 13.1 and Eqs. (13.7) and (13.8).

Solve: From Table 13.1, the average machining energy per unit volume of material removed is 1.4 $hp \cdot min/in^3$. Then, Eq. (13.7) gives the machining power as

$$P_{\text{machining}} = 1.4(0.062) \left[hp \cdot min \, / \, in^3 \right] \times \left[in^3 \, / \, min \right] = \textbf{0.087 hp}.$$

Now, Eq. (13.8) gives the cutting torque as

$$T_{\text{cutting}} = P_{\text{machining}}/(2\pi N) = 0.087/(2\pi \times 500) \, [hp]/[rad \, / \, min] = 2.8 \times 10^{-5} \, hp \cdot min.$$

Since 1 hp = 33,000 $ft \cdot lbf/min$, the cutting torque is

$$T_{\text{cutting}} = 2.8 \times 10^{-5} \, (33,000) \, [hp \cdot min] \times [ft \cdot lbf \, / \, (hp \cdot min)] = \textbf{0.92 ft} \cdot \textbf{lbf}.$$

13.5.1.2 Drilling

Drilling is the process of using a cutting tool called a ***drill bit*** to produce round holes in materials such as wood or metal. A common twist drill shown in Fig. 13.4 (the one sold in hardware stores) has a point angle of 118 degrees. This angle works well for a

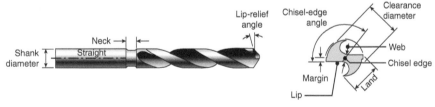

FIGURE 13.4

The characteristics of a common twist drill.

---

[5] Since there are 2π radians in 1 revolution (360 degrees), you can convert N in revolutions per minute (RPM) to radians per minute simply by multiplying it by 2π. Also, note that radians are dimensionless and do not appear in the final dimensions of the answer.

A typical drill press.

wide array of drilling operations. A steeper point angle (say, 90 degrees) works well for plastics and other soft materials. A shallower point angle (say, 150 degrees) is suited for drilling steels and other tough materials. Drills with no point angle are used in situations where a blind, or flat-bottomed hole, is required. Fig. 13.5 shows a typical drilling machine, called a ***drill press***.

Creating holes in a workpiece is a relatively common requirement. Drills are particularly good for making deep holes. The **feed**, f, of a drill is the distance the drill penetrates the workpiece in 1 revolution of the drill. The axial speed of the drill is the product of its feed and rotational speed, N, or

$$V_{\text{drill}} = fN \tag{13.9}$$

The cross-sectional area of a drill with a diameter D is $\pi D^2/4$, and the MRR of a drill is the product of its cross-sectional area and its axial speed, or

$$MMR_{\text{drill}} = (\pi D^2/4)fN \tag{13.10}$$

Eqs. (13.7) and (13.8) can also be used to calculate the drilling power and torque.

Example 13.3

We need to drill a 10. mm hole in a piece of titanium alloy. The drill feed is 0.20 mm/revolution and its rotational speed is 600. RPM. Determine

1) The material removal rate.
2) The power required.
3) The torque on the drill.

Need: MMR_{drill}, $P_{machining}$, and $T_{machining}$.

Know: $D = 10.$ mm, $f = 0.20$ mm/rev, and $N = 600.$ RPM.

How: The MRR is given by Eq. (13.10). The power and torque are given by Eqs. (13.7) and (13.8). The machining energy per unit volume can be found in Table 13.1.

Solve: 1. Eq. (13.10) gives

$$MMR_{drill} = \left(\pi D^2/4\right) fN$$

$$= \left(\pi 10^2/4\right)(0.2)(600.) \left[mm^2\right] \times [mm/rev] \times [rev/min]$$

$$= 9400 mm^3/min = \mathbf{160\ mm^3/s}.$$

2. From Table 13.1 for titanium alloys, the machining energy per unit volume of material removed is 3.5 W·s/mm³. Then, Eq. (13.7) gives

$$P_{machining} = 3.5(160) \left[W\cdot s/mm^3\right] \times \left[mm^3/s\right] = \mathbf{560\ W}.$$

3. From Eq. (13.8), the torque on the drill is

$$T_{machining} = P_{machining}/(2\pi N) = 560/(2\pi \times 600.) \; [W]/[radians/min]$$

$$= 0.15\ W\cdot min, = 0.15(60) \; [W\cdot min] \times [s/min] = 9.0\ W\cdot s = 9.0\ (J/s)s$$

$$= 9.0\ J = \mathbf{9.0\ N\cdot m}.$$

Eqs. (13.5) and (13.6) can also be used to determine the time it takes to drill the hole to a specific depth. For example, if we were required to drill the hole in Example 13.3 to a depth of 50. mm, then the drill feed rate would be $FR = fN = 0.20(600.)$ [mm/rev]×[rev/min] = 120 mm/min, and the time required to drill the hole 50. mm deep would be $t_{machining} = L/FR = 50./120$ [mm]/[mm/min] $= 0.42$ min, or 25 s.

13.5.1.3 Milling

Milling is carried out on a machine tool called a ***milling machine*** used for the shaping of metal and other solid materials. It has a cutting tool that rotates about the spindle axis similarly to a drill. However, a drill moves along only one axis, but on a milling machine, the cutter and workpiece move relative to each other, generating a tool path along which material is removed along three axes. Often the movement is achieved by moving the workpiece while the cutter rotates in one place, as shown in Fig. 13.6. Milling machines may be manually operated, mechanically automated, or digitally automated via CNC (computer numerical control).

Milling machines (Fig. 13.7) can perform a vast number of operations, some of them with quite complex tool paths, such as slot cutting, planning, drilling, and routing.

Fig. 13.6 illustrates the basic shape and operation of a milling cutter. The cutting speed, $V_{milling}$, of a cutter of diameter D rotating at N RPM is

$$V_{milling} = \pi DN \tag{13.11}$$

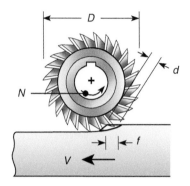

FIGURE 13.6

The characteristics of a common slab milling cutter.

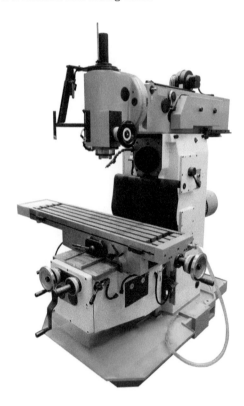

FIGURE 13.7

A typical milling machine.

The feed rate of the cutter is the rate at which its axis moves along the workpiece and can be calculated from the feed per tooth f, the rotational speed of the cutter N, and the number of teeth on the cutter n as

$$FR = fNn \tag{13.12}$$

The MRR for a workpiece of width w, a depth of cut d, and a cutter feed rate FR, is

$$MRR_{\text{milling}} = wdFR \tag{13.13}$$

The time required to mill a length L of material at a feed rate, FR, is

$$t_{\text{milling}} = L/FR \tag{13.14}$$

Example 13.4

A flat piece of cast iron 4.0 inches wide and 10. inches long is to milled with a feed rate of 20. inches/min with a depth of cut of 0.10 inches. The milling cutter rotates at 100. RPM and is wider than the workpiece. Determine
1) The material removal rate.
2) The machining power required.
3) The machining torque required.
4) The time required to machine the workpiece.

> **Need:** MMR_{milling}, $P_{\text{machining}}$, $T_{\text{machining}}$, and t_{milling}.
> **Know:** The material is cast iron with a length of $L = 10.$ inches and a width of $w = 4.0$ inches. The feed rate $(FR) = 20.$ in/min, $N = 100.$ RPM, and the depth of the cut $(d) = 0.10$ inches.
> **How:** Eq. (13.13) provides the MRR. Eqs. (13.7) and (13.8) can be used to find the cutting power and torque, and Eq. (13.14) gives the machining time.
> **Solve: 1**. From Eq. (13.13) we get

$$MRR_{\text{milling}} = wdFR = (4.0)(0.10)(20.) \; [\text{inches}] \times [\text{inches}]$$
$$\times \; [\text{inches} / \text{min}] = \textbf{8.0 in}^3/\textbf{min}.$$

> **2**. From Table 13.1 for cast iron, we find that the machining energy per unit volume of material removed is 1.2 hp·min/in³. Eq. (13.7) gives the machining power as
> $$P_{\text{machining}} = 1.2(8.0) \; [\text{hp} \cdot \text{min} / \text{in}^3] \times [\text{in}^3 / \text{min}] = \textbf{9.6 hp}.$$
> **3**. From Eq. (13.8) the torque on the milling cutter is

$$T_{\text{machining}} = P_{\text{machining}}/(2\pi N) = 9.6/(2\pi \times 100.) \; [\text{hp}]/[\text{radians} / \text{min}] = 0.015 \; \text{hp·min}$$

$$= 0.015(33,000) \; [\text{hp} \cdot \text{min}] \times [\text{ft} \cdot \text{lbf} / \text{hp} \cdot \text{min}] = \textbf{500 ft·lbf}$$

Eq. (13.6) gives the machining time as

$$t_{\text{machining}} = L/FR = 10./(20.) \; [\text{in}]/[\text{in} / \text{min}] = 0.75 \; \text{min} = \textbf{45 s}.$$

13.5.2 Additive processes

Additive processes involve material being added, such as in joining (welding, soldering, gluing, etc.), rapid prototyping, stereolithography, 3D printing (see Fig. 13.10), and the application of composite layers of resin and fiber. Additive machining processes add material to a base object to create complex shapes. This is far less expensive than cutting an intricate product from a solid block of material by the subtractive process.

13.5.3 **Continuous processes**

Continuous processes involve a product that is continuously produced, such as in the extrusion of metals and plastics and the pultrusion of composites. **Extrusion** (Fig. 13.8) is a continuous process of manufacture used to create objects of a fixed cross-sectional profile by pushing or drawing material through a die of the desired cross-section. In extrusion, a bar or metal or other material is forced from an enclosed cavity through a die orifice by a force applied by a ram. The extruded product has the desired, reduced cross-sectional area and a good surface finish, so that further machining is not needed. Extrusion products include rods and tubes with varying degrees of complexity in cross-section.

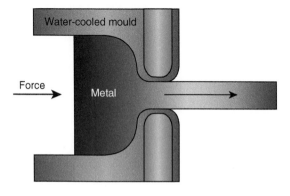

FIGURE 13.8

An extrusion process.

Pultrusion is a similar continuous process of manufacturing materials with constant cross-section except the material is **pulled** through a process or a die to its final shape (Fig. 13.9). Pultrusion is the only continuous manufacturing process available for obtaining a high-quality composite profile, with high mechanical properties. Pultruded products are normally composed of high-performance glass or carbon fibers embedded in a polymer matrix.

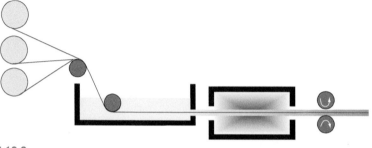

FIGURE 13.9

A typical pultrusion process.

13.5.4 Net shape processes

A **net shape** process occurs when the output is at or near its final shape, such as stamping, forging, casting, injection molding, blow molding, 3D printing, and thermoforming.

Stamping is a metalworking process in which sheet metal is formed into a desired shape by pressing or punching it on a machine press. Stamping can be a single-stage operation, where only one stroke of the press produces the desired form, or could occur through a series of these stages. The most common stamping operations are piercing, bending, deep drawing, embossing, and extrusion.

Casting is a manufacturing process by which a liquid material is poured into a mold containing a cavity of the desired shape and allowed to solidify. The mold is then opened to complete the process. The casting process is subdivided into two distinct subgroups: expendable and nonexpendable mold casting. Casting is most often used for making complex shapes that would be otherwise difficult or uneconomical to make by other machining methods.

Injection molding is a manufacturing process used to make parts by injecting molten plastic at high pressure into a mold. Injection molding is widely used for manufacturing a variety of parts, from the smallest gears to entire automotive body panels. Injection molding is the most common method of production today and is capable of tight tolerances. The most commonly used thermoplastic materials are polystyrene, polypropylene, polyethylene, polycarbonate, and polyvinyl chloride (PVC, commonly used in extrusions such as pipes, window frames, and the insulation on wiring). Injection molding can also be used to manufacture parts from aluminum, zinc, or brass in a process called *die casting*. The melting points of these metals are, of course, much higher than those of plastics, but it is often the least expensive method for mass producing small metal parts.

Blow molding is a manufacturing process in which hollow plastic parts are formed. The blow molding process begins with melting the plastic and forming it into a tubelike piece with a hole in one end, into which air can be injected. It is then clamped into a mold and air pressure pushes the plastic into the mold. Once the plastic has cooled and hardened, the mold opens up and the part is ejected.

3D Printing is a new material additive manufacturing process that builds a three-dimensional object from a computer-aided design model, usually by successively adding material layer by layer. There are many different 3D printing processes such as material jetting, binder jetting, powder bed fusion, and sheet lamination. Fig. 13.10 illustrates a typical 3D printer.

Thermoforming (Fig. 13.11) is a manufacturing process used with plastic sheets. Plastic sheets or film is converted into a finished part by heating it in an oven to its forming temperature, and then stretching it on a mold and cooling it. A thermoform machine typically utilizes vacuum to draw the plastic onto the mold in the forming process.

FIGURE 13.10

Computer-controlled 3D printing operation.

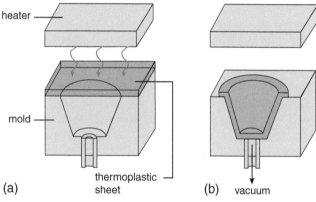

FIGURE 13.11

Thermoforming a drinking cup. A flat plastic sheet is placed over the mold and heated in illustration A. In illustration B, the sheet is drawn in to the mold with a vacuum.

Example 13.5

The energy needed to manufacture a reusable cup is significantly greater than that required to manufacture a disposable cup, as shown in the table. For a reusable cup to be an improvement over a disposable cup on an energy basis, we have to use it multiple times. Determine the number of uses required for the energy per use of a reusable cup to become less than for a disposable cup (this is called the *breakeven point*) if it takes about 180 kJ of energy to wash a single cup in a commercial dishwasher. (Data from Professor Martin B. Hocking at the University of Victoria, Canada.)

Note: since the values used in this example are whole numbers, we do not need to be concerned about significant figures in the calculations.

Energy required to manufacture a cup	
Cup type	**kJ/cup**
Ceramic	14,000
Plastic	6300
Glass	5500
Paper	550
Foam	190

Need: The breakeven point for a reusable cup.

Know: Manufacturing energy from the given table and that it takes about 180 kJ of energy to wash a single cup in a commercial dishwasher.

How: If a cup is used 10 times, then each use costs 1/10 of the manufacturing energy. If it is used 100 times, then each use costs just 1/100 of the manufacturing energy. However, to reuse a cup it has to be cleaned, and the energy required for each cleaning needs to be added to its manufacturing energy. The energy required per cup use can be computed from:

Energy per use = Wash energy + (Manufacturing energy)/(Number of uses)

Solve: Since it takes about 180 kJ of energy to wash a single cup in a commercial dishwasher, the total amount of energy per cup use is given by the preceding equation. So how many times would you have to use a ceramic cup to "breakeven" with the manufacturing energy used in producing a foam cup?

$$\text{Energy per use} = 190 \text{ kJ/cup (Foam)}$$

$$\text{Wash energy} = 180 \text{ kJ}$$

$$\text{Manufacturing energy} = 14,000 \text{ kJ/cup (Ceramic)}$$

Then, $190 = 180 + 14{,}000/N$, where $N =$ number of uses (and washes) of the ceramic cup. Solving for N gives $N = 1400$. Therefore, you would have to use and wash the ceramic cup 1400 times to equal the manufacturing energy of a single foam cup.

The table that follows shows how the energy per use of the three reusable cups declines the more you use them.

Number of times a reusable cup must Be used to break even with a disposable cup		
Reusable cup	Made of paper	Made of foam
Ceramic	38	1400
Plastic	17	630
Glass	15	550

These results are extremely sensitive to the amount of energy the dishwasher uses for cleaning each cup. The energy calculation for the dishwasher, requiring 180 kJ/cup per wash, is barely less than the manufacturing energy of the foam cup, 190 kJ/cup. If even a slightly less energy efficient dishwasher is used, then the reusable cups would never break even with the foam cup.

In situations where cups are likely to be lost or broken and thus have a short average lifetime, disposable cups are the preferred option *if* energy usage is the main criterion.

13.6 Modern manufacturing

What makes a good manufacturing process? Today, the overall performance of a company is often dictated by the design of its manufacturing process and facility. It turns out that four key elements are important in designing a good manufacturing process.

The first key element is the ***movement of materials***. A well-designed facility results in efficient material handling, small transportation times, and short queues. This, in turn, leads to low work-in-process levels, effective production management, decreased cycle times and manufacturing inventory costs, improved on-time

delivery performance, and higher product quality. The second key item is *time*. The setup time plus process time from order to shipping is extremely important in meeting customer demands and keeping a constant flow of product. The third key element is *cost*. Material, labor, tooling, and equipment costs must all be monitored and controlled. The fourth key element is *quality*. Customers today demand high-quality products, so deviations from design specifications must be kept to a minimum.

Modern manufacturing philosophies such as JIT manufacturing, flexible manufacturing, lean manufacturing, and life cycle manufacturing contain all these elements. They are discussed in detail in the following material.

13.6.1 Just-in-time manufacturing

JIT manufacturing is easy to understand. Things are planned to occur *just in time*. For example, consider your activities today. You left your room *just in time* to walk to your first class. That class ended *just in time* for you to go to your next class, which ended *just in time* for you to have lunch. JIT processes may seem easy, but achieving them in a manufacturing environment is often very difficult.

In a manufacturing process, parts should arrive at the factory just in time to be distributed to the work stations, and they should arrive at a work station just in time to be installed by a worker. Finally, the factory should produce finished products just in time to be handed to a waiting customer. This eliminates any inventory of parts and products, and in theory, JIT does not need *any* inventory of raw materials, parts, or finished products.

JIT originated in Japan with the Toyota Motor Company, and JIT was initially known as the *Toyota Production System*. After World War II, the president of Toyota wanted to compete aggressively with American automobile production. What he found was that one American car worker could produce about nine times as much as a Japanese car worker. By further studying the American automobile industry, he found that American manufacturers made large quantities of each item (parts, engines, car model, etc.). American car manufacturers also stocked all the parts needed to assemble their cars.

Toyota felt that this would not work in Japan because the small Japanese market wanted small quantities of many different car models. Consequently, the company developed a production system that provided parts to the workers only when they were needed (i.e., just in time). It analyzed the concept of "waste" in general terms that included wasted time and resources as well as wasted materials. It subsequently identified the following **seven wasteful activities** that could be minimized or eliminated:

1. **Transportation** (moving products that are not required to perform the processing).
2. **Inventory** (all components, work-in-progress, and finished product not being processed).
3. **Motion** (workers or equipment moving more than is required to perform the processing).

4. **Waiting** (workers waiting for the next production step).
5. **Overproduction** (production ahead of demand).
6. **Over processing** (due to poor product design).
7. **Defects** (the wasted effort involved in inspecting and fixing defects).

A number of Japanese jargon terms are associated with JIT that are in common use today. For example, **Andon** refers to trouble lights that immediately signal to the production line that there is a problem, and the production line is stopped until the problem is resolved. In the Toyota system, the Andon light indicating a stopped production line is hung from the factory ceiling so that it can be clearly seen by everyone. This raises the profile of the problem and encourages rapid attention to a solution. However, when General Motors instituted an Andon light, U.S. workers were reluctant to take the responsibility for stopping a production line, and defective products were produced. General Motors later solved this problem by allowing workers to signal that they had a problem without stopping the production line.

The JIT philosophy involves the elimination of waste in its many forms, a belief that ordering and stocking costs can be reduced, and always striving to progress through continuous improvement. The elements of a JIT process typically include the following:

- Meet regularly with the workers (daily/weekly) to discuss work practices and solve problems.
- Emphasize consultation and cooperation with the workers rather than confrontation.
- Modify machinery to reduce setup time.
- Reduce buffer stock.
- Expose problems, rather than have them covered up.

It is not necessary to apply JIT to all stages of a process. For example, we could keep large stocks of raw material but operate the production process internally in a JIT fashion (hence eliminating work-in-progress stocks).

13.6.2 Flexible manufacturing

A flexible manufacturing system (FMS) is a manufacturing system that contains enough flexibility to allow the system to rapidly react to production changes. This flexibility is generally considered to fall into two categories. The first is **machine flexibility**. This allows the system to be changed to produce new product types and to change the order of operations executed on a part. The second is **routing flexibility**. This consists of the ability to use multiple machines to perform the same operation on a part as well as the system's ability to absorb large-scale changes, such as in volume, capacity, or capability.

Most FMS systems consist of three main systems: a **material handling system** to optimize the flow of parts, a **central control computer** that controls material movement, and the **working machines** (often automated CNC machines or robots).

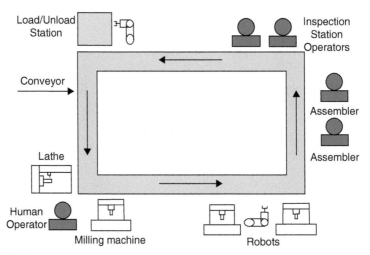

FIGURE 13.12

A flexible manufacturing system.

The use of robots in manufacturing industries has many benefits. Each robotic cell is connected with a material handling system, which makes it easy to move parts from one robotic cell to another. At the end of processing, the finished parts are routed to an automatic inspection cell and removed from the FMS. This process is illustrated in Fig. 13.12.

The main advantage of an FMS is its high flexibility in managing manufacturing resources like time and effort to manufacture a new product. The best application of an FMS is found in the mass production of small sets of products.

13.6.3 Lean manufacturing

Lean manufacturing is the optimal way of producing goods through the removal of waste and implementing flow, as opposed to batch processing. Lean manufacturing is a generic process management philosophy derived mostly from Toyota and focuses mainly on reduction of the seven wastes originally identified by Toyota (see Section 13.6.1, on JIT).

Lean manufacturing is focused on getting the right things, to the right place, at the right time, in the right quantity to achieve perfect work flow while minimizing waste and being flexible and able to change. All these concepts have to be understood, appreciated, and embraced by the workers who build the products and own the processes that deliver the value. The cultural and managerial aspects are just as important as, and possibly more important than the actual tools or methodologies of production itself. Lean manufacturing tries to make the work simple enough to understand, to do, and to manage.

The main principles of lean manufacturing are zero waiting time, zero inventory, internal customer pull instead of push, reduced batch sizes, and reduced process times.

13.6.4 Life cycle manufacturing

Today the term *green* is synonymous with environmental sustainability. Sustainability is the development and application of processes or the use of natural resources that can be maintained indefinitely. As public environmental awareness increased, manufacturing facilities began examining how their activities were affecting the environment.

To understand the environmental impact of a manufacturing process, you need to examine all the inputs and outputs associated with the entire existence of the product. This "cradle-to-grave" analysis is called **life cycle analysis**, or simply LCA. LCA normally ignores second-generation impacts, such as the energy required to fire the bricks used to build the kilns used to manufacture the raw material. The concept of conducting a detailed examination of the life cycle of a product or a process in response to increased environmental awareness on the part of the public, industry, and government is relatively recent.

Just as living things are born, age, and die, all manufactured products have an analogous life cycle. Each stage of a product's development affects our environment, from the way we use products to the way we dispose of them when we are finished with them (see Fig. 13.13). Looking at a product's life cycle helps engineers understand the connections between the Earth's natural resources, energy use, climate change, and waste disposal. Product life cycle analysis focuses on the processes involved in the entire production system, including the following:

- The design and functionality of the product.
- The extraction and processing of raw materials.
- The processes used in manufacturing.
- Packaging and distribution.
- How the product is used by the customers.
- Recycling, reuse, and disposal of the product.

Engineers use life cycle analysis during product design and development and manufacturers use it during the production stage. However, LCA has its greatest potential to reduce environmental impacts during the design stage, when 70% of the total product cost is determined (see Fig. 13.14).

Life cycle analysis uses detailed environmental and energy estimates for the manufacture of a product, from the mining of the raw materials to production and distribution, end use and possible recycling, and disposal. LCA enables a manufacturer to measure how much energy and raw materials are used; and how much solid, liquid, and gas waste is generated at each stage of a product's life.

FIGURE 13.13

Life cycle manufacturing.

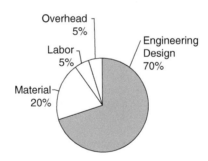

FIGURE 13.14

The influence of engineering design on the cost of manufacturing.

13.7 Variability and six sigma

Unintentional variability is a fact of life. As a simple example, it's instructive to have the students in the class each measure the same thing with the same instrument. We

Table 13.2 Student measurements of a pencil's length in mm.

Students 1–5	Students 6–10	Students 11–15
21.2	22.7	21.0
21.0	20.1	21.2
20.7	20.9	20.8
21.7	20.9	21.2
21.7	21.5	21.2

find that there will be slightly different results. For example, if you have 15 students in your class measure a brand new unsharpened pencil with the same ruler, you will get something like that shown in Table 13.2.

Note that the measured lengths are not all the same. One student measured the pencil's length as 22.7 mm, a number much greater than all the other measurements. Another student measured the same pencil's length as 20.1 mm, which is also quite a bit smaller than most of the other measurements. These limits are called **outliers** and are quite common due to inattention or misreading of the measuring instrument.

Which reading is correct? Any, none, or some combination? Most of us would agree that the average or mean defined as the total of all the measurements divided by the number of measurements has less bias than does any individual measurement.

Let us analyze what we have measured. The mean or average length of these 15 ($N = 15$) measurements is $\bar{x}$ where

$$\bar{x} = \frac{1}{N}(x_1 + x_2 + x_3 + ...) = \frac{1}{15}(21.2 + 21.0 + 20.7 + \cdots + 21.2) = 21.2 \text{ mm}$$

Or, more generally, $\bar{x} = \frac{1}{N}\sum_{i=1}^{N} x_i = 21.2$ mm

So, you say the pencil is 21.2 mm long. But is it? Suppose we drop the outlier of 22.7 mm. Then the average length of the pencil becomes 21.1 mm. Is that correct? Or should we drop the other outlier of 20.1 mm, giving a new average of 21.3 mm. Again, is that correct?

Let us try something else. Subtract each measurement from the average (mean value) of 21.2 mm, as shown in Table 13.3.

If you add all these "deviations," they will total 0.0 since this is what we mean by an average. Now plot these deviations as a frequency curve (called a histogram) with the frequency of each measurement on the vertical axis and the measured values on the horizontal axis, as in Fig. 13.15.

The histogram shows that the most frequent measurements are centered around zero so that many, but not all, of the students in the class did think the actual length of the pencil was 21.2 mm.

Table 13.3 Mean value ($\bar{x} = 21.2$ mm) of pencils minus measured lengths.

$\bar{x}$ − Measured length	$\bar{x}$ − Measured length	$\bar{x}$ − Measured length
0.0	−1.5	0.2
0.2	1.1	0.0
0.5	0.3	0.4
−0.5	0.3	0.0
−0.5	−0.3	0.0

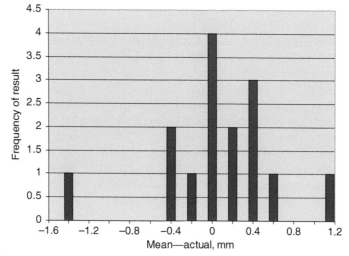

FIGURE 13.15

Histogram of deviations from mean.

But, is ignoring all those who did not measure exactly 21.2 mm the best we can do? Suppose we square these deviations—some of which were positive and some which were negative—so that each squared term is then positive. An individual squared term is still a poor measure of all the data points. We need a measure that eliminates the bias of a single measurement just as the mean reduced the bias of a single number. So, add all the squares and take the square root of them then divide by the number of points.[6]

$$\sigma = \sqrt{\frac{1}{N}\sum_{i=1}^{N}(\bar{x}-x_i)^2} \qquad (13.15)$$

---

[6] For mathematical reasons, if we use all N points of data to determine the mean of the data, we should divide by $N-1$ rather than N. The proper use of statistical methods requires that the number of samples should be "large", say $N \geq 30$. Then whether we divide by N or $N-1$ should make no difference.

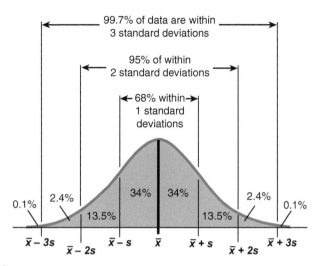

99.7% of data are within
3 standard deviations

95% of within
2 standard deviations

← 68% within→
1 standard
deviations

0.1%   2.4%   34%   34%   2.4%   0.1%
       13.5%   13.5%

$\bar{x}-3s$ $\bar{x}-2s$ $\bar{x}-s$ $\bar{x}$ $\bar{x}+s$ $\bar{x}+2s$ $\bar{x}+3s$

FIGURE 13.16

Normal distribution curve.

The Greek letter sigma σ defined in Eq. (13.15) is called the **standard deviation** of the data. In our case, $\sigma = 0.6$ mm, but what does that signify?

A typical statistical analysis of manufacturing variables starts with the histogram in Fig. 13.15. A curve is then produced which shows the number of standard deviations between the mean value and a process control limit (see Fig. 13.16). The vertical axis is the "probability" of the variable's occurrence and the horizontal axis is known as the **Z-score** or **Process Sigma Level**, $\sum$, which is defined by the difference between the variable's mean value $\bar{x}$ and a sample point x_i all divided by the standard deviation, σ

$$Z\text{-score} = \text{Process Sigma Level} (\Sigma) = \frac{1}{\sigma}(\bar{x} - x_i) \tag{13.16}$$

The **Z-score** is just the number of standard deviations between the sample mean $\bar{x}$ and a measurement x_i or a process control upper or lower limit. (Note: the process sigma level $\sum$ is NOT the same as the standard deviation sigma, σ.)

The theoretical equation describing the frequency versus standard deviation (or F vs. Z) curve is

$$\text{Frequency} = F = \frac{1}{\sigma\sqrt{2\pi}}\exp(-0.5Z^2) \tag{13.17}$$

The curve defined by Eq. (13.17) is called a **normal distribution or Gaussian distribution**.

Why go to these lengths to plot this curve in F - Z coordinates? The reason is that, with this coordinate system, the area under the curve (taken out to $Z = \pm\infty$) is 1.

Table 13.4 Manufacturing sigma values.

Sigma level or Z-score	Errors per million products	Percentage of product within sigma value (%)
±1	690,000	68.27
±2	308,000	95.45
±3	66,800	99.73
±4	6210	99.994
±5	230	99.9999
±6	3.4	99.9999998

This means that there is 100% probability that a measurement is somewhere under this curve. The utility of this type of analysis is that we can predict the amount of manufacturing errors as shown in Table 13.4.

Example 13.6

Convert the numerical data in Table 13.3 to Z-score values. Then, plot its frequency (F) versus Z curve. (This should be like the curve in Fig. 13.16.)

Need: The frequency of occurrence for a normal distribution of pencil length measurements.

Know–How: Use Eq. (13.15) to calculate σ, (13.16) to calculate Z and (13.17) to calculate frequency F, then plot the $F - Z$ curve.

Solve: Use a spreadsheet to calculate and plot the results.

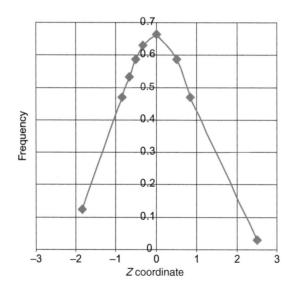

A "Six Sigma"[7] process means you are rejecting parts for which $Z \geq \pm 6$. This means that very few parts are out of tolerance, since at these values the area under the tails of the normal error curve is very small (about 3.4 defective parts per million parts produced). This can have a profound effect on manufacturing costs. Note that a 6-sigma process is equal to a Z-score of 6.

Example 13.7

If a process produces parts whose length dimension in not constant, but varies according to a bell-shaped normal distribution with mean value of $\bar{x} = 63.8$ mm and standard deviation $\sigma = 4.2$ mm, what is the length dimension range of the parts that fall between plus or minus of the following values:

a) one standard deviation ($\pm 1\sigma$) of the mean value,
b) two standard deviations ($\pm 2\sigma$) of the mean value, and
c) three standard deviations ($\pm 3\sigma$) of the mean value.

Need: The length dimension range of parts whose length are between $\pm 1\sigma$, $\pm 2\sigma$ and $\pm 3\sigma$ of the mean length.

Know: From Table 13.4, 68.27% fall between $\pm 1\sigma$, 95.45% fall between $\pm 2\sigma$, and 99.73% fall between $\pm 3\sigma$.

How—Solve:

a) If we measure one standard deviation on either side of the mean, we get

$$63.8 - 4.2 = \textbf{59.6 mm} \text{ and } 63.8 + 4.2 = \textbf{68.0 mm.}$$

We know from Table 13.4 that 68.27% of the total area under the normal bell-shaped curve is over the interval of $\pm 1\sigma$. This means that 68.27% of the parts have a length between 59.6 and 68.0 mm.

b) Now measure two standard deviations from the mean,

$$63.8 - 2(4.2) = \textbf{55.4 mm} \text{ to } 63.8 + 2(4.2) = \textbf{72.2 mm}$$

From Table 13.4, we find that 95.45% of the length of the part fall within this interval.

c) Finally, measure out three standard deviations from the mean,

$$63.8 - 3(4.2) = \textbf{51.2 mm} \text{ to } 63.8 + 2(4.2) = \textbf{76.4 mm.}$$

From Table 13.4 we find that 99.73% of the parts length fall within this interval of 51.7 and 76.4 mm. That leaves only 0.135% of the parts that are shorter than 51.2 mm and 0.135% of the parts that are longer than 76.4 mm.

In manufacturing, all parts produced must meet certain tolerance limits set by the engineering designer. These tolerances provide lower and upper Z-score values for acceptable parts. Example 13.8 illustrates this principle.

---

[7] Six Sigma is a registered service mark and trademark of Motorola, Inc. Motorola reported over $17 billion in savings from Six Sigma as of 2006.

Example 13.8

The design specification for the parts in Example 13.7 is required to fall between 70 and 74 mm. What range of standard deviations of the parts, with the mean and standard deviation given in Example 13.7, meet this requirement?

Need: Standard deviation range of parts described in Example 13.7 fall between 70 and 74 mm.

Know: The parts have a bell-shaped normal distribution with mean value of $\bar{x} = 63.8$ mm and standard deviation $\sigma = 4.2$ mm.

How: Use Eq. (13.16) to determine the lower and upper Z-scores for these specifications. This specifies the sigma range of acceptable parts.

Solve: The Z-score for the lower limit of 70 mm, according to Eq. (13.16), is determined by subtracting the mean of 63.8 and then dividing the by the standard deviation of 4.2:

$$(\text{Z-score})_{\text{lower}} = (70 - 63.8)/4.2 = \mathbf{1.48}$$

This tells us that for a part to have a length greater that the lower limit, it must be at least 1.48 standard deviations above the mean. Converting the upper limit of 74 mm into a Z-score gives:

$$(\text{Z-score})_{\text{upper}} = (74 - 63.8)/4.2 = \mathbf{2.43}$$

This tells us that for a part to have a length less that the upper limit, it must be at least 2.43 standard deviations above the mean. Consequently, to meet the design specifications the part lengths must fall between 1.48 and 2.43 standard deviations from the mean.

Summary

In this chapter, we reviewed four major manufacturing processes—**additive, subtractive, continuous,** and **net shape**. We also investigated the basics of **just in time, flexible, lean,** and **life-cycle** manufacturing.

Statistical methods are important for the control of manufacturing processes because the manufactured parts are never exactly as specified. Their dimensions vary about some **mean value**, and the variability follows Gauss's **normal distribution**. The most important variable that describes the consistency of a manufacturing process is the **standard deviation**, σ, of the process. It tells us the number of rejects in a normal distribution of manufactured objects. The key variable is the **Z-score**. For a Z-score between ± 1 σ, 68.27% of the parts are within specifications; for a Z-score between ± 2 σ, 95.45% are within specifications; and for a Z-score between ± 3 σ, 99.73% are within specifications.

Exercises

1. Discuss the building materials use by the three little pigs (straw, sticks, and bricks). Why were they chosen? Why did they fail? What was the environmental impact?

2. A casting process involves pouring molten metal into a mold, letting the metal cool and solidify, and removing the part from the mold. The solidification time is a function of the casting volume and its surface area, known as *Chvorinov's Rule*:

$$\textbf{Solidification Time} = K \times (\textbf{Volume} \div \textbf{Surface Area})^2$$

where K is a constant that depends on the metal. Three parts are to be cast that have the same total volume of 0.015 m³, but different shapes. The first is a sphere of radius R_{sph}, the second is a cube with a side length L_{cube}, and the third is a circular cylinder with its height equal to its diameter ($H_{\text{cyl}} = D_{\text{cyl}} = 2R_{\text{cyl}}$). All the castings are to be made from the same metal, so K has the same value for all three parts. Which piece will solidify the fastest? The table that follows gives the equations for the volume and surface area of these parts.

Object	Surface area	Volume
Sphere	$4\pi R_{sph}^2$	$(4/3)\pi R_{sph}^3$
Cube	$6L_{cube}^2$	L_{cube}^3
Cylinder	$2\pi R_{cyl}^2 + 2\pi R_{cyl}H_{cyl} = 6\pi R_{cyl}^2$	$\pi R_{cyl}^2 H_{cyl} = 2\pi R_{cyl}^3$

(**Ans.** The cube will solidify the fastest and the sphere will solidify the slowest.)

3. For a particular casting process, the constant K in Chvorinov's equation in Exercise 2 is 3.0 s/ mm² for a cylindrical casting 150 mm high and 100 mm in diameter. Determine the solidification time for this casting.

4. Tool wear is a major consideration in machining operations. The following relation was developed by F. W. Taylor for the machining of steels:

$$VT^n = C$$

where V is the cutting speed; T is the tool life; the exponent n depends on the tool, the material being cut, and the cutting conditions; and C is a constant. If $n = 0.5$ and $C = 400$ mm/min$^{0.5}$ in this equation, determine the percentage increase in tool life when the cutting speed is reduced by 50%. (**Ans. The tool life increases by 300%.**)

5. For a particular machining operation $n = 0.6$ and $C = 350$ mm/min$^{0.5}$ in the Taylor equation given in Exercise 4. What is the percentage increase in tool life when the cutting speed V is reduced by (a) 25%, and (b) 74%?

6. Using Taylor's equation in Exercise 4, show that tool wear and cutting speed are related by

$$T_2 = T_1 (V_1/V_2)^{1/n}$$

7. A shaft 150. mm long with an initial diameter of 15. mm is to be turned to a diameter of 13.0 mm in a lathe. The shaft rotates at 750. RPM in the lathe and the cutting tool feed rate is 100. mm/min. Determine
   a) The cutting speed.
   b) The material removal rate.
   c) The time required to machine the shaft.

8. Repeat the calculations in Example 13.2 for a magnesium alloy being machined at 700 RPM.

9. Determine the machining power and cutting torque required to machine a magnesium shaft if the material removal rate, MMR, is 0.1 in³/min and $N = 900$. RPM.

10. Determine the machining time required to turn a 0.2 m long shaft rotating at 300 RPM at a tool feed of 0.2 mm/rev.

11. A lathe is powered by a 5 hp electric motor and is running at 500. RPM. It is turning a 1.0-inch cast iron shaft with a depth of cut of 0.035 in. What maximum feed rate can be used before the lathe stalls?

12. A 2.0-inch diameter carbon steel shaft is to be turned on a lathe at 500. RPM with a 0.20 inch depth of cut and a feed of 0.030 in/rev. What minimum horsepower and torque must the lathe have to complete this operation?

13. Suppose the material used in Example 13.3 is an aluminum alloy and the drill rotational speed is 500. RPM. Recalculate the material removal rate, the power required, and the torque on the drill.

14. A drill press with a 0.375-inch diameter drill bit is running at 300. RPM with a feed of 0.010 in/ rev. What is the material removal rate?

15. You need to drill a 20. mm hole in a piece of stainless steel. The drill feed is 0.10 mm/revolution and its rotational speed is 400. RPM. Determine
   a) The material removal rate.
   b) The power required.
   c) The torque on the drill.

16. Repeat the calculations in Example 13.4 for stainless steel instead of cast iron. Use the same material dimensions and operating conditions, and compare the machining times for the two materials.

17. A flat piece of cast iron 100. mm wide and 150. mm long is to be milled with a feed rate of 200. mm/min with a depth of cut of 1.0 mm. The milling cutter rotates at 100. RPM and is wider than the workpiece. Determine
    a) The material removal rate.
    b) The machining power required.
    c) The machining torque required.
    d) The time required to machine the workpiece.

18. A milling operation is carried out on a 10.-inch long, 3.0-inch wide slab of aluminum alloy. The cutter feed is 0.01 in/tooth, and the depth of cut is 0.125 inches. The cutter is wider than the slab and has a diameter of 2.0 inches. It has 25 teeth and rotates at 150. RPM. Calculate (a) the material removal rate, (b) the power required, and (c) the torque at the cutter.

19. A part 275. mm long and 75. mm wide is to be milled with a 10-toothed cutter 75 mm in diameter using a feed of 0.1 mm/tooth at a cutting speed of 40. $\times 10^3$ mm/min. The depth of cut is 5.00 mm. Determine the material removal rate and the time required to machine the part.

20. In the face-milling operation shown in the figure that follows, the cutter is 1.5 inches in diameter and the cast iron workpiece is 7.0 inches long and 3.75 inches wide. The cutter has eight teeth and rotates at 350. RPM. The feed is 0.005 in/tooth and the depth of cut is 0.125 inches. Assuming that only 70% of the cutter diameter is engaged in the cutting, determine the material removal rate and the cutting power required.

End mill

21. In Example 13.5, we did not take into account recycling or disposal costs or benefits. Do you think his conclusions would change if these were included? Plot the net energy per use versus the number of uses for the reusable and disposable cups.

22. The theoretical equation describing the frequency versus standard deviation (or F vs. Z) curve is can be written as a combination of Eqs. (13.16) and (13.17) as

$$F = \frac{1}{\sigma\sqrt{2\pi}}\exp\left(-0.5Z^2\right) = \frac{1}{\sigma\sqrt{2\pi}}\exp\left(\frac{1}{2}\left(\frac{(x-\bar{x})}{\sigma}\right)^2\right)$$

or in spreadsheet script: exp(−0.5*((x − x̄)/σ)`2)/(σ*sqrt(2*pi).

   Plot a normal distribution curve for the set of 100 numbers from 0 to 99. (Hint: Use the Chart —Column graphical representation. Observe its shape and calculate its mean and standard deviation. See Exercise 23.)

23. Students often are confused by the difference between a normal distribution curve, as in Exercise 22, and a *random* distribution. Excel has a variable, rand(), that generates random

numbers between 0 and 1. (See the Help menu for how to use it.) Plot a curve for the set of random numbers from 0 to 99. (**Hint:** Use the Chart–Column representation. Observe its shape and calculate its mean and standard distribution.)

24. Three-hundred widgets are manufactured with the following normal error distribution statistics: mean $\bar{x} = 123$ mm; standard deviation $\sigma = $ (a) 23 mm, (b) 32.5 mm, and (c) 41 mm. How many units measure less than 140. mm?

25. We wish to fit steel rods (mean, $\bar{x} = 21.2$ mm; $\sigma = 0.20$ mm) into the jig shown below to make sure we can meet specifications. The process requires that 97.5% of all rods are to be within specifications. (Presume that no rod is so short that it cannot pass specifications.) What is the appropriate size L mm of the jaws of the jig? (Hint: 95% of normally distributed data are found within $\pm 2\sigma$ from the mean and therefore 2.5% of the rods are too large.)

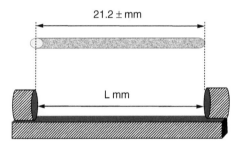
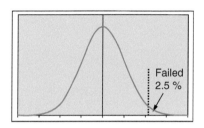

26. The jig in the Exercise 25 is OK, except now too many undersized rods are now passing specifications. The manufacturing section of your company wants to eliminate all rods that are 0.2 mm or more undersized. What fraction now passes specifications?

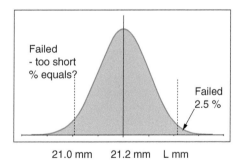

27. You perform 100 tests on one of two prototypes of the X15-24 line of products and a colleague performs another 100 tests on the other prototype.

Test set statistic	Your test	Colleague's test
Mean, $\bar{x}$	125.7	123.5
Standard deviation, σ	1.50	13.3

Your data are circles and your colleagues are crosses in the figure that follows.

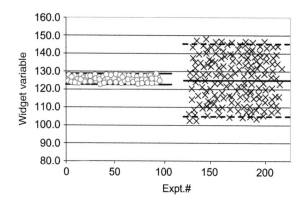

Your colleague notes that, because the means of the measurements are virtually identical, there is no difference between the two prototypes and therefore hers should be used because it will be cheaper to manufacture (which you concede). What counterarguments can you muster?

28. Quality control is more personal when the products are hand grenades and it's your turn to learn how to throw one. You understand that, once you pull the pin, if you hold on too long, you might be severely injured or killed. However, if the fuse burns too long after you have thrown it, the enemy might have time to pick it up and throw it back at you, with results similar to holding it too long. Suppose the fuse burn time is designed to be 4.00 s with a standard deviation of 0.2 s and the grenade has been made to Six Sigma standards. What is the variability you can expect on the time for it to detonate after pulling its safety pin?

29. You are a new quality control engineer at a company that manufactures bottled drinking water. All the bottles and the filling water are checked hourly to make sure that there are no contaminants. Monday morning you notice that, for some reason, the water that was used to wash the bottles before filling was not tested over the weekend. Now you have several carloads of product ready to be shipped. What do you do? (Use the Engineering Ethics Matrix.)
    a) Have the shipment destroyed and start filling the order over again.
    b) Test random samples for the shipment and if they pass send the shipment.
    c) Tell your supervisor and let him or her decide what to do.
    d) Since the wash water has never been contaminated in the past, do nothing and release the shipment.

30. As a young engineer, you have been told not to trust the machinists who work for you. They make up excuses for bad parts that are not true because they are lazy or incompetent. One day a machinist tells you his milling machine is "out of calibration" and that he cannot make parts to specification unless the machine is repaired. What do you do? (Use the Engineering Ethics Matrix.)
    a) Replace him with a more competent machinist.
    b) Ask the machine shop supervisor to check the milling machine to see if there is a problem.
    c) Ask the machinist to show you exactly what he means by demonstrating the problem.
    d) Tell your supervisor you need a raise if you have to work with idiots like this.

Materials engineering

14

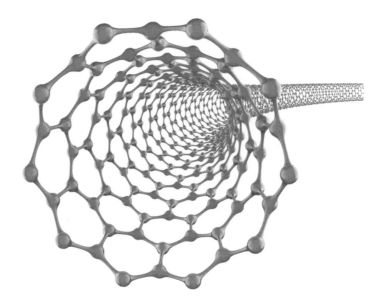

Source: koya79/iStockphoto.com

14.1 Introduction

Engineers select the materials that are to be used in their designs. Should a refrigerator case be made of steel or plastic? Should armor plate be a single sheet of steel or a lighter layered composite alloy? Should a transistor be made out of germanium or silicon? Should a telephone cable be made of copper or fiber optic glass? Should an artificial hip joint be made out of metal and, if metal, should it be titanium or stainless steel, or would a polymer composite be better? These materials selection problems are further examples of "*constraint optimization*". An engineer finds the solution that best meets given need while also satisfying a set of constraints.

Exploring Engineering. https://doi.org/10.1016/B978-0-443-13541-5.00028-3

14.2 Choosing the right material

The constraint to be optimized when choosing the right material for an engineering application might be cost, weight, or performance or some index such as minimizing weight × cost with a requirement of a particular strength. The constraints (also called design requirements) typically involve such words as **elastic modulus** (which is closely related to **stiffness**), **elastic limit**, **yield strength** (sometimes abbreviated as plain "strength"), and **toughness**. While we all have loose ideas of what is meant by these terms, it is necessary to precisely express them as engineering variables. Those variables must, in turn, contain the appropriate numerical values and correct units.

To help develop appropriate variables, numbers, and units, this chapter will introduce a new tool: the **stress–strain diagram**. It is a tool for defining elasticity, strength, and toughness as engineering variables, as well as a tool to extract the numerical values of those variables in the selection of appropriate materials.

Working through the examples and problems in this chapter will enable you to:

1) Define **material requirements**
2) Relate two important classes of materials—**metals and polymers**
3) Understand the **internal microstructure of materials** that can be **crystalline and/or amorphous**
4) Use a stress–strain diagram to express materials properties in terms of the **five engineering variables:** stress, strain, elastic limit, yield strength, and toughness
5) Use the results determined for those properties to carry out materials selection.

Humankind's reliance on the properties of materials for various applications, from weapons to clothing to shelter, has been around since the dawn of the ascendancy of our species.

Fig. 14.1 illustrates how various materials have evolved over time.[1,2] The earliest were naturally occurring elements and compounds, followed by the method of trial and error, culminating in today's modern materials designed by systematic investigation. Our interest here is to discover the principles behind new materials. Notice how modern materials are classified by type: metals, polymers, composites, and ceramics. For the sake of brevity, we will confine ourselves to just the first two of these classes: metals and polymers.

Generally, metals are strong and dense. Metals also reflect light and conduct heat and electricity. On the other hand, polymers, popularly known as "plastics," are relatively weak, sometimes opaque and sometimes transparent, and generally do not

---

[1] The Neolithic Revolution started around 10,000 BCE in the Fertile Crescent, a boomerang-shaped region of the Middle East where humans first took up farming. Shortly after, Stone Age humans in other parts of the world also began to practice agriculture. Civilizations and cities grew out of the innovations of the Neolithic Revolution.

[2] Ashby, M. F., Technology of the 1990s: Advanced Materials and Predictive Design. Phil. Trans. R. Soc. Lond., A322, 393, (1987).

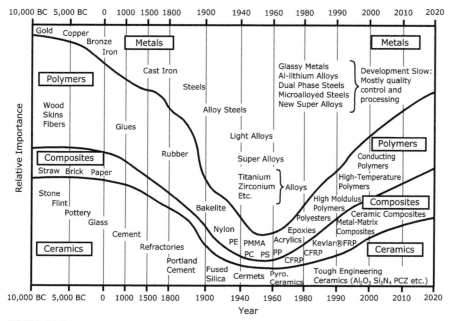

FIGURE 14.1

Materials since the Neolithic Revolution,[1] about 10,000 years **BCE**.

Reprinted by permission of the Royal Society (London) and of Professor Ashby.[2]

conduct heat or electricity. We will shun the term *plastic* in favor of *polymer* in describing these materials, since, as we shall see, the word *plastic* is reserved to describe a particular behavior in materials.

14.3 Strength

What are the reasons for a material's strength? Material properties are directly related to their molecular properties and hence to the properties of their constituent atoms. The only fundamental example we will attempt to calculate is the breaking strength of a material, such as a piece of pure iron, as determined by its molecular structure. This calculation represents the upper limit to its strength. All failures at lower strength values are due to defects in the material or its structure.

We will make a quick review of material structure such as whether a material is **amorphous** (meaning no structure observable at the microscopic level, which is 10^{-6} m and smaller) or whether it is **crystalline** (which means it consists of definite microstructures that can be seen under a low power microscope or even by the naked eye).

Suppose you clamped the ends of a piece of pure iron and pulled it as shown in Fig. 14.2 until it eventually breaks.

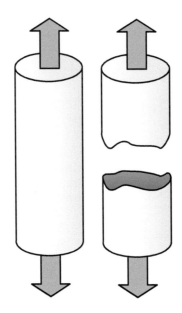

FIGURE 14.2

Failure under tension.

A straight pull like this is called **tension**. If the nominal cross-sectional area of the break is known, and the number of atoms in that plane is also known, one can, in principle, calculate the force to separate the atoms at the break zone and also the work required (i.e., force × distance) to break it.

In a crystal of iron, the geometry is particularly simple. A **crystal** contains a repeated arrangement of the atoms that make up the structure. Fig. 14.3 shows the arrangement of the atoms in a perfect crystal of iron. If the crystals are large enough

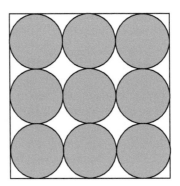

FIGURE 14.3

Structure of an iron crystal containing nine atoms. The centers of the atoms are 0.234×10^{-9} m apart.

(i.e., there are trillions and trillions of atoms all arranged in the same pattern), they can often be seen with low-power optical microscopes, and their atomic structure can be deduced using X-rays.

The crystal structure will arbitrarily extend in every direction in this idealized model (as in a single crystal). At the fracture plane shown in Fig. 14.3, this works out to contain the planar projection of 1.8×10^{19} atoms per m^2. All we need to know now is the strength of each bond and we will have a model of the theoretical strength of this material. We can estimate this quite easily by doing a simple thought experiment: Imagine you started with a lump of iron and then you heated it until it evaporated (just like boiling water but much hotter). The total energy absorbed represents the energy to break all the bonds in the solid iron and make individual atoms detach from the solid. This turns out to be about 6.6×10^{-19} J per atom of iron. We will use this as a measure of the bond strength of all the atoms in the original piece of iron. Therefore, the *work* per unit area required to fracture a piece of iron is just about $(1.8 \times 10^{19}$ atoms per $m^2) \times (6.6 \times 10^{-19}$ J per atom of iron$) = 12$ J/m^2. This sounds quite modest.

Calculating the force required to break the material is another matter. Recall that work is force × distance. So, we need to consider how far do the atomic planes in iron need to be pulled apart to consider them fully separated. Atomic forces fall off very rapidly with distance and a reasonable estimate of the minimum distance required to produce fracture is 0.1−0.2 nm (recall that a nm is a nanometer, which is 10^{-9} m) *beyond* the equilibrium center-to-center separation of the individual iron atoms (0.234 nm), as shown in Fig. 14.4. At that additional separation, the individual iron atoms will no longer interact, and fracture has occurred.

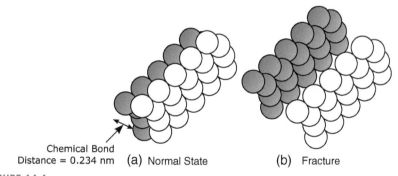

Chemical Bond
Distance = 0.234 nm   (a) Normal State   (b) Fracture

FIGURE 14.4

Before and after view of fracture along an atomic plane in iron.

The origin of the attractive force in metals is found in their atomic structure. Metals have free (negatively charged) electrons inhabiting the spaces between (positively charged) metal ions (see Fig. 14.5). This electronic structure provides strong forces between the iron atoms and also explains why it is easy to get electrons to flow in metal wires.

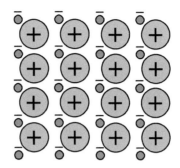

FIGURE 14.5

Metallic bonds.

We will equate the work done in separating the atoms beyond their equilibrium positions to a force × an *assumed* average distance of 0.15 nm of separation. Using the 12 J/m^2 fracture work per unit area calculated above,

$$\text{Work/Area}\left(12 \text{ J/m}^2\right) = \text{Force/Area}\left(\text{N/m}^2\right) \times \text{Distance}\left(0.15 \times 10^{-9}\text{m}\right)$$

$$\text{Therefore, } \frac{\text{Force}}{\text{Area}} = 80. \times 10^9 \text{ N/m}^2$$

The units N/m^2 have the name **Pascal** (abbreviated as **Pa**), but since we are dealing with very large numbers here, we use the *giga* prefix of 10^9. Our answer is therefore about 80 GPa (G being the prefix for *giga*). This force per unit area is equivalent to piling 80 billion apples on top of a 1 × 1 m tray here on Earth, and it is a much larger force/area than observed in practice, which is about 450 MPa (or 0.45 GPa). Why was our estimate so far below laboratory test values? It's because real crystals have a finite size with atomic level defects that are usually randomly oriented. Fig. 14.6 shows a polished piece of copper as viewed under a

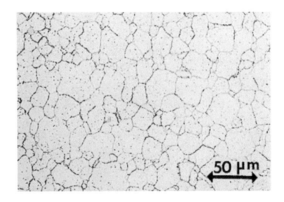

FIGURE 14.6

Polished copper showing individual crystals.

Courtesy of the Copper Development Association, Inc.

microscope. The individual random crystals and crystal grain boundary defects of copper are obvious.

Fracture may occur between crystals at their relatively weak **grain boundaries** rather than within the crystals, resulting in lower fracture forces than the previous mathematical model.

In addition, not all materials are crystalline. Some have a completely random **amorphous** structure. Many materials may also consist of mixed phases such as crystals in a matrix of amorphous material.

The other class of materials we will deal with is **polymers**. Polymers are repeating chains of small molecular groups (the word *polymer* means "many molecular pieces"). Generally, polymers are made of organic chemical links.[3] For example, the common polymer polyethylene consists of long chains of ethylene molecules written as -[CH_2 - CH_2]- strung together, each dash representing the net attraction of electrons between two adjacent carbon atomic nuclei (see Fig. 14.7). Each carbon atom also has two off-axis hydrogen atoms and one on-axis C - C bond. Roughly speaking, the strength of a C - C bond is about 10% that of an iron-to-iron bond, so you might think that our model apparently implies a force/area of about $(0.1)(80\ GPa) = 8\ GPa$ as the upper limit to break a typical polymer bond. However, this is not true.

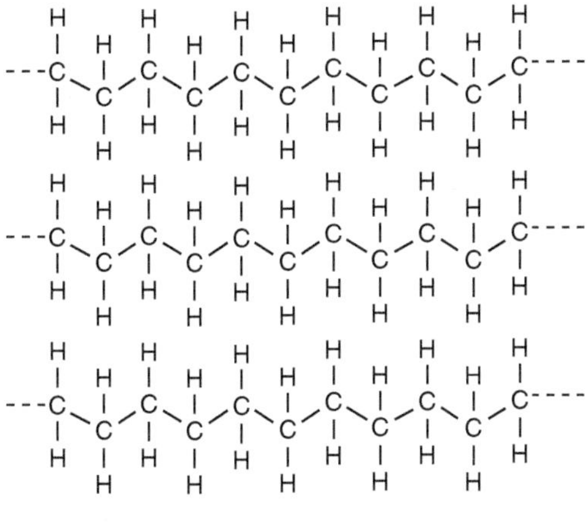

FIGURE 14.7

Regular array of polyethylene molecular chains.

What you need in this case is not the force between individual carbon atoms but that between adjacent polymer chains. While there are many kinds of possible molecular interactions, for the case of polyethylene, the kinds of forces are lumped under the name "van der Waals" forces. They are quite weak, about 0.1% of the

---

[3] "Organic" meaning based on carbon, hydrogen, and a few other atoms such as nitrogen, oxygen, and sulfur.

strength of the metal bonds, or about 0.08 GPa. The upper limit for the strength of a polymeric material might therefore be of this magnitude.

14.4 Defining materials requirements

Consider the choice of material for a car's bumper. Defining material requirements for a bumper begins by understanding what a bumper is and what it does. A bumper is a structure often integrated into the car's chassis; typically, it is 1−2 m wide, 0.1−0.2 m high, and 0.02−0.04 m thick and is attached to the front and rear of the car about half a meter above the ground.

Fig. 14.8 shows a "conceptual bumper" for which a material will be selected. A bumper does two main jobs:

1) It survives undamaged at a very low-speed (1−2 m/s) collision.[4]
2) It affords modest protection for the car, though sustaining major damage to itself, in a moderate-speed (3−5 m/s) collision.

What does an engineer mean by terms such as *collision*, *survive undamaged*, and *protection for the car*? Translation of such language into engineering variables, units, and numbers begins with the now familiar concept of energy conversion.

A collision is a rapid process of energy conversion. A moving car is carrying translational kinetic energy. A collision rapidly converts that TKE into **"elastic"** and **"plastic"** energy of the materials of construction. Elastic energy is a form of stored potential energy, while plastic energy is converted, via internal collisions of the atoms in the material, into heat. In a properly designed bumper, these two processes of energy conversion will occur in the bumper—not in the car itself and certainly not in the passengers.

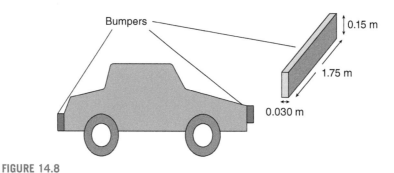

FIGURE 14.8

Conceptual bumpers.

---

[4] Federal regulations call for protection against damage to the vehicle in a 2.50 mph (1.12 m/s) collision; while this might seem to be a modest requirement, remember that cars are heavy!

In a really low-speed collision, the energy conversion should leave the bumper looking as it did before the collision. In a moderate-speed collision, the bumper may be damaged or destroyed, but it should absorb and release the energy in a way that it leaves the car and the passengers undamaged.

A wide range of bumper designs can achieve these low- and moderate-speed energy conversion objectives. To select a design from these possibilities, an engineer first picks dimensions for a bumper. For simplicity, consider a bumper to be a rectangular solid object 1.75 m × 0.15 m × 0.030 m = 0.0079 m³ (width, height, and thickness, respectively). For a given level of protection, the engineer will then seek the lightest material that allows a bumper of these dimensions to meet the design requirements.

Why does the engineer seek the lightest material? In the days of metal bumpers, a bumper provided a significant part (about 5%) of the mass of the car. The higher the mass of the car, the lower the gas mileage. Conversely, anything that reduces mass will increase gas mileage. So all other things being equal, a car with lighter bumpers will have a higher gas mileage than a car with heavier bumpers. Our optimization problem will be choosing the material that will give the lightest bumper of given dimensions that will do a bumper's job.

Solving that optimization problem begins by using the principle of energy conservation to translate the design requirements from words into variables, units, and numbers.

Example 14.1

Estimate the total energy that a 1.00×10^3 kg car has to absorb in a 2.50 mph (1.12 m/s) collision with a perfectly rigid wall. If all of this is absorbed by the bumper, what is the "specific" energy absorbed (i.e., energy per m³ of bumper volume) given its dimensions of 1.75 m × 0.15 m × 0.030 m?

Need: Energy absorbed ____ in J and the specific energy absorbed ____ in J/m³.

Know: Speed of car, dimensions of bumper.

How: The conservation of energy provides the energy absorbed, and the energy absorbed divided by the volume of the bumper gives the specific energy absorbed.

Solve: The bumper must absorb the total TKE of the vehicle from a 1.12 m/s collision. The vehicle's TKE is

$$TKE = \frac{1}{2}mV^2 = \frac{1}{2} \times 1.00 \times 10^3 \times 1.12^2 \text{ [kg]} \times \text{[m/s]}^2 = \textbf{6.27} \times \textbf{10}^2 \textbf{ J}$$

and the bumper's volume = 1.75 m × 0.15 m × 0.030 m = 0.0079 m³. Therefore, the **specific energy** absorbed by bumper = energy absorbed/volume of bumper = $6.27 \times 10^2 \times 10^{-6}/0.0079$ [J]×[1/m³]×[MJ/J] = **0.079 MJ/m³**.

Notice if we wanted 5.00 mph (2.24 m/s) protection, the total energy absorbed goes to about 2.5 kJ and the specific energy absorbed goes to 0.32 MJ/m³, and at 35.0 mph (15.6 m/s) the corresponding numbers are 0.12 MJ and 1.6×10^2 MJ/m³.

Is there a candidate bumper material that can absorb these amounts of energy elastically and rebound to its original dimensions? Or will there be sustained damage to the bumper or, even worse? For us, the choice of a bumper material boils down to a choice between a polymer and a metal. A polymer offers light weight at some sacrifice of strength. A metal offers high strength but with a penalty of higher weight. Which should an engineer choose? Answering these questions begins by considering

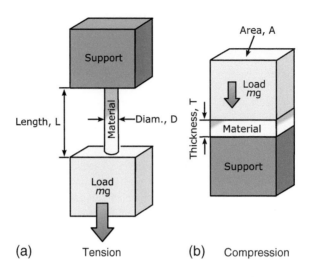

FIGURE 14.9

(a and b) Basic material tests.

the types of materials that nature and human ingenuity offer. Making that choice requires applying a new tool: the **stress–strain diagram**.

The stress–strain diagram is a tool for using measurements of two new variables, **stress** and **strain**, to quantify the terms **modulus of elasticity**, **elastic limit**, **plasticity**, **yield strength**, and **toughness**. Two basic experiments are shown in Fig. 14.9. Materials are either stretched under **tension**, as in Fig. 14.9a, or put into **compression**, as in Fig. 14.9b. Add more weight, mg, in either of the situations and the wire stretches or the plate compresses accordingly. In this way, one can plot a diagram of the stretch in L or compression in T versus the applied load. The response is considered **elastic** if the material returns to its initial dimensions when the load is removed.

Fig. 14.10 is a composite plot of four elastic experiments. In tension, the long rod (or wire, etc.) shown in curve (b) stretched in proportion to its load twice as much as did the shorter one shown in curve (a). The compression shown in curve (d) is half that shown in curve (c) and is also proportional to its load. This is an expression of **Hooke's law**.

$$(L - L_0) \propto L_0 \frac{mg}{A} \tag{14.1}$$

Notice that we have used a convention that tension is a positive load so that stretching is a positive response to it, and the opposite is true for compression and the contraction response to it. But why do we have to plot these data as clumsily as in Fig. 14.10? Eq. (14.1) is our clue that we can be more inclusive as well as more general. We write it as:

$$\frac{mg}{A} \propto \frac{(L - L_0)}{L_0} \quad \text{or} \quad \frac{mg}{A} = E \frac{(L - L_0)}{L_0} \, (\text{thus defining } E) \tag{14.2}$$

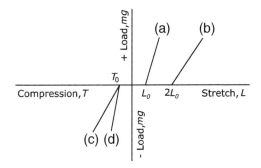

FIGURE 14.10

Elastic response of a material to tension and to compression: (a) Tension, initial length L_0, diameter D; (b) tension, initial length $2L_0$, diameter D; (c) compression, initial thickness T_0, load area A; (d) compression, initial thickness T_0, load area $2A$.

The fractional stretching or compression is the change in length divided by its original length, $(L - L_0)/L_0$. This is called the **strain**. If the material stretches in tension, the strain is positive, and if the material is compressed, the strain is negative. We use the Greek letter epsilon (ε) to denote strain. Note that strain is a dimensionless number, since it is the quotient of two quantities with the length dimensions that in SI units are meters/meter but just as well could be inches/inch.

When the force causing the strain is divided by the applicable cross-sectional area we get the **stress** in the material, and denoted it by the Greek letter sigma (σ). Again, a tensile stress is positive, and a compressing stress is negative. Stress has SI units of N/m^2 (or Pa) and English units of lbf/in^2 (or psi). The constant E in Eq. (14.2) is called **Young's modulus** or the **Elastic modulus**.

Hooke's law can now be written as follows:

$$\sigma = E\varepsilon \qquad (14.3)$$

Eq. (14.3) is the way that engineers use Hooke's law, and it should be committed to memory in this form once it is understood.

Example 14.2

When a mass of 1000. kg of steel is carefully balanced on another piece of steel of depth 1.00 m, height 0.15 m, and thickness 0.030 m, and sits on a rigid base, the length of the middle piece of steel contracts by 0.000010 m.

a. Is the middle piece of steel in tension or compression?
b. What is the stress in the middle piece of steel?
c. What is the strain in the vertical direction of the middle piece?

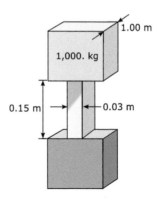

Need: (a) Type of stress, (b) stress _____ N/m², and (c) strain _____ fraction.
Know: Mass of the object compressing the piece = 1000. kg. Initial thickness of sample is
0.15 m and it suffers a contraction of 1.0×10^{-5} m when the load is applied.
How: By definition, stress = force/area, and force = weight = mg. The supporting cross-
sectional area is $1.00 \times 0.030 \text{ m}^2 = 0.030 \text{ m}^2$. By definition,

$$\text{strain} = \text{(change in length)/(initial/length)}$$

Solve: (a) The middle piece is in compression.
(b) **Stress, σ** = weight/area = $-(1000. \times 9.81)/0.030$ [kg] × [m/s²] × [1/m²]

$$= -3.27 \times 10^5 \text{ N/m}^2 = -3.27 \times 10^5 \text{ Pa.}$$

The negative sign indicates compression.
(c) Since **strain** is the change in length divided by the initial/length or

$$\varepsilon = -1.0 \times 10^{-5}/0.15 \text{ [m]/[m]} = -6.7 \times 10^{-5}$$

ε is dimensionless, and the negative sign indicates compression.

Suppose one made a large number of measurements like those described in
Example 14.2 but with differing loads. Suppose one then plotted the results on a
graph, with strain as the horizontal axis and stress as the vertical axis. For each
axis, the negative direction would be compression, and the positive direction would
be tension. The resulting graph is called a **stress–strain diagram**. It captures in one
diagram three of the most important properties of a material: elasticity, strength, and
toughness.

The stress–strain diagram of an idealized steel looks like Fig. 14.11. It also defines some new terms: **yield stress** (or **yield strength**), **yield strain**, and **plastic
deformation**.

The **yield stress** is the maximum stress at the limit of elastic behavior in which
the specimen returns to its initial length when the load is removed. In the case shown
in Fig. 14.11, it is 200. MPa. The **yield strain** is the strain corresponding to the
elastic yield stress. But plastic deformation causes a *permanent* change in the properties of the steel when it passes beyond its elastic limits. Stretching here is called
plastic deformation. For this idealized steel, compressive properties mirror the tensile ones.

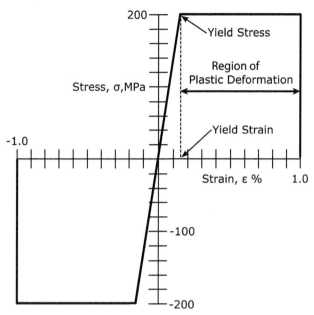

FIGURE 14.11

Simplified stress–strain curves for an idealized steel.

Compare the yield stress in Fig. 14.11 of 200 MPa to the theoretical failure value for iron calculated earlier at 80 GPa. In other words, our sample of steel is failing at a stress about 200 MPa or a factor of about 400 lower than its theoretical maximum. Presumably, this failure is initiated at inclusions or at crystal boundaries rather than by pulling apart the rows of atoms of iron in the steel.

The elastic modulus E defined in Eq. (14.3) is just the slope of the stress–strain curve. In Fig. 14.11, it is $E = 200$ MPa/0.0015 $= 130$ GPa.

Another variable of interest is **toughness**. In ordinary speech, people sometimes confuse it with (yield) stress or strength. To an engineer, however, strength and toughness have very different meanings. A material can be strong without being especially tough. One example is a diamond. It can be subjected to great force and yet return to its original shape, so it is therefore very strong. Yet, a diamond can be easily shattered and therefore is not very tough. A material can also be tough without being strong. An example is polycarbonate, a kind of polymer used in football helmets. It is relatively easily dented and therefore not very strong. Yet, it absorbs a great deal of energy per unit mass without shattering and is therefore very tough.

In Fig. 14.11, the stress–strain curve for strains greater than the yield strain is shown as a horizontal line parallel to the strain axis. In actual stress–strain diagrams, the shape of that portion of the curve is more complicated. This region is called the **plastic** region, and when it is described with a horizontal line, it is called a **perfectly**

plastic region. This overall model is called **perfectly elastic**, **perfectly plastic**. It's a useful simplification of how real materials behave.

The horizontal line in the perfectly plastic region extends out to a **maximum strain**. At that strain, the material "fails". The total strain at failure is as the sum of the elastic strain and the plastic strain. The important point is that a wire stretched beyond its maximum strain breaks. A plate compressed to its maximum strain shatters or spreads. The **toughness** of a material is defined as the area under the portion of the stress–strain curve that extends from the origin to the point of maximum strain.

Because this area (shaded in Fig. 14.12) is the result of stress, measured in N/m^2, and of strain, measured in m/m, toughness has the units of $N{\cdot}m/m^3$, or J/m^3 (that is, energy or work per unit volume). It is usually measured in a dynamic experiment involving a strike by a heavy hammer. Toughness is therefore the preferential criterion to be used to predict material failure if a sudden load is applied.

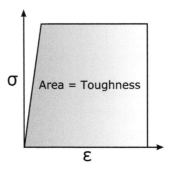

FIGURE 14.12

Toughness, defined as the work done until failure, represents the ability of the material to absorb energy.

Example 14.3

Given the stress–strain diagram for steel in Fig. 14.11, determine (**a**) its modulus under compression, (**b**) its yield strength under tension, (**c**) its toughness under tension to yield, and (**d**) its toughness under tension to 1.0% strain.

Need: E = _____ GPa under compression, σ = _____ MPa at yield, and toughness at yield and 1.0% strain = _____ MJ/m^3.

Know: Stress/strain curve in Fig. 14.11; toughness is area beneath the stress/strain curve.

How: Hooke's law $\sigma = E\varepsilon$, and use stress/strain curves.

Solve: a) Need slope in compression region for σ/ε curve.

  $E = -200.$ [MPa]/-0.0015 [m/m] = **130 GPa**.

  b) Under tension, the **yield stress** is $\sigma =$ **2.00 MPa**.

  c) Toughness under tension up to the yield stress is equal to the triangular area, toughness at yield $= \frac{1}{2} \times 0.0015$ [m/m] $\times 200.$ [MN/m^2] = **0.15 MJ/m^3**.

  d) **At 1% strain**, add rectangular area: $(0.01-0.0015) \times 200.$ = 1.7 MJ/m^3. Therefore, the **toughness** = 0.15 + 1.7 = **1.9 MJ/m^3**.

The no-damage elastic portion of the curve is only a small fraction of the total protection afforded by a car's bumper. You can also see that plastic deformation, which will manifest itself in permanent deformation or crushing, affords a much better place to dissipate energy than does the purely elastic behavior. Of course, whether a particular bumper design can distribute this energy uniformly and avoid exaggerated *local* deformation is up to the skill of the engineer.

We are now in a position to see if a metal bumper made of steel satisfies our main constraint—what does the stress—strain diagram tell us about its ability to absorb the energy of a collision yet return to its original shape?

Example 14.4

Can the car's bumper from Example 14.1 made of steel absorb the TKE of the 2.50 mph (1.12 m/s) collision assumed there? Or will it plastically deform? What if it is traveling at 5.00 mph (2.24 m/s)? Also, what is the weight of the steel bumper if its density is 7850 kg/m^3? Assume the mechanical behavior is described by the stress—strain diagram of Fig. 14.11.

Need: Energy to be absorbed in bumper during collision = ____ MJ/m^3 compared to energy of collision. Also, weight of bumper = ____ N.

Know: From Example 14.1: TKE = 6.27×10^2 J at 2.50 mph; specific energy absorbed = 0.079 MJ/m^3; steel bumper volume = 0.0079 m^3; density of steel, = 7850 kg/m^3. At 5.00 mph, TKE = 2.5 kJ and the specific energy absorbed is 0.32 MJ/m^3. Also, from Example 14.3, the elastic toughness = 0.15 MJ/m^3 and plastic toughness is 1.8 MJ/m^3 at 1.0% strain.

How: Compare specific energy to be absorbed to the capability of the bumper.

Solve: Assume the entire surface of the bumper contacts the wall simultaneously (i.e., the force on the bumper is uniform over its entire area).

At 2.50 mph, the specific energy to be absorbed is 0.079 MJ/m^3 of steel. Steel can absorb up to 0.15 M J/m^3, so it will elastically absorb this amount of energy—again if uniformly applied. If so, the bumper will bounce back to its original shape.

At 5.00 mph, the specific energy to be absorbed goes to 0.32 MJ/m^3 of steel. The bumper will plastically deform since 0.32 MJ/m^3 > 0.15 MJ/m^3 but will not fail since 1.9 MJ/m^3 > 0.32 MJ/m^3.

The weight of this bumper is mg = density × volume × g

$$= 7850 \times 0.0079 \times 9.81 \ [kg/m^3] \times [m^3] \times [m/s^2] = \mathbf{610 \ N}$$

or the weight of a physically fit student!

14.5 Materials selection

The previous section showed how the stress—strain diagram provides the information needed to determine if a material satisfies the design requirements. The question now becomes: Is there a material of less weight than steel that also satisfies the design requirements? To be specific, let us consider a polymer, a material that is much less dense than steel. Can it compete with steel in the bumper application? Again, our tool is the stress—strain diagram. Fig. 14.13 and Table 14.1 are for a polycarbonate.

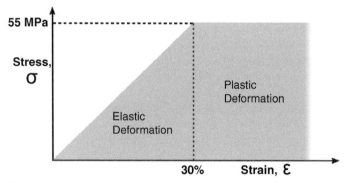

FIGURE 14.13

Stress and strain for a polycarbonate.

Table 14.1 Properties of a polycarbonate.

Elastic yield, MPa	Strain yield %	Density, kg/m³
55.	30.	1300.

Note: The properties shown in this table are that of a very common polycarbonate used in crash helmets, CDs, DVDs, and bottles, as well as car bumpers.

Notice that the polycarbonate it is not as "strong" as steel because its yield strength is considerably less, but it can be stretched much further while elastically returning to its original strength. Let us repeat Example 14.4 using this material for the bumper.

Example 14.5

If the car's bumper in Example 14.1 is made of the polycarbonate described by Fig. 14.13, can it absorb the TKE of the low-speed 2.50 mph (1.12 m/s) collision, or will it plastically deform? What if it is traveling at 5.00 mph (2.24 m/s)? What is the weight of this bumper if its density of polycarbonate is 1300 kg/m³? Assume the compressive stress–strain diagram mirrors the tensile stress–strain diagram in Fig. 14.13.

Need: Energy to be absorbed in bumper after collision = _____ MJ/m³ compared to energy of collision and the weight of a polycarbonate bumper.

Know: From Example 14.1, at 2.50 mph, TKE = 6.27×10^2 J and the specific energy absorbed = 0.089 MJ/m³. The bumper volume = 0.0079 m³.

How: Compare specific energy to be absorbed to capability of material to absorb it. Need to calculate the latter from the σ, ε diagram.

Solve: Calculate the elastic toughness by looking at Fig. 14.13. It is given at yield by the triangular area, $\frac{1}{2} \times 0.30$ [m/m] $\times$ 55. [MN/m²] = 8.3 MJ/m³.

At 5.00 mph, TKE = $\frac{1}{2} mV^2 = \frac{1}{2} \times 1.00 \times 10^3 \times 2.24^2$ [kg]×[m/s]² = **25×10^2 J**, and specific energy absorbed is

$$= 25 \times 10^2 \times 10^{-6}/0.0079 \text{ [J]}\times[1/m^3]\times[MJ/J] = \textbf{0.32 MJ/m}^3.$$

Since this is much less than 8.3 MJ/m³. Thus, this bumper can survive elastically. The weight of this bumper is mg/g_c = density × volume × g

$$= (1300 \times 0.0079) \times 9.81/1 \text{ [kg/m}^3]\times[m^3]\times[m/s^2] = \textbf{98. N}$$

or much less than the weight of a student!

So, although polymer is not as "strong" as steel, its ability to compress further without breaking gives it adequate toughness to do the job. In addition, polymers are substantially less dense than steel. So, the polymer is the winner. In the last 2 decades, polymers have almost entirely replaced steel as the material from which automobile bumpers are made.

Of course, there are still plenty of options for the materials engineer to design the preferred configuration of the bumper. For example, the metal bumper may be manufactured with springs to absorb the impact. There is surely enough material in our steel bumper that we could use some of it in the form of coil springs, as in Fig. 14.14.

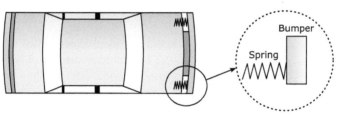

FIGURE 14.14

Mechanical design also influences the choice of materials.

A steel spring can be stretched or compressed much further than a flat piece of the same material. It's as if we traded the material for one with a lower modulus (less steep curve) and a much larger strain to yield. Furthermore, there is now plenty of room to design for the randomness of crashes—that is, how the vehicles will interact with an immoveable object such as a pole as indicated in Fig. 14.15.

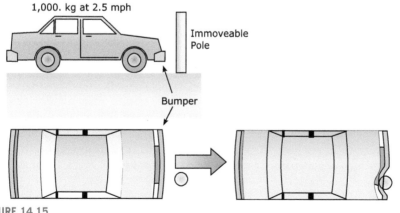

FIGURE 14.15

Not all crashes are head-on. The material may yield locally in low-speed impacts.

This chapter contains just as a glimpse into the most basic considerations that would go into a material design. Of course, materials engineers have become very clever at rearranging the macroscopic properties of materials as they need them. In the car bumper case, the polymer may be backed with a second one that is in the form of a foam—the advantage being its strain at yield is mediated by the thousands of gas-filled bubbles within its foam structure.

14.6 Properties of modern materials

This discussion of selecting a material for a bumper gives a feel for the engineer's problem of material selection. However, elasticity, strength, and toughness, though very important, are not the whole story. In the decision whether to use silicon or germanium to make transistors, the electronic and thermal properties of materials play a key role. In the selection of copper versus fiberoptic glass, information-carrying capacity becomes crucial. In the selection of steel versus polymers for refrigerators, considerations of appearance, manufacturability, and corrosion resistance become important. In the selection of a material for a hip joint transplant, compatibility with the human body becomes an essential materials requirement.

In addressing these varied requirements, 21st century materials go far beyond the traditional categories of metals, polymers, and the other classes of materials shown in Fig. 14.1. A wide range of "composite" materials now combine the properties of those original categories. Some are created by embedding fibers of one material within a matrix of a second material. This makes it possible to combine, for example, the high strength of a thin carbon fiber with the high toughness of a polymer matrix.

One of the newest areas of materials engineering are materials whose structure has been engineered at the nanometer scale ($1 \text{ nm} = 10^{-9} \text{ m}$).[5] These materials are called **nanomaterials** and are important because the unique properties of materials at the nanoscale enable engineers to create materials and devices with enhanced or completely new characteristics and properties.

Many engineers consider nanotechnology to be the next industrial revolution because it is predicted to have enormous social and economic impact. Companies have already introduced nanotechnology in several consumer products. Examples of nanomaterials already on the market include nanoscale titanium dioxide used in some cosmetics and sunscreens, nanoscale silica being used as dental fillers, and nanowhiskers used in stain-resistant fabrics. Nanoclays and coatings are being used in a range of products from tennis balls to bikes to cars to improve bounce, strengthen high-impact parts, or render material scratchproof. Some nanomaterials

---

[5] As small as a nanometer is, it's still relatively large compared to atoms. Atoms, which are the basic building blocks of materials, are on the order of $0.1-0.5$ nm in size. An atom's nucleus is much smaller again—about 0.00001 nm.

even are based upon biological models. Fig. 14.16 illustrates the use of carbon nanotubes.

FIGURE 14.16

Carbon nanotubes.

Dr. Michael De Volder Institute for Manufacturing, University of Cambridge, UK.

Nanomaterials are literally tailored molecule by molecule. This makes it possible, for example, to provide materials with precise combinations of electronic and optical properties. In one sense, however, the emergence of these new composites is actually a matter of going "back to the future." The original materials used by humanity some 10,000 years ago, such as wood, stone, and animal skins, are complicated natural composites. Fig. 14.1 summed up this long-term history of humanity's materials use. Even with the new challenges of the 21st century, however, the traditional properties of elasticity, strength, and toughness will continue to play a central role in materials selection. As they do, the stress—strain diagram, the best representation of these properties, remains a crucial tool for the 21st century engineer.

Summary

Materials engineers develop and specify materials. They do so using constrained optimization. Thus one property, such as performance or weight or cost, is to be optimized (maximized or minimized), subject to meeting a set of constraints (design requirements).

Materials selection begins by quantifying design requirements. It proceeds by searching the major classes of candidate materials, metals, polymers, composites, and so on for candidates capable of meeting design requirements. The candidates are then subjected to further screening, based on such characteristics as the **strain** they exhibit in response to **stress**. This particular characteristic is plotted on a graph called the **stress—strain diagram**. This stress—strain or σ - ε

diagram can be then used to determine key properties of the material: **modulus of elasticity, yield strength, plastic deformation, and toughness**. These properties, expressed in the appropriate units, provide the basis (the "know") for solving particular constrained optimization problems of materials selection.

Exercises

The figures below depict the situations described in Exercises 1–6.

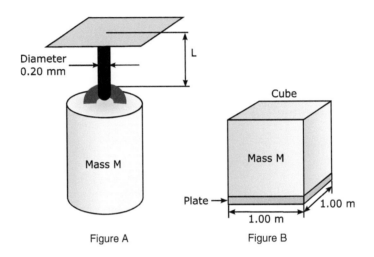

Figure A   Figure B

1. A mass of $M = 1.0$ kg is hung from a circular wire of diameter 0.20 mm as shown in Figure A. What is the stress in the wire? (**Ans. 0.31 GP.**)
2. If the wire in Figure A is stretched from 1.00 to 1.01 m in length, what is the strain of the wire?
3. In Figure B, when a block of metal 1.00 m on a side is placed on a metal plate 1.00 m on a side, the stress on the plate is 1.00×10^3 N/m^2. What is the mass of the metal cube? (**Ans. 102 kg.**)
4. Suppose the plate described in Exercise 3 and Figure B was 0.011 m thick before the cube was placed on it. Suppose that placing the cube on it causes a compressive strain of 0.015. How thick will the plate be after the cube is placed on it? (**Ans. 0.011 or unchanged to three significant figures.**)
5. Suppose the wire in Figure A is perfectly elastic. When subjected to a stress of 1.00×10^4 Pa, it shows a strain of 1.00×10^{-5}. What is the elastic modulus (i.e., Young's modulus) of the wire?
6. A plate of elastic modulus 1.00 GPa is subjected to a compressive stress of -1.00×10^3 Pa as in Figure B. What is the strain on the plate? (**Ans. -1.00×10^6.**)

For Exercises 7–9, assume that a silicone rubber has this stress–strain diagram.

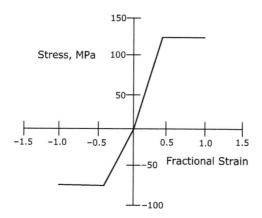

7. What is the yield strength under compression of the silicone?
8. A flat saucer made of silicon rubber has an initial thickness of 0.0050 m. A ceramic coffee cup of diameter 0.10 m and mass 0.15 kg is placed on a plate made of the same silicon rubber. What is the final thickness of the plate beneath the cup, assuming that the force of the cup acts directly downward and is not spread horizontally by the saucer?

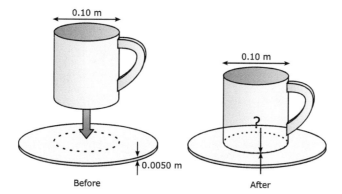

Before   After

9. What is the maximum number of such coffee cups that can be stacked vertically on the saucer and not cause a permanent dent in the plate? (**Ans. 4.3 × 10⁵ cups**—a tough balancing act)
10. A coat hanger is made from polyvinyl chloride (PVC). The "neck" of the coat hanger is 0.010 m in diameter and initially is 0.10 m long. A coat hung on the coat hanger causes the length of the neck to increase by 1.0×10^{-5} m. What is the mass of the coat? The stress—strain diagram is provided. (**Ans. 1.3 kg.**)

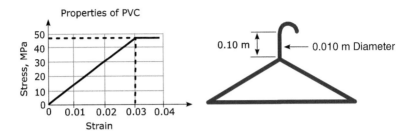

Properties of PVC

11. How many coats having the same mass as those in exercise 10 must be hung on the hanger to cause the neck to remain permanently stretched (i.e., plastically deformed) after the coats are removed?

For Exercises 12–15, imagine the idealized situation as shown in Figure A for an artillery shell striking an armor plate made of an (imaginary) metal called "armory." Figure B is the stress–strain diagram of armory. Assume the material diagram shows armory strained to failure.

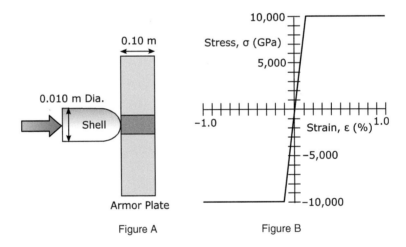

Figure A   Figure B

12. Suppose the shell has mass 1.0 kg and is traveling at 3.0×10^2 m/s. How much TKE does it carry?
13. Assume that the energy transferred in the collision between the shell and the armor plate in Exercise 12 affects only the shaded area of the armor plate beneath contact with the flat tip of the shell (a circle of diameter 0.0010 m) and does not spread out to affect the rest of the plate. What is the energy density delivered within that shaded volume by the shell? (**Ans.** 5.7×10^9 J/m³.)
14. Which of the following will happen as a result of the collision in the Exercise 13? Support your answer with numbers.
    a. The shell will bounce off without denting the armor plate.
    b. The armor plate will be damaged but will protect the region beyond it from the collision.

c. The shell will destroy the armor plate and retain kinetic energy with which to harm the region beyond the plate. (**Ans. a**—but it is uncomfortably close to yielding and to a permanent set to the armor; an armored tank crew will at least get quite a headache.)

15. In exercise 14, what is the highest speed the artillery shell can have and still bounce off the armor plate without penetrating it?

Exercises 16–19 involve the situation depicted below. Consider a "micrometeorite" to be a piece of mineral that is approximately a sphere of diameter 1.0×10^{-6} m and density 2.0×10^3 kg/m^3. It travels through outer space at a speed of 5.0×10^3 m/s relative to a spacecraft. Your job as an engineer is to provide a micrometeorite shield for the spacecraft. Assume that if the micrometeorite strikes the shield, it affects only a volume of the shield 1.0×10^{-6} m in diameter and extending through the entire thickness of the shield.

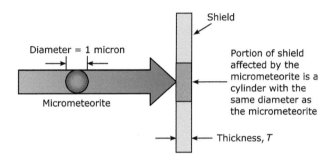

16. Using the stress–strain diagram for steel in Fig. 14.11, determine the minimum thickness a steel micrometeorite shield would have to be to protect the spacecraft from destruction (even though the shield itself might be dented, cracked or even destroyed in the process).

17. Using the stress–strain properties for a polymer (Fig. 14.13 and Table 14.1), determine whether a sheet of this polymer 0.10 m thick could serve as a micrometeorite shield if this time the shield must survive a micrometeorite strike without being permanently dented or damaged. Assume the properties of the polymer are symmetric in tension and in compression. (**Ans. Yes**, it will survive unscathed.)

18. Suppose you were required to use a micrometeorite shield no more than 0.01 m thick. What would be the required toughness of the material from which that shield was made if the shield must survive a micrometeorite strike without being permanently dented or damaged?

19. Suppose a shield exactly 0.01 m thick of the material in exercise 18 exactly met the requirement of surviving without denting or damage at a strain of -0.10, yet any thinner layer would not survive. What are the yield stress and elastic modulus of the material? (**Ans.** $\sigma_{\text{Yield}} = -33$ **MPa**, $E = 0.33$ **GPa.**)

20. Your company wants to enter a new market by reverse-engineering a popular folding kitchen step stool whose patent has recently expired. Your analysis shows that by simplifying the design and making all the components from injection-molded PVC plastic you can produce a similar product at a substantially reduced cost. However, when you make a plastic prototype and test its performance you find that it is not as strong and does not work as smoothly as your competitor's original stool. Your boss is anxious to get your design into production because he/she has promised the company president a new high-profit item by the end of this quarter. What do you do? (Use the Engineering Ethics Matrix.)

    a. Release the design. Nearly everything is made from plastic today, and people do not expect plastic items to work well. You get what you pay for in the commercial market.

    b. Release the design but put a warning label on it, limiting its use to people weighing less than 150 pounds.

    c. Quickly try to find a different, stronger plastic that can still be injection molded and adjust the production cost estimate upward.

    d. Tell your boss that your tests show the final product to be substandard and ask if he/she wants to put the company's reputation at risk. If he/she presses you to release your design, make an appointment to meet with the company president.

21. As quality control engineer for your company, you must approve all material shipments from your suppliers. Part of this job involves testing random samples from each delivery and making sure they meet your company's specifications. Your tests of a new shipment of carbon steel rods produced yield strengths 10% below specification. When you contact the supplier, they claim their tests show the yield strength for this shipment is within specifications. What do you do? (Use the Engineering Ethics Matrix.)

    a. Reject the shipment and get on with your other work.

    b. Retest samples of this shipment to see if new data will meet the specifications.

    c. Accept the shipment, since the supplier probably has better test equipment and has been reliable in the past.

    d. Ask your boss for advice.

Mechanical engineering

Source: Maxx-Studio/Shutterstock.com

15.1 Introduction

Mechanical engineering is one of the most versatile disciplines in the broad field of engineering. Virtually every aspect of life is touched by mechanical engineering. If something moves or uses energy, a mechanical engineer probably was involved in its design, testing, and production. Mechanical engineers also use economic, social, environmental, and ethical principles when they create new systems and products.

15.2 Mechanical engineering

Mechanical engineering consists of the fields of **thermal design** and **machine design** (see Fig. 15.1). Mechanical engineers use the principles of thermal and

Exploring Engineering. https://doi.org/10.1016/B978-0-443-13541-5.00034-9

339

machine design in the development of systems such as cars, trucks, aircraft, ships, spacecraft, turbines, industrial equipment, robots, heating and cooling systems, medical equipment, and much more. They also design the automated machines that mass-produce these products. In this chapter, we will explore the basic elements of the six subtopics shown in Fig. 15.1. We begin by defining the meaning of each of the subtopics.

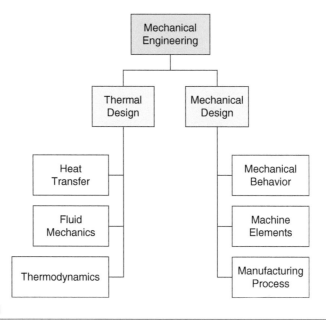

FIGURE 15.1

The structure of mechanical engineering.

15.2.1 Thermal design

Thermal design is the study of heat transfer, fluid mechanics, and thermodynamics.

- **Heat transfer** has applications in heat exchanger design, engine cooling, air conditioning, and cooling of electronics.
- **Fluid mechanics** includes fluid behavior in applied areas such as piping networks, hydraulic systems, bearings, turbo-machineries, and airfoils.
- **Thermodynamics** is the study of the conservation of energy in applied areas such as engine and refrigeration cycles, power plants, and combustion.

15.2.2 Machine design

Machine design is the study of mechanical behavior, machine elements, and manufacturing processes.

- **Mechanical behavior** includes statics, dynamics, strength of materials, vibrations, reliability, and fatigue.
- **Machine elements** are basic mechanical parts of machines such as gears, bearings, fasteners, springs, seals, couplings, and so forth.
- **Manufacturing processes** include areas such as computerized machine control, engineering statistics, quality control, ergonomics, and life cycle analysis.

In this chapter, we will explore the basic elements of these six areas of mechanical engineering.

15.3 The elements of thermal design

We talk about "heat" as though it is a material substance. The phrase "*heat transfer*" implies that some "thing" is being moved from one place to another. Many years ago, heat was thought to be a colorless, odorless, weightless fluid. This led to the concept that heat *flows* from one object to another. It was later discovered that heat is not a physical fluid substance. Heat is really just another form of energy, called *thermal energy*. Nevertheless, the phrase *heat flow* is still in common use today.

15.3.1 Heat transfer

When two objects at different temperatures come into thermal contact, they exchange thermal energy (heat). The hotter object gives thermal energy to the colder body until their temperatures are equal, at which point they are said to be in a state of "thermal equilibrium". Heat transfer is such a large and important mechanical engineering topic that most engineering programs have at least one required course in it.

Heat transfer equations are usually written as a heat transfer *rate* (i.e., heat per unit time). To determine the total amount of thermal energy transferred as a system undergoes a process from one equilibrium state to another you need to multiply the heat transfer rate by Δt, the time required for the process to occur.

In engineering, **heat transfer** refers to the thermal energy that passes between objects either by **thermal conduction** in which energy is transferred by molecular vibration, by **thermal convection** in which a fluid moves between regions of different temperature, or by **thermal radiation**, in which energy is transmitted by electromagnetic radiation between surfaces of different temperature. These three heat transfer modes are shown in Fig. 15.2 and described below.

15.3.1.1 Thermal conduction

The basic equation for the rate of steady state conduction heat transfer is called **Fourier's Law of Conduction**. For steady state conduction heat transfer through a flat plate or wall as shown in Fig. 15.3, Fourier's law is:

$$\left(\dot{Q}_{\text{cond}}\right)_{\text{plane}} = k_t A\left(\frac{T_{\text{hot}} - T_{\text{cold}}}{\Delta x}\right) \tag{15.1}$$

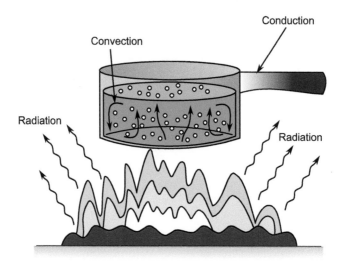

FIGURE 15.2

The three modes of heat transfer.

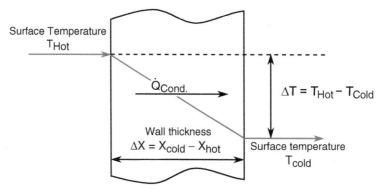

FIGURE 15.3

Thermal conduction through a wall.

where $\dot{Q}_{cond}$ is the conduction heat transfer *rate*, k_t is the **thermal conductivity** of the material, A is the cross-sectional area perpendicular to the heat transfer direction, and Δx is the thickness of the plate or wall. The algebraic sign of this equation is such that a positive $\dot{Q}_{cond}$ always corresponds to heat transfer from the hot to the cold side of the plane or wall.

Example 15.1

The glass in the window of your dorm room is 2.00 ft wide, 3.00 ft high, and 0.125 in thick. If the temperature of the outside surface of the glass is 21.0°F and the temperature of the inside surface of the glass is 33.0°F, what is the rate of heat loss through the window?

Need: $\dot{Q}_{cond} = ?$

Know: $T_{hot} = T_{inside} = 33.0°F$, $T_{cold} = T_{outside} = 21.0°F$, $A = 2.00 \times 3.00 = 6.00\ ft^2$, and

$$\Delta x = 0.125\ in \times [1.00\ ft\ /\ 12.0\ in] = 0.0104\ ft.$$

How: The heat loss rate can be calculated from Eq. (15.1).

Solve: From Table 15.1, the thermal conductivity of the glass is $k_t = 0.45\ Btu/(h \cdot ft \cdot R)$, and then Eq. (15.1) gives

$$(\dot{Q}_{cond})_{loss} = k_t A \left(\frac{T_{hot} - T_{cold}}{\Delta x} \right) = 0.45 \left[\frac{Btu}{hr \cdot ft \cdot R} \right] \times 6.00 [ft^2] \left(\frac{33.0 - 21.0}{0.0104} \right) \left[\frac{R}{ft} \right]$$

$$= 3120\ Btu\ /\ h$$

Note that the temperature difference, $T_{hot} - T_{cold}$, has the same numerical value regardless of whether Fahrenheit or Rankine temperatures are used because

$$T_{hot} - T_{cold} = (460. + 33.0\ R) - (460. + 21.0\ R) = 33.0 - 21.0 = 12.0\ R = 12.0°F$$

Table 15.1 Thermal conductivity of various materials at 20°C.

Material	Thermal conductivity k_t	
	B t u/(h•ft·R)	W/(m·K)
Air at atmospheric pressure	0.015	0.026
Engine oil	0.084	0.145
Liquid water	0.343	0.594
Window glass	0.45	0.78
Mercury	5.02	8.69
Aluminum	118.0	204.0
Copper	223.0	386.0

[1] **Btu** (British thermal unit) = 1055 J, which is the amount of heat needed to raise the temperature of 1 lbm of water by 1°F.

15.3.1.2 Thermal convection

Convective heat transfer occurs whenever an object is either hotter or colder than a surrounding *fluid*. Convection only occurs in a moving liquid or a gas, never in a solid. The basic equation for the rate of convection heat transfer is known as **Newton's Law of Cooling**,

$$\dot{Q}_{conv} = hA(T_\infty - T_s) \tag{15.2}$$

where $\dot{Q}_{cond}$ is the convective heat transfer *rate*, h is the convective **heat transfer coefficient**, A is the surface area of the object being cooled or heated, T_∞ is the **temperature of the surrounding fluid**, and T_s is the **surface temperature** of the object. In English units, h is expressed in units of $Btu/(h•ft^2•R)$, and in SI units, it is $W/(m^2•K)$. The algebraic sign of Newton's law of cooling is positive for $T_\infty > T_s$ (heat transfer into the object) and negative when $T_\infty < T_s$ (heat transfer out of the object).

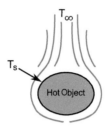

FIGURE 15.4

Natural convection around a hot object.

There are two types of convection—natural convection and forced convection. **Natural convection** is produced by density differences in a fluid produced by temperature differences in the fluid (e.g., as in "hot air rises"). Global atmospheric circulation and local weather phenomena (including wind) produce convective heat transfer. Fig. 15.4 illustrates natural convection around a hot object.

In **forced convection**, the fluid is in motion produced by a pump or a fan. It is one of the main forms of heat transfer use by engineers because large amounts of thermal energy can be moved efficiently by forced convection. It is used in heating and air conditioning systems, electronics cooling, and in numerous other technologies. Forced convection is used in designing *heat exchangers*, in which one fluid stream is used to heat or cool another fluid stream. Fig. 15.5 illustrates a flat plate being cooled by forced convection from a fan.

Table 15.2 provides values for the convection heat transfer coefficient h, under various conditions.

FIGURE 15.5

Forced convection over a flat plate by a cooling fan.

Table 15.2 Typical values of the convective heat transfer coefficient.

Type of convection	Convective heat transfer coefficient, h	
	$Btu/(h \cdot ft^2 \cdot R)$	$W/(m^2 \cdot K)$
Air, free convection	1−5	2.5−25
Air, forced convection	2−100	10−500
Liquids, forced convection	20−3000	100−15,000
Boiling water	500−5000	2500−25,000
Condensing water vapor	1000−20,000	5000−100,000

Example 15.2

A new computer chip with a surface area of 1.00 cm^2 generates 10.0 W of heat. Determine the convective heat transfer coefficient needed to keep the temperature of the chip less than 20.0°C above the environmental temperature. Can the chip be cooled with air, and will it require free or forced convection?

Need: The convective heat transfer coefficient, h.

Know: $A = 1.00 \text{ cm}^2$, $\dot{Q}_{conv} = -10.0$ W (the sign is negative because the chip is losing heat), and $T_\infty - T_s = -20.0°\text{C}$ (the surface temperature, T_s, is greater than the environmental temperature $T\infty$ here).

How: Determine h from Eq. (15.2) and compare it to the values in Table 15.2.

Solve: Eq. (15.2) is $\dot{Q}_{conv} = hA(T_\infty - T_s)$ so

$$h = \dot{Q}_{conv} / A(T_\infty - T_s) = (-10.0 \text{ [W]}) / \left\{ (1.00 \text{ [cm}^2\text{]})(1 \text{ [m]}/100 \text{ [cm]})^2(-20.0 \text{ [°C]}) \right\}$$

$$= 5000 \text{ W/m}^2 \cdot \text{K (to one significant figure)}$$

Table 15.2 has this value of h in the forced convection range for liquid.

15.3.1.3 Thermal radiation

Radiation heat transfer includes all electromagnetic radiation. Infrared, ultraviolet, visible light, radio and television waves, microwaves, X-rays, gamma rays, and so on are all forms of radiation heat transfer. If an object with a surface temperature T_s is emitting or receiving thermal radiating energy to or from its surroundings at temperature T_∞, then the rate of thermal radiation is expressed by the **Stefan–Boltzmann Law of Radiation**,

$$\dot{Q}_{rad} = \varepsilon A \sigma \left(T_\infty^4 - T_s^4 \right) \tag{15.3}$$

where $\dot{Q}_{rad}$ is the radiation heat transfer *rate*, ε is the dimensionless emissivity (the hotter object is said to *emit* energy, while the colder object *absorbs* energy) of the object (a **black** object is defined to be any object whose emissivity is $\varepsilon = 1$, and a perfectly reflecting object has an emissivity of 0), A is the surface area of the object, and σ is the **Stefan–Boltzmann constant**,

$$\sigma = 5.69 \times 10^{-8} \text{ W/m}^2 \cdot \text{K}^4 = 0.1714 \times 10^{-8} \text{ Btu/h} \cdot \text{ft}^2 \cdot \text{R}^4$$

Note that this Eq. (15.3) contains temperatures raised to the fourth power. This means that *absolute* temperature units (Rankine or Kelvin) must **always** be used in this equation.

Fig. 15.6 illustrates how solar radiation can be used to heat water for household use, and Table 15.3 lists values for the thermal emissivity values for various materials.

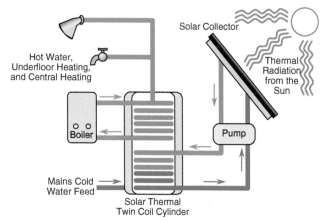

FIGURE 15.6

Solar water heater using thermal radiation from the sun.

Table 15.3 Emissivity values of various materials.

Material	Temperature		Emissivity ε (dimensionless)
	°C	°F	
Aluminum	100	212	0.09
Concrete	21	70	0.88
Flat black paint	21	70	0.90
Flat white paint	21	70	0.88
Glass	21	70	0.95
Iron (oxidized)	100	212	0.74
Wood	21	70	0.90

Example 15.3

The ancient Egyptians were said to be able to make ice in the desert by exposing a tray with a thin layer of water to a cloudless black night sky. Some say that the night sky temperature is 3 K (which is the background temperature of the universe.) However, the night sky is full of air that radiates back at us, so the generally accepted value of the temperature of a cloudless night sky is about $-70°C$ or 203 K. The emissivity of the water is $\varepsilon = 0.9$ and the surface area of the tray is 0.250 m^2. Determine how long it will take to freeze water initially at $0°C$ if 385 kJ must be removed by radiation.

Need: Time to freeze water initially at $0°C$.

Know: $\dot{Q} = \frac{dQ}{dt} = \frac{\Delta Q}{\Delta t}$ so $\Delta t = \frac{\Delta Q}{\dot{Q}}$

How: From Eq. (15.3), we know that $\dot{Q}_{rad} = \varepsilon A \sigma \left(T_\infty^4 - T_s^4\right)$, $\sigma = 5.69 \times 10^{-8} \text{ W/m}^2 \cdot \text{K}^4$, and $\Delta Q = -385 \text{ kJ}$ (negative because heat must be removed), $T_\infty = 203 \text{ K}$, and $T_s = 273 \text{ K}$.

Solve: $\dot{Q}_{rad} = \varepsilon A \sigma (T_\infty^4 - T_s^4) = (0.9)(0.250 [m^2])(5.69 \times 10^{-8} [W/(m^2 \times K^4)]) (203^4 - 273^4 [K^4]) = -49.4$ W Then, $\Delta t = (-385 \times 10^3 [J])/(-49.4 [J/s]) = 7800$ s $=$ **2.17 h**. Note that we must use absolute temperatures in Eq. (15.3) because the temperatures are raised to the fourth power.

15.3.2 Fluid mechanics

A *fluid* can be either a liquid or a gas, and like solid mechanics that separates the study of statics (fixed solids) from dynamics (bodies moving under forces), fluid mechanics deals with **fluid statics** and **fluid dynamics**. Liquids are normally incompressible, whereas gases are compressible. Fluids are characterized by various physical properties, such as the following:

- **Density** is the mass per unit volume, and is measured in kg/m^3 or lbm/ft^3.
- **Pressure** is the force per unit area and is measured in pascals (1 Pa $= 1 N/m^2$) or in psi (1 psi $= 1$ lbf/in^2). Normal air pressure at sea level is 101,325 Pa $= 14.696$ lbf/in^2 (usually rounded to 101.3 Pa and 14.7 psi).
- **Viscosity** is the fluid's resistance to flow and is measured in units of $N \cdot s/m^2$ or in $lbf \cdot s/ft^2$.

Table 15.4 summarizes these fluid mechanic units.

Pressure is measured either relative to the atmospheric pressure, called *gage* pressure, or as an absolute pressure. For example, atmospheric pressure at the surface of the Earth is 0 N/m^2 (or 0 psi) gage, but it is 101.325 kN/m^2 absolute pressure. Therefore,

$$p_{absolute} = p_{gage} + p_{atmosphere} \tag{15.4}$$

Fluid flow can be either laminar or turbulent. **Laminar** flow occurs at low speeds when a fluid flows in parallel layers that slide past another like playing cards. In laminar flow, the motion of the fluid is very orderly, moving in straight lines parallel to the walls.

Turbulent flow is characterized by chaotic fluid motion. The fluid undergoes irregular fluctuations, or mixing, and the speed of the fluid at a point is continuously changing. Common examples of turbulent flow are blood flow in arteries, fluid flow

Table 15.4 Fluid mechanics units.

Property	Symbol	Units	
Pressure	p	N/m^2 (pascal)	lbf/in^2 (psi)
Speed	V	m/s	ft/s
Density	ρ (Greek "rho")	kg/m^3	lbm/ft^3
Viscosity	μ (Greek "mu")	$N \cdot s/m^2$	$lbf \cdot s/ft^2$

in pipelines, convective motions in the atmosphere and in ocean currents, and the flow of fluids through pumps and turbines.

The variable that determines which type of flow is present is the **Reynolds number** (Re), defined as

$$\text{Re} = \frac{\rho V D}{\mu} \qquad (15.5)$$

where ρ and μ are the fluid's density and viscosity, respectively, and V and D are a fluid speed and a characteristic distance (for example, for fluid flowing in a pipe, V is the average fluid speed, and D is the pipe diameter). The Reynolds number is dimensionless, as shown by its units:

$$\text{Re} = \frac{\rho V D}{\mu} = \left[\frac{\text{kg}}{\text{m}^3}\right]\left[\frac{\text{m}}{\text{s}}\right][\text{m}]\left[\frac{\text{m}^2}{\text{N}\cdot\text{s}}\right] = \left[\frac{\text{kg}}{\text{m}^3}\right]\left[\frac{\text{m}}{\text{s}}\right][\text{m}]\left[\frac{\text{m}^2\cdot\text{s}^2}{\text{kg}\cdot\text{m}\cdot\text{s}}\right] = \text{dimensionless}$$

Fluid flow in a pipe is **laminar** for Reynolds numbers up to 2000. Beyond a Reynolds number of 4000, the flow is completely **turbulent**. Between 2000 and 4000, the flow is in transition between laminar and turbulent.

Example 15.4

Show that the Reynolds number is dimensionless when Engineering English units are used, and calculate the Reynolds number of water flowing at an average speed of 6.75 ft/s in a pipe with an inside diameter of 3.25 inches. The density of the water is 62.4 lbm/ft^3 and its viscosity is 2.35×10^{-5} lbf·s/ft^2.

Need: The units of the Reynolds number in Engineering English units and the Reynolds number of water flowing at 3.75 ft/s inside a 3.25 inch pipe.

Know: Expression (15.5) gives the Reynolds number as Re = ρVD/μ.

How: In Engineering English units, $[\rho] = [\text{lbm}/\text{ft3}], [V] = [\text{ft}/\text{s}], [D] = [\text{ft}]$, and

$$[\mu] = \left[\text{lbf}\cdot\text{s}/\text{ft}^2\right].$$

Solve: Notice that since these units contain both lbm and lbf, we will need to introduce g_c at some point. Substituting into the Reynolds number equation, we get the following:

$$[\text{Re}] = \frac{\left[\dfrac{\text{lbm}}{\text{ft}^3}\right]\left[\dfrac{\text{ft}}{\text{s}}\right][\text{ft}]}{\left[\dfrac{\text{lbf}\cdot\text{s}}{\text{ft}^2}\right]} = \left[\frac{\text{lbm}}{\text{ft}^3}\right]\left[\frac{\text{ft}}{\text{s}}\right][\text{ft}]\left[\frac{\text{ft}^2}{\text{lbf}\cdot\text{s}}\right] = \left[\frac{\text{lbm}\cdot\text{ft}}{\text{lbf}\cdot\text{s}^2}\right]$$

Since this is not dimensionless, we need to divide by $g_c = 32.2$ bm·ft/lb·fs^2 to get

$$\frac{[\text{Re}]}{[g_c]} = \frac{\left[\dfrac{\text{lbm}\cdot\text{ft}}{\text{lbf}\cdot\text{s}^2}\right]}{\left[\dfrac{\text{lbm}\cdot\text{ft}}{\text{lbf}\cdot\text{s}^2}\right]} = [0]$$

For water flowing at 6.75 ft/s inside a 3.25 inch pipe, the Reynolds number is

$$\text{Re}/g_c = (62.4)(6.75)(3.25)(1/12) / \left[(2.35 \times 10^{-5})(32.2)\right] = 1.51 \times 10^5 \text{ [dimensionless]}.$$

15.3.2.1 Fluid statics

For an incompressible fluid as rest, the pressure difference between two points is just the weight per unit volume of the fluid between these points, or

$$p_2 - p_1 = \frac{\rho g(z_2 - z_1)}{g_c} \tag{15.6}$$

where $z_2 - z_1$ is the difference in height between the two points, g is the acceleration of gravity (9.81 m/s^2 or 32.2 ft/s^2), and $g_c = 32.2$ lbm·ft/lbf·s^2 in the Engineering English units, and 1 (a dimensionless integer) in SI units. Eq. (15.6) accurately predicts how the pressure varies with depth in lakes and oceans, but not the atmosphere. Fig. 15.7 shows the pressure profile of the Earth's atmosphere. Note that it is not linear as predicted by Eq. (15.6) because the atmosphere is compressible and Eq. (15.6) is only accurate for incompressible fluids (i.e., liquids). For compressible fluids, the relationship between pressure and height is not linear.

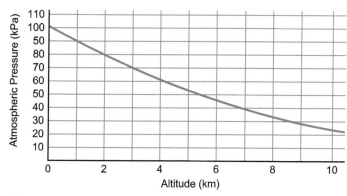

FIGURE 15.7

Atmospheric static pressure profile. We live comfortably only at the top left of this curve.

Example 15.5

Find the gage and absolute pressure under the surface of the ocean at a depth of 300. m. The density of salt water is 1025 kg/m^3.

Need: $p_{absolute}$ and p_{gage} at 300. m under the sea.
Know: The density of sea water is $\rho = 1025$ kg/m^3.
How: Eq. (15.4) gives $p_{absolute} = p_{gage} + p_{atmosphere}$.
Solve: From Eq. (15.6) with $g_c = 1$ and dimensionless in the SI system of units, we have

$$p_{gage} = p_{300m} - p_{atmosphere} = \frac{\rho g \left(z_{300m} - z_{surface} \right)}{g_c}$$

$$= \frac{(1025 \ [\text{kg/m}^3])(9.81 \ [\text{m/s}^2])(300. - 0 \ [\text{m}])}{1} = 3.02 \times 10^6 \ \text{kg/(m·s}^2) \ \text{gage}$$

$$= 3.02 \times 10^6 \ \text{N/m}^2 \ \text{gage} = 3020 \ \text{kN/m}^2 \ \text{gage} \qquad = 3020 \ \text{kPa gage}$$

Then, the absolute pressure at 300. m is:

$$p_{300m} = \left(p_{300m} - p_{\text{atmosphere}}\right) + p_{\text{atmosphere}} = 3020 \; \left[\text{kN/m}^2\right] + 101.3 \; \left[\text{kN/m}^2\right]$$
$$= 3020 \; \text{kN/m}^2 \; \text{absolute (to 3 significant figures)}$$

15.3.2.2 Fluid dynamics

An incompressible fluid in steady motion can be modeled between points 1 and 2 in a flow by the **Bernoulli Equation**:

$$\frac{p_1 - p_2}{\rho} + \frac{V_1^2 - V_2^2}{2g_c} + \frac{g}{g_c}(z_1 - z_2) = 0 \tag{15.7}$$

where p is the fluid pressure, V is the fluid speed, z is the fluid height, and ρ is the fluid density.

Also, g is the acceleration of gravity (9.81 m/s² or 32.2 ft/s²), and $g_c = 32.2$ lbm·ft/lbf·s² in the Engineering English units, and 1(an integer) in SI units. This form of the equation is approximate only since it neglects any frictional pressure losses due to the fluid's viscosity.

Example 15.6

Water flows through the nozzle shown below. The average water speed at the inlet is $V_1 = 20.0$ ft/s. If the pressure difference between the inlet and the outlet of the nozzle is $p_1 - p_2 = 100.$ lbf/ft², determine the outlet water speed.

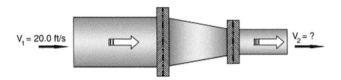

The density of the water is 62.4 lbm/ft³.
Need: The speed the nozzle's outlet water, $V_2 = ?$
Know: The inlet water speed is $V_1 = 20.0$ ft/s, $p_1 - p_2 = 100.$ lbf/ft², and the density of water is 62.4 lbm/ft³. The nozzle is horizontal, so $z_1 - z_2 = 0$.
How: Use Eq. (15.7).
Solve: From Eq. (15.7), we have:

$$\frac{p_1 - p_2}{\rho} + \left(\frac{V_1^2 - V_2^2}{2g_c}\right) + \frac{g}{g_c}(z_1 - z_2) = \frac{100. \; \left[\text{lbf/ft}^2\right]}{62.4 \; \left[\text{lbm/ft}^3\right]} + \left(\frac{20.0^2 - V_2^2 \; \left[\text{ft}^2/\text{s}^2\right]}{2(32.2 \; \left[\text{lbm·ft/lbf·s}^2\right])}\right)$$
$$+ 0 = 0$$

Solving for V_2 gives: $V_2 = \mathbf{22.4}$ **ft/s.**

15.3.3 Thermodynamics

Thermodynamics is the study of energy and its conversion from one form into another form. The efficient use of natural and renewable energy sources is one of the most important technical, political, and environmental issues of the 21st century. For those reasons, thermodynamics can be used in ways that improve the lives of people around the world.

15.3.3.1 Thermodynamic system

In thermodynamics, we are interested in the energy flows in and out of **systems**, and we carry out our analysis using **system diagrams**. The "system" is the item we are analyzing and everything else is called the system's **surroundings**. There are three types of systems: isolated, closed, and open. An **isolated system** cannot exchange mass or energy with its surroundings. A **closed system** can exchange energy, but not mass, with the surroundings. An **open system** can exchange mass and energy with the surroundings. Fig. 15.8 illustrates these different systems.

15.3.3.2 The first law of thermodynamics

The **first law of thermodynamics** states that energy is conserved, meaning that energy cannot be created or destroyed. But energy *can* be converted from one form of energy (e.g., chemical) into another form of energy (e.g., heat). There are many different forms of energy—kinetic, potential, thermal, magnetic, chemical, and so forth. Modern society benefits from a variety of energy conversion technologies. For example, an automobile's engine converts the chemical energy of gasoline into the kinetic energy of the vehicle.

In addition to the common forms of energy, we also have something called **internal energy**, designated by the symbol U. The first law of thermodynamics tells us that for a closed system the internal energy, U, can change only if energy moves

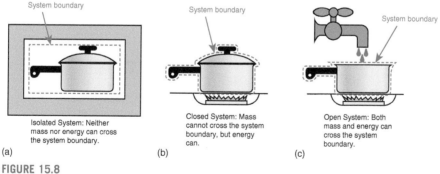

(a) Isolated System: Neither mass nor energy can cross the system boundary.

(b) Closed System: Mass cannot cross the system boundary, but energy can.

(c) Open System: Both mass and energy can cross the system boundary.

FIGURE 15.8

An illustration of isolated, closed, and open systems.

in or out of the system in the form of heat, work, or changes in the system's kinetic or potential energy. For any process that converts the system from *state 1* into *state 2*, the change in internal energy, $U_2 - U_1$ is determined by an energy balance on the system:

$$Q - W = U_2 - U_1 + m\left(\frac{V_2^2 - V_1^2}{2g_c}\right) + \frac{mg}{g_c}(Z_2 - Z_1) \qquad (15.8)$$

where the quantity Q is the *heat added to the system* during the change of system state and W is *the work done by the system* during the change of system state.[1] The two remaining terms in Eq. (15.8) are the changes in kinetic and potential energy of the system, and $g_c = 32.2$ lbm·ft/lbf·s^2 in the Engineering English units, and 1 (an integer) in SI units. Eq. (15.8) is the mathematical form of the first law of thermodynamics for a closed system. Note that U, Q, and W must all have the same units; Btu, ft·lbf, or joules (J).

The heat transfer term, Q, in Eq. (15.8) consists of any of the three modes that were discussed in Section 15.3.1, conduction, convection, and radiation. The work term, W, in Eq. (15.8) can also have several modes, mechanical work, electrical work, electrochemical work, and so forth.

Example 15.7

A 0.100 kg steel ball is dropped vertically from a height of 30.0 m and strikes the ground. The initial and final speeds of the ball are both zero. If there is no heat transfer or work involved in the process, determine the change in the ball's internal energy.

Need: $U_2 - U_1$ for the ball.
Know: Q and W for the ball are both zero, $V_1 = V_2 = 0$, and $m = 0.100$ kg.
How: Use Eq. (15.8) to find $U_2 - U_1$ for the ball.
Solve: Substituting the known values into Eq. (15.8), we get

$$Q - W = 0 = U_2 - U_1 + 0 + \frac{(0.100 \text{ [kg]})(9.81 \text{ [m/s}^2\text{]})}{1}(0 - 30.0 \text{ [m]})$$

or $U_2 - U_1 = 29.4 \text{ kg·m}^2/\text{s}^2 = 29.4 \text{ (kg·m/s}^2\text{)·m} = 29.4 \text{ N·m} = 29.4$ J

The example below contains an error in the solution. The error could be in the equations, units, or computation. Can you find the error(s)?

Example 15.8: What's wrong with this solution?

A 5.00 lbm is book sitting 3.00 feet from the floor on the edge of a desk. The book is pushed off the desk and falls to the floor. Determine the change in the book's internal energy right after the book hits the floor if there is no work done or heat transfer?

---

[1] This is due to historical convention. Heat entering a system is positive and heat leaving a system is negative, but work has the opposite sign convention. Thus, work leaving a system (i.e., the system doing work) is positive but work entering a system (i.e., something doing work on the system) is negative. The usage developed because engineers wanted to know how much power was developed for how much coal was needed to run their steam engines and they made these terms both positive.

Need: Where the initial potential energy of the book went.
Know: $Z_1 = 3.00$ ft and $Z_2 = 0$ ft, $W = 0$, $Q = 0$, and $V_2 = V_1 = 0$.
How: The answer can be found from Eq. (15.8).
Solve: Start with Eq. (15.8)

$$Q - W = (U_2 - U_1) + m\left(\frac{V_2^2 - V_1^2}{2g_c}\right) + \frac{mg}{g_c}(Z_2 - Z_1) = 0$$

Then solve for the change in internal energy

$$U_2 - U_1 = 0 - m\left(\frac{V_2^2 - V_1^2}{2g_c}\right) - \frac{mg}{g_c}(Z_2 - Z_1) = 0 - \frac{mg}{g_c}(Z_2 - Z_1)$$

$$= -\frac{(5.00 \text{ lbm})(32.2 \text{ ft/s}^2)}{2(32.2 \text{ lbm}\cdot\text{ft/lbf}\cdot\text{s}^2)}(0 - 3.00 \text{ ft}) = 7.50 \text{ ft}\cdot\text{lbm}$$

Is this answer correct?[2] Why?

---

[2] The correct answer is 15.0 ft·lbf.

15.4 The elements of machine design

Any form of design is basically a decision-making process. The design process involves developing a plan to meet a need or solve a problem. For example, if you are designing something as simple as a chair, some of the things you will need to consider are as follows:

1. The type of chair (e.g., an easy chair, an office chair, or a chair for a dining room table)
2. The person using the chair (e.g., an adult or a child)
3. The material used in the construction of the chair (wood, metal, plastic, or a combination of materials)
4. The strength and cost of the material used to make the chair
5. The shape and artistic appeal of the chair

In this section, we are concerned with the elements of machine design. In mechanical engineering, a "machine" is any mechanical system that transforms or transmits energy to do useful work.

15.4.1 Mechanical behavior

When engineers select the materials that are to be used in the machines and parts they design, they need to know how these materials will behave under reasonable operating conditions. This subject is covered in detail in Chapter 14 of this book with a very short review below.

The criterion to be optimized when choosing materials for an engineering application might be cost, or weight, or performance, or some index such as minimizing weight × cost with a requirement of a specific strength. The constraints (also called design requirements) typically involve such words as **elastic modulus** (also called **stiffness**), **elastic limit**, **yield strength**, and **toughness**. While we all have loose ideas of what is meant by these terms, it is necessary to precisely express them as engineering variables. Those variables must, in turn, contain the appropriate numbers and correct units.

A design engineer must be able to:

1. Define material requirements.
2. Understand classes of materials such as metals and polymers.
3. Understand the internal microstructure of materials that can be crystalline and/or amorphous.
4. Use stress–strain diagrams to express materials properties in terms of five engineering variables: stress, strain, elastic limit, yield strength, and toughness.
5. Use the results determined for those variables to carry out materials selection.

For example, if a child drops a plastic toy onto the floor, the impact should leave the toy looking like it did before it was dropped. So, as the toy's designer, how do you chose a plastic that is tough enough to survive a child's play, but not so expensive as to be unaffordable?

Example 15.9

Estimate the total energy that a 0.09 kg plastic toy car has to absorb in a 1.00 m/s collision with a perfectly rigid floor. If all of this energy is absorbed in the toy, what is the "specific" energy absorbed (i.e., energy per m^3 of volume) if the toy car has a volume of 5.69×10^{-5} m^3?

Need: Energy absorbed and the specific energy absorbed.

Know: Speed of impact, dimensions of toy.

How: The conservation of energy provides the energy absorbed, and the energy absorbed divided by the volume of the toy gives the specific energy absorbed.

Solve: The toy car must **absorb** the total kinetic energy of the impact from a 1.00 m/s collision with the floor. The vehicle's kinetic energy is

$$\tfrac{1}{2}\,mv^2 = \tfrac{1}{2} \times 0.09 \times 1.00^2 \; [\text{kg}][\text{m/s}]^2 = 0.045 \text{ J}$$

The toy's volume $= 5.69 \times 10^{-5}$ m^3. Therefore, the specific energy absorbed $=$ energy absorbed/volume of toy

$$= 0.045 \; / \; (5.69 \times 10^{-5}) \; [\text{J}][1/\text{m}^3] = 790 \text{ J}/\text{m}^3$$

With this information, the toy designer could survey the physical properties of various commercial plastics and choose one that meets his/her needs.

15.4.2 Machine elements

Machine elements are basic mechanical parts used as the building blocks of most machines. They include shafts, gears, bearings, fasteners, springs, seals, couplings, and so forth. In this section, we are going to focus on the most commonly used machine element, gears.

Gears are used when engineers must deal with rotary motion and rotational speed. Rotational speed is defined in two ways, the more familiar being N, the *revolutions per minute* (RPM) of a wheel. There is also a corresponding "scientific" unit of rotational speed in terms of circular measure, radians/s. Its symbol is the Greek lowercase letter omega (ω). There are 2π radians in a complete circle. Hence, $N = 60 \times \omega/2\pi$ [s/minute] [radians/s] [revolution/radian] = RPM; conversely:

$$\omega = 2\pi N/60 \text{ (in radians / s when } N \text{ is in RPM)} \tag{15.9}$$

Angular speed ω is also directly related to **linear speed** v. Each revolution of a wheel of radius r covers 2πr in forward distance per revolution. Therefore, at N RPM the wheel's tangential speed is $v = 2\pi rN/60 = r\omega$ (in m/s if r is in meters). Hence,

$$v = r\omega \tag{15.10}$$

Automobile engines can have rotational speeds of 500–7000 RPM (maybe 10,000 RPM in very high-performance engines), but we have vehicles that are moving at speeds of, say, 100. km/h (62.1 mph). If the tire outer diameter is 0.80 m (radius of 0.40 m), what is the wheel's rotational speed when the vehicle is moving at 100. km/h?

In our current case, the wheels are rotating at a circular speed corresponding to the formula $r\omega = v = (100.\text{ km/h}) \times (1000\text{ m/km}) \times (1\text{ h/3600 s}) = 27.8$ m/s. Hence, $\omega = v/r = 27.8/0.40 = 70.$ radians/s or $70. \times 60/2\pi = 670$ RPM. Somehow, the rotational speed of the engine must be transformed into the rotational needs of the wheels. How can these two different speeds of rotation be reconciled? It is done by a mechanism called a transmission. A manual transmission is made of several intermeshing toothed gears. These gears are simply a wheel with a toothed circumference (normally on the outside edge) as shown in Fig. 15.9.

A gear *set* or gear *cluster* is a collection of gears of different sizes, with each tooth on any gear having exactly the same profile as every other tooth (and each gap between the teeth being just sized to mesh). The teeth enable one gear to drive another—that is, to transmit rotation, from one gear to the other. *Note:* A simple gear pair as in Fig. 15.9a *reverses* the rotational direction of the driven gear from that of the driving gear. You need at least three gears in a set of simple gears as per Fig. 15.9b to transmit in the same direction as the original direction.

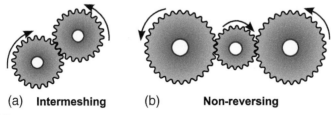

(a) **Intermeshing**   (b)   **Non-reversing**

FIGURE 15.9

(a) Intermeshing gears and (b) nonreversing gears.

The **gear ratio** (GR) of a gear train is the ratio of the angular speed of the input gear to the angular speed of the output gear. It is easier to think in terms of N, the RPM, rather than in terms of angular speed ω, in radians per second, so that the GR of a simple gear train is:

$$\text{Gear Ratio (GR)} = \frac{\text{Input rotation}}{\text{Output rotation}} = \frac{N_1}{N_2} = \frac{d_2}{d_1} = \frac{t_2}{t_1} \qquad (15.11)$$

in which "d" stands for diameter and "t" stands for the number of teeth per gear. In other words, to make the output (driven gear) turn faster than the input (driving gear), we need a GR less than 1 and must choose an output (driven gear with diameter d_2) that is smaller than the input (driving gear with diameter d_1). To make the output (driven gear) turn slower than the input (driving gear), we need a GR greater than 1 and must choose an output (driven gear with diameter d_2) that is larger than the input (driving gear with diameter d_1).

Example 15.10

A 6.00 cm diameter gear is attached to a shaft turning at 2000. RPM. That gear in turn drives a 40.0 cm diameter gear. What is the RPM of the driven gear?

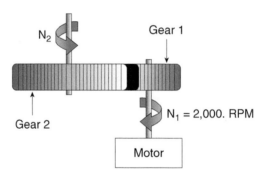

Need: RPM of driven gear, N_2.
Know: Speed of driving gear $N_1 = 2000.$ RPM. Radius of driving gear (r_1) is 3.00 cm. Radius of driven gear (r_2) is 20.0 cm.
How: Use Eq. (15.11): $\frac{N_2}{N_1} = \frac{d_1}{d_2}$
Solve: $N_2 = 2000.$ [RPM] $\times 6.00/40.0$ [cm /cm] $= \mathbf{3.00 \times 10^2}$ **RPM**.

Compound gear sets—sets of multiple interacting gears on separate shafts, as shown in Example 15.10, can be very easily treated using the gear ratio concept introduced above.

What is the gear ratio for the full set of compound gears? Note that in Fig. 15.10 gears 2, 4, and 6 are *driven* and 1, 3, and 5 are *drivers*; in addition, some of these gears are connected by internal shafts so that these turn at a common speed. These relationships make it very easy to analyze compound gear trains.

$$\text{GR} = \frac{\text{Product of diameter or number of teeth of DRIVEN gears}}{\text{Product of diameter or number of teeth of DRIVING gears}} \qquad (15.12)$$

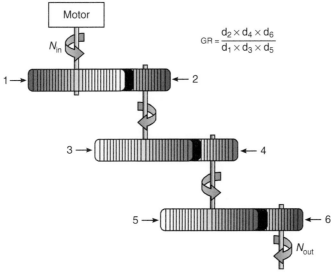

$$GR = \frac{d_2 \times d_4 \times d_6}{d_1 \times d_3 \times d_5}$$

FIGURE 15.10

Compound gear train.

Example 15.11

A 70.0 RPM motor is connected to a 100-tooth gear that couples in turn to an 80-tooth gear that directly drives a 50-tooth gear. The 50-tooth gear drives a 200-tooth gear. If the latter is connected by a shaft to a final drive, what is its RPM? We will go straight to **solve**.

Again, use a sketch to help visualize the problem.

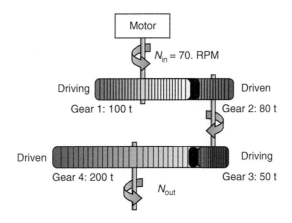

Need: The rotational speed in RPM of the final drive.

Know: The motor speed is 70.0 RPM, the driving gears have 100 and 50 teeth respectively, and the driven gears have 80 and 200 teeth.

How: Using Eq. (15.12).

Solve: From Eq. (15.12), we have

$$\text{Overall GR} = N_{in}/N_{out} = \frac{\text{Product of number of teeth on "driven" gears}}{\text{Product of number of teeth on "driving" gears}}$$

Hence, $N_{in}/N_{out} = (t_2 \times t_4)/(t_1 \times t_2) = (80 \times 200)/(100 \times 50) = 3.19$, and therefore

$$N_{out} = 70.0/3.19 = \textbf{21.9 RPM}$$

Note that the number of gear teeth is a "counted" integer, and thus has infinite significant figures.)

Torque (or twisting moment) is also a variable in gear analysis. Gears not only change the rotation speed, but they also change the torque on the axle. Eq. (15.13) shows that the torque T varies inversely with speed:

$$\frac{T_2}{T_1} = \frac{N_1}{N_2} = \frac{t_2}{t_1} = \frac{d_2}{d_1} = \frac{r_2}{r_1} \tag{15.13}$$

Therefore, to apply high torque to a shaft, you use large gears turned slowly by small intermeshing gears. However, you can use gears to achieve a desired combination of torque and RPM. For example, in a car with manual transmission, you can first use one set of gears to provide the wheels with high torque and low RPM for initial acceleration (first gear). Then you can shift to another set of gears providing the wheels with lower torque and higher RPM as the car speeds up (second gear). Then you can shift to a third gear combination that offers low torque and high RPM for cruising along a level highway at 65 miles per hour. In a modern automobile, there may be four, five, or six forward gears.

15.4.3 Manufacturing processes

This subject is covered in detail in Chapter 13 of this book and only briefly reviewed here. Virtually everything that we use at home, at work, and at play was manufactured. Manufactured goods are everywhere—aircraft, bicycles, electronics, coat hangers, automobiles, refrigerators, toys, clothing, cans, bottles, cell phones, etc. Manufacturing processes involve machining, drilling, milling, welding, extrusion, blow molding, and thermoforming.

Manufacturing plays a vital role in the U.S. economy. However, it has undergone major changes over the past several decades. Today, manufacturers use sophisticated technology to produce a variety of useful products. They have increased worker productivity, reduced the size of their workforce, and increased output. Previously, many industries required only a high school diploma for their production-related jobs, but manufacturing jobs today require workers with greater skills.

Manufacturing in the 21st century is far more than assembly and production. It also involves a broad spectrum of other functions such as research and development, engineering, sales and management, warehousing and storage, and shipping, just to touch on a few of modern manufacturing's key product life cycle components (see Fig. 15.11).

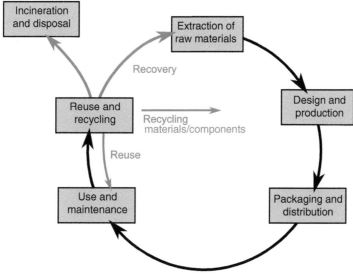

FIGURE 15.11

Life cycle manufacturing.

Life cycle manufacturing activities are critically important and add value to products by:

1) Providing a "Green" manufacturing environment.
2) Creating competitive advantage through innovation, short delivery times, customer satisfaction, and so forth.
3) Utilizing software components and communication connectivity.
4) Enabling mass customization of products for unique market segments through increased understanding of customer groups and on-demand production responses.

Summary

In this chapter, we have explored the mechanical engineering fields of thermal design and machine design. Thermal design is composed of the subjects of **heat transfer, fluid mechanics, and thermodynamics**. Machine design is composed of the subjects of **mechanical behavior, machine elements, and manufacturing processes**. Each of these subjects is so important to a mechanical engineering that a student will have at least one course in each of these six subjects.

Exercises

1) If the heat conduction rate through a 3.00 m^2 wall 1.00 cm thick is 37.9 W when the inside and outside temperatures are 20.0 and 0.00°C respectively, determine the thermal conductivity of the wall.

2) If a 15.0 × 3.00 m wall 0.100 m thick with a thermal conductivity of 0.500 W/m·K is losing 75.0 W of heat, determine the inside wall temperature if the outside wall temperature is 0.00°C.

3) An engineer needs to transfer heat at a rate of 50.7 W through a wall 0.15 m thick and 3.00 m high. If the thermal conductivity of the wall is 0.085 W/m·K and the inside and outside temperatures of the wall are 20.0 and 10.0°C, respectively, how long must the wall be in order to satisfy the heat transfer rate requirement?

4) The equation for the conduction heat transfer rate through a circular tube of length **L** is

$$\dot{Q} = 2\pi L k_t \frac{T_{inside} - T_{outside}}{\ln(r_{outside}/r_{inside})}.$$

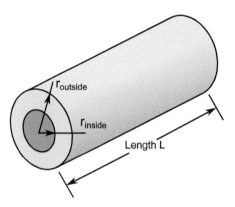

where ln() is the natural logarithm. A 35.0 foot long copper pipe with an inside diameter of 0.920 inches and an outside diameter of 1.08 inches has an inside temperature of 122°F and an outside temperature of 118°F. Determine the heat transfer rate through the pipe. The thermal conductivity of copper can be found in Table 15.1.

5) A cooking pot on a stove has a surface area of 0.01 m². The surface temperature of the pot is 70.0°C and the temperature of the surrounding air in the kitchen is 20°C. The convective heat transfer coefficient is 12.0 W/(m²·K). Determine the convection heat transfer rate from the pot to the kitchen air.

6) An electric heater provides 1500. W of heat to a room. A fan blows air over the heating element with a forced convection heat transfer coefficient of 275 W/(m²·K). The temperature of the heating element is 300.°C and the surface area of the heating element is 0.01 m². Determine the air temperature in the room under these conditions.

7) The convective heat transfer rate from a computer chip is 3.00 W. The surface temperature of the chip is 28.0°C and the air temperature is 18.0°C. Determine the convective heat transfer coefficient for the chip.

8) An electric circuit board is to be cooled by forced convection through cooling fins. 100. W of heat needs to be removed from the circuit board from a maximum temperature of 40.0°C to the room air at 20.0°C. The forced convection heat transfer coefficient is 355 W/(m²·K). Determine the required surface area of the fins.

9) A flat plate at 200.°C is painted flat black. It radiates heat into a room whose walls are at 18°C. If the surface area of the plate is 0.210 m², what is the rate of heat radiated into the room? The thermal emissivity of flat black paint can be found in Table 15.3.

10) The surface area of an average naked human adult is 2.2 m² and the average skin temperature is 33°C (91°F). The emissivity of human skin is 0.95. Determine the radiation heat transfer rate

from the human to the walls of a room if they are at 22°C (7°F). This is why a naked person feels chilly at room temperature.

11) A 4.00 inch diameter glass sphere in a vacuum chamber contains a chemical reaction that produces heat at a rate of 1.00×10^5 Btu/hr. The sphere radiates this to the inside walls of the vacuum chamber that are at 85.0°F. What is the surface temperature of the glass sphere?

12) The heat exchanger on a space craft has a surface temperature of 70°F, and emissivity of 0.83, and radiates heat to outer space at 3.00 K. If the heat exchanger must remove 375 W from the space craft, what should its surface area be?

13) What is the Reynolds number of water flowing at 3.50 ft/s through a 3.00 inch diameter pipe? The density and viscosity of the water are 62.4 lbm/ft^3 and 2.03×10^{-5} lbf·s/ft^2 respectfully.

14) You are required to keep the flow of air in a 6.00 inch diameter duct in the laminar flow region. What is the maximum average speed that the air can have before it becomes turbulent? The density and viscosity of the air are 0.074 lbm/ft^3 and 3.82×10^{-7} lbf·s/ft^2.

15) Fig. 15.7 is a plot of atmospheric pressure versus height. If you draw a straight line from the starting point to the ending point of this line and determine the slope of this line you will be able to use Eq. (15.6) to determine the "average" density of the Earth's atmosphere over this height range. What is this average density?

16) If the pressure required to crush a new submarine is 11,700 kN/m^2 absolute, what is the maximum depth it can reach in the ocean? See Example 15.5 for the properties of sea water.

17) Water slowly flows through a vertical cylindrical pipe 30.0 m in height. The density of the water is 998 kg/m^3. Determine the pressure change between the inlet (bottom) and outlet (top) of the pipe.

18) What is the exit speed from the nozzle in Example 15.6 if the fluid flowing through the nozzle is water instead of air? The density of water is 998 kg/m^3.

19) How much heat is produced by a stationary incandescent 100. W electric light bulb that has a constant internal energy after it has been illuminated for 1.00 h?

20) A 10.00 g bullet is fired horizontally at a rigid wall. The speed of the bullet just before it hits the wall is 867 m/s and it is completely stopped by the wall. If there was no heat transfer or work done by the bullet when it hit the wall, what was its initial kinetic energy and what happened to this kinetic energy after impact?

21) If you put 127 J of work into throwing a 0.123 kg rock vertically into the air without changing its internal energy, how high will it go?

22) If you drop a book from your desk onto the floor 0.983 m below the desk, with what speed does it hit the floor? There is no work or heat transfer done during this process and the book's internal energy does not change during the fall.

23) If the pedal gear in the illustration below has 60 teeth and the sprocket has 30 teeth, what is the gear ratio?

Bicycle Gears

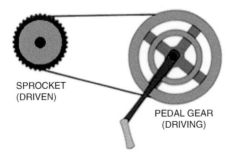

SPROCKET
(DRIVEN)

PEDAL GEAR
(DRIVING)

24) If the pedal in the illustration for Exercise 23 revolves once, how many times does the sprocket gear revolve?

25) The diameter of the sprocket in Exercise 23 is 3.25 inches. What is the diameter of the pedal gear?

26) If the 6.00 cm diameter gear in Example 15.10 has 30 teeth, how many teeth does the 40.0 cm gear have?

27) If the torque produced by the motor in Example 15.11 was 875 N m, what would be the torque at the output of the gear train?

28) What is the overall gear ratio of the 3-gear design show below and what is the RPM of gears B and C?

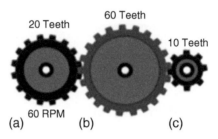

29) This concerns the case of the American Society of Mechanical Engineers (ASME) versus Hydrolevel Corporation.[3] In 1971, the engineering firm of McDonnell and Miller, Inc. requested an interpretation of the ASME Boiler and Pressure Vessel ("BPV") Code from the ASME Boiler and Pressure Vessel Codes Committee. T.R. Hardin, chairman of the ASME committee and employee of the Hartford Steam Boiler Inspection and Insurance Company in Connecticut, wrote the original response to McDonnell and Miller's inquiry. ASME's interpretation was used by McDonnell and Miller salesmen as proof of Hydrolevel's noncompliance. Subsequently, Hydrolevel never acquired sufficient market penetration for sustaining business, and eventually went bankrupt.

Hydrolevel then sued McDonnell and Miller, the Hartford Steam Boiler Inspection and Insurance Company, and ASME for restraint of trade. Hydrolevel's lawyers argued that two key ASME Subcommittee members acted not only in the self-interest of their companies, but also in violation of the Sherman Antitrust Act. The litigation against ASME went all the way to the Supreme Court where the case was settled for $4.75 million in favor of Hydrolevel.

Please prepare a one-paragraph response to the following questions:

a. How could McDonnell and Miller have avoided the appearance of a conflict of interest?

b. What was T.R. Hardin's responsibility as chairman of the BPV Code Heating Boiler Subcommittee? How could he have handled things differently to protect the interests of ASME?

30) You are an engineer who has experience as an expert witness.[4] You have been contacted by two attorneys working on the same case. The first attorney wants to hire you as an expert witness to support an inventor who claims to have developed a device that operates in a cycle and delivers 1000 Btu while consuming only 320 Btu per cycle. The second attorney is wants to hire you as an expert witness against the same inventor for allegedly defrauding investors with the invention described by the first attorney.

Please prepare a one-paragraph response to the following questions:

a. Whose expert witness would you prefer to be? Why?

b. In deciding which attorney to help as an expert witness, what ethical questions will you ask?

---

[3] Abstracted from http://ethics.tamu.edu/ethics/asme/asme1.htm

[4] Abstracted from http://wadsworth.com/philosophy_d/templates/student_resources/0534605796_harris/cases/Cases/case64.htm

Nuclear engineering

16

16.1 Introduction

This chapter describes one alternative to today's dominant fuels for electrical energy production, coal, oil, and natural gas, all of which produce copious amounts of CO_2, which is infamous as a greenhouse gas. We also overwhelmingly use hydrocarbon fossil fuels for our motor vehicles. Nuclear power can substitute as the source of electrical power but cannot directly make transportation fuels.

There are about a thousand nuclear reactors located around the world of which about 500 are in large electrical power generating plants. They have various sizes and perform various functions. The smallest are research reactors with minimal power output; at the other end of the scale, the largest nuclear power reactors are

Exploring Engineering. https://doi.org/10.1016/B978-0-443-13541-5.00036-2

designed to produce about 1300 MW[1] of electricity. The United States has over 100 nuclear power reactors producing about 17% of its electrical power requirements— and without the concomitant production of CO_2.

There have been nuclear reactor incidents/accidents worldwide of varying seriousness. At least four accidents have produced significant concerns well beyond the borders of the nuclear plant that had the accident. In historical order, Windscale/ Sellafield (UK), Three Mile Island (U.S.), Chernobyl (Ukraine), and Fukushima (Japan). Table 16.1 lists the total radioactive releases of these major accidents relative to Three Mile Island.

Table 16.1 Relative radioactive releases following major nuclear accidents.

Reactor accident	Radioactive release relative to Three Mile Island
Windscale/Sellafield	About 1500
Chernobyl	About 10,000,000
Fukushima	200,000,000 to 400,000,000

It's worth noting that the release from Three Mile Island (the worst nuclear accident in U.S. history) produced much less radiation release than the others because its **secondary containment** (that is, its outer steel/concrete enclosure) held. The others did not (and, in the case of Chernobyl, it had no secondary containment to begin with).

This chapter is intended to inform as well as to instruct. It is organized with three objectives: (A) to present the basic physics of nuclear reactions, (B) explain what occurs in nuclear power plants, and (C) explain the origin and implications of nuclear waste materials. The reader should then be in a position to make an *educated* assessment of whether possible radioactive releases from nuclear power plants under severe accidents offset the global effects of CO_2 release from fossil energy sources.

16.1.1 Nuclear fission

An analysis starts with the basic nuclear physics of **fission**, a word meaning "splitting" of an atom by electrically neutral **neutrons**. Normal chemistry depends on the motion of a local electrons belonging to atoms, but **nuclear** reactions such as fission

---

[1] If this size nuclear reactor runs for more than 300 days/year, it produces about 10^{10} kWh of electricity while producing revenue of about $2,000,000 *per day*. A reactor of this size will serve a million homes.

are too energetic (perhaps several million times so) for meaningful interactions with electrons.

Fissile atoms can be fissioned, that is, split by neutrons; for reasons that will become apparent, we are particularly interested in fissile atoms that can be readily split by very low energy neutrons.

The tiny nucleus of an atom is where virtually all of its mass resides including its positively charged **protons** and its uncharged **neutrons**. The major atom of interest in fission is uranium-235, written as ^{235}U or as $^{235}_{92}U$; its mass is due to 235 particles consisting of 143 neutrons plus 92 protons.

An **isotope** is an atom with the same nuclear charge as another isotope; it is therefore the same chemical substance but of different mass due to more or to fewer neutrons within its nucleus. U-235 is one of several uranium **isotopes** occurring to the extent of 0.72% in natural uranium, the rest being another isotope U-238.

Thus, the superscript indicates the **atomic mass** of the atom's nucleus and the subscript (which is called the **atomic number**) indicates its number of protons. Hence, $^{235}_{92}U$ has 92 protons plus 143 neutrons. Similarly, the isotope $^{238}_{92}U$ has 146 neutrons but still just 92 protons. In the nucleus, neutrons and protons are collectively called **nucleons**.

The uranium isotope $^{238}_{92}U$ is relatively stable toward neutrons but $^{235}_{92}U$ will readily absorb a low energy neutron momentarily fusing with it making its nucleus too unstable to stay together so it will fission with a variety of possible outcomes with these typical fission fragments:

$$^{1}_{0}n + ^{235}_{92}U \Rightarrow ^{141}_{56}Ba + ^{92}_{36}Kr + 3^{1}_{0}n \tag{16.1}$$

$$^{1}_{0}n + ^{235}_{92}U \Rightarrow ^{90}_{38}Sr + ^{143}_{54}Xe + 3^{1}_{0}n \tag{16.2}$$

$$^{1}_{0}n + ^{235}_{92}U \Rightarrow ^{131}_{53}I + ^{100}_{39}Y + 5^{1}_{0}n \tag{16.3}$$

The **fission products** in these particular nuclear reactions are isotopes of barium, krypton, strontium, xenon, iodine, and yttrium, respectively, all appearing on the right hand side of the expressions. It is best to write a neutron as $^{1}_{0}n$ instead of just n indicating it has a mass of one atomic unit and also the neutron has no charge and thus its subscript is '0'.

We need to pay attention to two **conserved** quantities in nuclear reactions: the number of nuclear charges (protons) and the total number of neutrons and/or total nucleons. That sounds complicated but it's not. Table 16.2 shows how to think about nuclear reactions (it is actually quite close to how ordinary chemical reactions are balanced).

Table 16.2 Some common nuclear reactions.

Nuclear reaction	Nuclear particle	LHS neutrons	LHS protons	Nuclear fragments	RHS neutrons	RHS protons
(16.1)	$^{235}_{92}U$	$143 + 1 = \mathbf{144}$	**92**	$^{141}_{56}Ba\ /\ ^{92}_{36}Kr$	$85 + 56 + 3 = \mathbf{144}$	$56 + 36 = \mathbf{92}$
(16.2)	$^{235}_{92}U$	$143 + 1 = \mathbf{144}$	**92**	$^{90}_{38}Sr\ /\ ^{143}_{54}Xe$	$52 + 89 + 3 = \mathbf{144}$	$54 + 38 = \mathbf{92}$
(16.3)	$^{235}_{92}U$	$143 + 1 = \mathbf{144}$	**92**	$^{131}_{53}I\ /\ ^{100}_{39}Y$	$78 + 61 + 5 = \mathbf{144}$	$53 + 39 = \mathbf{92}$

LHS and RHS = left and right hand sides, respectively, of the reaction.

Notice that each side of the reaction balances both the number of proton charges and the number of nucleons. Any deficiency in the number of nucleons in the fissioned nuclei is balanced by the number of free neutrons produced. Be careful: these conservation laws do *not* imply that the total masses from the LHS and RHS are also exactly equal. In fact, using precision values for the masses of the fission products show that a small amount of mass is missing in each nuclear reaction (see next section).

A self-sustaining nuclear reaction occurs when sufficient neutrons are produced on the RHS of the fission reactions to supply the next round of LHS reactions with sufficient neutrons. In practice, we need more than just enough neutrons per generation shown on the LHS because of losses of neutrons to parasitic side reactions or other losses. In practice, in water-cooled nuclear reactors, the net number of neutrons to sustain a **chain reaction** is about 2.4 per generation. Thus, of the ~2.4 neutrons in say generation 1, just one (per reaction) must make it to be fissioned in the next generation. If we can maintain one net neutron in the next generation and in subsequent generations, we have a self-sustaining nuclear reaction: if less than one net neutron, the nuclear reaction will die down. Of course, if there is more than 1 neutron/reaction in generations 2, 3, 4, ... we have the makings of a nuclear explosion.

Some boron compounds are added to some nuclear reactors because they have a huge ability to absorb neutrons. The favored isotope of boron is $^{10}_{5}B$; it is converted to isotopes of lithium under neutron irradiation, a strategy that is useful in control of a nuclear power reactors.

16.1.2 Nuclear energy

Einstein's special relativity theory quantitatively shows that the missing fission mass, Δm, in a nuclear reaction is converted to energy according to:

$$\Delta E = \Delta mc^2 \tag{16.4}$$

where c is the speed of light, 3.00×10^8 m/s (186,000 miles/s). In a typical fission reaction, the deficiency of mass missing from the right-hand side of the reaction is about 0.2 **atomic mass units** (abbreviated as amu) where **1 amu** $= 1.660540 \times 10^{-27}$ kg.

Example 16.1

Convert the missing mass of 0.200 amu per fission reaction into kg.
    Need: 0.200 amu of mass = _____ kg.
    Know: 1 amu $= 1.66 \times 10^{-27}$ kg
    How—Solve: Thus, **0.200 amu** $= 0.200 \times 1.66 \times 10^{-27}$ kg $= \mathbf{3.32 \times 10^{-28}}$ **kg**.
        Surprisingly, this is not much mass.

Energy units based on a single electron's charge are more convenient to use for reactions at the single atom level. The charge on one electron is 1.602×10^{-19} coulombs. If an electron is accelerated by $+1.000$ V, its energy is 1.602×10^{-19} coulomb $\times$ volts $= 1.602 \times 10^{-19}$ joules $= 1$ **eV** (electron volt) of energy. Mega is 10^6 so 1 MeV $= 10^6$ eV, a convenient unit when dealing with nuclear reactions.

Example 16.2

Convert the missing mass of 0.200 amu per fission reaction into MeV/fission using Einstein's mass-energy equation.

Need: 0.200 amu of mass = ___ MeV/reaction.

Know: An eV is 1.60×10^{-19} joules and thus 1.00 MeV is 1.60×10^{-13} J.

How: From Example 16.1, 0.200 amu $= 3.32 \times 10^{-28}$ kg and

Solve: Then, using Einstein's equation:

$$\Delta E = \Delta m c^2 = 3.32 \times 10^{-28} \times \left(3.00 \times 10^8\right)^2 \, [\text{kg}][\text{m/s}]^2 = \mathbf{2.99 \times 10^{-11} \, J/fission}$$

Since 1 MeV is 1.60×10^{-13} J, then

$$\Delta E = 2.99 \times 10^{-11}/1.60 \times 10^{-13} \, [\text{J/fission}][\text{MeV/J}] = \mathbf{187 \, MeV/fission.}$$

The energy stated in joules is not a particularly convenient magnitude to remember; it's far easier to work in MeV. The result of this calculation is about right for any of the fission reactions 16.1, 16.2, and 16.3 in Table 16.2, although a more accurate average number is about 200 MeV/fission. Most of the 200 MeV/fission energy ends up as heat inside a nuclear reactor which is then converted into electricity. 200 MeV/fission is easy to remember, but how big is it?

Example 16.3

Compare fission energy to a chemical reaction such as the combustion of a kilogram mole of ethanol whose heat of combustion is 1.25×10^6 kJ/kmol.

Need: Compare a typical fission reaction to the combustion of ethanol, which has a combustion heat of 1.25×10^6 kJ/kmol.

Know: Fission reactions produce about 200 MeV, which in joules, is about (200 MeV/ fission) $\times$ (1.60×10^{-13} joules/MeV) $= 3.2 \times 10^{-11}$ J/fission.

How: There are Avogadro' number of fissions/kmol, or 6.02×10^{26} fissions/kmol.

Solve: The fission reaction scales to

$$3.2 \times 10^{-11} \times 6.02 \times 10^{26} \, [\text{J/fission}][\text{fissions/kmol}]$$

$$= 1.9 \times 10^{16} \, \text{J/kmol} = \mathbf{1.9 \times 10^{13} \, kJ/kmol}$$

The heat of combustion of ethanol is 1.25×10^6 kJ/kmol, but the energy from a missing fission mass of 0.200 amu is 1.9×10^{13} kJ/kmol—*a difference of about 10 million.* **This is the energy source for both nuclear reactors and for nuclear bombs.**

16.2 **Nuclear power reactors**

The way a nuclear reactor works starts with a fuel pellet. Most reactors use uranium oxide pellets although some countries use a mixture of uranium oxide and plutonium oxide, the latter containing fissile $^{239}_{93}Pu$. These pellets fit tightly into zirconium[2] metal tubes; fuel rods are these tubes filled with fuel pellets and which are then fitted into subassemblies (Fig. 16.1).

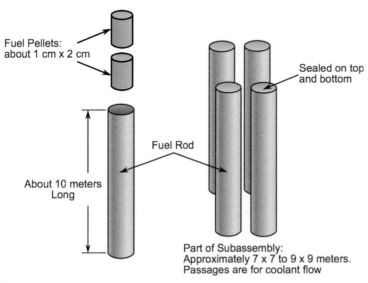

FIGURE 16.1

Nuclear fuel rods.

Fig. 16.2 shows a completed subassembly. Notice the technician has minimal protective wear because unirradiated nuclear fuel is only very slightly radioactive.

A rod bundle is then inserted into a rectangular metal sleeve called a subassembly. The next step is to insert the subassemblies into the shell of the nuclear reactor. The shell is a massive pressure vessel that contains the hot water and steam circulating within the reactor. Finally, a coolant such as water is forced into the subassemblies from the bottom to remove the heat generated by nuclear fission.

---

[2] Zirconium metal is resistant to damage by nuclear bombardment inside the reactor.

FIGURE 16.2

Nuclear fuel subassemblies.

US Nuclear Regulatory Commission. http://www.nrc.gov/images/reading-rm/photo-gallery/20071114-045.jpg

Two of the most common types of reactors are schematically shown in Fig. 16.3: they are a Pressurized Water Reactor (PWR) and a Boiling Water Reactor (BWR) that are collectively known as "light water reactors (LWRs)", light water being ordinary water. In nuclear reactors, water is either boiled to steam directly or indirectly in a steam generator and the steam then passed through a steam turbine that turns an electrical generator.

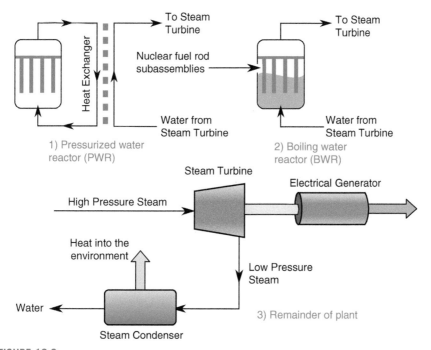

FIGURE 16.3

Nuclear power reactors.

PWRs operate at a substantial pressure of about 16 MPa (2500 lbf/in^2) and BWRs at approximately one half that. They both have efficiencies of about 33% meaning that about 33% of the heat from the nuclear fuel is captured into electricity and the remaining 67% is lost as waste heat in the steam condenser. An important issue therefore is that nuclear reactors are often located near large sources of water so they can condense the turbine exhaust steam. Note that the balance of the plant that produces electric power is the same for either kind of LWR.

Nuclear power plants are used as "base load", meaning they are run flat out for as long as possible because they are expensive to install (so you want to run them 24/7 to spread their capital costs as much as possible) but they are cheap to run primarily because so little fuel[3] is used.

---

[3] A 1000 MW power reactor produces about 100 tons/*year* of highly radioactive spent fuel; a coal-fired power plant of the same capacity needs about 8000 tons/*day* of coal producing about 800 tons/*day* of solid waste and about 9000 tons/*day* of CO_2.

16.3 Neutron moderation

Water in LWRs has two purposes. The first is the more obvious—removing the nuclear energy produced as a result of fission as heat and delivering that heat as steam to a steam turbine—generator. But there is also a much more compelling nuclear effect of water that is responsible for the ultimate reactivity of the nuclear chain reaction. When a neutron collides with a proton (i.e., a hydrogen atom less its electron), kinetic energy is efficiently exchanged between them. This is important because fresh neutrons from the fission process called **fast neutrons** have an average energy of about 2 MeV (so they move at about 5% the speed of light) but they are slowed by collisions with protons to "**thermal energies**" or as "**thermalized neutrons**" and behave rather like atoms in an ideal gas. Their energy is in the form of kinetic energy, $\frac{1}{2}mV^2$. Neutron thermal energies are only a few hundredths of an eV compared to ~2 MeV for fast neutrons.

Example 16.4

What is the average energy of an atom in an ideal monatomic gas at 20.°C and 350.°C? Assume the average energy of an ideal gas is kT in which T is the absolute temperature and k is Boltzmann's constant, 8.617×10^{-5} eV/K.

   Need: Average energy of an atom in a gas at 20.°C and 350.°C.
   Know: A gas atom's or molecule's average energy is kT and $k = 8.617 \times 10^{-5}$ eV/K.
   How: Convert temperature to Kelvin. $T = 20. + 273 = 293$K and $T = 350. + 273 = 623$K.
   Solve: Atom's energy $= kT = 8.617 \times 10^{-5} \times 293 = $ **0.0252 eV at 20.°C** and at **350°C**,
      $8.617 \times 10^{-5} \times 623 = $ **0.054 eV**

   Which kind of neutron causes the faster rate of nuclear fissions? It turns out that the probability of a fission reaction is about 500—1000 times greater for thermal neutrons than for fast ones. For this reason, we need only modest enrichment of the fissile components in the fuel so water moderated nuclear reactors need only 2%—4% enrichment of ^{235}U compared to its naturally occurring concentration of about 0.7%.

   A good demonstration of the effects of collisions between neutrons and protons is demonstrated by a Newton's Cradle, shown in Fig. 16.4. Two steel balls of the same mass are suspended on hooks by a piece of string. One is hanging vertically and the other is pulled away and released so that it hits the stationary ball full on. If the two balls are both hard and 100% elastic, *all* of the kinetic energy of the swing ball is transferred to the previously stationary one. In Fig. 16.4, ball A hits ball B and it swings exactly to a position mirroring ball A's initial condition while ball B stops completely. Of course ball B is then free to repeat the motions with the ultimate number of collisions determined by various sources of friction.

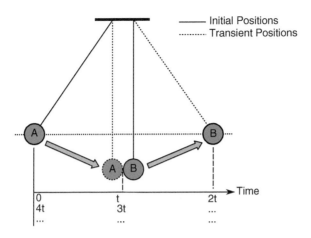

FIGURE 16.4

Newton's cradle.

In a perfectly elastic system (one that loses no kinetic energy after the rebound), the swing amplitude and the incremental time to swing between collisions remains fixed. The time intervals are given as successive multiples of "t" in Fig. 16.4.

When a neutron hits a proton head on, the neutron will behave the same way—it will lose all of its kinetic energy and come to a stop with the proton taking all of the original kinetic energy. This process is called **elastic scattering**. Of course, in practice, only a few neutrons will hit protons head on; more likely they will give a glancing blow. A fuller analysis of the scattering process shows that about 17 successive collisions with protons are enough to slow the fast neutron to thermal energies. The overall process is known as **neutron moderation** and the substance doing it (in this case the hydrogen nuclei of water molecules) is the **moderator**.

When neutrons hit larger nuclei than protons, say the nucleus of a carbon atom (which contains six neutrons and six protons), it requires many more collisions (say a 100 or so) to slow down a fast neutron. If a neutron elastically hits a still heavier nucleus, it simply bounces off.

This process is known as **"thermalization"** because it ultimately slows the 2 MeV fast neutrons produced by fissioning to thermal speeds characteristic of the protons in the moderator.

Example 16.5

Suppose a fast neutron in a nuclear reaction has initial energy E_{init} before thermalization. After thermalization, it has final energy E_{final}. The number of collisions to thermalize a fast neutron in a water-moderated reactor is p given by

$$p = \frac{1}{\xi} \ln \left(\frac{E_{init}}{E_{final}} \right) \tag{16.5}$$

The atomic mass of the struck nucleus is A. If A < 10, then $\xi = 1.00$, and if A > 10, then $\xi = \frac{2}{A+2/3}$. In this example, the struck nucleus is hydrogen (A = 1), so $\xi = 1.00$.

Element	A	ξ
H	1	1
C	12	0.158
Na	23	0.084
Fe	56	0.035
U-238	238	0.008

The logarithmic term in Eq. (16.5) is called the **logarithmic energy decrement**. How many collisions does it take for neutrons in a water-moderated reactor to slow to thermal energies at 20.°C and at 350°C?

Need: The number of collisions to thermalize a fast neutron given its initial energy is about 2 MeV.

Know: From Example 16.4, at 20.°C, a thermal neutron has energy of 0.0252 eV and at 350°C an energy of 0.054 eV, $\xi = 1.00$.

How: Use the logarithmic energy decrement method, Eq. (16.5).

Solve: At 20.°C, $E_{init}/E_{final} = 2.0 \times 10^6/0.025 = 8.0 \times 10^7$
and at 350°C, $E_{init}/E_{final} = 2.0 \times 10^6/0.054 = 3.7 \times 10^7$.
Thus, at **20.°C**, $p = \ln(8.0 \times 10^7) = $ **18 collisions**.
and at **350°C**, $p = \ln(3.7 \times 10^7) = $ **17 collisions**.

16.4 How does a nuclear reactor work?

In order to determine how a nuclear reactor manages to balance a lot of factors, we are going to break down the reactor core physics into several manageable pieces. The object is getting a nuclear reactor core that achieves self-perpetuating "criticality", the state in which a reactor exactly produces the correct amount of neutrons from one generation to the next generation.

16.4.1 Fissile nuclei

As we have already seen, the collisions of fissile nuclei with neutrons breaks down certain heavy and unstable atoms destroying a small amount of mass in each reaction and also leaving two fragments of lighter atoms. Fissile atoms include U^{233}, U^{235}, Pu^{239}, and Pu^{241} with the prime example being the fissioning of U-235. The fission product spectrum shown in Fig. 16.5 is very important in determining how we treat the waste products of nuclear reactions because they may be very radioactive, and some are very abundant. Notice that there are two major peaks in the fission product spectrum.

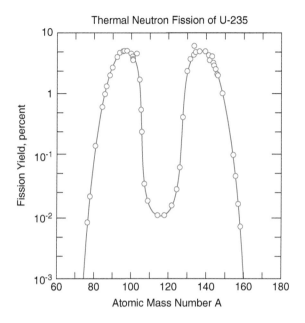

Thermal Neutron Fission of U-235

FIGURE 16.5

Fission product yields by thermal fission.

16.4.2 Fertile nuclei

Fertile atoms will convert to fissile atoms by neutron collisions followed by a radio-active decay sequence of which the most important product is plutonium, e.g.,

$$^{238}_{92}U \rightarrow ^{239}_{94}P$$

How did two additional positive charges in the nuclei (see subscripts) come about? The radioactive decay of $^{239}_{92}U$ to $^{239}_{93}Np$ and subsequent decay to $^{239}_{94}Pu$ each produces a byproduct of negatively charged electrons, which accordingly raises the number of positively charged protons by two. A simple view is that neutrons break down to produce a proton and an electron, the latter being called beta decay.

Pu-239 is an important atom because it is a fissile atom and not simply a fertile one. It is used in nuclear reactors as an additive to uranium fuel and also it is formed internally from fertile U-238, the latter being far and away the most abundant atom in a reactor core.

There is only 2%–4% directly fissionable U-235 in the reactor core. Fortunately, the remaining ~97% of the core uranium is fertile U-238. As such, it provides extra fuel beyond that loaded by fissioning of the initial U-235 so that the conversion of fertile atoms is a significant contributor to the performance of nuclear reactors.

16.5 The four factor formula

For a nuclear reactor to work continuously, you must make the proper number of neutrons during fission reactions, not too many and not too few. To understand what this statement means, we will consider how to calculate the number of neutrons by breaking down the problem into simpler pieces. The factors that need to be considered on sustaining a self-perpetuating neutron cycle are as follows:

- The fast fission factor, $\varepsilon = \frac{\text{Total neutrons from fast and thermal fission neutrons}}{\text{Fission neutrons produced from thermal fissions alone}}$

- Resonance escape probability, $p = \frac{\text{Number fission neutrons slowing to thermal energies}}{\text{Total of fast neutrons}}$.

- The thermal utilization factor, $f = \frac{\text{Thermal neutrons absorbed into fuel}}{\text{Total of thermal neutrons absorbed}}$

- The thermal fission factor, $\eta = \frac{\text{Neutrons produced by fission}}{\text{Neutrons absorbed by fuel etc.}}$

When you multiple these factors together, they will give the infinite multiplication factor, k_∞.

$$k_\infty = \frac{\text{Neutrons in current generation}}{\text{Neutrons in preceding generation}} = \eta \times f \times p \times \varepsilon \qquad (16.6)$$

If $k_\infty = 1$ in an infinitely large reactor, it is critical. If $k_\infty \leq 1$, it is subcritical and if $k_\infty \geq 1$, it is supercritical. For a finite sized reactor, you need $k_\infty > 1$ to account for leakages. Eq. (16.6) is called the Fermi Four factor equation after the great Italian physicist Enrico Fermi, who built the world's first nuclear reactor (under the sports stadium at the University of Chicago). The equation can be derived by chasing the fate of the first round of neutrons through the next generation as in Fig. 16.6.

Eq. (16.6) is only a guide that strictly works only for an infinitely large reactor; it's a good first estimator of a nuclear reactor's behavior as to whether you have a viable design.

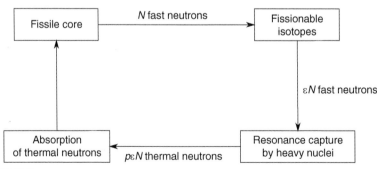

FIGURE 16.6

Neutron cycle in a nuclear reactor.

Example 16.6

Consider a nuclear reactor containing 95% of a graphite moderator and 5.0% fissile fuel $^{235}_{92}U$. The values to be used in Eq. (16.6) are $\eta = 1.931$, $\varepsilon = 1.000$, $f = 0.982$, and $p = 0.648$.

a. What is the value of k_∞?
b. How much neutron leakage can you afford with this design?
c. Why is $\varepsilon = 1.000$ for this design?

Need: The value of k_∞, the percent of neutrons lost to leakage, and why $\varepsilon = 1.000$.
Know: Reactor factors, $\eta = 1.93$, $\varepsilon = 1.000$, $f = 0.982$, and $p = 0.648$.
How: Use the Four Factor Formula to calculate k_∞ and from that the loss of neutrons.
Solve:

a. $k_\infty = \eta \times f \times p \times \varepsilon = 1.931 \times 0.982 \times 0.648 \times 1.000 = \textbf{1.229.}$
b. To get $k = 1.000$ for a critical finite core, $k/k_\infty = 1.000/1.229 = 0.814$ and thus **leakage = 1.000 − 0.814 = 0.186 = 18.6%.**
c. Why is $\varepsilon = 1.000$? Look at the core design—95% moderator and 5.0% fissile fuel. In the immediate vicinity of the thermal fission, there are 19 moderator nuclei for each fissionable one so the fast neutron that is formed will immediately slow to thermal speeds and **the fast fission factor $\varepsilon = 1.000$.**

16.6 Fission products and nuclear waste

All three nuclear reactions shown in Eqs. (16.1−16.3) produce fission products. Of these, iodine-131 and strontium-90 are notorious because they dangerously radioactive elements that can enter the food chain.

Nuclear radiation can negatively affect living creatures. Iodine-131, for example, behaves chemically and biochemically as ordinary iodine, natural iodine being entirely composed of a single nonradioactive isotope, I-127. Biochemically, any isotope of iodine has the same affinity for the thyroid gland. Radioactive I-131 can thus move through the food chain into the human thyroid and cause throat cancer. But it has a short **half-life**[4] of only 8 days. So, if milk is exposed to I-131, it will become radioactively contaminated. But it can be made into cheese and aged for a few months to make it wholesome again.

Radioactive strontium-90 also behaves chemically and biochemically as its nonradioactive cousin Sr-88. Further, since Sr is in the same periodic table group as calcium, it can fool the body into absorbing it into bone; if Sr-90 is absorbed into the body, it can produce serious illness (including fatal bone cancers). Sr-90 has a half-life of 29 years and represents a serious hazard if it leaks from its containment.

Unfortunately, the above two radioactive isotopes are not the only radioactive species generated in nuclear reactions. In the fissioning process, hundreds of nuclear reactions occur beyond those expressed as Eqs. (16.1−16.3). The fissioning process

---

[4] Meaning 50% of the atoms of the iodine-131 decay after just 8 days and 75% after 16 days etc.

does not split atoms uniformly into equal amounts of radioactive nuclei but forms a broad spectrum of fission products.

What is happening during fissioning is that the uranium nucleus becomes unstable when it absorbs an extra neutron. It flies apart as quickly as it can and forms a smorgasbord of fission products. Fig. 16.5 neatly summarizes the amounts of fission products that are formed as well as the concentrations of essentially all of their radioactive radioisotopes. Notice the ordinate, yield percent, is logarithmic and covers four orders of magnitude. Smaller amounts of other fission products species are not shown because they are only produced in parts per million (or even much less) and are too dilute to show. Still some of the most persistent radioactive isotopes are technetium-99, neptunium-237, and plutonium-242, all with half-lives measured in hundreds of thousands of years.

In the fission products' curve, there are two prominent peaks with concentrations between about 1% and 8%. These peaks contain the most abundant of fission products. One broad peak occurs at a mass number of about 95 ± 10 (which includes Sr-90) and the other at about 140 ± 10 (which includes I-131).

Example 16.7

How many years does it take to reduce the initial activity of Sr-90 to 1% of its original radioactivity? Assume the rate of radioactive decay of an isotope follows an exponential decay in time:
$C = C_0 \exp\left[-0.693\left(\frac{t}{t_{1/2}}\right)\right]$. The half-life $t_{1/2}$ is defined as when 50% of the isotope has decayed.

Need: Time in years to reduce the radioactivity of Sr-90 to 1%.
Know: Half-life, $t_{1/2} = 29$ years.
How: The rate of radioactive decay follows an exponential decay in time:

$$C = C_0 \exp\left[-0.693\left(\frac{t}{t_{1/2}}\right)\right] \qquad (16.7)$$

where C is the final number of atoms of Sr-90 and C_0 is the original amount written in the same units.
Solve: $C = C_0 \exp\left[-0.693\left(\frac{t}{t_{1/2}}\right)\right]$ in which $C/C_0 = 1\% = 0.01$

Therefore, $\ln(0.01) = -\frac{0.693t}{t_{1/2}} = -4.605$ and $t = \frac{-4.605 \times 29}{-0.693} = $ **193 years**.

One can see that the management of a radioactive accident site (as has befallen the Japanese at Fukushima) is a long-term affair. Most of the high-activity fission products have half-lives of 30–50 years. Chernobyl in the Ukraine, the site of one of the worst nuclear accidents in history, has been totally abandoned due to radiation contamination.

16.6.1 Nuclear waste, the Achilles Heel of nuclear power

The spectrum of fission products is the ultimate weakness of nuclear power because so many of these products are highly radioactive and many are biologically active as well. Fission products are an enormous burden that, if and when they leak, can cause expense (clean up costs), illness, and death. The effects of nuclear radiation depend on the types of radiation emitted as well on the chemical and biochemical properties of the materials. For example, radioactive xenon Xe-133 has a half-life of 5 days, but, importantly, xenon is a rare gas meaning that it does not form any chemical bonds and is chemically and biochemically inert. In general, it is less dangerous, whereas I-131, with a similar half-life, can enter the human thyroid gland with devastating consequences.

16.6.2 Nuclear accidents

When a 1000 MW nuclear reactor is operating at full power, it will produce the entire radioactive spectrum described in Fig. 16.5. If such a reactor shuts down, intentionally or not, the radioactive burden of this reactor at shut down is about 10^{20} Bq, where 1 Bq (Becquerel) is 1 radioactive disintegration per second. The accident at Chernobyl *released* somewhere in the range 5×10^{18} Bq and Three Mile Island about 5×10^{11} Bq. The releases at Fukushima was about 9.4×10^{17} Bq.

There is another very important phenomenon that accompanies fission products. It is that radioactive decay is an exothermic event, which is harder to deal with as the reactors get larger. When a nuclear power reactor is shut down by inserting neutron absorbing rods into the reactor's core, it does not *fully* shut down. For example, a 1000 MW nuclear reactor at the instant of shutdown continues immediately after shutdown to produce about 180 MW of heat from the radioactive decay of the fission products. After 1 h of shutdown, this falls to about 45 MW of heat and after a day to 15 MW, which is still a large amount of heat. This radioactive heat is just a not just minor nuisance; if not continuously cooled, the cladding of the fuel rods will melt and red-hot cladding metal will chemically react with water producing explosive hydrogen. In the absence of cooling, the fuel pellets will also melt (at $\sim 2860°C$) and destroy the pressure vessel that had contained now volatile fission products. Most of these things happened during the accidents at Three Mile Island in the United States, at Chernobyl in the Ukraine, and at Fukushima in Japan.

Example 16.8

Statistics of ^{90}Sr fission product in a large power reactor core:

Breakdown accounting	Result
Fuel in nuclear core as UO_2	More than a 100,000 kg
Enrichment of ^{235}U	3% or 3000 kg
At reactor shutdown, assume total fission products $\sim 6\%$ of ^{235}U	180 kg
Yield of ^{90}Sr is about 4.5% at shutdown	8 kg of ^{90}Sr

How many disintegrations per second will the Sr initially undergo? What about after 193 years of storage?

Need: ^{90}Sr disintegrations/s: (A) On shutdown, (B) after 193 years of storage.

Know: $t_{1/2} = 29.$ years and radioactivity drops to 1% after 193 years (from Example 16.6). The atomic mass of ^{90}Sr $= 89.9$ kg/kmol. Avogadro's number is 6.02×10^{26} particles/kmol.

How: Differentiating Eq. (16.7) gives the radioactive decay equation:

$$\frac{dC}{dt} = -\frac{0.693C}{t_{1/2}} \tag{16.8}$$

with $C = C_0$ at $t = 0$. The required units are disintegrations of atoms of ^{90}Sr per second. Converting $t_{1/2} = 29$ years to seconds $= 9.1 \times 10^8$ s.

Solve: Converting exactly 8 kg of ^{90}Sr into atoms

$$= (8./89.9) \times 6.02 \times 10^{26} \, [\text{kg}][\text{kmol/kg}][\text{atoms/kmol}] = 5.4 \times 10^{25} \text{ atoms of} ^{90}\text{Sr}.$$

Eq. (16.8) gives the rate of disintegrations of ^{90}Sr atoms/s as

$$= 0.693 \times 5.4 \times 10^{25}/(9.1 \times 10^8) = \mathbf{4.1 \times 10^{16} \text{ atoms of} ^{90}Sr} \text{ that decay every second.}$$

After 193 years, this falls to 1% of the original activity or **4.1 × 10^{14} atoms/s**, which is still not small!

16.7 Is nuclear power a viable renewable energy source?

Uranium fuel is assuredly a renewable energy source. Not only are there significant reserves of the ore (known as "pitchblende"), but, with better fuel cycles than the common ^{235}U/LWR combination, there is a large potential supply of fissile materials. Uranium-235 is used in conjunction with the ^{238}U, which is roughly 97%—98% of the total uranium in nuclear reactor fuel. Some neutrons are absorbed in this material eventually making plutonium-239, which is also fissile, and which will sustain additional nuclear reactions much as U-235 does. If this ^{239}Pu is chemically separated from the rest of the fission products, it can be used as additional nuclear fuel when it is reloaded into a fresh batch of fuel. In fact one can, in principle, extract the ^{239}Pu repeatedly and end up with 250% of the original ^{235}U fuel value!

In addition, LWRs are neutron hogs because there is a significant likelihood that protons in water will parasitically absorb some of the spare neutrons making **heavy hydrogen** (called deuterium) by neutron absorption, i.e.,

$$^1_0 n + ^1_1 p \Rightarrow ^2_1 d \tag{16.9}$$

where d is its nucleus with twice the mass of ordinary hydrogen.

There are other reactor schemes such a graphite moderated reactors that are so sparing with neutrons you can recover up to 95% of the original fuel as ^{239}Pu in each complete fuel cycle. Further, this fuel can be chemically recovered and it will deliver 20 × the original ^{235}U fuel worth by reprocessing. **Heavy water** (deuterium) is used to moderate Canadian reactors. Its advantage is that it is frugal with parasitic neutron absorption compared to ordinary water.

There are also "fast" reactors that produce more uranium than they consume. They have no moderator but rely on an efficient capture process for fast neutrons.

These can in principle recycle all of the ^{238}U as ^{239}Pu (^{238}U being $\sim 99 \times$ as abundant as ^{235}U).

Finally, thorium 232 and uranium 233 can be used as a nuclear fuel (although no one has done this commercially). Thorium is available as beach sand in India!

Levels of uranium in the range of 1–10 ppm are widely scattered in topsoil. In fact, soil is slightly radioactive due to the presence of both uranium and thorium. Uranium is also present in coal ash so that a coal-fired power plant will emit thousands of times more radioactivity in particulates up their smoke stacks than will a nuclear power plant.

Summary

Nuclear fission reactions can sustain a nuclear **controlled chain reaction** and will produce both large quantities of thermal energy and **neutrons**. The key step is **moderation** of **fast** neutrons produced by **fission**. This occurs in LWRs when neutrons collide with protons in water molecules that are there for the dual purposes of providing a coolant and a moderator. A necessary variable in the design of a nuclear reactor is that the **infinite multiplication factor k_∞** is > 1 and, with neutron losses accounted for in a real reactor, $k = 1$ to maintain the proper cycling of neutrons.

An important and unfortunate consequence of fission reactions are **fission products** that are not only dangerously **radioactive** (and for a long time) but produce **decay heat**, the latter capable of causing meltdown of both the nuclear fuel and the nuclear containment system. Loss of containment has the potential of releasing large quantities of radioactivity into the biosphere.

The prime advantage of nuclear power is that it does *not* produce a CO_2 effluent, and thus does not tip the atmosphere into a long warming period. Does this outweigh its disadvantages which include long-term spent fuel custodianship and the possibility of more nuclear meltdowns.

Exercises

Some useful conversions:

1.00 MeV is 1.60×10^{-13} J.

1 fission reaction produces 2.0×10^2 MeV.

Boltzmann's constant, 8.617×10^{-5} eV/K

Avogadro's number $= 6.02 \times 10^{23}$ entities per mol or 6.02×10^{26} entities per kmol.

1. Write down possible fission reactions (conserving charge and nucleons) for $^{92}_{36}Kr$, $^{142}_{56}Ba$, $^{92}_{38}Sr$ and $^{140}_{54}Xe$ (krypton, barium, strontium and xenon, respectively).

2. **Find the error in this nuclear reaction:** It describes a process for nuclear **fusion**. (Fusion is the opposite of fission, because it's the addition of a nuclear particle to a nucleus to make a heavier element than the starting materials.)

$$^2_1H + ^3_1H \Rightarrow ^4_2He + 2^1_0n$$

Why is tritium a *particularly* dangerous radioisotope?

3. How much uranium-235 is used in grams in producing 1.0 MWD (i.e., a megawatt-day) of energy given 2.0×10^2 MeV/fission?
4. In the previous question, approximately how much of this uranium has been *annihilated* (in mg).
5. **Find the Error in This Solution**: What is the average energy for an ideal gas of neutrons in a graphite-moderated reactor at 750.°C? Note: An ideal gas's average energy is kT with $k = 8.617 \times 10^{-5}$ eV/K.

   Need: Average energy for an ideal gas of neutrons in a graphite-moderated reactor at 750.°C.

   Know-How: An ideal gas's average energy is kT with $k = 8.617 \times 10^{-5}$ eV/K.

   Solve: $E = kT = 8.617 \times 10^{-5} \times 750. = 0.065$ eV.
6. **Find the Error in This Solution**: What's the logarithmic energy decrement for a neutron that thermalizes from 2.0 MeV to 0.1 eV?

   Need: The logarithmic energy decrement for a neutron between 2. MeV and 0.1 eV.

   Know-How: The logarithmic energy decrement is defined as LED $= \log\left(\frac{E_{\text{Init}}}{E_{\text{Final}}}\right)$

   Solve: LED $= \log\left(\frac{2. \times 10^6}{0.1}\right) = 7.3$
7. How many collisions does it take for neutrons in a graphite-moderated reactor to slow to thermal energies at 20.°C and at 350.°C?
8. How many times can you reuse a charge of ^{235}U/^{238}U fissile fuel if you reprocess the fuel when it is spent and recover all of the fissile Pu-239 there? Assume that 60% of the original fissile charge of U-235 was indirectly converted to Pu-239 from U-238. (**Hint:** It's a geometric series if all the Pu-239 converted from the original U-235 is reused.)
9. What is the ultimate fuel value of ^{235}U in a light water reactor ($x = 60\%$) and in a graphite-moderated reactor ($x = 95\%$)? (**Hint:** See Problem 8).
10. How much heat in watts is generated by radioactive ^{90}Sr in a nuclear core at shutdown and at 193 years later? Assume that each radioactive decay of ^{90}Sr produces 0.545 MeV/disintegration and that the initial amount of ^{90}Sr is 8.0 kg. Does ^{90}Sr meaningfully add to the decay heat in a nuclear meltdown?
11. The total decay heat on shutdown is about 180 MW for a large power reactor. What's the average fission product half-life *on shutdown* if there are 500. kg of fission products and each radioactive decay releases 0.50 MeV? Assume the average molecular mass of the fission products is 130. kg/kmol.
12. A heterogeneous reactor, as the name suggests, has nonuniform properties, e.g., a graphite moderator slab adjacent to a fissile $^{235}_{92}U$ slab. Without doing the arithmetic can you deduce arguments why this might increase the net thermal neutron flux? **Hint**: Sketch the expected neutron distribution, both fast and thermal, under the sketch below. What might this do to the resonance escape probability?

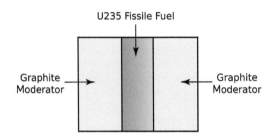

13. The world's first nuclear reactor was designed and built by Enrico Fermi in 1942. It consisted of a checkerboard of three dimensional alternating blocks of a graphite moderator and fissile U-235. The fuel was natural grade uranium containing just 0.7% U-235 because no one had yet solved the problem of how to enrich it beyond its natural state.

    The resonance escape probability for an infinite (no neutron loss) homogeneous core is about 0.648. Predict k_∞ if $\varepsilon = 1.000$, $\eta = 1.931$, $f = 0.750$ and $p = 0.648$. If you calculate $k_\infty \leq 1.000$, how did Fermi get it to achieve criticality?

14 **Find the Error in This Solution**: You have a radioactive sample consisting of 10^{25} atoms. It is decaying at a rate of 10^{21} per second. What is the residual activity after 24 h?

    Need: Residual activity of a radioisotope after 24 h.

    Know: Original amount is 10^{25} atoms decaying at a rate of 10^{21} per second.

    How: The original amount present divided by the rate gives the time to complete the decay.

    Solve: Time to decay to zero is $10^{25}/10^{21}$ [atoms][atoms/s] $= 10^4$ s or **2.8 h**. Since it was all decayed after 2.8 h, it was also completely decayed after 24 h as none was left.

15. How much radiation in Becquerel's (radiation particles/s) is released by the burning of 8000 metric tons/day of coal in a power plant? (The low-level radiation is there because uranium is slightly radioactive and widespread in minerals of all types including coal.)

    Assume coal contains 10.0% by weight of ash. If the coal contains 10.0 parts per million (ppm) of $^{238}_{92}U$, the ash contains 100. ppm of $^{238}_{92}U$. The half-life of $^{238}_{92}U$ is 4.468 billion years. There are eight other radioactive isotopes per disintegration of $^{238}_{92}U$.

16. Milk, which is easily contaminated by I-131, can be made into cheese and aged for a few months so it will become wholesome again. How long should you wait?

17 A particularly long-lived isotope is ^{237}Np with a half-life of 2.2 million years so that its radioactivity is long lasting. How many Bq does 1.00 mg of it produce?

18. What does it mean when a nuclear reactor is to be moderated by graphite and cooled by helium? Can you *sketch* such a reactor?

19. While not concerned with nuclear power, the analytical tools you have mastered can be used to give a lower limit on *the age of the Earth*. U-238 decays through multistep step radioactive cascade to Pb-206, a stable isotope of lead. It is dominated by the first step in the decay which has a half-life of 4.47 billion years.

    The measurement depends on small zircon crystals (ZrSiO$_4$) which can dissolve uranium compounds (but they will not dissolve lead componds). Liquid zircons will freeze at about 2500°C trapping any available uranium in them. This uranium will eventually decay to stable Pb-206, which is then trapped within the zircon crystal. In one zircon crystal, the ratio of U-238/Pb206 was determined to be 0.888. When was the Earth last molten?

20. You are a nuclear engineer in the now defunct Soviet Union, which was a heavy-handed dictatorship tolerating neither criticism nor behavior contrary to the official line. You are shown a proposed design for the Chernobyl reactor that is destined to explode in 1986 causing about 30 immediate deaths and inflicting thyroid damage on thousands (perhaps tens of thousands) of other victims (particularly the young). You immediately see that the reactor has no secondary containment and, perhaps as bad, the last 60 cm. of the control rods are coated with a heavy layer of a graphite moderator. You point out to the engineer-in-chief that a sudden insertion of withdrawn control rods all the way into the core of the reactor will not shut it down as expected, but will *increase* the nuclear reactions by virtue of the extra moderation via the graphite coating. Furthermore, lacking a secondary containment vessel, there could be disastrous widespread consequences. The engineer-in-chief takes you aside and says the reactor is designed to be economical. The design must stand essentially as is. He warns you to be careful because, if you bring it up again, the Soviet oversight authorities may send you to jail. What should the engineer do? (Use the Engineering Ethics Matrix and the NSPE Code of Ethics for Engineers.)

    a) Work within the system to get the best reactor he can under the circumstances.

b) Still insist on changes to make the reactor safer including changes he has been warned to ignore.

c) Resign and seek a new job, a job that will surely be a serious demotion given the Soviets' aversion to explicit or implicit criticism.

21. You are a nuclear engineer designing the nuclear reactor complex in Fukushima, Japan. It contains six virtually identical reactors. Japan is of course a democracy but one imbued with a culture of respect for ones elders. During a design review for the nuclear complex, you notice that all the secondary cooling systems have a common design based on diesel-electric generators to run the cooling pumps. You realize that a common failure mode exists. Anything that knocks out one generator could do the same to all of the reactors. Very politely you explain your concerns to the engineer-in-chief who tells you the design has been fixed as a compromise between absolute safety and safety based on an extremely unlikely event. And, based on the virtually zero likelihood of a common mode failure, the design stands. Remember, in Japan, consensus is very important. Are you OK with the engineer-in-chief's decision? What should the engineer do? (Use the Engineering Ethics Matrix and the NSPE Code of Ethics for Engineers.)

a) Work within the system to get the best reactor he can under the circumstances

b) Still insist on changes to make the reactor complex safer including changes he has been warned to ignore.

c) Resign and seek a new job, a job that could be a demotion given the Japanese aversion to explicit or implicit criticism of their collective decision-making.

Interdisciplinary engineering fields

2.1

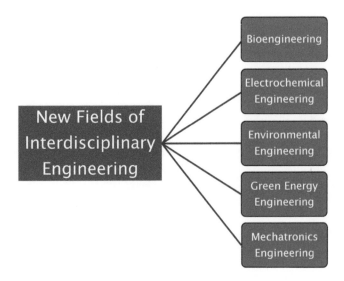

We can expect many new fields of engineering to emerge as technology advances in the 21st century. Some of these fields are apparent today. Fields such as micro-electro-mechanical systems (MEMS), mechatronics (robotics), nanotechnology, synthetic biology, and smart technologies (cars, structures, buildings, etc.) are all emerging engineering fields. New engineering fields are often multidisciplinary and hold enormous opportunities for future engineers. The following chapters are devoted to four specific emerging engineering fields.

Bioengineering combines the analytical and experimental methods of engineering with biology and medicine to produce a better understanding of biological phenomena and develop new therapeutic techniques and devices. Students with bioengineering degrees may work as biomedical engineers to develop new medical techniques, medical devices, and instrumentation for hospitals or individual patients.

Electrochemical Engineering is a combination of electrical engineering and chemical engineering. While the subject involves some mature technologies, such as the synthesis of chemicals, refining metals, the development of new batteries and fuel cells, electroplating, etching, and corrosion resistance, it also holds a central position in the development of new generations of hybrid cars as well as being the core technology for all-electric vehicles of the future.

Environmental Engineering is the branch of engineering concerned with protecting people from adverse environmental effects, such as pollution, as well as improving environmental quality. Environmental engineers work to improve recycling, waste disposal, public health, and water and air pollution control. They study the effect of technological advances on the environment and address environmental issues such as acid rain, global warming, ozone depletion, water pollution, and air pollution from automobile exhausts and industrial sources.

Green Energy Engineering is energy produced from sources that are environmentally friendlier (or "greener") than fossil fuels (coal, oil, and natural gas). Green energy includes all renewable energy sources (including solar, wind, geothermal, biofuels, and hydropower), and by definition, should also include nuclear energy (though there are many environmentalists who oppose the idea of nuclear energy as green energy because of the nuclear waste issues, accidents, and its harmful environmental effects).

Mechatronics is a new interdisciplinary engineering field that combines elements of mechanical, computer, and electrical engineering in product design and manufacturing. It includes robotics and other modern electromechanical devices. The **Internet of Things** (IoT) is the internetworking of physical devices containing electronics, sensors, actuators, software, and network connectivity that allow these devices to collect and exchange data. Many of the smart components associated with the Internet of Things are mechatronic. The development of the IoT is forcing mechatronics engineers, designers, practitioners, and educators to research the ways in which mechatronic systems and components are perceived, designed, and manufactured. This allows them to face up to new issues such as data security, machine ethics, and the human–machine interface.

Bioengineering

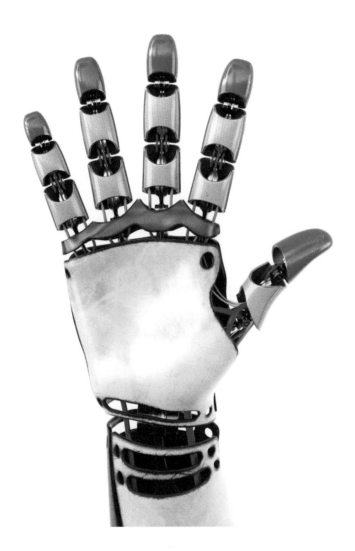

3D printed hand. https://techcrunch.com/2016/06/26/the-future-of-3d-printed-prosthetics/.

Exploring Engineering. https://doi.org/10.1016/B978-0-443-13541-5.00015-5

17.1 Introduction

Bioengineering (or biomedical engineering) is the application of engineering, life sciences, and applied mathematics to define and solve problems in biology, medicine, health care, and other fields that deal with living systems. Some examples of bioengineering include the design and development of:

- Devices that substitute for damaged body parts such as hearing aids, cardiac pacemakers, and synthetic bone and teeth.
- Artificial kidneys, hearts, heart valves, blood vessels, arms, legs, hips, knees, and other joints.
- Medical imaging techniques (ultrasound, MRI, CT, and others).
- Engineered organisms for chemical and pharmaceutical manufacturing.
- Blood oxygenators, dialysis machines, and diagnostic equipment.

Bioengineering is a relatively new discipline that combines many aspects of traditional engineering fields, such as chemical, electrical, and mechanical engineering as well as a number of subspecialties, including biotechnology, biological engineering, biomechanics, biochemical, and clinical engineering. Each of these fields may differ slightly in its focus of interest, but all are concerned with the improvement of human life.

17.2 What do bioengineers do?

Bioengineers have a wide variety of career choices. Some work alongside medical practitioners, developing new medical techniques, medical devices, and instrumentation for manufacturing companies. Hospitals and clinics employ clinical engineers to maintain and improve the technological support systems used for patient care. Engineers with advanced bioengineering degrees can perform biological and medical research in educational and government research laboratories.

Many bioengineers help people by solving complex problems in medicine and health care. Some bioengineering areas combine several disciplines, requiring a diverse array of skills. Digital hearing aids, implantable defibrillators, artificial heart valves, and pacemakers are all bioengineering products that help people combat disease and disability. Bioengineers develop advanced therapeutic and surgical devices, such as a laser system for eye surgery and a pump that regulates automated delivery of insulin.

In genetics, bioengineers try to detect, prevent, and treat genetic diseases. In sports medicine, bioengineers develop rehabilitation and external support devices. In industry, bioengineers work to understand the interaction between living systems and technology. Government bioengineers often work in product testing and safety, where they establish safety standards for medical devices and other consumer products. A biomedical engineer employed in a hospital might advise on the selection and use of medical equipment or supervise performance testing and maintenance.

In biocommunication, bioengineers develop new communication systems that enable paralyzed people to communicate directly with computers using brain waves. Through bioinformation engineering, they are exploring the remarkable properties of the human brain in pattern recognition and as a learning computer. Through biomimetics, bioengineers try to mimic living systems to create efficient designs. These areas have already developed far beyond the material in this chapter, but their basic themes are similar.

By engineering analysis of the situations to which living matter might be exposed and by characterizing, in engineering terms, the remarkable properties of living matter, knowledge is gained that improves safety and health. That understanding can be used to make life safer and healthier. Among the tasks undertaken by bioengineers is the design of safety devices, ranging from football helmets to seat belts to air bags; the development of prosthetic devices for use in the human body, such as artificial hip joints; the application of powerful methods for imaging the human body, such as computed axial tomography (CT/CAT scanning) and magnetic resonance imaging (MRI); and the analysis and mitigation of possible harmful health effects on humans subject to extreme environments, such as the deep sea and outer space.

In this chapter, we introduce a few simple descriptions of human anatomy and of the effects of large forces on hard and soft human tissues. We also learn (1) why head injuries can be devastating, (2) why vehicle collisions can kill, (3) how to make a first approximation of the likelihood of damage during collisions to the human body using a fracture criterion, (4) how to predict the injury potential of a possible accident using a criterion that we call "stress-speed-stopping distance-area" (**SSSA**), and (5) how to apply two other criteria for the effect of deceleration on the human body (the 30. g limit and the Gadd Severity Impact parameter).

17.3 Biological implications of injuries to the head

In many automobile accidents, the victims suffer severe head and neck injuries. Some of the accidents directly cause brain trauma, and others cause neck injuries. The kind of accident caused by severe overextension of the neck and associated tissues is called **whiplash**. Head and neck injuries are all too common in accidents where people suffer **high g decelerations** (meaning decelerations of many times that of gravity). What this means is that the victim absorbs very high decelerations on impact with another vehicle or with a stationary object. Still other victims are injured within the vehicle when they contact interior components of the passenger compartment, such as the dashboard and the windshield.

We first take a brief look at human anatomy from the neck up to understand what forces can do to neurological function. Fig. 17.1 shows the essentials of the skull, the brain, the spinal column, and the spinal cord.

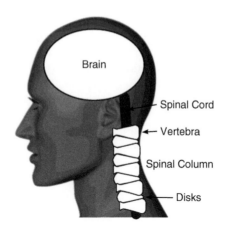

FIGURE 17.1

Part of the human nervous system.

The skull protects the brain, which floats inside the brain in a fluid-like layer. The base of the brain connects to the spinal cord through the spinal column, which provides protection for the spinal cord. The spinal column consists of individual bony vertebras that surround the all-important spinal cord. The spinal cord is the essential "wiring" that takes instructions from the brain to the various bodily functions. The vertebrae are separated by cartilaginous matter known as *disks* to provide flexibility and motion. Injury to the disks can result in severe pain and, in the case of injuries to the spinal cord per se, in paralysis. Injury to the spinal cord can lead to very severe bodily malfunctions while injury to the brain can cause a number of physical deteriorations or death. This system, as efficient as it is, can suffer a number of possible injuries during high g decelerations.

Consider the injuries that are illustrated in Fig. 17.2. The brain moves relative to the skull in its fluid-like layer when experiencing high g's. In **hyperflexion**, the skull moves forward relative to the brain, causing damage to the occipital lobe (the back of the brain), and in **hyperextension**, it moves in the opposite direction relative to the brain, causing damage to the frontal lobe. Further, damage can occur to the basal brain, a potentially devastating injury, since it may interrupt the nerve connections to the spinal cord. Also, there can be fractures to the vertebrae and extrusion or rupture of the protective disks.

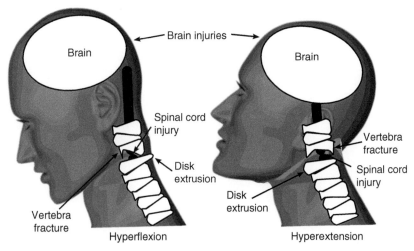

FIGURE 17.2

Brain injuries due to extreme whiplash.

17.4 **Why collisions can kill**

What is it about collisions that can kill or severely injure people? The challenge of this chapter is to express the answer using engineering variables. By using the right numbers and units for the values of these variables, criteria can be identified that distinguish a potentially fatal crash from one where the victims walk away.

One key variable in the bioengineering analysis of automobile safety is deceleration. Just as the "zero-to-sixty" time of acceleration is a measure of a car's potential for performance, the "sixty-to-zero" time of deceleration is a first measure of an accident's potential for injury. But, while the zero-to-sixty time is typically measured in seconds, the sixty-to-zero time may be measured in milliseconds, that is, in thousandths of a second.

Rapid deceleration causes injury because of its direct relation to force. As discussed in the section on Newton's Second Law (Chapter 4), force is directly proportional to positive or to negative acceleration (i.e., deceleration). And, as discussed in Chapter 14, force divided by area is stress. As further discussed in that chapter, the stress–strain curve gives the yield strength of a material. Suppose that the biologic material is flesh, bone, or neurons (nerve cells found in the brain, spinal column, and local nerves). If their yield strength is exceeded, serious injury or death follows.

An engineering forensic analysis of collision injury combines the effects of acceleration with the properties of materials. So, a first step in characterizing collisions is to determine the deceleration that occurs. We already have a tool for visualizing this: the v–t (or velocity – time) graph.

Example 17.1

A car traveling at 30.0 miles per hour (13.4 m/s) runs into a sturdy stone wall. Assume the car is a totally rigid body that neither compresses nor crumples during the collision. The wall "gives" a distance of $D_S = 0.030$ m in the direction of the collision as the car is brought to a halt. Assuming constant deceleration, calculate that deceleration.

   Need: Deceleration = _____ m/s².

   Know—How: First sketch the situation to clarify what is occurring at the impact. Then, use the $v-t$ graph to describe the collision. A positive slope of the $v-t$ graph indicates "acceleration," and a negative slope indicates "deceleration."

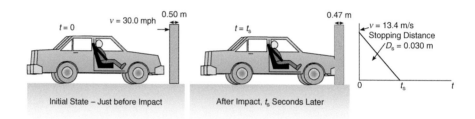

Initial State – Just before Impact   After Impact, t_s Seconds Later

   Solve: We first need to calculate the stopping time, t_s, to decelerate from 30.0 mph (13.4 m/s) to zero. If the deceleration is constant, we know that stopping distance is equal to the area under the $v-t$ graph or $D_S = \frac{1}{2} \times v \times t_s$, where t_s is the stopping time. So,

$$D_S = \frac{1}{2} \times 13.4 \text{ [m/s]} \times t_s \text{ [s]} = 0.030 \text{ m}$$

Therefore, $t_s = 2(0.03)/13.4 = 0.06/13.4 = 0.0045$ s, and the slope of the $v-t$ graph is

$$\Delta v/\Delta t = (0 - 13.4)/(0.0045 - 0) = -2980 = -\ 30.\times 10^2 \text{ m/s}^2.$$

The negative sign indicates this is the **deceleration** rate.

   What does this deceleration mean? How high is it? One good way to understand its damaging potential is to compare it to the acceleration of gravity. This nondimensional ratio is $a/g = 30.\times 10^2/9.8 = 310$ g's (to two significant figures).

The use of acceleration in terms of the number of g's can be quite helpful. Referring to Chapter 4, what is the weight of a mass m on earth? It is mg. What is the force on a mass accelerating by an amount a? It is ma. But, let us simultaneously multiply and divide that force by g; that is,

$$F = ma \times (g/g) = mg \times (a/g) = \text{weight } (mg) \times \text{number of } g\text{'s } (a/g) \qquad (17.1)$$

so that the number of g's = F/mg. Note that the number of g's = a/g is dimensionless.

   In the impact described in Example 17.1, your head would weigh 310 times its usual weight (and the same for the rest of your body). An average human head has a mass of about 4.8 kg, so it will experience a force of about 4.8 [kg] × 9.8 [m/s²] × 310 = 15,000 N or about 3300 lbf! Your entire body will experience a similar force 310 times its normal weight.

   You can also calculate the forces involved by kinetic considerations as opposed to these basically kinematic considerations. The force × distance to stop is equal to the kinetic energy that is to be dissipated. Let us assume the driver is belted into the car and therefore decelerates at the same rate as the car.

Example 17.2

The car and its belted driver of Example 17.1 suffer the same constant deceleration of 310 g's or 30. $\times$ 10^2 m/s^2. What is the force the driver experiences during the collision if his mass is 75 kg? Assume the car is a totally rigid body that neither compresses nor crumples during the collision.

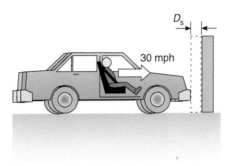

Need: Force stopping 75 kg driver on sudden impact = _____ N.
Know: The driver's weight is 75 [kg] $\times$ 9.8 [m/s^2] = 740 N and deceleration rate is 310 g.
How: $F = ma = mg \times a/g$.
Solve: $F = 740 \times 310 = \mathbf{2.3 \times 10^5}$ **N** (or about 52,000 lbf!).

17.5 The fracture criterion

Under what conditions does a force cause living material to fail, that is, to crack, or break into pieces, or lose its ability to contain or protect fluid or soft structures? As we saw in Chapter 14, a material breaks when the stress on it exceeds its ultimate strength. Stress and strength are variables with units (N/m^2) that can be calculated from a description of the loading and the material's stress–strain diagram.

One aspect of the loading is the applied force, which was found in Example 17.2. Another aspect is the area over which that force is applied. This is needed to calculate the stress on the material in question.

Fig. 17.3 shows an illustration of a bone. Bone is a natural composite material that consists of a porous framework made of the mineral calcium phosphate, interspersed with fibers of the polymeric material collagen. The calcium phosphate gives bone its stiffness, and the collagen provides flexibility. Fig. 17.3 shows the internal structure of the body's all-important long bones, which have a hard outer layer, spongy interior layers with axial channels built from bone cells known as *osteoblasts*, and a central core of marrow that is responsible for renewing blood supply and other important cells. Bone is "anisotropic," which means that its properties are not the same in all directions. For example, it has different mechanical properties along its axis than it has perpendicular to the axis.

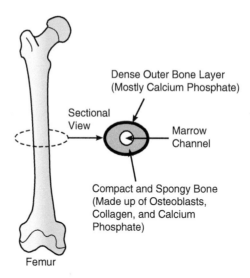

FIGURE 17.3

Structure of the body's long bones.

The properties of bone can be approximated by a stress—strain diagram as in Fig. 17.4. Notice bone is more elastic along (that is, in the longitudinal direction) the bone's length than crosswise (that is, in its transverse direction). These properties are due to the dense outer bone layer being tough with the compact and spongy interior bone providing surprisingly high flexibility. This model is a simplification of real bone behavior and is modeled here as if it were a perfectly elastic and perfectly plastic material. The corresponding data are shown in Table 17.1.

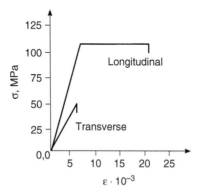

FIGURE 17.4

Bone mechanical properties.[1]

---

[1] Adapted from A. H. Burstein and T. M. Wright, *Fundamentals of Orthopedic Biomechanics* (Baltimore, MD: Williams and Wilkins, 1994), p. 116.

Table 17.1 Approximate mechanical properties of human bone.

Direction	Elastic yield, MPa	Young's modulus, GPa	Strain at yield, %	Strain at fracture, %	Toughness at yield and fracture, MJ/m^3
Longitudinal	110	16.	0.7	2.0	0.4 at yield and 1.8 at fracture
Transverse	60.	9.0	0.7	0.7	0.2 at yield and fracture

Bone, like other materials, may fail in several ways:

1. If bone is subjected to a local stress greater than its yield strength, it takes on a deformation.
2. If that stress exceeds the bone's ultimate stress, the bone will break.
3. If the energy absorbed exceeds the toughness at yield × volume of affected bone, the bone takes a permanent set.
4. If the energy absorbed exceeds the toughness at fracture × volume of affected bone, the bone breaks.

In general, bone failure modes 1 and 2 apply most to the situation in which the load is applied slowly. For a suddenly applied load, it is usually the toughness criterion that applies.

Most bones that fail in vehicular accidents fail due to transverse stress rather than longitudinal stress[2] (for example, the ribs hitting the steering wheel if their owner is not protected by a seat belt or air bag). However, whiplash injuries may produce neck bone failure in the longitudinal direction. In general we prefer not to permanently distort bone. Therefore, the elastic yield condition is the more conservative situation for a victim of an accident.

In another version of the accident described in Examples 17.1 and 17.2, the driver is unbelted, so the skull hits and dents the dashboard. How does the stress on the skull compare with the elastic yield point of the bones in the skull? That depends on the area of the skull over which the force is applied. For a given force, the smaller the area, the higher is the stress.

Example 17.3

This is an accident similar to that in Examples 17.1 and 17.2, except the 75 kg driver is not wearing a seat belt. Assume the full weight of the driver causes his/her skull to dent the dashboard to a depth of 0.030 m. Assume the area of contact between the driver's skull and the dashboard is $0.030 \times 0.030 \text{ m} = 9.0 \times 10^{-4} \text{ m}^2$, and that the properties of skull bone are those given in Table 17.1. Will the collision *fracture* the driver's skull?

---

[2] Think of snapping a dead tree limb across your knee. The break occurs on the side away from your knee as fulcrum. That layer of branch is being *stretched* and it fails mostly in tension.

Need: Is the yield strength of the skull exceeded? (Yes/No)

Know: Since the head suffers the same deceleration of 310 g given $D_S = 0.03$ m, we can use the results of Examples 17.1 and 17.2. Consequently, the force on impact of the 75 kg driver on her/his skull is 2.3×10^5 N.

How: Stress is force per unit area, and from Table 17.1, the yield strength of the bone is 60. MPa.

Solve: Compare 60. $\times 10^6$ [N/m^2] with the applied stress on impact. That stress is $(2.3 \times 10^5)/(9.0 \times 10^{-4})$ [N]/[m^2] = 2600×10^6 N/m^2 which is considerably greater than 60. $\times 10^6$ N/m^2. So, yes, the bone fractures, and the brain is seriously damaged. Unfortunately, this driver is fatally injured.

17.6 The stress-speed-stopping distance-area (SSSA)

Let us generalize the results of Example 17.3. Collisions kill or injure because they involve a high rate of deceleration. That deceleration causes bodily contact with inflexible material, and the result is experienced as a force. The smaller the area over which that force is applied, the higher is the stress on the bone or tissue. The higher the localized stress, the greater is the likelihood that the material fails in that region.

We can express this insight as a relationship among the variables. Consider first the generalized form of the $v-t$ diagram of a collision with a constant deceleration rate. Suppose a body of mass m is subjected to a constant deceleration a and stops in a distance D_S. Then, from Newton's Second Law, the force experienced by the body is $F = ma$, and $a = v/t$, so $F = mv/t$. For a constant deceleration, the stopping distance is: $D_S = \frac{1}{2}vt$, so $t = 2D_S/v$. Then, the force experienced by the body is

$$F = ma = mv^2/2D_S \tag{17.2}$$

Now suppose that this force is applied to an area A on the head or body with a hard surface. Then, the stress can be calculated from Eq. (17.2) as

$$\text{Stress} = \sigma = F/A = mv^2/2AD_S \tag{17.3}$$

We call this relationship the **SSSA criterion**. It states that, technically, it's not enough to say that "speed kills." What kills is the combination of high speeds, short stopping distances, *and* small contact areas!

Strategies for reducing the severity of a collision follow from these insights. By decreasing speed, increasing the stopping distance, and increasing the area of application of the force, the effects of the collision on living tissue can be reduced.

The presence of the v^2 term in the numerator of Eq. (17.3) indicates that decreasing speed is a highly effective way of decreasing collision severity. Cutting speed in half reduces the stress of a collision to one fourth its previous value (if everything else remains the same).

How might the other two terms in the SSSA criterion be brought into play? Application of the area term is simple in concept, though more difficult in practice. The larger the area of contact between body and surroundings during a collision, the smaller is the stress. This is one motivation behind the air bag. It is a big gas-filled cushion that can increase the area of contact by a large factor. Of course, that cushion must be deployed rapidly enough to be effective in a collision. Only by conquering these two technical challenges—deployment in a few milliseconds when appropriate, nondeployment at all other times—did engineers turn the air bag into a valuable safety tool.

However, the area portion of the SSSA criterion is only part of the air bag story. A seat belt does indeed increase the area over which force is applied (compared to the forehead) and, as important, reduces the probability of an impact of the head with the dashboard. But that increase would not by itself account for the great life-saving and injury-reducing potential of seat belts. A more important part of that potential involves that other term in the denominator, the stopping distance, D_S.

A more practically important contribution of seat belts is to attach the driver to a rigid internal passenger shell while the rest of the car shortens by crumpling. The high effectiveness of the seat belt as a safety device has been achieved only in combination with the design of a car, which significantly crumples during a collision. Let us now assume that the car has such a *crumple zone*. This is illustrated in Fig. 17.5.

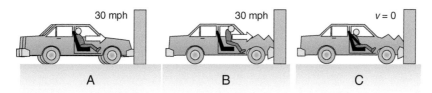

FIGURE 17.5

Seat belts and crumple zones.

Illustration A shows the situation of driver, car, and wall during a 30. mph collision. The car rapidly crumples to a stop several milliseconds later, as shown in illustration B. But the driver is still moving forward at the speed of 30. miles an hour. It is the collision of the driver's head and the windshield that brings the driver to a stop in a short distance, on the order of a few hundredths of a meter. By contrast, when the driver is wearing a seat belt, the body stops with the car, as shown in illustration C.

In the last case, the driver gets the full benefit of the crumple of the car, which is on the order of a half to 1 m. This hundredfold increase in stopping distance results, according to the SSSA criterion, in a reduction in stress to one hundredth of its previous value. This can be the difference between life and death.

Example 17.4

A car traveling at 30.0 miles per hour (13.4 m/s) runs into a rigid stone wall. Assume the car's crumple zone results in a stopping distance of 0.60 m. Assume that the 75 kg driver is wearing a seat belt. Assume that the area of contact of seat belt and body is A = 0.040 m × 0.30 m = 0.012 m². Determine the maximum stress and the value of g the driver's body experiences.

   Need: Maximum stress on the driver's body _____ N/m² and the number of g's.

   Know–How: We could simply repeat the analysis of the previous examples, but the SSSA criterion provides a shortcut.

Solve: Stress, $\sigma = mv^2/2AD_S$. Substitute into the SSSA formula:

$$\sigma = 75. \times 13.4^2/(2 \times 0.012 \times 0.60) \text{ [kg] [m/s]}^2 \left[1/m^2\right][1/m]$$

$= 9.4 \times 10^5$ **N/m²** (which is less than the elastic yield of bone in Table 17.1). The number of g's is calculated from Eqs. (17.1) and (17.3) as

Number of g's $= F/mg = \sigma A/mg = 9.4 \times 10^5 \times 0.012/(75 \times 9.8)$ [N/m²][m²]/[kg] [m/s²] $= 15$ g's.

Previously it was 310 g's, so survivability has been greatly enhanced.

To sum up, here is an answer to our original question, expressed in engineering variables and units. The principal way seat belts save lives is by attaching the driver securely to the inner shell of the car, enabling the driver to take advantage of the car's crumple zone of about 0.5−1 m. During a collision from 30. miles an hour, that strategy restricts deceleration to less than about 150 m/s², resulting in stresses on the body that are less than about 1.0 MPa and almost surely less than enough to break bones.

17.7 Criteria for predicting effects of potential accidents

The preceding sections established that the effects of rapid deceleration on human bone and tissue may produce serious injury. We just indicated one way to address that issue by translating deceleration into force and force into stress. A comparison of maximum stress experienced by the body with the strength of bone provides a first criterion for the prediction of accident severity.

The effects of acceleration and deceleration on the human body go far beyond the potential to break bones. These effects range from the danger of blackouts for pilots experiencing very high acceleration or deceleration to the bone loss and heart arrhythmia experienced by astronauts exposed to the microgravitational forces of space flight for extended periods of time.

Just how many g's can the human body stand? As a first approximation, engineers drew on a wide range of experiences, such as those of pilots and accident victims, to arrive at the following criterion:

The human body should not be subjected to more than 30. g's.

At 30. g's and above, the damaging effects of acceleration or deceleration on the human body can range from loss of consciousness to ruptured blood vessels to concussion to the breaking of bone to trauma or death. This criterion still serves as an initial rule of thumb for the design of safety devices.

Example 17.5

Does a driver without a seat belt who experiences a 30.0 mile per hour (13.4 m/s) collision and is stopped in 0.10 m by a padded dashboard exceed the 30. g criterion? Does the role of a seat belt in taking advantage of the crumple zone to increase collision distance to 0.60 m meet the 30. g criterion?

Need: Deceleration without seat belt = _____ (less than/equal/more than 30. g?)

Deceleration with seat belt = _____ (less than/equal/more than 30. g?)
Know–How: From Equation (17.2), $F = ma = mv^2/(2D_S)$; therefore, in g's,

$$F/mg = ma/mg = a/g = mv^2/(2mgD_S) = v^2/(2gD_S)$$

Solve: Case 1: $D_S = 0.10$ m, $a/g = (13.4^2)/(2 \times 9.8 \times 0.10)$ [m/s]^2/m/s^2][m] = 92 g's which is
more than 30. g's.
Case 2: $D_S = 0.60$ m, $F/mg = (13.4^2)/(2 \times 9.8 \times 0.60) = 15$ g's
which is **less than 30. g's.**

Consistent with our previous analysis, the combination of seat belt and vehicle crumple zone reduced the driver's deceleration from a probably fatal 92 g's to a probably survivable 15 g's.

However, experiences under extreme conditions, such as high-speed flight, soon revealed that the 30. g's criterion is seriously incomplete. In some cases, humans have survived accelerations far above 30. g's. In other cases, accelerations significantly below 30. g's caused serious injury.

The missing element in the simple 30. g's criterion is **time.** Deceleration has many effects on human tissue, ranging from destruction at one extreme to barely noticeable restriction of blood flow at the other. Each of these effects has a characteristic time interval needed to take effect. Very high decelerations, measured in hundreds of a g, can be survived *if* the exposure is short enough. Deceleration at moderate g levels, on the other hand, can prove fatal if the exposure is long enough.

In Fig. 17.6, the axes are expressed in terms of the **log$_{10}$** of the variables. Not only can we get a scale covering several orders of magnitude by this neat trick, we also can infer something important about the relationship. In fact, it is linear, not between the actual variables g and t_S, but between $\log_{10} g$ and $\log_{10} t_S$.

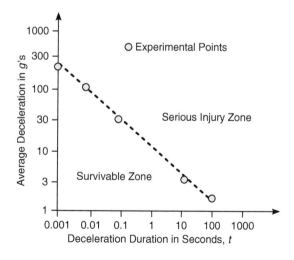

FIGURE 17.6

Injury criterion based on the deceleration in g's and the deceleration time, t.

S.A. Berger et al., Introduction to Bioengineering (Oxford: Oxford University Press, 1996).

The dotted line represents a *rough boundary* between accelerations likely to cause serious injury or death (above and to the right of the line) and survivable accelerations (below and to the left of the line). Notice that the previous "30. *g*'s" criterion is the point on that line corresponding to a duration of 0.1 s (the right order of magnitude for the duration of an automobile accident).

Figure 17.6 can also be expressed as a formula. A straight line on a log-log plot corresponds to the number on the x-axis raised to a certain power. In this case, the formula for the line dividing serious injury from survivability is

$$a = (0.002\ t_S)^{-0.4} \tag{17.4}$$

with *a* expressed in the nondimensional units of *g*'s and with t_S expressed in seconds. The slope of the log/log graph is −0.4. This is *not* a basic scientific law. Rather, it is an empirical relationship summing up the net effects of many physical and biological properties. It is, however, a useful guide. It is more convenient by rearranging it into the form

$$a^{2.5}\ t_S = 500 \tag{17.5}$$

We can generalize Eq. (17.5) by introducing the **Gadd Severity Index** (GSI).[3]

$$\mathbf{GSI} = a^{2.5} t_S \tag{17.6}$$

where t_S is in seconds and *a* is in *g*'s.

The GSI in Eq. (17.6) is a simplified numerical index that comes from assuming either constant deceleration or that a meaningful average deceleration can be measured. To use this equation, first, calculate the GSI, then compare the result to the number 500. If the GSI is greater than 500, there is a serious danger of injury or death.

Example 17.6

Calculate the GSI for a driver without a seat belt who experiences a constant deceleration in a 30.0 miles per hour (13.4 m/s) collision and is stopped in 0.050 m by the dashboard.

Need: GSI = _____ (a number).

Know−How: From Example 17.1, we know that, if deceleration is constant, the stopping distance is equal to the area under the $v−t$ graph or $D_S = \frac{1}{2} \times v \times t_s$, where t_s is the stopping time. So, then, $t_s = 2D_s/v = 2 \times 0.050/13.4 = 0.0074$ s. Therefore, the deceleration is: $a = \Delta v/\Delta t = (0 - 13.4)/(0.0074 - 0) = -1800$ m/s^2, or $a = 180$ *g*'s (to two significant figures and neglecting the negative sign indicating it's a deceleration).

Solve: GSI $= a^{2.5}\ t_S = 180^{2.5} \times 0.0074 = 3200$, which is much greater than 500. This suggests that severe injury or death is likely.

---

[3] The Gadd Severity Index was introduced in C. W. Gadd, in 1961. "Criteria for Injury Potential." National Research Council Publication #977 (Washington, DC: National Academy of Sciences, 1961), pp. 141−145.

Example 17.7

Calculate the GSI for a driver with a fastened seat belt who experiences a constant deceleration in a 30.0 miles per hour (13.4 m/s) collision and is stopped by a 1.0 m crumple zone.

Need: GSI = _____ (a number).

Know–How: From the development of Eq. (17.2), the stopping time can be calculated from
$$t_S = 2D_S/v = 2(1.0)/13.4 = 0.15 \text{ s, and the deceleration is}$$

$$a = \Delta v/\Delta t = (0 - 13.4)/(0.15 - 0) = -89. \text{ m/s}^2, \text{ or } \boldsymbol{a = 89/9.8 = 9.1 \text{ } g\text{'s.}}$$

Solve: GSI $= a^{2.5} t_S = 9.1^{2.5} \times 0.15 = \textbf{37}$, which is much less than 500.

This suggests that severe injury or death is very unlikely.

17.8 Healthcare engineering

Healthcare engineering is an interdisciplinary field that uses engineering methods to improve the welfare of people. Healthcare engineering is a fast-growing field where engineers design and develop devices, systems, and procedures to address issues such as the continuing increase in healthcare costs, the quality and safety of patients of all ages, and the management of diseases by utilizing electronics, robotics, miniaturization, material engineering, and other fields.

Healthcare engineers can be engineers from any engineering field such as biomedical, chemical, civil, computer, electrical, industrial, materials, mechanical, and so on who are engaged in supporting, improving, and advancing all aspects of healthcare through engineering methods. Healthcare engineering covers topics such as the following:

Assistive technology, which includes designing and developing treatment equipment to assist individuals who are disabled as the result of birth defects, injury, illness or aging. Engineering activities may involve an all-inclusive approach or one focused on a specific organ, body system, disease state or disability, with the goal of developing devices and sensors to help individuals with disabilities carry out daily activities.

Medical Imaging, which includes the imaging of the human body and its functions. It includes visualization, processing, rendering, analysis, and interaction with techniques used in radiology, such as MRI.

Patient Informatics, which refers to patient electronic medical records, how they are generated, stored, transmitted, and how a doctor accesses this information.

Healthcare systems, in which engineers are involved in the organizations, agencies, facilities, information systems, management systems, financing mechanisms, and all trained personnel engaged in delivering healthcare.

Example 17.8

A case study in healthcare engineering[4]

Lower-limb prosthetics often require a liner material to ensure mechanical transmission of forces while ensuring comfort and preventing skin damage. Made of silicone rubber these liners lead to sweating which, when coupled with friction, causes skin deterioration and infection. This is one of the most common reasons given by patients for why they choose not to wear prostheses.

The healthcare engineering challenge is to design and develop liner material that will not only measure forces in the prostheses but also be comfortable and prevent skin damage as a wearer's body changes size and shape throughout the day.

The design goal is to create breathable 3D printable silicone liners that contain active cooling channels and be able to sense stresses, and then feed the data back to the wearer.

---

[4] This case study is from University College London, see https://www.ucl.ac.uk/healthcare-engineering/case-studies/2018/sep/smart-prosthetic-liner-monitors-wearers-temperature-and-actively-cools

Summary

Bioengineering applies the methods of engineering to living bodies, organs, and systems. This chapter illustrated just one aspect of bioengineering, the use of biomechanics and engineering analysis of biomaterials to improve safety. In this chapter, we learned why collisions can kill, how to make a first approximation of the likelihood of collision damage to the human body using a fracture criterion, how to predict the injury potential of a possible accident using the **SSSA** criterion, how to apply two criteria for the effect of deceleration on the human body (the **30. g's limit** and the **Gadd Severity Impact** parameter), and how to analyze bioengineering problems.

Exercises

Exercises 1 and 2 concern the following situation: A car is traveling 30.0 mph and hits a wall. The car has a crumple zone of zero, and the passenger is not wearing a seat belt. The passenger's head hits the windshield and is stopped in the distance of 0.10 m. The skull mass is 5.0 kg. The area of contact of the head and the windshield is 0.010 m^2. Assume direct contact and ignore the time it takes the passenger to reach the windshield.

1. Provide a graph of $v-t$ of the collision of the skull and the windshield, then graph the force experienced by the skull as a function of time. (**Ans. See the following figure.**)

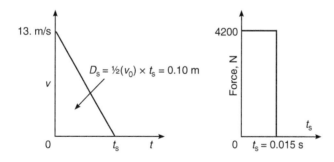

2. If the compressive strength of bone is 3.0×10^6 N/m^2, does the collision in the previous exercise break the skull?

    Exercises 3 and 4 concern an experiment in the 1950s, when Air Force Colonel John Paul Stapp volunteered to ride a rocket sled to test the resistance of the human body to "g forces." The sled accelerated from 0 to 625 miles per hour in 5.0 s. Then, the sled hit a water brake and decelerated in 3.0 s to a standstill. Assume that Stapp was rigidly strapped into the sled and that he had a mass of 75 kg.

3. Prepare $v-t$ and $F-t$ graphs of Stapp's trip, and compute the "g forces" he experienced in the course of acceleration and deceleration. (**Ans. 5.7 g's, −9.5 g's, see the following figure.**)

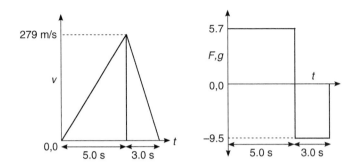

4. Using the force versus time graph (Fig. 17.6) for human resistance to g forces, predict whether Stapp suffered serious injury in the course of his record-breaking trip.

5. A tall person sits down on a sofa to watch TV. Assume that the center of gravity of the person falls 1.0 m with constant gravitational acceleration in the course of sitting down. The sofa compresses by 0.05 m. Assume constant deceleration. Determine the g forces experienced by the person in the course of this sitting down. (**Ans. 20. g's**, so be kinder to couch potatoes!)

Exercises 6 and 7 concern an infant's rear-facing safety seat as illustrated next.

6. A rear facing child safety seat holds a child of mass 12. kg rigidly within the interior of a car. The area of contact between the seat and the child is 0.10 m^2. The car undergoes a 30. mph collision. The car's crumple zone causes the distance traveled by the rigid interior to be 1.0 m. Give the stress experienced by the child's body in terms of a fraction of the breaking strength of bone assuming an infant's bone breaks at a stress of 10. MN/m^2. (**Ans.** As a fraction of breaking stress, 1.08×10^3, the bones should hold and the infant should be safe.)

7. A rear-facing child safety seat holds a child of mass 25. kg rigidly within the rigid interior of a car. The area of contact between the seat and the child is 0.10 m^2. The car undergoes a 30. mph collision. The car has no crumple zone, but a harness attached to the car seat stops it uniformly within a distance of 0.30 m. According to the Gadd Severity Index, will the child sustain serious injury or death?

8. Consider a parachute as a safety device. When the parachute opens, the previously freely falling person has typically reached a speed of about 50. m/s. The parachute slows to a terminal speed of about 10. m/s in 1.3 s. Approximating this set of motions by a constant deceleration, what is the maximum g experienced by the parachutist? (**Ans. -3.1 g's.**)

9. In the previous exercise, the force exerted by the parachute is spread by a harness in contact with $0.50 \, m^2$ of the parachutist, and the parachutist has a mass of 75 kg. What is the force per unit area (stress) experienced by the person during the deceleration?

10. The parachutist in the previous two exercises hits the ground (still wearing the parachute!) and is stopped in a distance of 0.10 m. If this final deceleration is constant, calculate the Gadd Severity Impact of the landing. (**Ans. GSI = 370.**)

11. A 75. kg person jumping from a 1.00×10^3 m cliff will reach a terminal speed of 50. m/s and uses a 1.00×10^2 m bungee cord to slow the descent. The bungee cord exerts a force F proportional to its extension, where F (in newtons) = (5.0 N/m) × (extension in m) and is designed to extend by 5.00×10^1 m in the course of bringing the user to a stop just above the ground. Is the maximum deceleration in g experienced by the falling person more or less than the maximum deceleration experienced by a parachutist undertaking the same leap (excluding landing forces)?

12. An air bag is designed to inflate very quickly and subsequently compress if the driver hits it. A collision uniformly stops a car from 30.0 mph to 0.0 and triggers the air bag. The driver is not seat belted and so hits the inflated air bag. This acts as a local "crumple zone" and consequently compresses by 0.20 m as his head is brought to rest. According to the Gadd Severity Index, does the driver suffer serious injury? Assume constant deceleration of the driver's head after hitting the air bag. (**Ans. GSI = 420**, and no, the driver should not suffer serious injury or death.)

13. Two 100. kg football players wearing regulation helmets collide helmet to helmet while each is moving directly at each other at 10.0 m/s and come to a near instantaneous (<1.0 ms) stop. The area of contact is $0.01 \, m^2$, and the helmets are each designed to provide a crumple zone of 0.025 m. What is the maximum stress exerted on each player? (**Ans. $4.0 \times 10^7 \, N/m^2$.**)

14. A designer of football helmets has two options for increasing the safety of helmets but, for economic reasons, can implement only one. One option is to double the area of contact experienced in a helmet-to-helmet collision. The other is to double the crumple distance experienced in a helmet-to-helmet collision. Which will be more effective in reducing the maximum stress? (Hint: Try the previous exercise first.)

15. A soccer player "heads" a wet 0.50 kg soccer ball moving directly toward him by striking it with his forehead. Assume the player initially moves his head forward to meet the ball at 5.0 m/s and the head stops after the ball compresses by 0.05 m during impact. Assume the deceleration of the head is constant during impact. Compute and comment on the calculated Gadd Severity Impact of heading a soccer ball under these conditions.

16. A car strikes a wall traveling 30.0 mph. The driver's cervical spine (basically, the neck) first stretches forward relative to the rest of the body by 0.01 m then recoils backward by 0.02 m as shown in the following figure. Assume the spine can be modeled by a material with a modulus $E = 10.$ GPa and a strength of 1.00×10^2 MPa. Does the maximum stress on the cervical spine during this "whiplash" portion of the accident exceed the strength of the spine? Assume a 0.15 m length of the cervical spine. (**Ans.** Stress on cervical spine is **greater than** its tensile strength.)

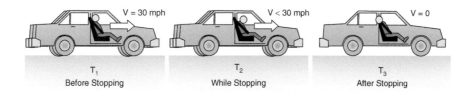

V = 30 mph	V < 30 mph	V = 0
T_1	T_2	T_3
Before Stopping	While Stopping	After Stopping

17. Which do you think has been more effective in reducing fatalities on American highways, seat belts or air bags? Give an engineering reason for your answer, containing variables, numbers,

and units. (Hint: Recall the SSSA formula previously developed and consider what other safety element is designed into a modern automobile.) Then, go on the Web and see if you were right.

18. As a bioengineer at the Crash Safety Test Facility of a major automobile company, you are asked to provide more data for the Gadd Severity Index (Fig. 17.6). Your boss suggests using live animals, dogs, and cats from the local pound, in hard impact tests and inspecting them for injury. You know their injuries will be severe or fatal, and using dogs and cats seems cruel. What do you do? (Use the Engineering Ethics Matrix format to summarize your conclusions.)

    a. Nothing. Live animals are used regularly in product testing, and besides, they will probably be killed in the pound anyway.
    b. Suggest using dead animals from the pound, since their impact injuries probably do not depend on whether or not they are alive.
    c. Suggest using human cadavers since you really want data on humans anyway.
    d. Suggest developing an instrumented human mannequin for these tests.

19. You are now a supervisor in the bioengineering department of a major motorcycle helmet manufacturer. Your engineers are testing motorcycle helmets manufactured by a variety of your competitors. Motorcycle helmets contain an inner liner that crushes on impact to decrease the deceleration of the head. This liner material is very expensive and can be used only once (i.e., once the helmet sustains a single impact it must be replaced). Your company has developed an inexpensive liner that withstands multiple impacts but is less effective on the initial impact than any of your competitors. The vice president for sales is anxious to get this new helmet on the market and is threatening to fire you if you do not release it to the manufacturing division. What do you do? (Use the Engineering Ethics Matrix format to summarize your conclusions.)

    a. Since your company has invested a lot of money in the development of this helmet, you should release it, and besides, if you do not, someone else will.
    b. Recommend continued testing until your company's product is at least as good as the worst competitor's product.
    c. Contact your company's legal department to warn them of a potential product liability problem and ask for their advice.
    d. Go over the vice president's head and explain the problem to the company's president.

20. During World War II, Nazi Germany conducted human medical experimentation on large numbers of people held in its concentration camps. Because many German aircraft were shot down over the North Sea, they wanted to determine the survival time of pilots downed in the cold waters before they died of hypothermia (exposure to cold temperatures). German U-boat crews faced similar problems. In 1942, prisoners at the concentration camp in Dachau were exposed to hypothermia and hypoxia experiments designed to help Luftwaffe pilots. The research involved putting prisoners in a tank of ice water for hours (and others were forced to stand naked for hours at subfreezing temperatures) often causing death.

    Research in the pursuit of national interests using available human subjects is the ultimate example of unethical bioengineering. Since the Nazi scientific data were carefully recorded, this produces a dilemma that continues to confront researchers. As a bioengineer today, should you use these data in the design of any product (such as cold-weather clothing or hypothermia apparatus for open heart surgery)? (Use the Engineering Ethics Matrix format to summarize your conclusions.)

    a. Since these experiments had government support and were of national interest at the time, they should be considered valid and available for scientific use now.
    b. Should you use these data, since similar scientific experiments have been conducted in other countries during periods in which national security was threatened and these data are not questioned today? Even the United States conducted plutonium experiments on unsuspecting and supposedly terminally ill patients (some of whom survived to old age!).
    c. This is just history and should have no bearing on the value or subsequent use of the data obtained.
    d. Do not use the data.

Electrochemical engineering

18

18.1 Introduction

Alternate energy sources are the hope for the future where we eliminate our heavy dependence on fossil fuel sources (coal, oil, and gas). Perhaps in the future we will use electrochemical technology systems to propel electric cars, improve electronic devices, and supplement solar energy systems.

18.2 Electrochemistry

Electrochemistry is the science behind **batteries** and **fuel cells** and, as such, plays an essential role in the development of novel power sources. You might be aware there is a heavy-duty battery in every automobile to activate the starter at the twist of the ignition key as well as for auxiliary uses in headlights, car radios, and so forth. But there are also small, familiar batteries that power flashlights, electronic devices, computers, and much more. In addition, special batteries are an integral part of "hybrid" and "electric" automobiles.

Exploring Engineering. https://doi.org/10.1016/B978-0-443-13541-5.00011-8

Batteries are intimately involved in various alternate power source concepts. For example, battery electrical storage units may be used in solar energy systems, such as small systems for individual houses or large ones for central power stations (Fig. 18.1).

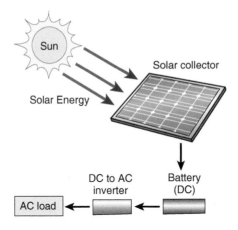

FIGURE 18.1

Solar power schematic.

A problem posed by such schemes is that they are subject to the variations in the weather as well as fluctuations in the demand for the electricity. Solar collectors make electricity when the sun is shining, but what happens if you don't have an immediate use for that power? Do you discard it even though this is a very wasteful thing to do?

And, what happens when you do need more power than you are producing at that moment? Suppose you use electric power for cooking the morning and evening meals. Your highest demand probably is in the early morning and evening before the sun comes up and after the sun goes down (or when the winds are quiet). Fig. 18.2 shows the problem.

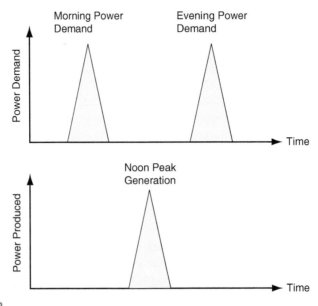

FIGURE 18.2

Mismatch of daily power production and need.

Obviously, what you want to do is to store the excess energy when it is abundant (noon in Fig. 18.2) and use it when it is required (early morning and evening in Fig. 18.2). Batteries are a relatively simple way to achieve this and individual house solar installations may rely on this tactic, known as **load leveling**, as shown in Fig. 18.3.

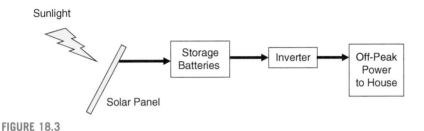

Sunlight

Solar Panel

FIGURE 18.3

Load leveling for solar energy. An inverter takes direct current (DC) from the batteries and converts it into alternating current (AC) for household use.

The other major topic of this chapter related to batteries is fuel cells. Fuel cells are basically continuously refueled batteries. Normal batteries, such as the battery in a car or a flashlight battery, have a finite amount of chemical energy stored in them. That is why batteries go "dead" after extended use—the potential energy stored in their electrodes and electrolyte (the chemical heart of a battery) has been extracted, and the equivalent amount of electrical energy has been consumed. A fuel cell does not suffer this limitation. It is constantly replenished with fresh chemicals and therefore with fresh chemical energy as more electricity is demanded. The high-energy feed chemicals (e.g., hydrogen gas or methanol) produce chemical products that are low energy, such as water or carbon dioxide.

Since chemistry is heavily involved, chemical engineers have been at the vanguard along with mechanical and electrical engineers who integrate such systems into useable energy resources.

Example 18.1

Between 10:00 A.M. and 2:00 P.M. on an otherwise cloudy day, the sun comes out and a solar-powered house produces and stores 20. kWh of electric energy. (Note: 1 kWh is the energy produced by 1 kW for 1 h and is equal to 3600 kJ.) The rest of the day it produces nothing. The house requires 1.0 kW from 10:00 A.M. to 6:00 P.M., and 2.5 kW between 6:00 P.M. and 8:00 P.M. Is a 10. kWh storage battery large enough to supply the house's electrical needs between 2:00 and 8:00 P.M.?

Need: Size of storage battery for load leveling.

Know: The house uses 1.0 kW except between 6:00 P.M. and 8:00 P.M., when it requires 2.5 kW. It stores 20. kWh between 10:00 A.M. and 2:00 P.M.

How: We need an energy balance relating the energy flow and the net energy stored. We make the assumption that no energy is stored until 10:00 A.M.

Solve: The easiest way to solve this kind of problem is to graph it.

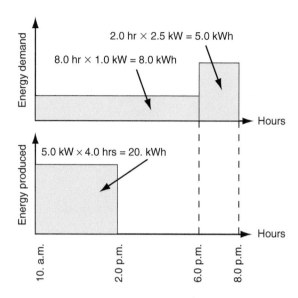

As the schematic shows, our house needs electricity at a rate of 1.0 kW for 4.0 h (4.0 kWh) from 2:00 P.M. until 6:00 P.M. and another 5.0 kWh between 6:00 P.M. and 8:00 P.M. for a total of **9.0 kWh**.

We store a net 20. − 4.0 = **16 kWh** of electrical energy between 10:00 A.M. and 2:00 P.M., much of which we have to dispose of, but our **10. kWh battery storage is large enough** for our needs. The additional 7.0 kWh may be sold to the electric power grid if the local utility allows it.

18.3 Principles of electrochemical engineering

Why does a battery work at all? Where do the electrons come from that carry the current in the external wires?

Many simple compounds are **ionic**, meaning they are held together by the balancing of positive and negative electrical charges. The perfect example is common salt, which can be symbolically written either as NaCl or as Na^+Cl^-. Ionic compounds are formed by the transfer of an electron from an electrically neutral atom or molecule to another electrically neutral atom or molecule. In a *crystal* of common salt, these charges hold the crystal lattice together and are called Columbic forces or ionic forces, which are nothing but the forces of electrostatics governed by Coulomb's law, expressed as follows:

$$F = \frac{e^2}{kr^2}$$

(18.1)

Here, F is the attractive force between two electrostatically opposite charges each of magnitude $e = 1.60 \times 10^{-19}$ coulombs and held apart at some distance, r. The constant k is the **dielectric constant** (or **permittivity**) of the medium in which the compound is immersed. This force is strongly attractive and is responsible for the crystalline structure of common table salt.

If an ionic crystal is immersed in water instead of in air, the relative value of k increases by a factor of ~ 80 and the corresponding force between atoms in the NaCl crystal is loosened by the same factor of ~ 80. The result is that positive ions of Na^+ and negative ions of Cl^- dissolve in water, and the ions drift apart. Thus, a solution of NaCl becomes a solution of two kinds of ions: Na^+ and Cl^- in equal proportion (and thus neutral overall). The trick in any electrochemical device is to physically separate these charges enough and create a voltage potential; this potential can then drive electrical charges through the external circuit.

Before we describe any electrochemical applications, it is necessary to define the terms **electrolyte** and **electrodes** which are further subdivided into **anodes** and **cathodes**. Fig. 18.4 illustrates these concepts.

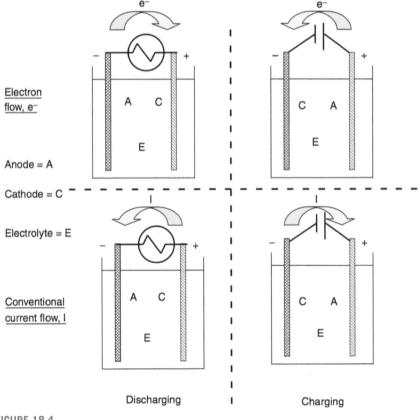

FIGURE 18.4

Some electrochemical terms.

This has been confusing ever since Ben Franklin arbitrarily decided that electric current flowed from the positive to the negative terminal. Much later, the British physicist J. J. Thompson discovered that negatively charged electrons carry the current in the opposite direction assumed by Franklin, that is, from negative to positive. It's much easier to think in terms of electrons than Franklin's current flows. For a discharging battery, the key concept is that the anodes churn out electrons and cathodes consume them. Conventional current during discharge flows into the anode (so it is marked as negative) and away from the cathode (so it is marked as positive).

18.4 Lead-acid batteries

As our first electrochemical example, consider today's lead-acid car batteries. The principles of their operation are conceptually similar to those for most electrochemical devices. The lead-acid battery was invented more than 150 years ago and has been much improved with detailed understanding of its internal chemistry.

One electrode is basically lead (chemical symbol Pb) plus several other metals added to improve its performance. The other electrode is lead coated with a lead oxide, known as litharge. The electrolyte is sulfuric acid (H_2SO_4) in water. The sulfuric acid solution in water may be thought of as dissociating into its constituent ions $2H^+$ and $SO_4^=$.

The discharge chemistry at the anode dissolves some lead and puts the subsequent positive lead ions into solution (since electrons are produced in the anode and flow into the external circuit).

Anode: The anode is made of "pure" lead. The principle anodic reaction is

$$Pb = Pb^{++} + 2e^- \qquad (18.2)$$

The lead ions Pb^{++} are doubly charged, combine with the sulfate ion from sulfuric acid, and mostly precipitate, since lead sulfate is only sparingly soluble in water:

$$Pb^{++} + SO_4^= = PbSO_4 \qquad (18.3)$$

The removal of sulfate ions from solution means they no longer exactly match the hydrogen ion concentration and the excess H^+ ions migrate across the electrolyte toward the cathode.

In the case of a lead-acid cell, the reactions near the anode produce a voltage of about 0.36 V and, near the cathode, another 1.69 V for a total cell voltage of about 2.0 V (see Fig. 18.5). There is also a voltage loss across the electrolyte, which we ignored and is referred to as *Ohmic losses*, because it is explained by Ohm's law. It occurs only when current is being drawn from the battery.

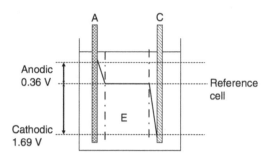

FIGURE 18.5

Cell potentials for lead-acid half cells.

Cathode: The cathode is mostly made of a sheet of lead coated with a paste of lead oxide (PbO_2). During discharge, the cathode receives electrons from the external circuit that, in turn, help to convert the lead oxide to its sparingly soluble sulfate, most of which stays on the cathode's surface. Overall, this represents the primary reactions:

$$PbO_2 + 4H^+ + SO_4^= + 2e^- = PbSO_4 + 2H_2O \qquad (18.4)$$

Both cathodic and anodic reactions remove lead ions and sulfate ions from the electrolyte and thus deplete the charge. The battery can be recharged several hundred times by applying a reverse voltage in excess of 2.0 V per cell. Eventually, the repeated solution and dissolution of the lead sulfate produces debris that falls to the bottom of the cell and fatally shorts it.

A car battery has six lead-acid cells in series and weighs about 25 kg. It can deliver 200–300 A at about 12 V for several minutes (see Table 18.1 for other energy storage parameters). The principal reason for its heavy weight is that it contains lead whose density is more than 11 times that of water. This fact is important for the future use of lead-acid batteries as energy storage devices.

Table 18.1 Measures of lead-acid battery efficiency.

Item	Value
Battery mass	25 kg
Energy content	3000 kJ
Mass energy storage density	120 kJ/kg
Volumetric energy storage density	250 kJ/L
Power	5 kW

Derived in part from: http://en.wikipedia.org/wiki/Lead-acid_battery and http://www.wdv. com/Hypercars/EAATalk.html

Example 18.2

Compare the statistics in Table 18.1 to that in 25 kg of gasoline. Comment on the prospects for all-battery driven cars in the future.

Need: Energy in 25 kg of gasoline and compare to Table 18.1.

Know: Combustion energy of gasoline is the same as its mass energy storage density, or about 46,500 kJ/kg (see Chapter 8).

How: We need the mass density of gasoline to compute its volumetric energy storage density. It is 740. kg/m^3 or 0.740 kg/L.

Solve: The energy content is $46{,}500 \times 25$ [kJ/kg][kg] $= \mathbf{1.2 \times 10^6\ kJ}$. For the volumetric energy storage density, $46{,}500 \times 0.740$ [kJ/kg][kg/L] $= \mathbf{34{,}400\ kJ/L}$.

Property	Lead-acid battery	Gasoline
Mass	25 kg	25 kg
Energy	3000 kJ	1.2×10^5 kJ
Mass energy storage density	120 kJ/kg	46,500 kJ/kg
Volumetric energy storage density	250 kJ/L	34,400 kJ/L
Power	5 kW	Typically >100 kW

In tabular form, you can easily see the challenge for the all-electric car that uses conventional batteries. Basically, a lead-acid battery is too heavy and has too little stored energy to compete with gasoline (hence, the compromise solution of hybrid gasoline/electric cars).

18.5 The Ragone chart

A neat way to compare the energy storage capability of an electrical storage device with its power producing capability is known as the Ragone[1] plot. It visually shows the energy storage and power producing characteristics of electrical storage devices. Fig. 18.6 is a pictorial representation of the logarithm[2] of energy storage density (expressed as Wh/kg) against the logarithm of power producing density (expressed as W/kg).

We want to create an electrochemical device that operates in the upper right-hand corner, so we can simultaneously maximize the power producing mass density and the energy storage mass density and make the smallest possible battery with the least materials. For example the Ragone plot indicates a lead-acid battery is more competitive for its energy storage capacity than for its power producing density compared to the NiMH (nickel metal hydride) battery.

---

[1] From David Ragone, who introduced the plot in 1962. http://en.wikipedia.org/wiki/Ragone_chart

[2] All logarithms are base 10 in this chapter.

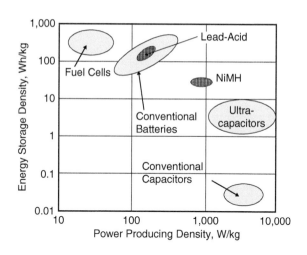

FIGURE 18.6

Ragone plot showing characteristics of some batteries and other electrochemical devices.

Modified from a graphic of Maxwell Technologies: http://www.maxwell.com

18.6 **Electrochemical series**

To introduce other batteries, we bring in one other useful concept arising from what we have already seen. The *electrochemical series* is defined by the relationship among half-cell reactions (such as developed for the lead-acid battery at each electrode) and put in order of their potentials measured against a standard cell called the *hydrogen electrochemical half-cell.*

Simple cells operate on principles that use this electrochemical series.

Example 18.3

In a Daniell cell, the electrolytes are $ZnSO_4(aq)$ with a Zn anode in its half-cell and $CuSO_4(aq)$ with a copper cathode in its half-cell. The two electrolytes are separated by an electroconducting porous and inert "salt bridge" that completes the path for the current flow while allowing the separation of ions from each electrolyte. Write down the reactions in each electrolyte and determine the voltage produced by the Daniell cell.

Need: Explanation of half-cell reactions in Daniell cell and the voltage produced.

Know: The electrochemical series table shown in Table 18.2.

How: Draw the cell and use the electrochemical series, knowing that the Zn half-cell is the anode.

Table 18.2 The electrochemical series.

Half-cell chemistry	Potential in volts
$Li^+ + e^- \leftrightarrow Li(s)$	−3.05 V
$Na^+ + e^- \leftrightarrow Na(s)$	−2.71 V
$Mg^{++} + 2e^- \leftrightarrow Mg(s)$	−2.37 V
$Zn^{++} + 2e^- \leftrightarrow Zn(s)$	−0.76 V
$Fe^{++} + 2e^- \leftrightarrow Fe(s)$	−0.44 V
$Ni^{++} + 2e^- \leftrightarrow Ni(s)$	−0.25 V
$2H^+ + 2e^- \leftrightarrow H_2(g)$	0.00 V (hydrogen half-cell is defined as zero)
$Cu^{++} + 2e^- \leftrightarrow Cu(s)$	0.34 V
$Cu^+ + e^- \leftrightarrow Cu(s)$	0.52 V
$Ag^+ + e^- \leftrightarrow Ag(s)$	0.80 V
$Pd^{++} + 2e^- \leftrightarrow Pd(s)$	0.95 V

(aq) means ions in aqueous solution, (s) means a solid, and (g) means a gas.

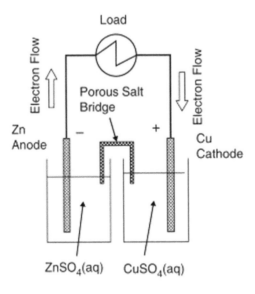

Solve: In the anodic electrolyte, Zn^{++} and $SO_4^=$ must be in aqueous solution in balance with each other:

$$ZnSO_4(aq) \leftrightarrow Zn^{++} + SO_4^=$$

But the anode is also dissolving and thus yields some locally extra Zn^{++} ions according to $Zn(s) \leftrightarrow Zn^{++} + 2e^-$ ($V = +0.76$ V).

Note that this is the reverse reaction of that in Table 18.2, so this half-cell reaction produces a positive voltage. We have also produced some additional Zn^{++} ions at a rate of one ion for every two electrons that leave the half cell. A corresponding number of Zn^{++} ions must also move across the salt bridge and be "neutralized" by a corresponding number of $SO_4^=$ ions.

The cathodic electrolyte must be electrically neutral, containing equal numbers of Cu^{++} and $SO_4^=$ ions. The cathodic reaction must be as written just as in Table 18.2: $Cu^{++} + 2e^- \leftrightarrow Cu(s)$ $(V = +0.34\ V)$.

Notice we have removed copper ions from solution; therefore, there must be a corresponding reduction in $SO_4^=$ ions in the electrolyte. They must move into the salt bridge to exactly counteract the Zn^{++} ions from the anodic side. Therefore, the cell potential is equal to $(+0.76\ V) + (+0.34\ V) = \mathbf{+1.10\ V}$.

Something else is going on that you might have spotted. In the anode, we dissolved one unit of solid zinc, $Zn(s)$, that became Zn^{++} ions; and in the cathode, we precipitated the corresponding number of ions of Cu^{++} ions as solid copper, $Cu(s)$. This is the type of reaction used in the electroplating industry. In that industry, solutions of metal ions are plated out to produce gold-, silver-, chromium-, copper-, and nickel-plated objects. In addition, corrosion engineers are interested in the opposite of plating—the dissolution of bulk metals. For example, the rusting of steel occurs in part because of the half-cell reactions $Fe(s) \leftrightarrow Fe^{2+} + 2e^-$ and $Fe(s) \leftrightarrow Fe^{3+} + 3e^-$.

Example 18.4

An electroplater wants to coat a 10.0 cm by 10.0 cm copper plate with 12.5 micrometers of silver $Ag(s)$, whose density is 10,500 kg/m^3. How many electrons must pass in the external circuit? How many coulombs are passed? If the plating takes 1200. s, what is the electrical current in amperes in the external circuit?

> **Need**: Number of electrons and the current flow to deposit 12.5 micrometers of silver onto 100. cm^2 $(100 \times 10^{-4}\ m^2)$ from a solution containing Ag^+ ions.
>
> **Know**: Atomic mass of Ag is 108 kg/kmol. Its density is 10,500 kg/m^3. Avogadro's number (N_{Av}) is 6.02×10^{23} atoms/mol or 6.02×10^{26} atoms/kmol. Current is the rate of flow of electrons, so
>
> $$1.00\ A = 1.00\ C/s \text{ and one electron carries } 1.60 \times 10^{-19}\ C.$$
>
> **How**: Sketch a credible electrochemical circuit using an electrolyte of silver nitrate (which is soluble in water producing equal numbers of Ag^+ and NO_3^- ions). Use a silver anode and the copper plate as a cathode. Finally, connect a battery with the polarity chosen as shown in the following figure.

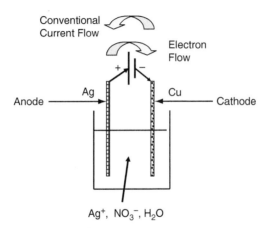

Solve: The reaction at the anode is $Ag(s) \rightarrow Ag^+ + e^-$, and the reaction at the cathode is $Ag^+ + e^- \rightarrow Ag(s)$; hence, one atom of silver dissolves at the anode and one atom of silver is deposited at the cathode. For each atom of silver dissolving at the anode and depositing at the cathode, one electron must circulate in the external circuit.

First, we need to know what the mass of silver is in the coating. We assume the deposited silver atoms fully pack with no internal pores. Then the mass of deposited silver is

Mass = Volume of the plating × Density of the plating

= Thickness of the plating × Area of the plating × Density of the plating

$= \left(12.5 \times 10^{-6} \text{ m} \right) \times \left(1.00 \times 10^4 \text{m}^2 \right) \times \left(10,500 \text{ kg/m}^3 \right) = 1.31 \times 10^{-3} \text{kg}$

Next, convert to kmol:

$$\text{kmol} = \frac{1.31 \times 10^{-3}}{108} [\text{kg}][\text{kmol} / \text{kg}] = 1.22 \times 10^{-5} \text{kmol}$$

Now, convert into the number of silver atoms using Avogadro's number:

$$\text{Number of atoms of Ag(s) deposited} = 1.22 \times 10^{-5} \times 6.02 \times 10^{26} \text{ [kg][atoms/kg]}$$
$$= 7.32 \times 10^{21}$$

This is equal to the number of electrons that flowed in the external circuit. Hence, we have used the services of **7.32×10^{21} electrons**.

We know that an ampere is the charge flowing/time; the electrical charge in this case is $7.32 \times 10^{21} \times 1.60 \times 10^{-19}$ [e⁻][coulombs/e⁻] $= 1.17 \times 10^3$ coulombs. This charge flows for 1200. s and hence the **current I** $= 1.17 \times 10^3/1200$. [coulombs]/[s] $= $ **0.975 A**.

18.7 Advanced batteries

Many batteries, such as the common "alkaline" battery, use a solid electrolyte rather than a liquid, but they operate on similar principles. Are there higher performance batteries available?

A widely available battery today is the NiMH battery. They are used both to propel hybrid cars and to start their auxiliary engines. Smaller versions of them are also used for long-lasting electronic devices.

In the anode, the overall principle reaction that occurs is

$$Ni(OH)_2 + OH^- \leftrightarrow NiO(OH) + H_2O + e^- \tag{18.5}$$

and the corresponding overall cathodic reaction is

$$M + H_2O + e^- = MH + OH^- \tag{18.6}$$

The basic advantage of this kind of battery is that it has both relatively high energy storage density and power producing density so, when packaged as a common D-size flashlight battery, it can supply more than 10 A for 1 h (which is more than 50 kJ). The Ragone plot, Fig. 18.6, shows its power performance is considerably better than a lead-acid battery. No wonder these batteries in various sizes are preferred for rechargeable use in higher-end electronic devices. For automotive hybrid use, similar sizes to D cells are packaged in a large array, thus simultaneously providing sufficient voltage and current capacity. But, the NiMH is being overtaken by the lithium-ion battery, whose overall chemistry is a shift of a Li ion, Li^+ from a compound of lithium, cobalt, and oxygen to lithium carbide. Li-ion batteries have "high" voltages (approaching 4 V compared to most batteries of 1.5−2 V) and still have a high energy storage density.

Whether NiMH or Li-ion (or the more common NiCd, pronounced "nyecad"), these batteries are two or three times better than lead-acid batteries in capacity and therefore will continue to find use in high-grade electronics and hybrid cars. None is likely to completely replace the gasoline engine, because the mass and volumetric energy densities are simply too low.

18.8 Fuel cells

If batteries with stationary solid or liquid electrolytes fail to completely replace gasoline engines, then why not use a constant flow of fresh replacement electrochemical fuel and continuously purge the used material? This is what a fuel cell does. One important class of fuel cells was invented after fundamental research by two GE scientists, Thomas Grubb and Leonard Niedrach, in the 1950s. The NASA moon program in the 1970s used these fuel cells. They almost led to disaster during the Apollo 13 moon shot, when internal poorly insulated electric wires shorted and sparked, resulting in a hydrogen/oxygen explosion.

The hydrogen fuel cell is the most promoted of all fuel cells for future transportation use, since it converts H_2 to H_2O *and nothing else* and thus would be a welcome relief to the air quality of urban areas currently relying on gasoline vehicles.

A schematic of the central feature of a fuel cell is shown in Fig. 18.7. The reactants, hydrogen and oxygen (actually air), flow to the anode and cathode, respectively, and hot water or steam, the sole product of reaction, is removed.

The key step of Grubb and Niedrach was to introduce the proton exchange membrane (PEM) based on a fluorinated polymer membrane. Niedrach showed how to impregnate the membrane with platinum, which is a catalyst for the oxidation of hydrogen by air at ambient temperatures. Hydrogen diffuses into the membrane,

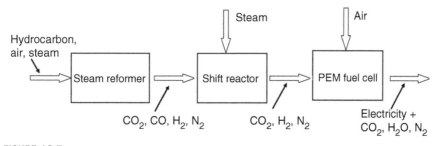

FIGURE 18.7

Principle of a proton exchange membrane.

where it is ionized. The resulting electrons flow from the anode, as shown in Eq. (18.7):

$$2H_2 \rightarrow 4H \rightarrow 4H^+ + 4e^- \tag{18.7}$$

The oxygen gas counter diffuses in the PEM and reacts to form water while absorbing electrons:

$$4H^+ + 4e^- + O_2 \rightarrow 2H^+ + 2OH^- \rightarrow 2H_2O \tag{18.8}$$

The actual construction of a fuel cell is quite complex (as schematically shown in Fig. 18.7), but this is not its principal challenge. The biggest problem for a hydrogen fuel cell is hydrogen.[3]

18.8.1 Fuel cells using novel fuels

While lots of electrochemical schemes can function as a fuel cell, one shows promise, because it uses a fuel that can be stored as a liquid—methanol, CH_3OH. The anode feed for such a cell is a mixture of methanol and steam onto a PEM membrane. With the appropriate catalyst on the PEM, the methanol is broken down, producing electrons and protons (H^+) as shown in Eq. (18.9):

$$CH_3OH + H_2O \rightarrow CO_2 + 6H^+ + 6e^- \tag{18.9}$$

The electrons are removed via the anode and the protons transport across the PEM as in a hydrogen cell. CO_2 is vented, a serious disadvantage for large-scale applications. At the cathode, the reaction is

$$^3/_2O_2 + 6H^+ + 6e^- \rightarrow 3H_2O \tag{18.10}$$

Overall, the reaction is

$$CH_3OH + ^3/_2O_2 \rightarrow CO_2 + 2H_2O \tag{18.11}$$

This cell operates at reasonable temperatures from 60 to 120°C. Its biggest disadvantage is that it produces CO_2. This is offset by the fact that an all-electric drive is roughly twice the efficiency of a gasoline powered car (thus about doubling its mpg). As a further use of these fuel cells, Fig. 18.8 shows a prototype methanol fuel cell designed to boost the recharging interval of laptop computers to 10 or more hours.

If we plot the electrical characteristics of fuel cells on a Ragone plot, they lie toward the upper left-hand corner, because their energy storage density is high (perhaps, 500 Wh/kg) while their power producing density is low (perhaps, 20–40 W/kg).

---

[3] It's a very flammable and explosive gas.

FIGURE 18.8

Fuel cells for laptops.

http://science.nasa.gov/headlines/y2003/images/fuelcell/notebook_med.jpg

Example 18.5

A PEM cell is fed with 100. mL/min of $H_2(g)$ and its stoichiometric equivalent of air. What is the cell voltage at zero current given the half cathodic cell produces 1.23 V, and what is the maximum possible electrical current produced by the cell?

Need: Hydrogen cell voltage and current capability.

Know: Anode half cell: $2H_2 \rightarrow 4H \rightarrow 4H^+ + 4e^-$ for which $V = 0.00$ V (see Table 18.2) and the cathode half-cell: $4H^+ + 4e^- + O_2 = 2H_2O$, for which $V = +1.23$ V. Avogadro's number is $N_{Av} = 6.02 \times 10^{26}$ molecules/kmol and the charge of an electron $= 1.60 \times 10^{-19}$ C.

How: Find the voltage from the half-cell potentials and current from the assumed 100% conversion of the energy in 100. standard mL/min of $H_2(g)$. Convert standard mL to moles using $pV = nR_uT$, where R_u is the **universal** gas constant, 8.31×10^3 J/kmol·K.

Solve: The **voltage** for the cell is found by algebraically adding the half-cell potentials $= 0.00 + 1.23$ V $= \mathbf{1.23}$ **V**.

So, **100.** mL/min of hydrogen requires **50.0** mL/min of O_2 for the reaction $H_2 + \frac{1}{2}O_2 = H_2O$. But the **amount of air** ($O_2 + 3.76$ N_2) needed is 4.76×50.0 mL (see Eq. 8.6 for the combustion equation) $= \mathbf{238}$ **mL/min.**

The net reaction uses 100. mL/min of H_2. The number of kmol/min of H_2 consumed is $\dot{n} = p\dot{V}/R_uT = (1.00 \times 10^5) \times (100.) (10^{-6})/(8.31 \times 10^3 \times 273)$ [N/m^2][mL/min] [m^3/ml]/[J/kmol·K][K] $= 4.41 \times 10^{-6}$ kmol/min.[4]

This contains $4.41 \times 10^{-6} \times 6.02 \times 10^{26} = 2.65 \times 10^{21}$ molecules/min of $H_2(g)$ or 4.42×10^{19} molecules/s flowing in. The maximum current is

$2 \times 4.42 \times 10^{19} \times 1.60 \times 10^{-19}$ [e^-/molecule] [molecule/s][C/e^-] $= 14.1$ A.

Note the factor of 2, since two electrons are needed for each molecule of H_2.

---

[4] The over-dot in $\dot{n}$ is used to denote rate, dn/dt.

If methanol is an "interesting" fuel for a fuel cell, can we use a hydrocarbon such as gasoline directly in a fuel cell? In due course, a fuel cell that could use gasoline directly could be advantageous, since the present infrastructure for gasoline could be used. An all-electric output gasoline fuel cell vehicle has the potential to double today's gasoline combustion engine mileage. A major scientific and engineering obstacle is to develop a low-temperature catalyst that readily breaks the sturdy chemical barriers between the C–C atoms that form most of the compounds in gasoline.

Using natural gas in a fuel cell is a complicated high pressure/high temperature affair as shown by a state-of-the-art reformer that produces 250 kW (335 HP) or enough for a large truck (Fig. 18.9). The final module is large compared to a diesel fuel engine and fuel tank for the same purpose.

FIGURE 18.9

A 250 kW fuel cell with a built-in natural gas reformer (Ballard power systems).

18.9 Ultracapacitors

The common **electrical capacitor** uses physical principles (instead of chemical ones) to store electrical energy in the form of electric charges. It works by charging a **dielectric** (which is essentially a nonconductor of electricity, such as a polymer or glass) by applying a voltage across it by sandwiching the dielectric material between two metal foils or plates. If the dielectric material has the appropriate structure, electrons can be drawn out of their normal orbits to induce net charges at the interface between the metal and the dielectric material. Common electrical capacitors have an energy storage density of less than 0.1 Wh/kg, which is too small for virtually all battery storage operations but is often useful for timing applications in electric circuits.

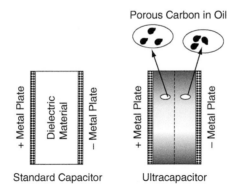

FIGURE 18.10

Comparison of conventional capacitor and ultracapacitor.

Ultracapacitors (sometimes called *supercapacitor*s) are a cross between a common electrical capacitor and a battery. Like common capacitors, they store electrical charges on surfaces between charged electrical conductors. Unlike regular capacitors, the charges accumulate in porous carbon immersed in oil; see Fig. 18.10.

A detailed description of ultracapacitors reveals that their charges are in double layers at interfaces, and so they are sometimes called *electric double-layer capacitors*. In ultracapacitors, porous carbon particles are the electrical storage medium, because they can have huge internal surfaces of $100-300 \ m^2/g$. If the carbon particles were nonporous, the carbon's external surfaces would total only a few cm^2/g.

Ultracapacitors are constructed with a separator and two compartments so + and − charges can be separated and collected on their external plates. There is a factor of about 10,000 in total charge if the charges can fill the internal pores of the carbon instead of just its external surface areas. This huge multiplier, compared to a standard capacitor, means the ultracapacitor has a correspondingly large energy storage capacity. Ultracapacitors have energy storage densities of $1-10 \ Wh/kg$, but their capability to deliver this energy as power can be as high as several thousand W/kg. Unlike batteries, the charge−discharge cycle has no *net* chemical changes and thus ultracapacitors should be capable of thousands of charge/discharge cycles.

The Ragone chart, Fig. 18.6, shows the areas of applicability of electrochemical batteries, fuel cells, and ultracapacitors. Ultracapacitors can deliver short pulses of high power at reasonable energy storage densities. The region of applicability for ultracapacitors is unique; the ability to deliver short bursts of energy for high-power applications coupled with their energy storage capability makes them suitable for small power demands, such as personal electronics.

Summary

Electrochemical systems are a part of most alternate energy schemes to store electrical energy for **load leveling** purposes. Batteries work because of the electrical nature of matter. Batteries all have **anodes** (as a source of electrons) and **cathodes** (a sink for them). The lead-acid battery is familiar because of its automotive uses but is limited by weight and its energy storage capacity.

Batteries are made up of two **half-cells**, the polarity of which is a result of the position of the half-cells in an electrochemical series. Such half-cells are also important in electroplating and in metal corrosion.

Modern battery systems, for example, NiCd, NiMH, and Li-ion, have higher energy storage density than traditional batteries but are marginal for a purely electric vehicle. They find application in electronic equipment that needs longer intervals between recharging. For longer-term or continuous use, a fuel cell is required. Practical fuel cells include the hydrogen **PEM** cell and, to a lesser degree, the PEM methanol fuel cell.

Both ultracapacitors and fuel cells fill power-energy niches not satisfied by standard battery systems. A **Ragone plot** offers a quick assessment of the region that a given electrochemical engineering technology best fills.

Exercises

Some useful unit conversions: $1.00\ kJ = 0.278\ Wh$; $1.00\ Wh = 3600\ kJ$, $1.00\ U.S.\ gallon = 3.79\ L$; 1.000 metric ton (tonne) $= 1000.\ Kg$; one electron charge $= 1.60 \times 10^{-19}$ coulombs; Avogadro's number $= 6.022 \times 10^{26}/kmol$; standard molar volume $= 22.4\ m^3/kmol$ or *equivalently* R_u is the universal (i.e., per kmol) gas constant, $8.31 \times 10^3\ J/K \cdot kmol$.

1. A vehicle has a 15. U.S. gallon gas tank and can be filled from empty in 60. s.
   a) What is the rate that power is transferred to the vehicle?
   b) If the vehicle is converted to an all-battery system, using a battery pack to replace the gas tank, what is the rate of power transferred to the batteries if it takes 4.0 h to charge?
   Assume the density of gasoline of $740\ kg/m^3$, its heat of combustion is $46{,}500\ kJ/kg$, and the battery's energy storage density is 525 kJ/L **(Ans. (a) 33. MW; (b) 2.1 kW)**

2. A windmill produces mechanical power according to this formula: $P = \frac{1}{2}\eta\rho V^3 A$, where η is its efficiency (assume $\eta = 60.\%$), $\rho =$ density of air ($1.00\ kg/m^3$), $V =$ wind speed in m/s (assume 5.0 m/s), and A is the cross-section of the mill that faces the wind (assume it is circular with a radius of 35.0 m).
   a) How many kW does the windmill produce?
   b) To load level, we have a battery storage device that can store 2.00 MWh. How long will we be charging it?
   c) If the energy storage density we can achieve is 125 Wh/kg, how big is this storage battery in tonnes?

3. We wish to store 2.00 MWh as an emergency power supply for a "big-box" store. If the gross energy storage density of the battery is 425 kJ/L, how big is the storage battery in m^3? If the density of the battery averages $2.5 \times 10^3\ kg/m^3$, is this reasonable compared to other batteries on the market? (Hint: See Fig. 18.6). **(Ans. a 2.60 meter cube, and "yes".)**

4. In a proposed hybrid car, the battery is "reinforced" by an ultracapacitor with the following characteristics: 5.0 Wh/kg and 5.0 kW/kg. If the car wants to draw on its ultracapacitor for 10. kW for 10.0 s, is it power or energy limited?

5. Will a Ragone plot confirm the conclusions of Exercise 18.4? Describe how by referencing Fig. 18.6. (Hint: How long does it take to discharge the ultracapacitor?)

6. Draw lines of constant discharge time on the Ragone plot. Interpret these lines. (Hint: Energy/power = time. On a log-log plot, since time is represented only to the first power, its slope is 1, which is represented as one ordinate decade to one abscissa decade.)

The figure that follows applies to **Exercises 7, 8, and 9**.

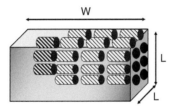

7. Design a battery pack composed of D cells (6.35 cm long × 3.18 cm diameter, each weighing 0.100 kg). The final package (see figure) must supply 42.0 V (a coming standard for all cars) at 30.0 A for 2.00 h. Each cell produces 3.0 V and has a *power producing density* of 125 W/kg. Estimate W and L for the package (ignore its energy storage density for now).

8. Design a battery pack composed of D cells (6.35 cm long × 3.18 cm diameter, each weighing 0.100 kg). The final package (see figure) must supply 42.0 V (a coming standard for all cars) at 30.0 A for 2.00 h. Each cell produces 3.0 V and has an *energy storage density* of 125 Wh/kg. Estimate W and L for the package (ignore its power producing density for now).

9. Design a battery pack composed of D cells (6.35 cm long × 3.18 cm diameter, each weighing 0.100 kg). The final package (see figure) must supply 42.0 V (a coming standard for all cars) at 30.0 A for 2.00 h. Each cell produces 3.0 V and has a *power producing density* of 125 W/kg **and** an *energy storage density* of 125 Wh/kg. Estimate W and L for the package.

10. Compare a battery made from the aqueous half cells: $Mg^{++} + 2e^- \leftrightarrow Mg(s)$ and $Cu^{++} + 2e^- \leftrightarrow Cu(s)$ with one made from $Fe^{++} + 2e^- \leftrightarrow Fe(s)$ and $Ag^+ + e^- \leftrightarrow Ag(s)$. What are the voltages and can you speculate on the relative weights of the batteries if the densities of $Mg(s)$, $Cu(s)$, $Fe(s)$, and $Ag(s)$ are 1,740, 7,190, 7,780, and 10,500 kg/m^3, respectively? (Hint: A significant part of the weight of an electrochemical cell is the weight of its electrodes.)

11. Consider a graph of the half-cell potentials in Table 18.2 versus the density of the electrode metal in the half-cell. Explain whether or not it is possible to design a cell that is both light and high voltage (say, more than 3 V).

12. An industrial fuel cell has to supply a continuous 20. MW to a perfectly efficient inverter at 123 V. Assuming it uses a hydrogen gas/air PEM system, what are the needed flows of H_2 and air to the cells in standard m^3/s? See the simplified schematic that follows. Assume each fuel cell produces 1.23 V.

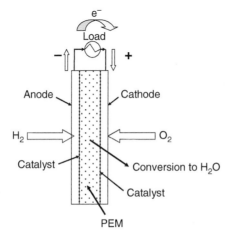

13. In the fuel cell configuration in Exercise 12, the actual voltage delivered by the cell varies with load; assume the voltage measured at full load is 0.75 V, so that the cell electrolytic efficiency is $0.75/1.23 = 61.\%$. How many cells do we need in series and what is the fate of this inefficiency?

14. If you run a fuel cell in reverse as an electrolysis cell, H_2O is split into its elementary components, $H_2(g)$ and $\frac{1}{2} O_2(g)$. This way we can make hydrogen and have it stored in case of the loss of outside electric power. The stored hydrogen can then be fed to our electrolysis unit acting as a fuel cell. If

   a) it takes 1.23 V to electrolyze water, what is the overall efficiency, defined as (Theoretical Power Required in MW)/(Actual Power Required in MW), of this scheme?

   b) the electrolysis requires 1.50 V, what then is its overall efficiency?

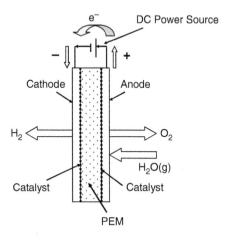

15. Solid oxide fuel cells use a unique electrolyte—a hard ceramic, based on a solid state solution of zirconium oxide "stabilized" by the element yttrium—which conducts oxygen ions when hot enough (about 1000°C). The anode is a Ni/ceramic material and the cathode is an exotic material, such as lanthanum strontium manganite. It works similarly to a PEM fuel cell. Air is applied to the cathode and hydrogen to the anode. It is about 60% efficient, and if it has an application, it will be to sophisticated central stationary power plants. Write down the anode and cathodic reactions; what is the theoretical voltage of such a cell?

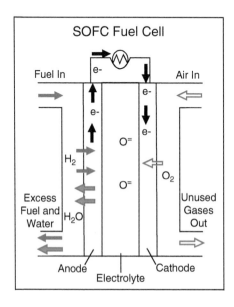

16. A methanol fuel cell is used as a battery for a laptop whose average electrical demand is 20. W at 10.0 V. Each cell produces 0.50 V. How much methanol in ml is needed each day? Methanol density is 792 kg/m^3.
17. A steam reformer is fed 2.00 L/s of methane, 0.952 L/s of air, and 1.00 L/s of steam. If it reaches its stoichiometric maximum extent of reaction, what is the composition of its product gases? Assume all quantities given in standard L/s (1.00 bar, 0°C basis). **(Ans. CO = 0.600 L/s, CO$_2$ = 0.400 L/s, H$_2$ = 5.00 L/s, and N$_2$ = 0.752 L/s)**
18. The same reformer as in the previous problem has a problem: the 3.00 L/s of H$_2$ is contaminated with 0.600 L/min of CO. Assume you can remove this contaminant down to 10.0 ppm of CO. What is the final composition of its product gases?
19. For the reformer in the two previous questions, what is its maximum electrical power production?

Comment: The next two problems are loosely based on experiences of Thomas Edison in developing a new battery, initially invented for use in automobiles and submarines. For details of the actual case, see Byron Vanderbilt's book, *Thomas Edison, Chemist.*

20. You are an engineer on a team developing a radically new battery invented by your immediate supervisor, a successful and famous inventor. Preliminary tests suggest that the new battery, if used in automobiles, will triple the range over the best existing battery. Detailed tests have not been made as to the safety or lifetime of the battery. But, the inventor urges immediate introduction to the market, using his personal reputation as a guarantee to customers of the value of the product. Only by testing it in actual use, the inventor argues, can any remaining "bugs" be identified and removed. You are concerned that customers will regard the inventor's confident statements as an implied warranty. You urge delay until safety and life tests can be done. The inventor, your immediate supervisor, rejects this approach. What do you do? (Use the Engineering Ethics Matrix.)
21. You are the CEO of a company that has developed a new fuel cell invented by one of your company's most creative engineers. After thoroughly testing the system for safety and reliability under a wide range of ordinary conditions, you arrange a sale to a customer operating under unusually stressful conditions (undersea exploration to unprecedented depths and pressures) not covered by your tests. As part of the contract, the customer takes full responsibility for any difficulties that arise under these conditions. In use, the fuel cell fails miserably in the application, due to an unanticipated pressure effect. The inventor, embarrassed by the failure, urges that you fully reimburse the customer for the sale price, keep the failure secret, and allow him to correct the problem at company expense. He argues that you are ethically obligated to do this to uphold the company's good name. What is the ethical way to respond to the inventor's request? (Use the Engineering Ethics Matrix.)

Environmental engineering

Source: *Zdenek Sasek/Shutterstock.com*

19.1 Introduction

Is the air safe to breathe? Is the water safe to drink? Are the fish we catch and the crops we grow safe to eat? Is our home and office safe from adverse chemical effects? Environmental engineering encompasses all of these concerns in its role of protecting people from harmful environmental effects.

Environmental engineering is the application of engineering principles to improve and maintain the environment to (a) protect human health, (b) protect nature's ecosystems, and (c) improve our environmentally related quality of life. It was originally a specialized field within civil engineering (called sanitary engineering) until the mid-1960s when the more accurate name *environmental engineering* was adopted. Today environmental engineering involves

- control of air, water, and noise pollution
- treatment and distribution of drinking water
- collection, treatment, and disposal of wastewater

Exploring Engineering. https://doi.org/10.1016/B978-0-443-13541-5.00009-X

- municipal solid-waste management and hazardous-waste management
- cleanup of hazardous-waste sites, and
- the preparation of environmental assessments, audits, and impact studies.

In this chapter, we provide material on environmental modeling tools, air and water quality and control, and solid waste management systems.

19.2 What do environmental engineers do?

Environmental engineers develop processes and equipment for the control of pollution of all kinds, the supply of clean water, and the disposal of waste. They protect public health by preventing disease transmission and preserve the quality of the environment by preventing the contamination and degradation of air, water, and land resources. They study the effect of technological advances on the environment by addressing issues such as acid rain, global warming, ozone depletion, air and water pollution and pollution from automobiles and industrial sources.

Environmental engineers typically do the following:

- prepare, review, and update environmental investigation reports
- design projects such as municipal and industrial wastewater treatment facilities
- research the environmental impact of proposed construction projects
- provide technical support for environmental remediation projects and for legal actions
- perform air and water quality-control checks, and
- inspect industrial and municipal facilities for environmental regulation compliance.

One of the most important responsibilities of environmental engineering is to prevent the release of harmful chemical and biological contaminants (also referred to as pollutants) into the air, water, and soil. This requires extensive knowledge of the chemistry and biology of potential contaminants as well as the processes that lead to their release. With this knowledge new processes can be designed or existing processes can be modified, to reduce or eliminate the release of pollutants.

19.3 How do we measure pollution?

Air, water, and soil pollution occurs when contaminants are introduced that have adverse effects on humans or the environment. Pollutants can originate from human activity or be naturally derived. They can take many forms such as chemicals, organic waste, pathogenic bacteria and virus, and even noise. The amount of a pollutant is defined by its **concentration** in mass per unit volume of the mixture in which it resides. For example, for a mixture with two pollutants A and B suspended in fluid C (either air or water), the concentration of pollutant A in mass per unit volume of the mixture is defined by

$$C_A = \frac{m_A}{V_{mixture}} = \frac{m_A}{V_A + V_B + V_c} \tag{19.1}$$

in which m_A is the mass of pollutant A and $V_A + V_B + V_C$ is the total volume of the mixture.

19.3.1 Pollutants in water

Concentrations of chemicals in water are typically measured in units of the mass of chemical (milligrams, mg or micrograms, μg) per volume of water (liter, L) or as parts per million (ppm) or parts per billion (ppb). Therefore:

For water, 1 ppm = 1 mg/L and 1 ppb = 1 μg/L.

One way to visualize one part per million (ppm) in water is to think of it as one cup of contaminate in a swimming pool, and one part per billion is about one drop of contaminate in a swimming pool.

Concentrations of chemicals in water may also be written as grams of contaminate per cubic meter of water, or as g/m^3. Since $1 \ m^3 = 10^3$ L, this is the same as grams per 1000 L or milligrams per liter (mg/L). Therefore:

For water, 1 g/m^3 = 1 mg/L = 1 ppm, and 1 mg/m^3 = 1 μg/L = 1 ppb.

19.3.2 Pollutants in soil

Concentrations of chemicals in soil are typically measured in units of the mass of chemical per unit mass of soil as mg/kg or μg/kg. Sometimes concentrations in soil are reported as parts per million (ppm) or parts per billion (ppb). Therefore:

For soil, 1 ppm = 1 mg/kg, and 1 ppb = 1 μg/kg.

19.3.3 Pollutants in air

Concentrations of chemical pollutants in air can either be reported as mass of the chemical per unit volume of air (mg/m^3), or as molecules of the chemical per million molecules in air (ppm). Since measurements in mg/m^3 depend on the volume of air, the atmospheric temperature and pressure affect the result. Consequently, chemical pollutants in air are typically reported at a pressure and temperature of 1 atm (101,325 N/m^2) and 25°C (298 K).[1]

---

[1] In chemistry, gas calculations are made at a Standard Temperature and Pressure (STP) of 0°C and 1 atm pressure (101,325 Nm^2), but air pollution calculations are normally made at 25°C and 1 atm pressure.

The molar volume (V_{molar}) is the volume of 1 mole of a gas at 1 atm pressure and 25°C. For an ideal gas, $pV = nRT$, so V_{molar} is

$$V_{molar} = V/n = RT/p = (8.3143 \text{ Nm/mol} \cdot \text{K}) \times (25 + 273 \text{ K}) / (101325 \text{ N/m}^2)$$

$$= 0.02445 \text{ m}^3/\text{mol} \times (10^3 \text{ L/m}^3) = 24.45 \text{ L/mol}$$

Therefore, at 25°C and atmospheric pressure, 24.45 L of carbon dioxide gas contains 1 mol (44 g) of CO_2 and 24.45 L of water vapor contains 1 mol (18 g) of H_2O. For this temperature and pressure, these two pollutant concentrations are related as follows:

$$\text{Concentration } (\text{mg/m}^3) = \text{Concentration (ppm)} \times \frac{M(\text{g/mol})}{24.45(\text{L/mol})} \qquad (19.2a)$$

and

$$\text{Concentration (ppm)} = \text{Concentration } (\text{mg/m}^3) \times \frac{24.45(\text{L/mol})}{M(\text{g/mol})} \qquad (19.2b)$$

where M is the molecular mass of the pollutant gas.

Example 19.1

The global average atmospheric carbon dioxide concentration in 2017 was 405 parts per million. What was its concentration in mg/m³?

Need: The concentration of CO_2 in mg/m³.

Know: The concentration on CO_2 is 405 ppm and the molecular mass of CO_2 is 44 g/mol.

How: Eq. (19.2a)

Solve: Using Eq. (19.2a), we have:

Concentration = $[405 \times 10^{-6}$ (i.e., ppm)]$(44 \text{ g/mol})(10^3 \text{ mg/g})/[(24.45 \text{ L/mol})$ $(10^{-3} \text{ m}^3/\text{L})] = 729 \text{ mg/m}^3$

19.4 The Mass Balance Equation

One of the most important tools available to an environmental engineer is the **Mass Balance Equation**. In its most general form, it accounts for accumulation of mass within defined space known as the control volume, as well as the creation or loss of a specific chemical mass that occurs because of a chemical reaction that takes place within the control volume.[2] This can be written as follows:

---

[2] Mass as a generic term cannot be created or destroyed (except in nuclear reactions), but the mass of specific chemical species can be created or destroyed in chemical reactions. For example, in the reaction $2H_2 + O_2 \rightarrow 2H_2O$, two molecules of hydrogen and one molecule of oxygen are destroyed while two molecules of water are created.

$$\left\{ \begin{array}{l} \text{The mass accumuled} \\ \text{inside the control} \\ \text{volume} \end{array} \right\} = \left\{ \begin{array}{l} \text{Total mass} \\ \text{entering the} \\ \text{control volume} \end{array} \right\} - \left\{ \begin{array}{l} \text{Total mass} \\ \text{leaving the} \\ \text{control volume} \end{array} \right\}$$

$$+ \left\{ \begin{array}{l} \text{The mass created or} \\ \text{destroyed inside the} \\ \text{control volume} \end{array} \right\}$$

which can be written mathematically as follows:

$$(m)_{CV} = (m_1 + m_2 + m_3 + ...)_{in} - (m_1 + m_2 + m_3 + ...)_{out} + G$$
$$= \sum (m)_{in} - \sum (m)_{out} + G \qquad (19.3a)$$

We can also write a **Mass Rate Balance Equation** as follows:

$$\left\{ \begin{array}{l} \text{The rate of mass} \\ \text{accumulation within} \\ \text{the control volume} \end{array} \right\} = \left\{ \begin{array}{l} \text{Total rate of mass} \\ \text{flowing into the} \\ \text{control volume} \end{array} \right\} - \left\{ \begin{array}{l} \text{Total rate of mass} \\ \text{flowing out of the} \\ \text{control volume} \end{array} \right\}$$

$$+ \left\{ \begin{array}{l} \text{The rate of mass created} \\ \text{or destroyed within the} \\ \text{control volume} \end{array} \right\}$$

This can be written mathematically as follows:

$$(\dot{m})_{CV} = (\dot{m}_1 + \dot{m}_2 + \dot{m}_3 + ...)_{in} - (\dot{m}_1 + \dot{m}_2 + \dot{m}_3 + ...)_{out} + \dot{G}$$
$$= \sum (\dot{m})_{in} - \sum (\dot{m})_{out} + \dot{G} \qquad (19.3b)$$

where $\dot{m} = dm/dt$ is the rate of change of mass, Σ is the summation symbol, and $\dot{G} = dG/dt$. Eqs. 19.3a and b can be applied to a gaseous, fluid, or solid mixture or to any one of its constituents.

Example 19.2

A Transfer Center receives 600. tons/day of solid waste (trash and recyclables) from a local community and 400. tons/day from another. If we assume 25% of this total is recyclables, and that at maximum capacity the local incinerator can burn 200. tons/day, how much solid waste should the Transfer Center plan on transporting to local landfills.

Need: Rate at which solid waste must be transported to local landfills.

Know: Mass flowrates into and out of the Transfer Center.

How: Define the system to be the Transfer Center, draw the control volume with inflows and outflows, and apply the Mass Rate Balance Eq. (19.3b).

Solve: First draw the control volume:

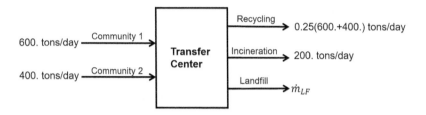

There are no chemical reactions here, so $\dot{G}_{waste} = 0$, and since there can be no accumulation of mass within the Transfer Center, Eq. (19.3b) reduces to the following:

$$(\dot{m})_{CV} = 0 = \sum (\dot{m})_{in} - \sum (\dot{m})_{out}$$

Substituting known values:

$$0 = (600. + 400.) - [0.250(600. + 400.) + 200. + \dot{m}_{LF}]$$

Solving for $\dot{m}_{LF}$ gives

$$\dot{m}_{LF} = 550. \text{ tons/day}$$

Often the goal of a treatment process is to reduce the amount of pollutant contained in a gaseous or fluid stream that has a **volumetric flowrate** Q, where $Q = dV/dt$ is the change in volume per unit time (e.g., m³/s). Here, it is convenient to measure the amount of pollutant in terms of its concentration in mass per unit volume (e.g., mg/m³). The relationship between the pollutant concentration C_p, the mass flowrate of the pollutant $\dot{m}_p$, and the volumetric flowrate Q_f of the transporting fluid (air or water) is given by

$$\dot{m}_p = C_p Q_f \tag{19.4}$$

which leads to the following alternative form of the Mass Rate Balance equation:

$$(\dot{m}_p)_{CV} = 0 = \sum (C_p Q_f)_{in} - \sum (C_p Q_f)_{out} + \dot{G}_p \tag{19.5}$$

If the volume of the pollutant is very small relative to the volume of water or air in which it is dissolved or suspended, then Eq. (19.5) can be written by assuming the concentration of water or air is $C = 1.0$ (or 100%) at all inlets and outlets. This provides another equation with which to solve for unknowns.

Example 19.3

A factory desires to release sulfur dioxide (SO_2) into the air through its smokestacks at the rate of 50.0 kg per day. The upwind concentration of sulfur dioxide in the air is 0.050 µg/m³ before reaching the smokestacks. For an average wind speed of 6.00 m/s, the estimated air flow rate into and out of the small town is 7.00×10^6 m³/s. Predict the average sulfur dioxide concentration in the air downwind from the smokestacks as it leaves the town.

    Need: The concentration of sulfur dioxide on the downwind side of the smokestacks.

    Know: The air flow rate, the concentration of sulfur dioxide upwind from the smokestacks and the rate at which sulfur dioxide emissions are released into the air.

How: Define the system to be the air mass above the town, draw the control volume with SO_2 inflows and outflows, and apply the Mass Rate Balance Eq. (19.5) to the SO_2.

Solve: First draw the control volume, showing the SO_2 inflows and outflows:

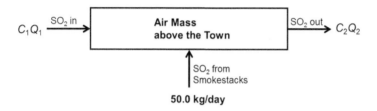

The smokestack SO_2 mass flow rate is

$$\left(\dot{m}_{SO_2}\right)_{\substack{\text{smoke} \\ \text{stack}}} = (50.0 \text{ kg/day}) \times (1000 \text{ g/kg}) \times \left(10^6 \text{ μg/g}\right) \times (1 \text{ day}/24 \text{ h})$$
$$\times (1 \text{ h}/3600 \text{ s}) = 5.79 \times 10^5 \text{ μg/s}$$

We assume there are no chemical reactions here, $\dot{G}_{SO_2} = 0$ and since there is no accumulation of SO_2 in the air mass above the town, then Eq. (19.5) simplifies to

$$\left(\dot{m}_{SO_2}\right)_{CV} = 0 = \sum \left(C_{SO_2} Q_{air}\right)_{in} - \sum \left(C_{SO_2} Q_{air}\right)_{out} = (C_1)_{SO_2}(Q_1)_{air} + \left(\dot{m}_{SO_2}\right)_{\substack{\text{smoke} \\ \text{stack}}}$$
$$- (C_2)_{SO_2}(Q_2)_{air}$$

Substituting values leads to

$$0 = \left[(0.05 \text{ μg/m}^3)(7.00 \times 10^6 \text{ m}^3/\text{s})\right] + 5.79 \times 10^5 \text{ μg/s} - (C_2)_{SO_2}$$
$$\times (7.00 \times 10^6 \text{ m}^3/\text{s})$$

Solving for the downwind SO_2 concentration gives

$$(C_2)_{SO_2} = 0.13 \text{ μg/m}^3$$

19.5 Air quality and control

We take air for granted, but we cannot live without it. The first air pollutants emitted by humans occurred when we learned to harness fire. Ancient Roman writers complained about the effects of wood smoke on health. In 535, Roman Emperor Justinian proclaimed the importance of clean air as a birthright. "By the law of nature, these things are common to mankind—the air, running water, the sea." Perhaps the first formal act of pollution regulation occurred around 1300 when King Edward II of England decreed that no coal could be burned in London while Parliament was in session.

19.5.1 Air quality

Air quality refers to the condition of the air within our surroundings. It should be clean, clear, and free from pollutants such as smoke, dust, smog, particulate matter, and gaseous impurities. Air quality is important for human health, vegetation, and crops, as well as aesthetic considerations (such as visibility) that affect appreciation of the natural beauty.

Air quality is determined by assessing a variety of pollution indicators. The **Air Quality Index** runs from 0 to 500 and indicates the amount of pollution in the air (see Table 19.1).

Table 19.1 Air quality index ratings.

Air quality index (AQI)	Air quality
0–50	Good
51–100	Moderate
101–150	Unhealthy for sensitive groups
151–300	Unhealthy
301–500	Dangerous

While air quality has improved significantly since the passage of the Clean Air Act in 1970, there are still many areas of the country where the public is exposed to unhealthy levels of air pollutants and sensitive ecosystems are damaged by air pollution. Poor air quality is responsible for an estimated 60,000 premature deaths in the United States each year. Costs from air pollution-related illness are estimated at $150 billion per year.

19.5.2 Air pollution control

Air pollution control involves methods to reduce or eliminate the emission of substances into the atmosphere that can harm the environment or human health. Air is polluted when it contains certain substances in concentrations high enough and for durations long enough to cause harm.

Most air contaminants originate from combustion processes. During the Middle Ages, the burning of coal for fuel caused significant air pollution problems in large European cities. Beginning in the 19th century, during the Industrial Revolution, the increasing use of fossil fuels intensified the severity and frequency of air pollution episodes. The production of automobiles and trucks had a tremendous impact on air quality problems in cities. It was not until the middle of the 20th century that meaningful and lasting attempts were made to regulate or limit emissions of air pollutants from stationary and mobile sources and to control air quality on both regional and local scales.

The pollutants known to contribute to urban smog and chronic public health problems include fine particulates, carbon monoxide, sulfur dioxide, nitrogen dioxide, ozone, and lead. Greenhouse gases include carbon dioxide, chlorofluorocarbons, methane, nitrous oxide, and ozone. In 2009, the U.S. Environmental Protection

Agency (EPA) ruled that greenhouse gases posed a threat to human health and could be subject to regulation as air pollutants.

The best way to protect air quality is to reduce the emission of pollutants by using cleaner fuels and more efficient combustion processes. Pollutants not eliminated in this way must be collected or trapped by appropriate air-cleaning devices before they can escape into the atmosphere. Airborne particles can be removed from a polluted airstream by a variety of physical processes. Table 19.2 describes the common types of equipment for removing fine particulates. Once collected, the particulates adhere to each other and can readily be removed from the equipment and disposed of.

There are also chemical methods for removing particles from the air. **Absorption** involves the transfer of a gaseous pollutant from the air into a contacting liquid that is either a solvent for the pollutant or can capture it by means of a chemical reaction. **Adsorption** is a surface phenomenon. The gas molecules are attracted to and held on

Table 19.2 Air pollution control devices.

Device	Description	Diagram
Cyclones	A cyclone removes particulates by causing an airstream to flow in a spiral path inside a cylindrical chamber. Large particulates move outward and slide down into a container at the bottom. Cyclones can achieve efficiencies of 90% for particles larger than about 20 μm in diameter.	
Scrubbers	Wet scrubbers wash particulates out of the airstream as they collide with tiny droplets in the spray. They can remove 90% of particulates larger than about 8 μm. Dry scrubbers spray dry chemicals into the air that cause the pollutant to chemically react and turn into a different substance that then falls out of the gas stream.	
Electrostatic precipitators	In an electrostatic precipitator, particles in the airstream are given an electric charge and are then attracted to and trapped on oppositely charged collection electrodes. An electrostatic precipitator can remove particulates as small as 1 μm with an efficiency exceeding 99%.	

Continued

Table 19.2 Air pollution control devices.—*cont'd*

Device	Description	Diagram
Baghouse filters	A baghouse contains many long narrow fabric bags suspended upside down in a large enclosure. Air is blown upward through the bottom of the enclosure and particulates are trapped inside the bags as air passes through them. A baghouse can remove very nearly 100% of particles as small as 1 μm.	

the surface of a solid. Activated carbon (heated charcoal) is one of the most common adsorbent materials and can remove gas with an efficiency exceeding 95%. **Combustion** can be used to convert gaseous hydrocarbon pollutants to carbon dioxide and water. Gaseous organic pollutants can be almost completely oxidized, with incineration efficiency approaching 100%.

Example 19.4

A clean baghouse filter has a mass of 90.0 g. After 24 h of operation, the filter plus the dust it contained has a mass of 90.1 g. The air flow rate through the filter was 2.83 m^3/min. What is the dust concentration in the air in μg of dust per m^3 of air?

Need: dust concentration in the air in μg of dust per m^3 of air.

Know: Air flow rate, mass of dust, and collection time interval.

How-Solve: The mass of the dust collected, $m_{dust} = (90.1 - 90.0 \text{ g}) \times (10^6 \text{ μg/g}) = 1.00 \times 10^5$ μg.

The total volume of air that passed through the filter is

$$V_{air} = (2.83 \text{ m}^3/\text{min}) \times (60 \text{ min/h}) \times (24 \text{ h}) = 408 \text{ m}^3$$

Then the dust concentration in the air is

$$m_{dust}/V_{air} = (1.00 \times 10^5 \text{ μg}) / 408 \text{ m}^3 = 245 \text{ μg of dust per m}^3 \text{ of air}$$

19.5.3 Global implications of air pollution[3]

The effects of pollutants in the air are not always about breathing in harmful chemicals. There can be other, sometimes unforeseen, consequences.

---

[3] Figures in this section from the Royal Society's online article: "The Basics of Climate Change". See also their article titled "Climate Change: a summary of the science", September 2010.

Most of the energy from the sun gets absorbed by the earth's surface, by clouds, by particles in the air, and gases in the atmosphere known as **greenhouse gases**. The amount of energy stored in the absorbing media determines its temperature. Thus, when the earth is in thermal equilibrium with the sun (i.e., incoming energy = outgoing energy), the average global temperature wants to remain constant. Changes to the amount of cloud cover, the amount of energy received from the sun, or the concentration of greenhouse gases in the atmosphere will cause the average global temperature to rise or fall until a new thermal equilibrium is attained. Such changes can be due to natural causes such as volcanic eruptions and small changes in the earth's orbit.

There is strong evidence indicating there has been a recent buildup of greenhouse gases in the atmosphere due to human activity, most notably carbon dioxide (CO_2). A plot of measured values of CO_2 concentration spanning the last 1000 years is shown in Fig. 19.1. Results prior to 1959 are based on sampling air trapped in Antarctic ice. These CO_2 levels are consistent with this new CO_2 being generated by human activity. For example, CO_2 produced after 1800 by the burning of coal, oil, and gas is enough to explain the sharp rise in CO_2 levels.

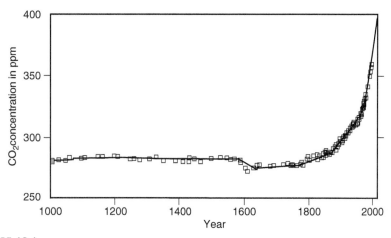

FIGURE 19.1

Variation in CO_2 concentrations in the atmosphere over the past 1000 years.

While greenhouse gases have been building up in the atmosphere, average global surface temperatures have also been increasing. Fig. 19.2 shows the average surface temperatures since 1880. As these results show, the average global surface temperature has increased by about 0.8°C since 1880 and has steadily increased over the last 3 decades to record levels. Predictions of future environmental consequences based on these continuing trends are worrisome. The 144 countries participating in the 2016 Paris Agreement announced that the world should limit the global increase in this century to 1.5°C (2.7°F).

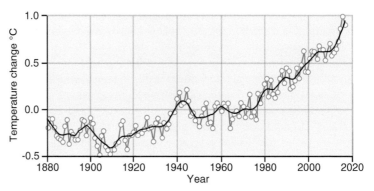

FIGURE 19.2

Average global surface temperature variation since 1880 relative to the average temperature from 1961 to 1990.

Do these small global temperature increases really matter? Yes, because this temperature rise is not spread evenly around the planet and it is undergoing one of the fastest changes in climate in the past 65 million years. The temperature increases in the Arctic where the climate is more sensitive to warming is already devastating glaciers. Even a small 2°C global temperature increase will produce changes in rainfall (more droughts and heat waves), snowfall, sea level, weather patterns (stronger hurricanes, more tornados, etc.), and other phenomena.

19.6 Water quality and treatment

The two primary goals of water treatment are (1) to produce safe, clear, good tasting, potable (i.e., drinkable) water and (2) to process wastewater so that it can be safely returned to the environment. In both cases, the volumes of water involved are huge and cost efficiency is paramount.

The complexity and details of the water treatment plant design will depend on the water quality at the source. For drinking water, the main sources are wells and surface water. Wells benefit from the natural filtering properties of soil but could still contain dissolved inorganics such as iron. Surface water (streams, lakes, reservoirs) comes in contact with human, animal, and plant life and thus should be expected to contain all sorts of organics, including microorganisms. Wastewaters from homes and industry pose an even tougher challenge as the starting quality is much worse and more variable.

19.6.1 Water quality

The level of impurities considered to be acceptable depends on the intended use of the water, with drinking water being at the top of the list in terms of tight limits on pollutant concentrations. Standards for treated wastewater that is passed back into the environment are different from those of drinking water and are regulated by the Clean Water Act in an effort to protect the environment, especially aquatic life. Anyone who has dealt with trying to stabilize the ecological environment of a small fish aquarium, without losing fish, understands the challenges this entails. These challenges can range from algae overgrowth and rising ammonia levels to lack of microorganisms (from overcleaning) for consuming organic waste and the adverse effects of chlorinated water.

Quantitative measures of water quality have been devised to track the state of the water as it passes through a water treatment plant. Some of the important ones are as follows:

- **Dissolved oxygen** (DO): needed to support aquatic life.
- **Biochemical oxygen demand**: an indirect measure of organic pollutants, determined by measuring the rate at which oxygen is depleted when microorganisms are introduced into a sample of the water.
- **Tests for pathogenic bacteria and virus**: *Escherichia coli*, hepatitis virus, and *Salmonella* which are harmful to humans.
- **Solids concentrations**: these solids could be suspended or dissolved, organic or inorganic, and can sometimes be identified by cloudiness or coloration of the water.
- **Nitrogen** and **phosphorus concentrations**: while undesirable in drinking water, they are valuable nutrients for aquatic life and crops.
- **Water hardness**: as measured by the concentration of calcium carbonate ($CaCO_3$). Higher concentrations cause soap to be less effective and produce mineral deposits in pipes.

Example 19.5

An industrial plant discharges water that contains trace amounts of lead into a nearby steam. Upstream from the plant, the concentration of the lead is 0.020 mg/L and the water flows at a rate of 10. m^3/s. Downstream from the plant, the concentration of lead must not exceed 0.050 mg/L. If the effluent from the plant is known to flow at a rate of 1.0 m^3/s, what is the maximum allowable concentration of lead in the plant's effluent?

Need: Concentration of lead in the water discharged to the stream.

Know: The upstream water flow rate and lead concentration and the downstream limit on lead concentration.

How: While this problem has elements in common with Example 19.3 (pollutant released into the environment), the flow rate of the effluent from the plant in this problem is large enough to significantly impact the steam's downstream flow rate. As a result, this problem has two unknowns, the concentration of lead in the water discharged from the plant and the stream's downstream water flowrate. This requires us to write two Mass Rate Balance Equations, one for the lead and another for the water.

Solve: Here we draw two control volumes, and we can apply the Mass Rate Balance Equation to each of them. The Mass Rate Balance Equation for water will be in terms of volumetric flowrates since the concentration of lead is very small so the concentration of water is effectively 1.0 (100%) at the inlet and the exit.

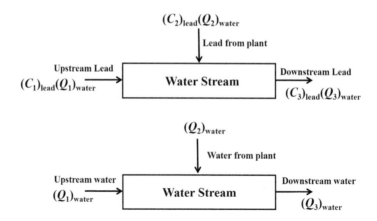

In the absence of chemical reactions ($\dot{G}_{\text{lead}} = \dot{G}_{\text{water}} = 0$) and no accumulation of mass within the control volume, the Mass Rate Balance Equation (19.5) for the lead and water reduces to

$$(\dot{m}_{\text{lead}})_{\text{CV}} = 0 = \sum (C_{\text{lead}} Q_{\text{water}})_{\text{in}} - \sum (C_{\text{lead}} Q_{\text{water}})_{\text{out}}$$

and

$$(\dot{m}_{\text{water}})_{\text{CV}} = 0 = \sum (C_{\text{water}} Q_{\text{water}})_{\text{in}} - \sum (C_{\text{water}} Q_{\text{water}})_{\text{out}}$$

Since the lead concentration is quite small, we can set $(C_1)_{\text{water}} = (C_2)_{\text{water}} = (C_3)_{\text{water}} = 1.0$, then the Mass Rate Balance Equation for water becomes

$$0 = (Q_1)_{\text{water}} + (Q_2)_{\text{water}} - (Q_3)_{\text{water}}$$
$$0 = 10. + 1.0 - (Q_3)_{\text{water}} \tag{19.6}$$

and then for lead[4]:

$$0 = (C_1)_{\text{lead}}(Q_1)_{\text{water}} + (C_2)_{\text{lead}}(Q_2)_{\text{water}} - (C_3)_{\text{lead}}(Q_3)_{\text{water}}$$
$$0 = (0.020\,\text{mg/L})(10.\,\text{m}^3/\text{s}) + (C_2)_{\text{lead}}(10\,\text{m}^3/\text{s}) - (0.050\,\text{mg/L})(Q_3)_{\text{water}} \tag{19.7}$$

We can now solve Eq. (19.6) for $(Q_3)_{\text{water}}$ and substitute the result in Eq. (19.7) to get $(C_2)_{\text{lead}}$ The results are given below.

The stream's downstream flow rate = $(Q_3)_{\text{water}} = 11.0\,\text{m}^3/\text{s}$.

Concentration of lead in the plant's effluent = $(C_2)_{\text{lead}} = 0.035\,\text{mg/L}$.

---

[4] Note that the volumetric flow rate of the pollutant (lead in this example) is always the same as the volumetric flow rate of the carrying fluid (water in this case).

19.6.2 Pretreatment of the water supply

Even backpackers hiking among seemingly pristine mountain streams are advised to purify water before drinking it. The available products for doing this offer some insight into the array of water purification tools available to environmental engineers. Methods available to hikers include boiling water for 1 min, chemically pretreating the water using iodine tablets or chlorine drops, water filters of various types, and UV light used in conjunction with a prefilter. All are effective at eliminating bacteria, viruses, protozoa, and solid particles to varying degrees, with boiling being the most effective. While boiling is the most effective, it is also costliest in terms of fuel, and therefore is not practical for large volumes of water.

The environmental engineer thus has a selection of strategies and processes from which to choose when designing a water treatment process. The exact size and sequence of these process units can be changed to meet the specific needs of the application. Some of the most common treatment process units, applicable to both drinking water and wastewater, are as follows:

- **Bar screen**—steel bars usually spaced one-inch (2.5 cm) apart to keep large debris from entering the inlet of the water treatment plant.
- **Comminutor**—motor-driven grinder that reduces the particle size of the grit and debris that makes it through the one-inch spaced bars of the bar screen.
- **Gravity settling tanks** (also referred to as clarifiers and sedimentation tanks)—gravity causes the higher density solid particles to settle at the bottom of the tank. The cross-section of a circular settling tank is shown in Fig. 19.3.

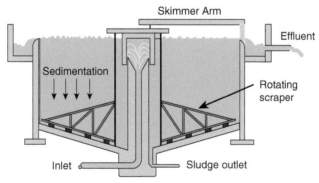

FIGURE 19.3

Circular settling tank design.

- **Aeration tanks/basins**—multiuse holding tank that usually involves adding DO.
- **Rapid sand filter**—composed of layers of sand and gravel, effluent from the settling tank seeps through the filter from above to remove solid particles from the water as shown in Fig. 19.4. The sand and gravel layers are sometimes seeded with carbon to remove dissolved organic material.
- **Suspended Growth Reactors**—Aerobic microorganisms, freely moving about in an aeration tank, are fed DO via rising air bubbles as they consume dissolved organic pollutants and metabolically transform them to CO_2, water, and other stable compounds.
- **Various chemical treatments**—including *softening* to lower the concentration of $CaCO_3$, *disinfection*, e.g., by chlorine injection to remove dangerous pathogens, and *absorption* which can remove trace amounts of organics.

FIGURE 19.4

Pressurized sand filter at the Niskayuna New York Treatment plant for extracting iron from drinking water. The illustration at left shows the gradation of grain size in the filter.

- **Oxidation ponds** (also called lagoons)—large shallow ponds on the output side of the water treatment plant. The pond is often shallow so that sunlight will induce the algae to produce oxygen, which in turn fuels the microorganisms as they consume the dissolved organics and metabolically transform them to CO_2 and water.

The water pretreatment plant shown in Fig. 19.5 was configured for treating surface water. In this design, disinfection and oxidation is followed by absorption, which removes dissolved organics using activated carbon particles. The addition of alum and polymer in the next stage encourages the solid particles to stick together, thus allowing them to overcome fluid forces and settle to the bottom of the

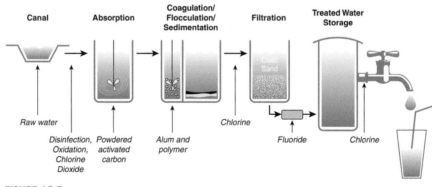

FIGURE 19.5

Water pretreatment plant design for producing drinking water from surface water.

sedimentation tank. The water then drops though a rapid sand filter to capture any remaining solid particles and organics. Before entering the filter, chlorine is added to prevent the return of any pathogens and in this design, fluoride, which can prevent tooth decay, is added upon exiting the filter.

19.6.3 Wastewater treatment

Treating wastewater that is clean enough to be returned to the environment is a complex process. For example, sludge, a thick smelly mixture of liquid and solid, must undergo its own treatment process before it is suitable for discharge to the environment. Microorganisms are used to eliminate some of the pollutants. The **aerobic** variety breathes in oxygen molecules while consuming dissolved inorganic pollutants and converting them to water and CO_2. In contrast, **anaerobic** microorganisms function without oxygen molecules, instead relying on their enzymes to break down organic matter into CO_2 and methane. They are used in **anaerobic digesters** to treat waste sludge. The methane generated from this process can be used as fuel to generate electricity.

In the case of the wastewater treatment plant of Fig. 19.6, the starting impurity could be domestic sewage and so the demands on the treatment process are huge. This type of wastewater treatment generally occurs in three treatment phases. The **primary treatment** removes the solid particles, beginning with the largest and ending with the smallest. Typically, this is accomplished in stages using screens, a comminutor, a grit chamber, and a primary settling tank, in which the settled mass is in the form of sludge. The **secondary treatment** then eliminates the dissolved organic pollutants using a suspended growth reactor. To sustain the population of microorganisms, the microorganism-bearing 'activated' sludge from the final settling tank is pumped back into the reactor. The **tertiary treatment** does the final polishing of the water quality and thus resembles the pretreatment process shown in Fig. 19.5.

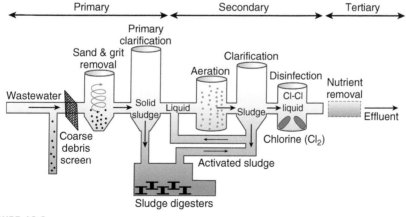

FIGURE 19.6

Wastewater treatment plant design.

University of Michigan Center for Sustainable Systems.

Example 19.6

The suspended growth reactor (or 'activated sludge unit') in the figure has been isolated with the effects of the adjoining process units replaced by their respective flow rates and solid concentrations, with only the water flowrate Q_2 being unknown. In the aeration tank, microorganisms are consuming dissolved organic pollutants and transforming them to solid particles via chemical reaction (thus the name reactor). At the same time, microorganisms could be multiplying, further increasing the concentration of solid matter in the tank. In the settling tank, most of the solid matter is removed from the processed water in the form of activated sludge. Some of this activated sludge is recycled back to the aeration tank. Determine the rate at which new solid matter is being generated in the aeration tank.

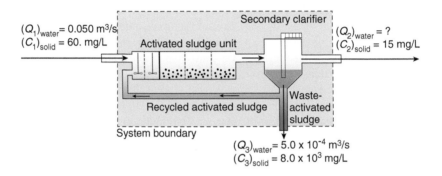

Need: the rate of solid mass generation ($\dot{G}_{\text{solid}}$)

Know: two of three flowrates, concentrations of solid matter at the inlets and exits.

How: Define the system to be the suspended growth reactor (the aeration tank plus the settling tank) and apply the Mass Rate Balance Eq. (19.5). With the solid flowrate $(Q_2)_{\text{solid}}$ unknown along with solid mass generation $\dot{G}_{\text{solid}}$, two Mass Rate Balance Equations will be needed.

Solve: Draw control volumes to set up Mass Rate Balance Equations for the solid matter and for the water. Note that the recycled activated sludge is not shown on the control volume drawings below because the piping that transports it never crosses the control volume boundary.

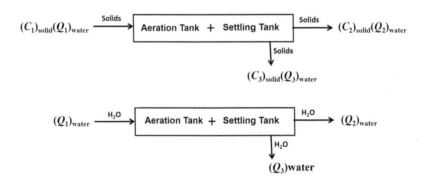

There is no net accumulation of mass within the control volume, but solid mass is being created within the aeration tank. Therefore, the Mass Rate Balance Equation (19.5) for the solid matter has a mass generation term $\dot{G}_{\text{solid}}$, while the equation for water does not:

$$(\dot{m}_{\text{solid}})_{\text{CV}} = 0 = \sum (C_{\text{solid}} Q_{\text{water}})_{\text{in}} - \sum (C_{\text{solid}} Q_{\text{water}})_{\text{out}} + \dot{G}_{\text{solid}}$$

The Mass Rate Balance Equation for the solid matter becomes:

$$0 = (C_1)_{\text{solid}} (Q_1)_{\text{water}} - (C_2)_{\text{solid}} (Q_2)_{\text{water}} - (C_3)_{\text{solid}} (Q_3)_{\text{water}} + \dot{G}_{\text{solid}}$$

$$0 = \left[(60.\,\text{mg}\,/\,\text{L})(0.050\,\text{m}^3\,/\,\text{s}) - 15(Q_2)_{\text{solid}} - (8.0 \times 10^3\,\text{mg}\,/\,\text{L})(5.0 \times 10^{-4}\,\text{m}^3\,/\,\text{s}) \right]$$

$$(1000\,\text{L}\,/\,\text{m}^3) + \dot{G}_{\text{solid}} \tag{19.8}$$

Since the solid concentration is quite small, we can set $(C_1)_{\text{water}} = (C_2)_{\text{water}} = (C_3)_{\text{water}} = 1.0$ and since $\dot{G}_{\text{water}} = 0$ the Mass Rate Balance Equation reduces to a balance of volumetric flow rates:

$$0 = (Q_1)_{\text{water}} - (Q_2)_{\text{water}} - (Q_3)_{\text{water}}$$

$$0 = 0.050\,\text{m}^3\,/\,\text{s} - (Q_2)_{\text{water}} - 5.0 \times 10^{-4}\,\text{m}^3\,/\,\text{s} \tag{19.9}$$

Finally, solving Eq. (19.9) for $(Q_2)_{\text{water}}$ and substituting the result into (19.8), yields:

$$(Q_2)_{\text{water}} = 0.050\,\text{m}^3\,/\,\text{s and } \dot{G}_{\text{solid}} = 1740\,\text{mg}\,/\,\text{s}$$

19.7 Solid waste management

The flow of domestic trash from source to endpoint is an entirely man-made creation. Common practice in the United States today is to create two separate streams of solid waste at the source, one for trash and another for recyclables (i.e., paper, cardboard, metal, certain types of plastic, glass). Once a week, specially marked trash and recyclable bins are rolled out for pick up by the collection trucks. The trash might be hauled to a **transfer station** or go straight to the local **landfill**. In contrast, the recyclables eventually end up at a **materials recovery facility** (MRF), where they are sorted for reusables.

At the MRF, the recyclables are placed on a conveyor belt and sorted by a combination of manual and automated methods. A typical sorting sequence is still very labor intensive as shown in Table 19.3 below.

Table 19.3 Recyclables sorting methods.

Material	Sorting method
Plastic wrappers and other trash	Sorted by hand
Corrugated cardboard	Sorted by hand, then baled
Glass	Sorted by rollers that sort and crush
Plastic and metal containers	Sorted using a rake-like separator
Various types of paper	Sorted by hand, then baled
Steel cans	Removed by rotating magnet then compacted and baled
Aluminum cans	Sorted by an eddy current separator, then compacted and baled

Example 19.7

A portion of the system of conveyor belts at an MRF is shown in the figure below. Recyclables enter the first conveyor at a rate of 100. tons/day. Then, at the intersection with a second conveyor, a mechanical rake diverts the plastic onto a second conveyor as 60.% (by mass) of the original recyclable stream, now mostly paper, continues down the first conveyor. The paper and plastic are sorted and removed from the conveyors. Finally, the trash that is not captured by the sorters falls into end-of-the-line hoppers 1 and 2 at the rates of 15 and 10. tons/day, respectively. Determine the mass removal rates at the paper and the plastic sorting stations.

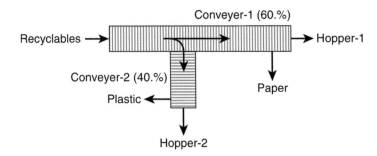

Need: mass removal rates at the paper and container sorting stations.

Know: throughput of 100. tons/day; 60.% of throughput remains on the first conveyor after passing through the rake separator; mass flow rates into hopper-1 is 15 tons/day and into hopper-2 is 10. tons/day.

How: define the control volume to consist of both conveyor belts and write the Mass Rate Balance Eq. (19.3b) for it. Then count unknowns and see where it takes you.

Solve: start by drawing the control volume, which in this case encloses both conveyor belts and the mass moving on them:

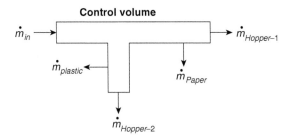

Mass neither accumulates nor is generated inside the control volume so that Eq. (19.3b) reduces to:

$$0 = \sum (\dot{m})_{\text{in}} - \sum (\dot{m})_{\text{out}}$$

For our control volume, this becomes

$$0 = \dot{m}_{\text{in}} - \left(\dot{m}_{\text{Paper}} + \dot{m}_{\text{Plastic}} + \dot{m}_{\text{Hopper}-1} + \dot{m}_{\text{Hopper}-2} \right)$$

$$0 = 100. - \left(\dot{m}_{\text{Paper}} + \dot{m}_{\text{Plastic}} + 15 + 10. \right) \qquad (19.10)$$

With two unknowns in Eq. (19.10), we still need one more equation. In a process like this one, comprised of individual process units (separator and two sorting stations, or any place where two or more streams of trash intersect), you can sometimes generate more equations by writing Mass Rate Balance Equations for the individual process units, provided we have some information about what is going in them. We know that downstream from the rake separator, 60.% of the throughput is on conveyor-1 and the other 40.% is on conveyor-2. With this information, we can do a Mass Rate Balance of the paper sorting station without introducing any additional unknowns.

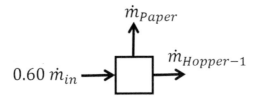

$$0 = 0.60 \, \dot{m}_{\text{in}} - \left(\dot{m}_{\text{Paper}} + \dot{m}_{\text{Hopper}-1} \right)$$

$$0 = 0.60 \, (100.) - \left(\dot{m}_{\text{Paper}} + 15 \right) \tag{19.11}$$

Eqs. (19.10) and (19.11) provide two equations in two unknowns. Solving them produces

$$\dot{m}_{\text{Paper}} = 45 \text{ tons/day and } \dot{m}_{\text{Plastic}} = 30. \text{ tons/day}$$

Some of the trash could be diverted to an **incinerator** to generate energy and reduce the volume of the trash by up to 90%, thus extending the life of the landfills. If organics such as food scraps and lawn trimmings can be isolated at the source, then **composting** (for use as fertilizer) becomes yet another option. Recent utilization of these waste management options in the United States is summarized below[5]:

- Landfilling, 52.5%
- Recycling, 25.8%
- Combustion with energy recovery, 12.8%
- Composting, 12.8%

The percentages can vary widely from region to region in the United States as a result of local fuel costs and resources. Contrast these numbers with Japan in which only 2% of its solid waste goes into a landfill and with China which converts 17% of its waste to energy by combustion.

As Table 19.4 suggests, recycling did not start to play a major role in solid waste management in the United Sates until the 1980s. The growing impact of recycling was

---

[5] EPA data from 2015; same reference as Table 19.4.

driven less by technological developments and economics (with the possible exception of paper) than by an increasing social awareness of the environmental benefits.

Table 19.4 Recycling rates as a percentage of available product waste in 2015.[6]

Waste	1970	1980	1990	2000	2014	2015
Paper	15%	21%	28%	43%	65%	67%
Glass	1%	5%	20%	23%	26%	26%
Metal	4%	8%	24%	35%	35%	34%
Plastic	Neg.	<1%	2%	6%	10%	9%
Yard trimming	Neg.	Neg.	12%	52%	61%	61%
Consumer electronics	NA	NA	NA	10%	42%	40%
Lead-acid batteries	76%	70%	97%	93%	99%	99%

19.8 Ethical considerations

A properly done environmental impact assessment depends not only on engineering skills, but also on National Society of Professional Engineers Code of Ethics (see Chapter 2 of this book). Ethical questions can arise in formulating decisions based on environmental assessment. Some of these questions are as follows:

- Is it ethical to limit resource extraction, with its associated environmental damage, by raising the resource price and thereby limiting its use to those who can afford it?
- Is it ethical to eliminate jobs in an area to protect the environment for a future generation?
- Is it ethical to deplete a resource so that future generations do not have it?
- Is it ethical to spend millions mitigating a high-consequence impact that is unlikely to occur?
- Is it ethical to destroy a watershed by providing logging jobs for 50 years?
- Is it ethical to close a lumber mill and eliminate jobs for an entire small community to save an old-growth forest?

The engineer should remember that none of these considerations are part of environmental assessment, nor are they addressed by the National Environmental Policy Act.[7] Addressing them is part of an engineer's assessment decision record and is ultimately the responsibility of the project's decision-makers.

---

[6] Table prepared by the EPA for their on-line article titled "National Overview: Facts and Figures on Materials, Wastes and Recycling", 2015.

[7] The National Environmental Policy Act is a United States environmental law that promotes the enhancement of the environment and established the President's Council on Environmental Quality (CEQ). The law was enacted on January 1, 1970.

Summary

Protection of the environment is a challenge that cannot be ignored. As the earth's population grows and industrializes, our finite natural resources are being consumed and contaminated at ever increasing rates. Even at low concentrations, pollutants can cause harm to humans and the environment. Units like **mg/L** for pollutants in water and **ppm** for pollutants in air are traditionally employed to express these concentrations concisely and to define limits that have been legislated and are enforced in this country by the EPA.

Mass Balance and Mass Rate Balance Equations are among the most fundamental of analytical tools available to the environmental engineer. They can be applied to any system with clearly defined boundaries, across which mass flows and within which mass can accumulate and/or is generated by chemical means.

Most **air treatment** methods such as **wet scrubbers** and **cyclones** are applied on a relatively small scale at the source. Most of the polluted air goes where the wind takes it. Some pollutants eventually leave the atmosphere, for example, SO_2 falls to the ground in the form of **acid rain** polluting plants and soil. Other pollutants, 'forever' chemicals like CO_2, continue to accumulate in the atmosphere, raising the risk of climate change.

The design of a suitable **water treatment process** for producing drinking water depends on the source and initial quality of the water. For example, groundwater from a well has already taken advantage of the natural filtering properties of soil to remove most pollutants. In this case, it may be enough to **disinfect** the water by injecting chlorine and to remove iron particles with **rapid sand filters**. However, wastewater from a sewage system will contain more diverse and concentrated forms of pollutants. Therefore, three stages of water treatment are employed. In the **primary stage**, wastewater flows through **screens**, **grit chambers**, and **clarifiers** to remove in succession smaller and smaller diameter particles. In the **secondary stage**, **aerobic microorganisms** fueled by oxygen in an **aeration tank** consume organic waste and transform it to water and CO_2 and other less toxic elements. **Sludge** is a byproduct of this biological process that must be separately processed before being released into the environment. The **tertiary** stage fine tunes the purity of the water to levels consistent with its intended use.

Solid waste is less about treatment and more about managing its flow. **Recycling**, **Composting**, **Incineration**, and **Anaerobic Digesters** (at landfills) are among the methods used to extract value from trash as an alternative to direct disposal at landfills.

Exercises

Problems for air quality and control

1. Which pollutants come from automotive sources?
    a) Hydrocarbons
    b) Nitrogen oxides

    c) Carbon dioxide

    d) Carbon monoxide

    e) All of the above

2. List two pollutants which result in air pollution produced by humans.

    a) Nitrogen and carbon dioxide

    b) Carbon monoxides and nitrogen

    c) Oxygen and carbon dioxide

    d) Carbon monoxides and nitrogen oxide

3. The air quality index that is considered to have the most serious health effects

    a) 25

    b) 60

    c) 140

    d) 455

4. How can Air Pollution harm the environment?

    a) Climate Change

    b) Ozone layer gets damaged

    c) Habitats get destroyed

    d) All of the above

5. A stack gas contains carbon monoxide (CO) at 25°C and 1 atm pressure and a concentration of 10. % by volume. What is the concentration of CO in ppm and in $\mu g/m^3$?

6. The exhaust from an idling automobile's engine at 25°C and 1 atm pressure contained nitrous oxide (N_2O) at a concentration of 60. ppm. What was the concentration of N_2O as (a) the percent by volume, and (b) in $\mu g/m^3$.

7. A dual-lane tunnel is proposed that is 3.0 km long, 12 m wide and 7.5 m high. During peak hours, the tunnel will be full of cars and their exhaust emissions will be uniform throughout the tunnel. The fresh air (wind) speed through the tunnel is $V = 1.0$ m/s. The most critical pollutants inside the tunnel are carbon monoxide (CO). The following data are available:

- CO emission rate inside tunnel = 1.5×10^{-4} mg/m^3 per m^2 of tunnel opening
- Ambient CO concentration = 20. mg/m^3
- Primary safety standard for CO = 40. mg/m^3

    a) Using the following equation, calculate the average concentration of CO inside the tunnel during peak hours.

$$C_{avg} = C_0 + \frac{EL}{VH}$$

    where C_0 is the ambient pollutant concentration entering the tunnel, E is the emission rate inside the tunnel, L is the length of the tunnel, V is the air speed through the tunnel, and H is the height of the tunnel.

    b) As an engineer, you are tasked to reduce the CO concentration to meet the primary safety standard and to provide good ventilation inside the tunnel. A typical way to achieve this is to install a fan in the tunnel to blow the pollutant out of the tunnel. Calculate the new air speed V inside the tunnel required to achieve safe CO conditions inside the tunnel.

Problems for water quality and treatment

8. Do a 20-min Internet search on how to set up and run a freshwater fish aquarium. What methods/products are used to (i) measure water quality, (ii) treat household tap water, (iii) add oxygen, (iv) control ammonia levels, (v) eliminate biological waste, (vi) remove suspended solid particulates and (vii) limit algae growth? Draw a sketch of an aquarium that illustrates (with appropriate explanatory labeling) the hardware and strategies commonly used. Then compare these methods with those used in the water treatment material given in this chapter.

9. Water flowing in the pipe of a water treatment plant is to be disinfected by injecting chlorine. If the water in the pipe has a volumetric flow rate of 570 gal/min, the upstream chlorine is zero

(C_{cl} = 0.0 mg/L) and the desired downstream concentration is 2.0 mg/L, determine the rate (in mg/s) at which chlorine should be added to the water.

10. Phosphorus from excess irrigation water is slowly leaking into a stream that has small farms located on both its left and right banks along one stretch. Too much phosphorus in the stream can overstimulate algae growth. The known phosphorus concentrations and flow rates entering this part of the stream are shown in the figure below. Determine the expected phosphorus concentration downstream from the farms.

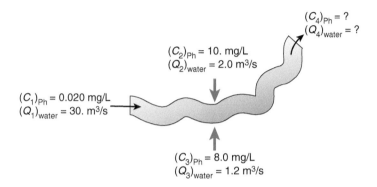

11. Iron particles are being removed from spring water using a rapid sand filter. The flow rate through the filter is 8100 L/min and the incoming and outgoing iron concentrations are 20. mg/L and 0.050 mg/L, respectively. Determine the rate at which iron solid particles are accumulating within the sand filter, in units of kg/min.

12. Top view of a storm sewer system is shown in the figure below. Maximum flow rates through the manhole covers during a worst-case storm are indicated. The flow eventually exits through the trunk line. Determine the maximum flow rates through each of the five connecting pipes shown in the figure by doing a control volume Mass Rate Balance around each of the five manhole covers.

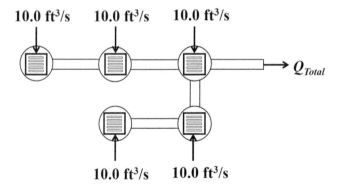

13. A lagoon on the backend of a water treatment plant receives the daily backwash from an array of rapid sand filters at the rate of 9.50×10^5 L/day. Time is given for the solid particles in the backwash to settle out in the lagoon before discharging the water to a local river. If the concentration of the solid particles in the backwash entering the lagoon is 150 mg/L and 30. mg/L when it is discharged into the river, determine the rate of accumulation of the solid particles in the lagoon, in units of kg/day.

Problems for solid waste management

14. A Transfer Center ships out 200. tons/day of trash for disposal. 80.% of this amount goes straight to the landfill. The other 20. % is directed to the local Incinerator where its total mass is first reduced by 90. % by combustion and then hauled to the same local landfill. Determine the rate at which mass from the Transfer Center accumulates in the landfill.

15. Sanitary landfills are an improvement over the landfills (or dumps) of the past since significant (and often expensive) measures are taken to protect the surrounding environment. A labeled schematic of the sanitary landfill in Columbia, Missouri, is shown in the figure below. It represents the next generation of landfills known as bioreactor landfills. Do a 30-min Internet search to the point where you can describe in detail the key features labeled in the figure and how it all works to protect the environment and generate value from trash. Present your description and list of Internet sources in one typed page.

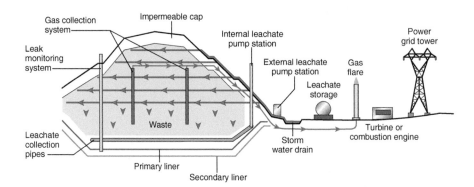

Ethics problems

16. Rachel is a newly hired environmental engineer at an industrial plant that discharges wastewater into a local river. She observes that safe handling and operating procedures are not being followed consistently and that this could lead to unacceptable levels of toxic chemicals being released into the environment. In response, she prepares and distributes a compilation of safety guidelines and proposes the purchase of detector alarms and materials handling equipment. Her immediate boss, the plant manager, says he does not have the money for the equipment and the process safety culture remains unchanged. What should she do?
    (a) Nothing, she did what she could.
    (b) Go over her boss's head and present her case to the higher ups.
    (c) Contact the EPA about the violations.
    (d) Threaten to quit, unless new safety measures are implemented.
    (e) Reduce her safety recommendations so at least some progress can be made

    Evaluate these alternatives using an ethics matrix before stating and explaining your choice of actions.

17. Anthony is an up and coming environmental engineer working in the federal government. He entered this field because of his passion for protecting the environment. Recently, his boss asked him to go to an annual meeting where the main item on the agenda was a confidential discussion of a new government initiative aimed at **reducing the limits** on a certain class of air pollutants. It turns out that the motivation for the reduced limits is to relieve the economic pressure on a specific industry, which runs counter to why Anthony entered this profession. He is considering the following options:

    (a) Inform the newspapers of the new air pollution limit reductions.
    (b) Send a memo up the chain of command detailing the risks involved in lowering the limits.
    (c) Say and do nothing in response, as the discussion was confidential.
    (d) Say nothing but start looking for a new job.

    Evaluate Anthony's options using an ethics matrix to determine the action that is the best compromise between his ethical responsibilities as a practicing engineer and his personally held ideals regarding protecting the environment.

Green energy engineering

20

Exploring Engineering. https://doi.org/10.1016/B978-0-443-13541-5.00021-0

20.1 Introduction

This chapter describes how nonfossil sources of energy may be substituted for today's dominant fossil fuels, which are coal, oil, and natural gas. We currently use these fuels for electric power production and for our vehicles.

In Chapter 8, we studied conventional chemical energy sources based on these fuels, and in Chapter 18, we studied electrochemical sources of stored energy and power. One of the things we learned is that a major combustion product from fossil fuel sources is carbon dioxide (CO_2), a gas implicated in global warming. Climate studies look at long range changes in weather patterns, such as previous historical earth warming and ice age cooling periods, of which there have been many in Earth's history. If and when either returns, the survival of humankind will be sorely tested.

There are both theoretical and experimental reasons to expect global warming due to CO_2 accumulation in the atmosphere. Of the radiation reaching us from the Sun, a small fraction has high-energy ultraviolet (UV) light content. While this is a small part of the total sunlight spectrum, it is absorbed by the atmosphere, preferentially by CO_2 molecules and by water vapor. A measured solar spectrum is shown as Fig. 20.1. The dips in the curve are due to absorptions by CO_2, H_2O, O_3 (ozone), and other constituents of the atmosphere. High-energy solar radiation occurs at short wavelengths (i.e., near the origin of this graph).

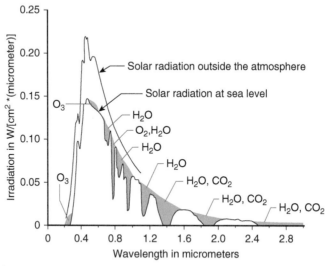

FIGURE 20.1

Solar irradiation as a function of wavelength.

The incoming photons from sunlight are measured as a function of their **wavelength**[1] (in micrometers, abbreviated μm); the longer the wavelength, the redder is the light and the lower its energy. The ordinate is in watts/cm^2 per micrometer. To get the total solar insolation amount of incoming sunlight, we must integrate for the area under the curve. In practice, we need to integrate only over the visible spectrum from its low-energy red end (or infrared (IR)) to its high-energy UV. By integrating the spectrum, we find that the Earth receives an average of 1.3 kW/m^2 on its surface area.

If none of the incident sunlight on Earth was absorbed in the atmosphere, we would not have very many warm evenings, because they are due in part to the retention of the absorbed heat in the atmosphere that accumulated during the sunny hours of the day. In addition, after striking the Earth, the resulting short-wave UV radiation is changed into long wavelength IR radiation, which is trapped by the lower atmosphere (hence, the term *greenhouse effect*).

The amount of CO_2 in the atmosphere has increased from the preIndustrial Age value of about 280 parts per million (ppm) to a current near 400 ppm, quite probably because we continue to emit some eight billion tons of CO_2 into our atmosphere every year. Many mathematical studies of the world's climate show a long-term heating trend in our atmosphere. We should now consider ways to produce both energy for electric power and transportation in ways that do not increase the CO_2 concentration in the atmosphere. Fig. 20.2 shows the overall pool of available energy resources.

---

[1] Individual light waves are sinusoidal. Their frequency (which is inverse to their wavelength) is proportional to their energy. High frequency (or short wavelengths) indicates energetic waves and vice versa.

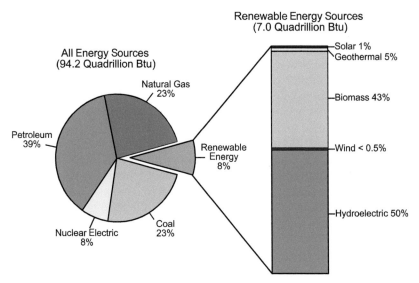

Totals may not equal sum of components due to independent rounding.

FIGURE 20.2

Current world energy resources. A Btu (British thermal unit) is the heat required to heat 1 lbm of water by 1°F.

Renewable energy resources are only 8% of the total available energy resources, according to Fig. 20.2. These include solar, wind, and hydroelectric generated power as shown in Table 20.1.

Table 20.1 Solar alternatives to fossil fuels.

Power methods	Ultimate source of power	Comments and rough cost per kW
Photovoltaics	Direct Sun	Relatively expensive, large potential for small applications (**$4750/kW**).
Solar thermal electric power plants	Direct Sun	May use the Sun directly to produce steam; expensive per unit of power (**$3150/kW**).
Solar thermal heat	Direct Sun	Direct household use; costs depend on specifics.
Windmills	Wind (which is caused by the Sun heating the earth)	Large installations are economical; limited sites available (**$1200/kW**).
Hydropower	Hydroelectric river dams and tidal power generators (which are another effect of solar energy moving water)	Many large facilities in the world; number of acceptable sites declining (**$1500/kW**).

The first two kinds of green power plants in Table 20.1 rely on immediate, bright sunlight. When the Sun goes down, the generated power drops to zero. In Chapter 18, we raised the question of how to deal with synchronizing solar output with immediate demand for such systems, since load leveling by batteries is expensive and inconvenient.

Wind is a form of solar energy because it is generated by air heated by the sun. For example, hot air rises and cool air sinks, and the interaction of these vertical winds with the Earth's rotation can transform them into horizontal winds. The energy of these winds can be captured by the blades of large windmills.

Hydropower is another form of solar energy. The sun evaporates water from the oceans into the atmosphere. Eventually, this water becomes rain or snow due to atmospheric cooling. If we are clever enough to catch the rain or the snowmelt behind a dam, we can release water from the dam when it is needed and convert its potential energy into electricity. Tidal flows also can be used in hydroelectric schemes.

Did you wonder why hydrogen has not been included in renewable green sources? After all, the by-product of the combustion of hydrogen is water *and only water*. It seems to be the perfect substitute for fossil fuels since hydrogen-fueled cars and fuels cells (Chapter 18) are well within our technological capabilities. The question is this: What is a viable source of hydrogen?

There are two industrial methods for making hydrogen: (1) by adding steam to partially burned hydrocarbons which also produces a large amount of CO_2, and (2) electrolysis, which is only used to make relatively small quantities of hydrogen and since it uses electricity, it too is a producer of CO_2.

20.2 Solar energy

Solar energy is the light and heat we receive from the Sun. The Earth receives 1.74×10^{17} W of solar radiation at the upper atmosphere. Approximately 30% is reflected back to space, while the rest is absorbed by clouds, oceans, and land masses.

Solar energy technologies include solar heating, solar photovoltaics, and solar thermal electricity, which can make considerable contributions to solving some of the energy problems the world now faces.

20.2.1 Photovoltaic power

Sunlight falls onto certain materials and electricity is produced. But how does this happen? To answer this question, we need to delve into a little quantum mechanics.

Einstein showed that an electron was produced only when the light exceeded a certain frequency. At longer wavelengths, there was zero current. This effect was immediately associated with Planck's postulate that sources of light emit in discrete frequencies, as measured by the famous Planck law equation:

$$E = h\nu \tag{20.1}$$

where E is the photon's (i.e., light wave's) energy, ν is its frequency (in cycles/s), and h is called *Planck's constant*, which is

$$h = 6.626 \times 10^{-34} \text{ J} \cdot \text{s}.$$

We can also express the energy of light by other variables (wavelength and frequency), using the traveling wave equation in which the speed of light is $c = 3.00 \times 10^8$ m/s:

$$\lambda \nu = c \qquad\qquad (20.2)$$

Example 20.1

The wavelength of the visible light is measured in micrometers, μm (10^{-6} m), and its spectrum is from the near ultraviolet (UV) at 0.4 μm to the near infrared (IR) at 0.70 μm. What are the corresponding energies and frequencies in joules and hertz (cycles/s) of near IR and UV photons?

Need: Frequencies and energies of near IR and near UV photons.

Know: Wavelengths of near IR and near UV are 0.70 and 0.40 μm.

How: Use Planck's law, Eq. (20.1) with $h = 6.626 \times 10^{-34}$ J·s and convert to frequency from Eq. (20.2) with $c = 3.00 \times 10^8$ m/s.

Solve: Eq. (20.1) gives $\nu = c/\lambda = (3.00 \times 10^8)/(0.70 \times 10^{-6})$ [m/s]/([μm] ×[m/μm])

$$= \mathbf{4.3 \times 10^{14}\ cycle/s} = \mathbf{4.3 \times 10^{14}\ Hz}\text{ for near IR photons}$$

and $\nu = c/\lambda = (3.00 \times 10^8)/(0.40 \times 10^{-6})$

$$= \mathbf{7.5 \times 10^{14}\ cycle/s} = \mathbf{7.5 \times 10^{14}\ Hz}\text{ for near UV photons}$$

Then, $E = h\nu$ or $E = (6.626 \times 10^{-34}) \times (4.3 \times 10^{14})$ [J· s][1/s] = $\mathbf{2.9 \times 10^{-19}\ J}$ for near IR photons, and for near UV photons, $E = \mathbf{4.1 \times 10^{-19}\ J}$.

In the photoelectric effect, a photon of light with energy less than a given frequency gives zero current, but at an incrementally higher frequency a current flows. A sufficiently energetic photon can thus produce a detectable electron. This is the origin of photovoltaic power.

A complication arises when we are not dealing with the photoelectric effect per se but with semiconductor solids that are relatively poor conductors of electricity. The situation is displayed in Fig. 20.3, which is known as an **energy level diagram**. On the left is a sketch of the basic photoelectric effect; an incoming photon stimulates the release of an electron only if its energy (i.e., frequency) is large enough. The analogous situation for photovoltaics is shown on the right. In photovoltaic solids, the number of energy levels is huge, because of the vast number of interconnected atoms. The energy levels smear into regions in which the electrons can exist. These regions are known as *valence* and *conduction bands*.[2] Electrons exist in conduction and valence bands and not in between them. If the bands overlap, as in metal conductors, electrons spontaneously flood the conduction band, leading to easy electricity flow. If the bands are so far apart, electrons cannot jump to the conduction band; the material is an insulator. In between are the **semiconductors**, whose properties are dominated by the properties of the band gap.

---

[2] The electrons trapped in these bands do what the names suggest, make chemical bonds (the valence band) and conduct electricity (the conduction band).

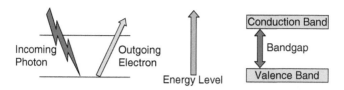

FIGURE 20.3

Photovoltaic basics.

Band gap values for four common semiconductors are given in Table 20.2, of which the most important is silicon. It is convenient to express the energy in the unfamiliar but useful units of the electron volt (eV), where an electron volt is the energy required to push one electron through 1 V. Since an electron's charge is 1.60×10^{-19} coulombs [C],

$$1\ eV = \left(1.60 \times 10^{-19}\ C\right) \times (1.0\ V) = \mathbf{1.60 \times 10^{-19}\ J}$$

Note that 1 [CV] = 1 [A·s] × [V] = 1 J

Crystalline silicon's band gap is 1.11 eV, meaning, among other things, that we would need to connect many such cells in series to reach a useful household level of 115 V.

Table 20.2 Semiconductor band gaps.

Material	Band gap (eV)
Crystalline silicon, c-Si	1.11
Amorphous silicon, a-Si	1.7
Germanium, Ge	0.67
Indium antimonide, nSb	0.17
Diamond, C	5.5

To promote a free electron in c-Si, we need incoming photons with energies ≥1.11 eV. Using Eq. (20.1) and Planck's constant in eV units, we can directly compare electron energies to the photon energies that produce them. In these units, Planck's constant is $h = 4.135 \times 10^{-15}$ eV s.

Example 20.2

What are the near IR and near UV photon energies in eV corresponding to wavelengths of (a) 0.40 and (b) 0.70 μm?

Need: The near IR and UV photon energies in eV.
Know: Planck's law $E = h\nu$ and Eq. (20.2), $\lambda \nu = c$, with $h = 4.135 \times e^{-15}$ eV s.
How: Combine Eq. (20.2) with Planck's law gives $E = hc/\lambda$.
Solve: $E = hc/\lambda = (4.135 \times e^{-15}) \times (3.00 \times 10^{8})/(0.4 \times 10^{-6})$ [eV·s][m/s]/[μm][m/μm]
= 1.240/0.4 eV = **3.1 eV** for near UV and **1.8 eV** for near IR.

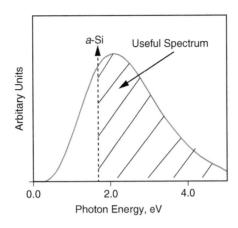

FIGURE 20.4

Band gap limitation on photon efficiency.

Fig. 20.4 shows the band gap of amorphous silicon in eV relative to the Sun's spectral energy (smoothed to ignore extraneous atmospheric absorbencies).

Since only photons with energy above the band gap contribute to the electric current, the smaller band gap in a germanium crystal theoretically has a larger efficiency than c-Si. Integration of the area under the spectral curve for energies larger than the band gap determines the possible photovoltaic efficiency. Photovoltaic devices have many factors that enter into their efficiencies. Even the "contamination" of c-Si by a-Si reduces the cell's efficiency given that c-Si is considerably harder to produce and thus more expensive than a-Si. A solar photovoltaic cell is shown in Fig. 20.5.

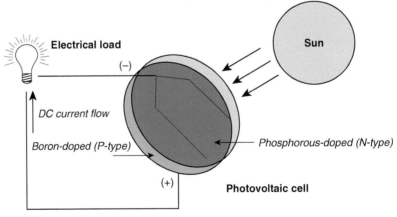

FIGURE 20.5

Photovoltaic cell.

This figure shows several factors other than band gap energy that are relevant to photovoltaic efficiency. Within the cell itself we have to consider diffusion of the photon-produced electrons as they flow toward the electrical collector plates on either side of the semiconductor's surface. Glass plates are also needed to protect the cells. Dopants (small quantities of active elements) are added to the semiconductors to modify their band structure. The cells may have many layers of semiconductor to optimize their efficiency. Actual a-Si photovoltaic cells have overall efficiencies of 10%–20%.

Example 20.3

If a single 100. cm^2 silicon cell produces a voltage of 0.500 V and a current of 2.00 A when exposed to sunlight with an intensity of 1000. W/m^2, determine the total number and area of these silicon cells required to produce 17,000 MW.

    Need: The total number of 100. cm^2 silicon cells and their total area.

    Know: The power from a single cell is 0.500×2.00 [V][A] = 1.00 [VA] = 1.00 W.

    How: The number of cells needed is the total power required divided by the power produced by a single cell.

    Solve: Therefore, to produce 17,000 MW = 17×10^9 W, we would need **17×10^9 cells**, which would have a total area of:

        Area = 17×10^9 [cells] [100 cm^2/cell][m/100 cm]^2 = **17×10^7 m^2** or about 66 square miles.

20.2.2 Solar thermal power plants

Standard fossil power plants use combustion heat to create high-pressure steam. The steam then drives a turbine connected to an electric generator that, in turn, produces the electricity. A similar conceptual solar thermal design is shown in Fig. 20.6. It concentrates focused sunlight onto tubes containing hot oil, which then produces high-pressure steam. The solar heat is concentrated about 75 times as the parabolic mirrors swivel to follow the maximum sunlight.

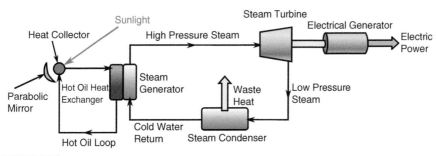

FIGURE 20.6

Solar thermal power plant.

As in fossil power plants, the high-pressure steam drives a turbine that then drives an electrical generator. The steam exiting the turbine is condensed back to liquid water, which is pumped back into the heat exchanger to be reconverted into high-pressure steam.

The theoretical amount of power produced by the turbine is the mass flow rate of the steam multiplied by the difference between the energy in the steam at the turbine's inlet and the turbine's exhaust. State-of-the-art solar thermal power plants are located in desert regions. An existing solar plant heats thermally stable synthetic oil to 400°C and produces up to 80 MW of electricity from a mirror field of nearly one-half million square meters.

Example 20.4

What is the gross efficiency of a solar power station that produces 80.0 MW for 10.0 h/day from an array field of parabolic mirrors covering 5.00×10^5 m^2 if the solar thermal flux is 1.30 kW/m^2?

Need: The efficiency of a solar thermal power plant.

Know: The power output is 80.0 MW for 10.0 h/day. Solar power is 1.30 kW/m^2, and the array field is 5.00×10^5 m^2.

How: Compare the power produced with the solar radiation of

$$5.00 \times 10^5 \text{ m}^2 \times 1.30 \text{ kW/m}^2 = 6.50 \times 10^5 \text{ kW}$$

Solve: Efficiency $= 100 \times 80.0 \times 1000/(6.50 \times 10^5)$ [%][MW][kW/MW]/[kW] = **12.3%**

Note that the United States can afford to use such large land areas in underpopulated southwestern deserts. But, Europe and Asia have larger population densities than the United States and this power plant could be considered an underutilization of land.

20.2.3 Solar thermal heating

Certainly, those of you who live east of the Rocky Mountains are familiar with the cold that winter brings. Do you know how your house is heated in winter? Do you know what fuel is used to keep your house warm? Did you realize your heating system has a CO_2 waste product[3]?

Household heating is an expensive and reoccurring expense. A typical northern suburban house consumes about 100×10^6 kJ per heating season. Provided that a typical winter's day is not too overcast, solar heating to supply this household heat would seem to be an obvious application. All that is needed is a strategically placed solar collector mounted perpendicular to the rays of the winter sun as in Fig. 20.7 (usually on a roof with a southern exposure). Because we need only relatively warm water for household heating, we can use simple tubes with water flowing inside to collect some of the incident solar radiation. Since black surfaces are

---

[3] If you have an electrically heated home, it too emits CO_2 but remotely at the power station.

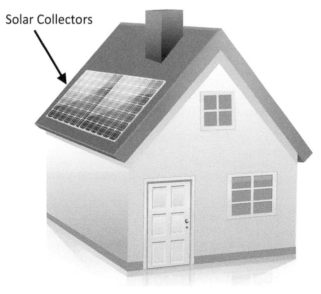

FIGURE 20.7

Solar heating.

very efficient at collecting solar radiation, the tubes are often blackened as well as being insulated with a glass cover.

Inside the house, we need a pump, thermal storage (such as drums of warm water to be called on when the sun is not shining) and standard heating coils to distribute the warm water throughout the house (see Fig. 20.8).

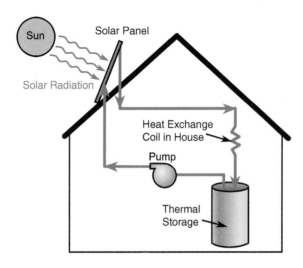

FIGURE 20.8

Domestic heating systems.

Example 20.5 Design of a solar thermal heating house

A typical suburban house in midwinter in a moderate climatic zone has a furnace capacity of about 20. kW that runs on a 50% cycle (meaning that it is on half the time). This means that, on average, about 10. kW is delivered to the house, and this is also the heat rate that needs to be supplied by solar heating.

If the solar radiation is 0.50 kW/m^2 and is directly delivered to the collection fluid in the roof panel, determine the following items:

a. The size of the required roof panels and is it possible to put panels of this size on a typical house roof?

b. The amount of solar heat that needs to be stored in a thermal storage tank for a winter night of 14 h if the house is maintained at 22°C and the hot fluid collected in the tank is at 45°C.

Need: **a.** Size of the solar collector = _____ m^2.
  b. Solar heat stored in thermal storage tank = _____kJ.
Know: Required heat rate is 10. kW, solar insolation rate is 0.5 kW/m^2, house temperature is 22°C, and the temperature of the hot fluid in the storage tank is 45°C.
How: Equate the solar insolation to the house's needs as measured by the furnace capacity.
Solve: **a.** Assume the solar panel area is A m^2. Then the collected heat rate is:
  Heat rate $= 0.50 \times A$ [kW/m^2][m^2] $= 10.$ kW. Therefore, $A = 10./0.50$ m^2 = **20. m^2**, and if panels are half as high as they are wide their size is $A = HW = 0.5W^2$ so $W = \sqrt{40.} =$ **6.3 m** and $H =$ **3.2 m**. These will fit on a typical house roof.
  b. After dusk, all of the energy needed by the house is to be recovered by the heat in the storage tank(s). This amount is 10. [kW]×14. [h] $= 140$ kWh $= 140\times3600.$ [kWh] [kJ/kWh] $= 5.0 \times 10^5$ kJ. Therefore, the amount of solar heat that needs to be stored in a thermal storage tank = **5.0 × 10^5 kJ**.

20.2.4 Windmills

Terrestrial wind is driven by solar heating in the atmosphere. Once winds are in accessible regions, their kinetic energy can be harvested, as it has been for centuries by windmills (see Fig. 20.9).

FIGURE 20.9

A Modern windmill.

The power output of a windmill is directly proportional to the circular area, A, swept by its rotors. Multiplying this area by the axial wind speed v determines the volumetric flow rate of air through the rotors, $\dot{Q}$, where:

$$\dot{Q} = Av \qquad (20.3)$$

After the passage of the wind through the windmill's rotors, the wind spreads out behind the windmill and has slowed after having some of its energy removed (see Fig. 20.10).

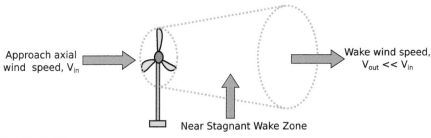

Approach axial wind speed, V_{in}

Wake wind speed, $V_{out} \ll V_{in}$

Near Stagnant Wake Zone

FIGURE 20.10

Fluid mechanics of a windmill.

Ideally, all the wind's power is converted to mechanical power:

$$P_{ideal} = \dot{Q} \times \frac{1}{2}\rho v^2 = \frac{1}{2}\rho v^3 A \qquad (20.4)$$

Since the ideal lossless state cannot be achieved, this equation must be corrected for the slow air flow behind the windmill's rotors. A detailed calculation shows that the actual power output of a windmill is:

$$P_{actual} = \frac{1}{2}C_p \rho v^3 A \qquad (20.5)$$

where C_p is the **power coefficient**, which has a theoretical maximum of $C_{p\text{-}max} = 0.593$ (or 59%). In this form, Eq. (20.5) is called **Betz's law** and the power coefficient is called the **Betz's limit**. In practice, Betz's limit is about 0.4–0.5 (or 40%–50%).

Example 20.6

A modern commercial windmill has a diameter of 100. meters and operates in winds of just 5.00 m/s. If the Betz limit is 0.50, what is the power produced by this windmill? Assume that the air density is 1.05 kg/m³. What wind speed is needed to achieve the windmill's design power output of 2.5 MW?

  Need: The power produced by a windmill of 100. meters diameter in a wind of 5.00 m/s with a power coefficient of 0.50. What wind speed corresponds to the design basis of 2.5 MW?

  Know: The Betz limit = 0.50, size of windmill, and air density is 1.05 kg/m³.

  How: Cross-sectional area of windmill is $A = \pi d^2/4 = 3.14 \times 100.^2/4 = 7.85 \times 10^3$ m².

  Solve: Using Eq. (20.5),

$$P_{\text{actual}} = \frac{1}{2}C_{\text{p}}\rho v^3 A = \frac{1}{2} \times 0.50 \times 1.05 \times 5.00^3 \times 7.85 \times 10^3 \ [\text{kg}/\text{m}^3][\text{m}/\text{s}]^3[\text{m}^2]$$

$$= 2.6 \times 10^5 \ [\text{N} \cdot \text{m}/\text{s}] = 2.6 \times 10^5 \ \text{W} = \textbf{0.26 MW}.$$

The design basis for this windmill is 2.50 MW. Since the power varies directly as the cube of the wind speed, **the required wind** is $(2.50/0.26)^{0.333} \times 5.00 = 10.6$ m/s = **11 m/s**.

Note that doubling the wind speed increases the power output by a factor of eight because the wind velocity is cubed in Eq. (20.5). Not surprisingly, windmills are located in areas where the wind is both constant and strong.

20.2.5 Hydropower

Water powered many manufacturing industries until the age of steam supplanted it. Instantaneous rainwater runoff was unreliable, but if the water was blocked and released when needed through a channel, it could be put to valuable use. Unlike heat cycles, hydropower uses potential energy and converts it to mechanical energy, a system that is very efficient (see Fig. 20.11).

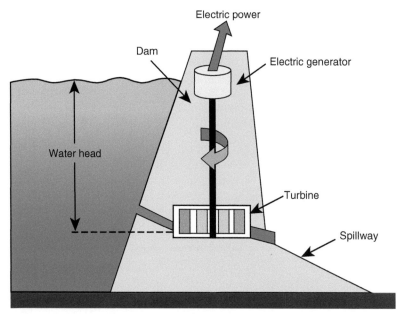

FIGURE 20.11

Hydroelectric power.

There are two key variables for a successful hydropower system: sufficient potential energy of the upstream water (the water head in Fig. 20.11) and the flow rate of water through the turbine. The flow rate of the water is determined by the geometry of the piping and the internals of the turbines. If the water volume flow rate is $\dot{Q}$ (ft^3/s or m^3/s), the power delivered to the turbine is

$$P_{\text{turbine}} = \frac{\rho g h \dot{Q}}{g_c} \tag{20.6}$$

where $g_c = 1$ (dimensionless) in MKS units, and $g_c = 32.2$ lbm·ft/(lbf·s^2) in Engineering English units, ρ (Greek rho) is the density of the water, and h is the height of the water above the turbine (called the *head*). In this formula, power is energy per unit time. The recovery of the potential energy by water turbines is more than 90% efficient. Hoover Dam, shown in Fig. 20.12, was built in the 1930s. It has 17 turbines with a combined capacity of 2080 MW; the water head there is 590 ft.

FIGURE 20.12

The Hoover Dam built in the 1930s.

Example: 20.7

A 100% efficient water turbine has a 590. foot water head and produces 161,000 HP. How much water in ft^3/s must flow through it?

Need: The water flow in ft^3/s to 100% efficient turbine system with a head of 590. feet of water. Turbine produces 161,000 HP.

Know: The water density is 62.4 lbm/ft^3, $g = 32.2$ ft/s^2, $g_c = 32.2$ lbm·ft/lbf· s^2 and 1.00 HP = 550. ft·lbf/s.

How: From Eq. (20.6), we have $P = \rho g h \dot{Q}/g_c$.

Solve: The water flow rate is $\dot{Q} = P g_c/\rho g h$, so

$$\dot{Q} = 161,000 \times 550. \times 32.2 / (62.4 \times 32.2 \times 590.)[\text{HP}]\left[(\text{ft}\cdot\text{lbf}/\text{s}) / \text{HP}\right]\left[\text{lbm}\cdot\text{ft} / \text{lbf} \cdot \text{s}^2\right]$$

$$/ \left\{\left[\text{lbm} / \text{ft}^3\right]\left[\text{ft} / \text{s}^2\right][\text{ft}]\right\}$$

$$= 2.41 \times 10^3 \text{ ft}^3/\text{s}.$$

Hoover Dam has 17 units of about the size described in this example, so that its total water throughput is about 41,000 ft^3/s. The new Three Gorges Dam in China has electricity production approximately 10 times larger than Hoover Dam and so requires roughly 10 times the water flow through its turbines.

The United States has about 7% of its electric power drawn from hydropower sources. While a decreasing number of sites are still available, hydropower is by far the most successful of green energy sources in the United States.

It should be noted that dams can fail with massive loss of life due to downstream flooding. A serious accident occurred in 2009 at a Russian 2000 MW hydroelectric plant due to a massive electrical failure that produced an explosion with severe loss of life of the plant's operating personnel.

20.3 Other green energy source

Solar energy in its various forms is not the only non-CO_2 power source. For example, OTEC, or Ocean Thermal Energy Conversion, was suggested for a floating power station that uses the differences in water temperature between deep and shallow zones as a thermal source to make electric power (but at very low efficiency). Another oceanic method relies on wave motion that is mechanically tapped for power. Yet another uses tidal power. The Rance Tidal Power station on the English Channel in the north of France produces 240 MW using tidal water and a dam to capture it. The trapped water is let down through a series of water turbines and thus is similar to standard hydroelectric power. Such systems may work well but there are few locations where they can be located.

In northern California, hot steam can be found near the Earth's surface. Using a power plant similar to that for solar thermal power (except that the heat source is geothermal steam), California produces about 1800 MW geothermally. But, in general, there is an obvious difficulty of finding suitable natural geothermal steam sources (Iceland being an exception where 25% of its electrical energy needs are met by geothermal power). Given that volcanic areas are the most suitable for geothermal energy, it is not surprising that acidic sulfurous gases accompany the production of geothermal power. These gases are smelly and toxic and very corrosive to the materials used to construct the power plant. For these reasons, these gases are collected and disposed of by reinjecting them back underground.

20.4 Sustainable engineering

Sustainable engineering is the process of designing devices and systems to use energy and resources at a rate that does not compromise the natural environment (i.e., sustainably). Every engineering discipline is engaged in sustainable design by using initiatives such as life cycle analysis (LCA), pollution prevention, and waste reduction.

In the 21st century, engineers need to be aware of potential short and long-term environmental impacts. Effects beyond the short-term that may not manifest themselves for decades are the main focus of sustainable design. For example, some materials previously used by engineers that were thought to be harmless are now known to be hazardous, such as asbestos in floor tiles, pipe insulation and shingles; lead in paint and pipes; and wall materials that could host toxic molds. These materials illustrate short-term design decisions that became long-term human health hazards.

Some design impacts may not occur until many years in the future. For example, the use of nuclear power to generate electricity may well be a sustainable design issue since the radioactive waste material will take many years to decay and become harmless to the environment. How can we design these systems so that they are environmentally safe for all time?

Sustainable engineering often begins with the choice of materials used in a product. For example, exhaust gases from the combustion of traditional hydrocarbon fuels are the major contributors to environmental problems such as global warming and air pollution. The use of renewable fuels is a promising option to reduce the environmental impact of fuel combustion. For example, ethanol-gasoline blends contain 10%–15% of the renewable fuel ethanol that are commonly in automotive fuel today.

The process of sustainable design involves

a) conducting an LCA,
b) prioritizing the most important problems found, and
c) finding the technologies and operations to address those problems.

Problems will vary by size, difficulty in treating, and feasibility. The most difficult problems are those that are small but expensive and difficult to solve.

20.4.1 Life cycle analysis

An LCA is a method of predicting the impact of a product, process, or activity, on the environment. It analyzes the effects of materials used, manufacturing methods, transportation and distribution systems, customer use, product maintenance, and final disposal. An LCA produces a complete cradle-to-grave history of a product as shown in Fig. 20.13.

FIGURE 20.13

Product life cycle.

Example 20.8

Describe the life cycle of your family car using words that describe the events involved.

    Solution:

1. Raw and recycled materials enter a factory at one end and our family car is driven out of the other end.
2. It is then shipped to a car dealer who sells it to my family.
3. We purchase the car and drive it to work and on family vacations.
4. It consumes fuel and pollutes the atmosphere, and it occasionally needs repairing.
5. After many years it breaks down and cannot be fixed anymore.
6. Our car is then sent to a salvage yard and any useful parts are stripped off.
7. The rest of our car is crushed and then recycled to manufacture a new car or other products.

The rapid adoption of plastics in many applications in recent years is due to their reliability, ease of manufacture, light weight, and affordability. However, with heightened awareness about energy and environmental concerns, plastic made from natural gas and oil are perceived by the public and decision makers as having negative impacts on sustainability.

An LCA of plastic packaging was carried out to evaluate the potential environmental effects of these materials. The study analyzed the energy requirements and greenhouse gas emissions of the following six general categories of plastic packaging: (1) caps and closures, (2) flexible beverage containers, (3) rigid containers, (4) shopping bags, (5) shrink wrap, and (6) other flexible plastic packaging. The study concluded that replacing all plastic packaging with nonplastic alternatives for these six types of packaging in 1 year in the United States would

- require 4.5 times as much packaging material by weight, increasing the amount of packaging used in the United States by nearly 55 million tons
- increase energy use by 80% (equivalent to the energy from 91 oil supertankers)
- result in 130% more global warming potential (equivalent to adding 15.7 million more cars)

However, there is an ongoing debate about whether paper or plastic is the better choice for the environment. Here is a look at plastic bag use and factors to consider when making the decision.

- Plastic bags were first introduced in 1977 and now account for four out of every five bags handed out at grocery stores.
- Americans make 2.3 trips to the grocery store each week. If people use 5–10 bags each time, that's between 600 and 1200 bags per person per year.
- Stores usually pay less than a penny per plastic shopping bag, and 3–4 cents per paper shopping bag.
- Plastic grocery bags consume 40% less energy to produce and generate 80% less solid waste than paper bags.
- Plastic bags can take 5–10 years to decompose. Paper bags take about a month to decompose.
- Paper bags are made from trees, which are a renewable resource. Most plastic bags are made from polyethylene, which is made from petroleum, a nonrenewable resource.
- 2000 plastic bags weigh 30 pounds, 2000 paper bags weigh 280 pounds.
- A packed standard-sized paper bag can hold up to 14 items but an average plastic bag often holds 5–10 items.
- When compared to grocery bags made with 30% recycled paper, polyethylene bags use less energy for manufacturing, less oil, and less potable water. In addition, polyethylene plastic grocery bags emit fewer global warming gases, less acid rain emissions, and less solid wastes.

20.4.2 Recycling

Recycling is the process of collecting and processing materials that would otherwise be thrown away as trash and turning them into new products. The benefits of recycling include the following:

- reducing the amount of waste sent to landfills and incinerators
- conserving natural resources such as timber, water, and minerals
- preventing pollution by reducing the need to use new raw materials
- reducing greenhouse gas emissions that contribute to global climate change
- creating new jobs in the recycling and manufacturing industries
- helping to sustain the environment for future generations

Recycling is an important part of developing sustainable products. As recycling becomes more socially acceptable, it is important to consider the potential of all the products we use. Even the humble newspaper can be as a valuable resource after its useful life. Of the stuff we threw away in the United States in 2012, about 54% was discarded in landfills, 12% was incinerated for energy, and 34% was recycled or composted.

Recycling materials such as steel, copper, brass, and other metals reduces pollution caused by the excavation and refining of raw materials, and it saves energy and money. It also means that valuable and irreplaceable materials will last longer. For example, it is estimated that in 40 years the world reserve of copper ore will be nearly depleted. Since copper is a vital component in electrical devices, using recycled discarded copper will extend the time needed for engineers to find alternative materials to replace copper.

The challenge of developing a sustainable future depends heavily on the ability of engineers to reduce cost, energy, materials, and natural resources in a structured and environmentally safe manner. In the conventional design process, engineers use functional analysis, brainstorming, design matrices, experiments, computer simulations, risk analysis, cost estimation, and customer feedback. Design constraints such as budget limits, time schedule, federal and state regulations, building codes, existing patents, and public policies are also included. Sustainable design differs from conventional design in that sustainability issues must be considered in all the design steps. This can be done through "life cycle" thinking that introduces compromises between environmental, economic, social, and technical decisions. These design decisions involve the four life cycle stages: material extraction and use, production and transportation, customer use, and product disposal.

Summary

Most of what is called **Green Energy** is direct and indirect manifestations of solar energy: **photovoltaic power**, **thermal power**, **solar heating**, **wind power**, and **hydropower**. Unfortunately, solar power in most forms is limited by the local conditions of clouds and seasonal variations in the sun's output. Wind power is likely the most important source of green power and is probably the best Green energy source.

Photovoltaic power may be from a free source, but it carries a lot of auxiliary systems. Its efficiency is reduced by the effect of its **band gap** on the useable fraction of the solar spectrum. Direct thermal power for electricity or for heat is only making small inroads into commercial use. Wind power is subject to **Betz limit** that reduces the possible electric power output. Hydropower is limited only by finding suitable real estate for collection and containment of the required water. All of this plus LCA is necessary for a sustainable future for the people on earth.

Exercises

1. Which value in the table that follows is the best approximation to the efficiency of a diamond-based photovoltaic: 10%, 55%, 75%, or 96%? (a) Why? (b) Why would diamond be an interesting semiconductor?

Material	Diamond	Semiconductor X	Semiconductor Y	Semiconductor Z
Band gap, eV	5.5	1.0	2.0	3.0
Efficiency				

2. How long does it take to heat a 1.00 m. deep swimming pool by 10.0°C if the sun is directly shining over it and there are no heat losses? Assume the water is well mixed. Assume the sunlight totals 1.30 kW/m². The heat capacity of water is $C = 4200.$ J/kg/°C, and its density is 1000. kg/m³. Ignore any heat losses from the pool. (**Ans.** 9.0 h)

3. Heat losses from a hot body at T_h in a cold environment at T_c generally follow a rate law in which the losses are proportional to the temperature difference between them, $(T_h - T_c)$. What is the steady state temperature of a circulating liquid in a roof-mounted solar heater that receives sunlight at a rate 1.3 kW/m² and loses heat according to 110. × $(T_h - T_c)$ W/m²°C? Assume the outside temperature is 15.0°C.

4. An industrial thermal plant uses a 15:1 concentrator to catch the Sun's rays. The heat losses from the heated fluid obey the relationship 110. × $(T_h - T_c)$ W/m²°C. What is the daytime temperature of the circulating fluid if the outside temperature is $T_c = 15.0°C$?

5. The photograph of a modern windmill, Fig. 20.9, shows a common three-bladed design. The speed of rotation of these blades is determined from a combination of the wind speed, the size and shape of the blades, the variable blade pitch relative to the wind, and the resistive load imposed by their coupling to the electrical generator. Assume the rotational speed for a 100. m. diameter windmill is 30.0 RPM. What is the tip speed of a blade? Can you also suggest two (or more) limitations on the maximum blade speed?

6. A large manufacturing company is selling a 1025. MW windmill array. In the location chosen, you expect a 10.0 m/s steady wind. If each windmill has a Betz limit of 0.46, how many windmills do you need if each has a diameter of 100. m? The air density is 1.050 kg/m³.

7. What is the cost for each windmill and its supporting infrastructure in the preceding exercise, given an installed cost of $1200. Per kW?

8. You work for an electric power utility and receive several bids for a 1025 MW wind farm, including one particular bid that claims to be able to deliver the specified power with only 400 windmills each with 100. m diameter blades. Your job, as chief engineer for the electric

power utility, is to decide if this bid will in fact deliver the contracted amount of power, 1025 MW. Assume the wind speed is 10.0 m/s and the air density is 1.05 kg/m^3.

9. A dam holds back a lake that is 50. km long and fills a V-shaped channel that is 200. m. deep by 100. m. across. How long does it take to empty in a severe time of drought if the outflow is steady at 10.0×10^3 m^3/s?

10. You want to build a dam on a large river and extract 5000. MW from it. You can take 10.0×10^3 m^3/s of water through the dam's spillways. The turbines you intend to use are 90.0% efficient. Most of the expense goes into the construction costs of the dam, so you would like to locate it where the backup lake will be as shallow as possible. What is this depth?

11. The motors and generators in some hydropower systems are reversible. That is, the water turbine can act as a pump and the generator as an electric motor to drive the pump. A power management scheme assumes that you have excess power some of the time and a deficiency at other times. Suppose you have a small dam and a 10.0 MW generator. During the demand period of 6.00 h, you can produce 10.0 MW, which essentially empties all the water from the impoundment behind the dam. During the night, 12.0 MW of excess power is available from the power grid. How many hours does it take to refill the dam if the pumping and the generator steps are both 100% efficient?

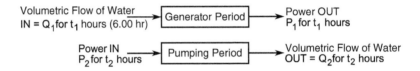

12. In northern France, the Rance tidal power plant captures the inflow from the English Channel behind a 750 m. long dam. The high tide average is 11. m. over a year. If the power produced peaks at 240 MW and its capacity factor is 0.40, what is the average energy produced? Assume the turbine is 90.0% efficient. (The *capacity factor* is the actual energy produced divided by the maximum energy if running at 100% rating.)

13. Some British engineers had an idea for a "slightly" tidal electric power concept. The Menai Strait in North Wales is located between the mainland and the small island of Anglesey. The channel between the island and mainland has a strong tide with a 5.0 foot average rise. The British considered damming it at its narrowest point (about 400. m. across). The dam was to hold large amounts of an ion exchange resin that would selectively leach natural uranium from the seawater, where it is present at a level of about 3.0 ppb (parts per billion). How much seawater in m^3 must flow through the ion exchange resin to make 1000. kg of natural uranium?

14. Write out six or more steps you think are in the life cycle of a newspaper. Start with trees and end with recycling and landfills.

15. In their book, *Time to Eat the Dog: The Real Guide to Sustainable Living*, the authors Brenda and Robert Vale compare the ecological footprints of pets with those of various lifestyle choices, like owning an SUV. For example, a medium size dog eats about 164 kilograms of meat and 95 kilograms of cereal in 1 year. It takes 43.3 m^2 of land to generate 1 kilogram of chicken and 13.4 m^2 to generate a kilogram of cereal per year. So that gives the dog an eco-footprint of about 8400 m^2 (for a big dog like a German shepherd, the figure is 11,000 m^2). An average SUV driven 6200 miles a year (about 10,000 km), uses 55.1 gigajoules (this includes the energy needed to both build and fuel it). 10,000 m^2 of land can produce approximately 135 gigajoules of energy per year, so the SUV's eco-footprint is about 4100 m^2—less than half that of a medium-sized dog. As well as consuming natural resources, cats and dogs devastate

wildlife populations, spread disease and add to pollution. Is there any good reason to own a dog?

16. You are the chief civil engineer for the county. In your professional capacity, you learn that two of the county's senior politicians each own a piece of land that is ideal to complete a new development. But only one piece of land is needed for the project and the two politicians are vying to sell their land to the development. Understandably perhaps, neither politician has revealed his personal interest in the development to the public. However, you know that neither came by his land recently nor dishonestly and both are asking the same price for their land. There is no difference to the development's costs whichever plan is adopted. None of this has been made public but a local newspaper reporter has filed court papers under the Freedom of Information Act (FOIA) for all of the county's information about the development. Both politicians ask you to redact data concerning their involvement in the development because it only looks like they are being dishonest while they are, in fact, not. (Use the Engineering Ethics Matrix and the NSPE Code of Ethics for Engineers.)
    a) Should you comply with the FOIA request "implicating" the two politicians?
    b) Should you comply with the FOIA request but leave out the part played by the two politicians, since neither is being dishonest?
    c) Should you approach either politician and suggest one or both should himself leak it to the local press?
    d) Should you take the initiative and reveal the land ownership in the regular monthly county meeting and explain the circumstance of the land deal?

17. You sold a house that you claimed is "certifiably" passive. You designed it with lots of windows, so that it is warm enough in winter due to solar gain through them. Unfortunately, you suddenly realize that, with so many windows, the interior gets too hot in summer, especially on the south side; indeed, the cooling energy costs for summer could even exceed heating energy costs for winter.

    You already signed a contract with the house buyer and may lose money if you have to make too many modifications. But you guaranteed only the winter's heating costs. You have several options. Which should you do? (Use the Engineering Ethics Matrix and the NSPE Code of Ethics for Engineers.)
    a) Proceed with the original design.
    b) Install smaller windows to reduce both the summer's and winter's solar gain, thus reducing the overall energy costs.
    c) Use higher quality windows on the southern exposures to limit summer heating and standard ones on the northern exposure.
    d) Add a summer thermal storage system to augment the winter's thermal storage system.
    e) Go back to the customer and ask to reopen the negotiation, explaining you cannot build the house for what was previously agreed to.

Mechatronics and physical computing* 21

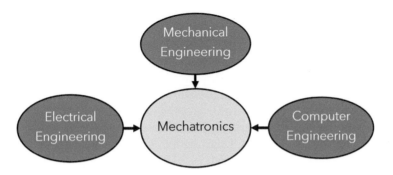

21.1 Introduction

What is Mechatronics? The word *mechatronics* is composed of *mecha* from mechanics and *tronics* from electronics. Mechatronics is a multidisciplinary field of study concerned with the design, analysis, and control of systems that combine mechanical elements with electronics components, including computers and microcontrollers. Some of the applications of mechatronics are show in Table 21.1.

21.2 Physical computing

Physical computing is an interactive physical system that can sense and respond to the world around it using software and hardware.[1] A physical computer senses its

---

* This chapter was written by Dr. David Hodgson, Associate Professor of Mechanical Engineering, Union College, Schenectady, New York.
[1] The term "physical computing" was introduced by Professor Dan O'Sullivan in 1991 to describe extending the physical inputs and outputs of the then-increasingly popular personal computer.

Exploring Engineering. https://doi.org/10.1016/B978-0-443-13541-5.00033-7

481

Table 21.1 Mechatronic examples.

Mechatronic examples	Typical applications
Household	Smart appliances (ovens, toasters, dish washers, clothes washer-dryer), air conditioning, heating, home security, etc.
Medical	Vision-guided surgery, prosthetic limbs, brain–computer interface for prosthesis, wheelchair control, etc.
Automotive	Autonomous vehicles, antilock brakes, active suspension, cruise control, air bags, engine control unit, intelligent highways, etc.
Entertainment	Amusement parks, high tech toys, theater production, movies, television, cameras (automatic focusing), CD players, etc.
Manufacturing	Industrial robots, machine tools, production line automation, product measurement devices, 3D printing, etc.
Military	Drones, underwater autonomous vehicles, smart weapons and munitions, exoskeletons, heat-seeking missiles, etc.

environment, processes that information, and then performs some action. It is a set of tools and practices that enable engineers to create electro-mechanical devices that can sense aspects of the physical world and effect changes to it. It uses sensors to detect a physical input, microcontrollers to reformat and process that input, and actuators to produce changes in response to that input.

Since a computer is central to physical computing, we have chosen to illustrate this chapter with the computational platform offered by Arduino, an open-source hardware and software company that designs and manufactures single-board microcontrollers and microcontroller kits for building digital devices and interactive systems (see: www.arduino.cc).

21.2.1 Actuators

An actuator is a device that converts a control signal (usually electronic) into mechanical motion. They have two basic motions, linear and rotary. Linear actuators provide straight line motion through push and pull functions. Rotary actuators provide turning or rotary motion, for example, to control the flow of fluid through valves. Table 21.2 describes the functions of several types of actuators.

21.2.2 Sensors

We live in a world of sensors. A sensor is any device that converts an external physical signal into a signal that can be interpreted by a computer. There are sensors in our homes, schools, offices, cars, and so forth that make our lives easier by turning on/off lights, detecting our presence, adjusting room temperature, detect smoke, open garage doors, and many other tasks. Table 21.3 illustrates the wide variety of sensors available today. All these sensors are used for detecting or measuring a physical property like temperature, pressure, position, and so forth.

Table 21.2 Types of actuators.

Type	Actuator function
Hydraulic	A hydraulic cylinder or hydraulic motor is used to produce mechanical motion. The motion can be either straight, rotating, or oscillating. The fact that liquids do not compress well makes a hydraulic actuator very powerful.
Pneumatic	A pneumatic cylinder or motor uses pressurized air to produce mechanical motion. They can make large linear or rotating motion with a small amount of air pressure. These actuators respond quickly and are good for stopping and starting motion.
Electrical	The most common kind of actuator is electrical that converts electrical energy into mechanical motion. AC, DC, or stepper rotating, oscillating or linear motors are common electrical actuators. Electrical solenoids are used to create push–pull motion.
Thermal	Another name for thermal actuators is "shape memory alloys" because these types of actuators use high power energy for commercial applications. Advantages of thermal actuators are that they are cost-effective, small, and lightweight.
Mechanical	Using pulleys, gears, rails, and chains, a mechanical actuator converts the rotating motion of physical objects into linear motion. One common example is rack and pinion steering systems used in automobiles.

Table 21.3 Types of sensors.

Temperature	Light	Infrared	Color	Fluid level	Touch
Pressure	Acceleration	Smoke	Humidity	Tilt	Motion
Proximity	Ultrasonic	Gas	Flow	Altitude	Distance

21.2.2.1 Temperature sensor

One of the most common sensors is a temperature sensor. In a temperature sensor, a change in temperature corresponds to change in a physical property like resistance or voltage. This change can then be correlated with the actual temperature. There are different types of temperature sensors like thermistors, thermocouples, resistive temperature devices, and so forth. Temperature sensors are used everywhere in computers, mobile phones, automobiles, air conditioning systems, and throughout industry.

21.2.2.2 Proximity sensors

A proximity sensor is a noncontact type sensor that detects the presence of an object. Proximity sensors can be implemented using different techniques such as optical, ultrasonic, Hall effect,[2] capacitive, etc. Some of the applications of proximity sensors are mobile phones, car parking sensors, ground proximity in aircraft, and so forth.

---

[2] When an electrical conductor is placed in a magnetic field perpendicular to this current, then a voltage (the Hall effect) will be generated across the conductor at right angles to both the current and the magnetic field.

21.2.2.3 Infrared sensors[3]

There are two types of infrared (IR) sensors, transmissive and reflective. In transmissive IR sensors, the transmitter and the detector are positioned facing each other so that when an object passes between them, the IR light beam is broken and the sensor detects the object. In reflective IR sensors, the transmitter and the detector are positioned beside each other facing the object. When an object comes in front of the transmitter, it reflects IR light to the detector to indicate the object is present. Different applications where IR sensors are implemented include mobile phones, robots, industrial assembly lines, and automobiles.

21.2.2.4 Ultrasonic sensor

An ultrasonic sensor is a noncontact type device that can be used to measure distance as well as velocity of an object. An ultrasonic sensor uses sound waves with frequencies greater than that of human hearing. Using the time of flight of a sound wave, an ultrasonic sensor can measure the distance to and velocity of an object.

21.3 Microcontrollers and Arduino

A microcontroller is a tiny microcomputer on a single integrated circuit. It contains a central processing unit (CPU), random access memory, special function registers, program and data read only memory, several input/output ports, and functions such as analog-to-digital converters (ADCs), digital-to-analog converters (DACs), timers, voltage reference, pulse width modulation (PWM), serial peripheral interface, universal serial bus (USB) port, Ethernet port, along with a host of other peripherals.

Microcontrollers usually must have low-power requirements since many devices they control are battery-operated. Microcontrollers are used in many consumer electronics, car engines, computer peripherals, test or measurement equipment, and are well suited for long-lasting battery applications. Microcontrollers are designed for embedded applications, in contrast to the microprocessors used in personal computers.

In an **8-bit**[4] microcontroller, the arithmetic logic unit performs the arithmetic and logic operations. An 8-bit integer can hold a value up to 255. Examples of 8-bit microcontrollers include Intel 8031/8051, PIC1x, and Motorola MC68HC11 families.

A **16-bit** microcontroller performs greater precision and performance compared to an 8-bit microcontroller. A 16-bit integer can hold a value up to 65,535 and it can

---

[3] Infrared waves occur at frequencies above those of microwaves and just below those of red visible light, hence the name "infrared."

[4] Bit is short for binary digit. A **bit** has a single binary value, either 0 or 1. 8-bit, 16-bit, 32-bit, and 64-bit all refer to a processor's word size. A "word" is the size of information it can place into a register and process.

automatically operate on two 16-bit numbers. Some examples of 16-bit microcontrollers are 16-bit Intel 8096 and Motorola MC68HC12 families.

The **32, 64, and 128-bit** microcontrollers use much larger values to perform the arithmetic and logic operations. These are used in automatically controlled devices including implantable medical devices, engine control systems, office machines, appliances and other types of embedded systems. Some examples are Intel/Atmel 251 family and PIC3x.

21.3.1 What is Arduino?

Small circuit boards that incorporate a microcontroller and the circuitry needed to easily supply power to and communicate with the microcontroller are called **microcontroller evaluation boards**. These evaluation boards are commonly referred to as microcontrollers even though they include more components than just the microcontroller. There are many different evaluation boards to choose from. Because of their popularity and the large amount of high quality online resources, we use Arduino boards (specifically the Uno) to demonstrate the application of microcontrollers in this chapter. The example programs and the implementation details presented are specific to the Arduino Uno; however, the concepts apply to many microcontroller-based boards available today.

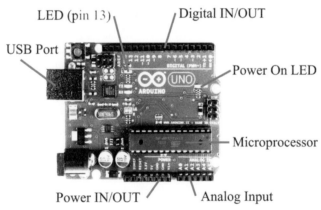

FIGURE 21.1

The Arduino Uno circuit board.

Arduino is an open-source programmable circuit board that can be integrated into a wide variety of projects. There are several different kinds of Arduino boards. The Arduino Uno board shown in Fig. 21.1 contains a programmable microcontroller that can respond to sensors and a large array of outputs actuators such as LEDs, motors and displays. The board contains a USB port to connect the board to a computer for programming, a power jack for a power supply or a battery, 14 pins numbered 0—13 that can be used as digital inputs or outputs, six analog input pins numbered A0-A5, and a reset button.

Arduino also refers to the programming language and integrated development environment (IDE) that is used to write, compile, and upload programs to an Arduino board. Microcontrollers were originally programmed only in assembly language, but various high-level programming languages, such as C, Python and JavaScript, are now in common use. The Arduino programming language is based on C/C++. Arduino programs are often referred to as *sketches*.

There are numerous Arduino sites on the Internet that provide introductory information and details for programming.

21.3.2 The basic structure of an Arduino program (called a "sketch")

The Arduino language is **CASE SENSITIVE**: a capital letter is not the same as a lower-case letter. For example, every Arduino program must contain a "setup()" and a "loop()" function, but it will not recognize the terms "Setup()" and "Loop()" or any other misspelled erroneously capitalized term (e.g., if you initially write code in a word processor it will capitalize any word it thinks begins a sentence).

The keyword **void** indicates that the functions that follow do not return a value. The code that is placed in the **setup()** function will be executed once when the program is initiated and the code placed in **loop()** will be executed over and over again.

The braces, { and }, indicate the beginning and ending of the block of code that is included in each function. Two forward slashes, //, indicate that the following text is a comment statement to be ignored when the program is compiled. Comments are used to explain how the program works and having well-commented code is extremely important. Often code needs to be changed either to add features or fix bugs. Raw code can often be very hard for a person to understand, even for the person that wrote it. Without good comments even relatively simple programs are hard to maintain. To clearly identify complete Arduino program code, it is always put in a textbox in this textbook. Fig. 21.2 shows the format of these functions.

```
void setup() {
    // Put your setup code here to run once.
}
void loop()
{
    // Put your main code here to run repeatedly.
}
```

FIGURE 21.2

A bare minimum Arduino program.

Example 21.1

Create a code that will cause the LED on the Arduino board to continuously blink ON for one second, OFF for one second. The Arduino board built-in LED is on digital pin 13.

> **Need:** The Arduino code to cause the LED on the Arduino board to continuously blink on and off with a one second interval.
>
> **Know—How:** Use the Arduino commands (e.g., www.arduino.cc/reference/en/)
>
> **Solve:** First, use the LED's digital pin number as an integer with the **int**, command as
>
> ```
> int ledPin = 13; // Note: Each command must end with a semicolon.
> ```
>
> Second, define the output pin connected to the LED with **pinMode()** using **void setup()** to start the program as:
>
> ```
> void setup() {
> pinMode(ledPin, OUTPUT);
> }
> ```
>
> Finally, turn the LED on and off with **void loop()** and two calls to **digitalWrite()**, one with HIGH to turn the LED on and one with LOW to turn the LED off. Use **delay(X)** to separate the on-off signal in X milliseconds. For a one second delay ($X = 1000$ ms), this part of the code is as follows:
>
> ```
> void loop() {
> digitalWrite(ledPin, HIGH);
> delay(1000);
> digitalWrite(ledPin, LOW);
> delay(1000);
> }
> ```
>
> Put these three code segments together to create the complete code for this operation.
>
> ```
> int ledPin = 13;
> void setup() {
> pinMode(ledPin, OUTPUT);
> }
> void loop() {
> digitalWrite(ledPin, HIGH);
> delay(1000);
> digitalWrite(ledPin, LOW);
> delay(1000);
> }
> ```

21.3.3 Variables and storing information

All programs need to store information for later use. In Arduino, variables must be declared before there are used. When declaring a variable in Arduino, it must be given a data type. Fig. 21.3 contains code that demonstrates how variables are declared and shows some simple manipulation of the information stored in variables.

The data type indicates what kind of information the variable is going to store. The two most common types of variables for basic programs are *integers* and *floats*. Floats are used to store noninteger numbers. Each variable only has a finite amount of memory that it can use which limits the size of the numbers that can be stored. In Arduino, basic integer variables are declared with the keyword **int**, are represented

```
int x=0;        // x is a global variable and can be used by all blocks.

const int y=3; // y is also a global variable, but its value cannot be changed by the program.

void setup() {

 int a;         // a is a local variable and only exists within the setup () block.

 x=2;           // Since x is a global variable its value can be changed here.

 a=x+3;         // a now equals 5.

}

void loop () {

 float b;       // b is a local floating-point number.

 b=y/x;         // b is now equal to 1.5.

}
```

FIGURE 21.3

Examples of data types and variable manipulation.

by 16 bits, and must be in the range $-32,678$ to $32,767$. Basic floating-point numbers are declared with the keyword **float**, are represented by 32 bits, and must be in the range from -3.4×10^{38} to $+3.4 \times 10^{38}$. Float variables have a precision of 6 decimal places.

Variables only exist within the block in which they are declared. This means that variables declared in **setup()** cannot be used in **loop()**, and vice versa. The exception to this is that variables declared at the top of the program are global variables and they can be used by all blocks within the program. The range of the program for which the variable exists is called the **scope** of the variable.

When a variable is declared, its value can be initialized but this is not required. In addition to setting the data type, other attributes can be assigned when declaring a variable. One common attribute is **const** which indicated that the variable is a constant and it cannot be changed later in the program. Constant variables are used to store parameters so they cannot be inadvertently changed.

21.4 Basic communication for microcontrollers

One of the main functions of microcontrollers is to interact with the outside world. In this section, we provide some examples of the different ways microcontrollers interact using the Arduino Uno.

21.4.1 Digital input—output

The most basic way microcontrollers interact is by sending and receiving single bits of information. As discussed in Chapter 10, a bit represents a piece of binary information. A bit is either HIGH/ON/TRUE/1 or LOW/OFF/FALSE/0. Microcontrollers send and receive bits by setting and sensing voltages on digital input/output pins. On the Arduino Uno, HIGH/ON/TRUE/1 is nominally represented by 5 V and LOW/OFF/FALSE/0 is nominally represented by 0 V. Since no device can produce a precise voltage, the voltage range of 2.7−5 V is considered HIGH/ON/TRUE/1 and a voltage range of 0−1.2 V is considered LOW/OFF/FALSE/0.

The 14 digital pins (pins 0−13) and the six analog input pins (A0−A5) on an Arduino Uno can be used for digital input and output. The program must configure the pin as either in input or an output before it can be used. The following functions are used for digital input and output:

- **pinMode(pin, mode)** where "pin" is the number of the pin to be configured and mode is the desired configuration. The pin "mode" can take one of three values INPUT, OUTPUT, and INPUT_PULLUP (see Section 21.5.1 for details about the INPUT_PULLUP mode)
- **digitalRead(pin)** where "pin" is the number of the pin to read. If the voltage on the pin is between 2.7 and 5 V, the digitalRead() function will return a 1; if the voltage on the pin is between 0 and 1.2 V, the digitalRead() function will return a 0. For voltages between 1.2 and 2.7 V, the digitalRead() function could return either a 0 or a 1.
- **digitalWrite(pin,value)** where "pin" is the number of the pin to be set and "value" is the value to set (either LOW/0 or 1/HIGH). When the pin is set to LOW, its voltage will be about 0 V and when the pin is set to HIGH, its voltage will be about 5 V.

The program in Example 21.1 shows an example of digital reading and writing.

21.4.2 Serial communication

Microcontrollers often include hardware that allows them to communicate large amounts of data more efficiently than by using simple digital input and digital output commands. Serial communication involves sending sequences of precisely timed bits. There are several commonly used serial communication protocols, and each of these protocols has very specific parameters that must be followed for there to be successful transmission of information. Fortunately, most microcontroller development environments include libraries that take care of the detailed specifics and make using serial communication easy for the programmer. The microcontroller in an Arduino Uno contains a Universal Asynchronous Receiver/Transmitter (UART). The UART reads and writes streams of bits (pulses of high and low voltage) and it also has a buffer where it can store information that it has not yet sent, or store information that it has received. In this way, the CPU of the microcontroller can execute other actions, while the UART is transmitting and receiving data. On an Arduino Uno the **receive (rx)** and **transmit (tx)** pins of the UART are connected to pins 0 and 1, respectively. The rx and tx pins are also connected to another microchip (an ATMega16U2) that converts the asynchronous serial to a signal that can be used on a USB. In this way, the Arduino board can communicate with a computer connected to the board with a USB cable.

Because pins 0 and 1 are used for serial communication, beginners should not use these pins for any other purpose. Having other devices or circuits connected to pins 0 and 1 may inhibit serial communication and among other things make it impossible for the board to be reprogrammed.

When two devices are using serial communication, they must both be configured to send and receive data at the same rate. The rate at which data bits are transmitted is the **baud rate**. To initiate the UART in an Arduino program, a **Serial.begin(rate)** command must be used where "rate" is the baud rate at which the UART will be configured to operate. A default baud rate of 9600 is often used. Most devices are able to communicate at least this fast so it is a safe value to use; however, higher baud rates can often be used to speed up the transmission of data. After the baud rate is set, information can be sent using the following commands:

- **Serial.print(data)** sends data to the monitor where "data" can be the name of a variable or a character string enclosed in quotation makes. For example, Serial.print("Hello") prints Hello on the monitor.
- **Serial.println(data)** sends the monitor a line feed before it prints the data.

21.4.3 The serial monitor

The Arduino IDE includes a serial monitor that can be used to communicate with a connected Arduino board. The monitor can be started by clicking on the magnifying glass icon in the upper right-hand corner of the Arduino IDE. Once the serial monitor is launched, the baud rate should be set to match the baud rate of the UART on the Arduino. The monitor will display the information that is sent through the serial port. Fig. 21.4 shows an example program and Fig. 21.5 shows the output it would generate on the serial monitor.

```
int x=0;
void setup () {
   Serial.begin(9600);  // Sets the baud rate to 9600.
   Serial.printin("In Setup");   // Prints the title "In Setup" followed by a line feed.
}
void loop () {
   Serial.print("x=");      // Sends the string inside the quotes (x=).
   Serial.println(x);       // Send the value of x and a line feed.
   x=x+1;                   // Adds 1 to x.
}
```

FIGURE 21.4

Example Arduino program.

```
In Setup
x=1
x=2
x=3
x=4
x=5
....
```

FIGURE 21.5

The output generated by the example program in Fig. 21.4.

21.4.4 Analog-to-digital and digital-to-analog conversion

Sometimes it is necessary to measure the voltage at pin rather than just classify it is as high or low. To do this, many microcontrollers include an ADC. ADCs vary in their input voltage range and precision. The ADC on an Arduino Uno had a default range of 0–5 V and it has a 10-bit precision. A 10-bit ADC converts an analog voltage to $2^{10} = 1024$ district values. The Arduino will convert a voltage that is between 0 and 5 V to an integer between 0 and 1023. The digital conversion is determined by dividing the input voltage by 5 V, then multiplying it by 1024 and truncating the decimals. For example, if the input voltage is 2V, the ADC will output $2/5 \times 1024 = 409$. On an Arduino Uno, analog readings can only be made on pins A0-A5. The Arduino command to read the voltage at a pin is:

- **analogRead(pin)** where "pin" is the pin number. The function will return a value between 0 and 1023.

Some microcontrollers can output a voltage between 0 and 5 V. This is done with a DAC. The Arduino Uno does not have a DAC, but it can approximate the effect of a DAC by creating a PWM signal. PWM signals turn on and off rapidly. The duty cycle (percent of the time the signal is on) of the pulse can be adjusted, which controls the average power output of the pin. Fig. 21.6 shows PWM signals with different duty cycles. Pins marked with a $\sim$ on an Arduino board can produce a PWM signal. On a Uno board, these are pins 3, 5, 6, 9, 10, and 11. The Uno has an 8-bit PWM resolution, so the duty cycle can be adjusted to $2^8 = 256$ different values. The command to output PWM signals on a pin is:

- **analogWrite(pin,value)** where "pin" is the pin number. The "value" is used to control the duty cycle of the PWM signal; it should be between 0 (always off) and 255 (always on).

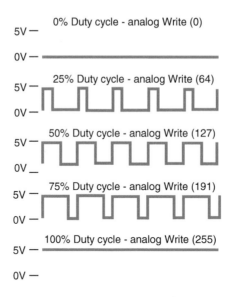

FIGURE 21.6

Pulse width modulation (PWM) signals with various duty cycles and the corresponding Arduino command to generate them.

From http://arduino.cc/en/Tutorial/PWM

Fig. 21.7 shows a simple program that reads the voltage at pin A0 and then sends a PWM signal to pin 3.

```
int Input_Voltage;
int Output_PWM;
void setup () {
  Serial.begin(9600);   // Starting serial communication with a baud rate of 9600.
        // Note that with analogRead and analogWrite the pin mode does not need to be set.
}
void loop () {
  Input_Voltage=analogRead(A0);   // Reads the voltage at pin A0 and converts it to an integer
                                   // between 0 and 1023.
  Output_PWM=Input_Voltage/4;     // Divides the input voltage by 4 converting for 10 bits to 8 bits.
  analogWrite(3, Output_PWM);     // Sends Output_PWM to pin 3.
                                   // Sending values to the serial monitor
  Serial.print ("Input_Voltage ");   // Prints title "Input_Voltage ".
  Serial.print (Input_Voltage);
  Serial.print (",  Output_PWM=");  // Prints more title",  Output_PWM=".
  Serial,print (Output_PWM);
}
```

FIGURE 21.7

A program that reads the voltage at pin A0 and then sends a PWM signal to pin 3.

21.5 Interfacing sensors and actuators

A transducer is a device that converts one type of signal to another. To interface a sensor or actuator to a microcomputer, we need an electronical transducer that has an input or an output that is voltage, current, or resistance.

A sensor is an electronic transducer that converts a physical quantity into an electrical signal. An actuator is an electronic transducer that converts electrical energy into a physical quantity and is an essential element of a control system.

Sensors are used to detect displacement, temperature, strain, force, light, and so forth. Almost all sensors require additional circuits to produce the voltage and current needed for analog-to-digital conversion. A thermistor changes its electrical resistance as a function of temperature, and a circuit is needed to produce a corresponding voltage, but a photodiode produces a current, so a circuit is needed to produce a voltage. Often, the term "sensor" includes both the transducer and the circuits needed to produce an output voltage.

21.5.1 Buttons and switches

Buttons and switches are the most basic devices that can be used as an input for a microcontroller. There are many different types of buttons and switches. The schematic for a normally open momentary contact pushbutton is shown in Fig. 21.8. When the button is pressed, a conductor contacts terminals A and B. When the button is not being pressed, a spring pushes the conductor off the terminals and the circuit is open.

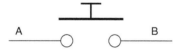

FIGURE 21.8

Schematic of spring loaded normally open momentary contact push button switch.

The state of a button can be read by a microcontroller by making a digital reading of the voltage at one of the terminals of the button. In Fig. 21.9a, terminal A of the button is connected to ground, while terminal B is connected to pin 3 of an Arduino Uno. When the button is pressed, there will be a direct connection between the ground and the pin 3. Making a digital reading of the voltage at pin 3 will result in a LOW/0/FALSE. However, when the button is not pressed pin 3 is not connected to anything. Its voltage is then "floating" and there is no way to know what value will be digitally read from the pin.

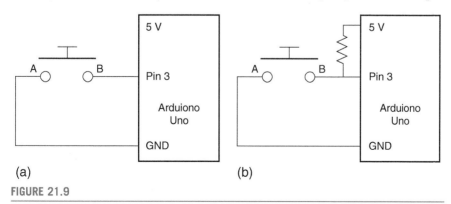

FIGURE 21.9

A pushbutton connected to an Arduino Uno (a) without a pull-up resistor and (b) with a pull-up resistor.

A pull-up resistor is used in Fig. 21.9b to eliminate the possibility of a floating pin. When configured as an input, the pins on a microcontroller have high impedance and little current flows into or out of them. When the button is not being pressed, there will be almost no current flowing through the resistor; therefore, both ends of the resistor will have the same voltage (5 V in this case). A digital reading at pin 3 will result in a HIGH/1/TRUE when the button is not pressed. When the button is being pressed, pin 3 will be directly connected to ground and current will flow through the resistor. A digital reading at pin 3 will result in a LOW/0/FALSE when the button is pressed.

Many microcontrollers including the one on the Arduino Uno have built in pull-up resistors that can be connected (or not) through programming. In an Arduino program setting, the mode of a pin to INPUT_PULLUP connects a pullup resistor to the pin. Fig. 21.10 shows a simple Arduino program that includes a pull-up resistor on pin 3. Fig. 21.11 shows sample output from the program.

```
void setup() {
  Serial.begin(9600);            // Staring serial communication.
  pinMode(3,INPUT_PULLUP);       // Internally connects a pullup resistor to pin 3.
}
void loop() {
                    // loop() repeatedly checks to see if the button is being pressed.
  If (digitalRead(3)=LOW)    // Pin 3 will be low when the button is being pressed.
  {
    Serial.println("The button is being pressed.");
  }
  else
  {
    Serial.println("The button is not being pressed.");
  }
}
```

FIGURE 21.10

Program using INPUT_PULLUP to connect a pull-up resistor to pin 3.

The button is not being pressed.

The button is not being pressed.

The button is not being pressed.

The button is being pressed.

The button is being pressed.

The button is being pressed.

The button is not being pressed.

FIGURE 21.11

Sample output of program in Fig. 21.10. The message displayed changes when the button is pressed and released.

21.5.2 Passive sensors

Sensors are broken into two broad categories. Passive sensors require no power input to operate, while active sensors require a power source to operate. Many passive sensors have resistances that change based on the input they are measuring. Most microcontrollers including the Arduino Uno cannot directly measure a resistance. As shown in Fig. 21.12, a simple voltage divider circuit can be constructed to detect changes in resistance. The analog voltage at the input will vary based on the relative magnitude of the fixed and variable resistor.

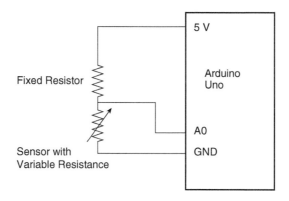

FIGURE 21.12

Example of how a circuit divider can be used to generate a variable voltage signal from a variable resistance sensor. When the resistance of the sensor goes up, the voltage at pin A0 will go up. When the resistance of the sensor goes down, the voltage at pin A0 will go down.

Example 21.2

A photocell has a resistance of 1 kΩ when exposed to light and 10 kΩ when it is in the dark. The photocell is used as the variable resistor in the circuit shown in Fig. 21.12, and a 10 kΩ resistor is used as the fixed resistor. Write an Arduino program that sends "It is light." to the serial monitor when the sensor is exposed to light and sends "It is dark." to the serial monitor when the photocell is not exposed to light.

Need: The Arduino code to display the state of the photocell on the serial monitor.

Know–How: Use the Arduino commands (e.g., www.arduino.cc/reference/en/)

Solution: When the photocell is exposed to light, the voltage at pin A0 will be the following:

$$V_{A0} = 5\ V \frac{1\ k\Omega}{1\ k\Omega + 10\ k\Omega} = 0.45\ V$$

When the photocell is not exposed to light, the voltage at pin A0 will be the following:

$$V_{A0} = 5\ V \frac{10\ k\Omega}{10\ k\Omega + 10\ k\Omega} = 2.5\ V$$

The analogRead() function is used to read the voltage at pin A0. When the photocell is exposed to light, the analogRead() function will return a value of:

$$\frac{0.45\ V}{5\ V} \times 1024 = 93$$

When the photocell is not exposed to light, the analogRead() function will return a value of:

$$\frac{2.5\ V}{5\ V} \times 1024 = 512$$

A value between 93 and 512 can be used as a threshold value. 300 is roughly in the middle of 93 and 512 so that is the value used in the example solution shown in Fig. 21.13.

```
void setup() {
  Serial.begin(9600);              // Starting serial communication.
                                   // pinMode does not need be set for analog writing and reading.
}
void loop() {
  if (analogRead(A0)<300)          // If the voltage at pin A0 is less than 2.93 volts.
  {
    Serial.println("It is light.");   // Prints title "It is light."
  }
  else
  {
    Serial.println("It is dark.");    // Prints title "It is dark."
  }
}
```

FIGURE 21.13

A solution to Example 21.2.

21.5.3 Active sensors

Active sensors require power to operate. This power can come from the microcontroller's power supply, but it can also come from an external power supply. The power is used to amplify and filter the raw signal from the transducer. Active sensors may also be able to communicate with a microcontroller using standard digital communication protocols rather than having an analog output for the microcontroller to measure.

Example 21.3

Proximity sensors of various kinds are quite common. One common type uses IR light. IR distance sensors consist of an IR LED and an IR phototransistor. Light from the LED bounces off of objects in front of the sensor and this light is then detected by the phototransistor. Some IR sensors produce an analog output voltage that varies based on the amount of IR light reflected back to the sensor from objects in front of it. Other IR sensors are designed to produce a digital output. Digital IR sensors output a high voltage (typically 5 V) if the amount of detected reflected light is above a threshold and output a low voltage (close to 0 Volts) otherwise.

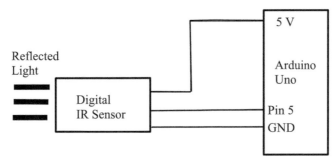

FIGURE 21.14

Circuit for a digital IR sensor.

The circuit shown in Fig. 21.14 shows a digital IR sensor that outputs 5 V when an object is within 10 cm of the front of the sensor and outputs 0 V when the object is further than 10 cm in front of the sensor. Write a program that sends "Object Detected" to the serial monitor when the sensor detects something within the 10 cm range and sends "No Object Detected" to the serial monitor when the sensor does not detect an object.

Need: The Arduino code to display the state of the IR sensor on the serial monitor.

Know: The circuit is wired as shown in Fig. 21.14.

How: Use the Arduino commands (e.g., https://www.arduino.cc/reference/en/)

Solution: See Fig. 21.15.

```
void setup() {
  Serial.begin(9600);    //Starting serial communication
  pinMode(5,INPUT);   //A pull-up resistor should not be used.
                      //The IR sensor will output 5 V or 0 V so a voltage divider is not needed.
}
void loop() {
  if (digitalRead(5)=HIGH)      //The output of the IR sensor is connected to pin 5.
  {
   Serial.println("Object Detected.");
  }
  else
  {
   Serial.println("No Object Detected");
  }
}
```

FIGURE 21.15

A solution for Example 21.3.

21.5.4 More complex output devices

Sometimes LEDs and speaker audible tones cannot convey enough information to the user of a device that contains a microcontroller. Even using seven-segment displays is limited to the number of pins on the microcontroller. Fortunately, many output devices contain built in logic circuits or a dedicated microcontroller that allow for sophisticated communication with an external microcontroller.

Liquid Crystal Displays (LCDs) can display one or more rows of characters. The most basic LCDs have 2 rows of 16 characters and can communicate with a microcontroller using only six pins (if the same amount of information were to be directly displayed on seven segment displays, $2 \times 16 \times 7 = 224$ output pins would be needed). Displays with preconfigured "serial backpacks" are also common; these can be controlled with just two microcontroller pins using standard serial communication protocols.

For speaker audio output, there are many devices available that are designed to simply interface with microcontrollers that will play a tone or a preloaded song.

21.5.5 Controlling higher power actuators

Some actuators require more current or a higher voltage than can be directly supplied by the microcontroller. When controlling such a device, some form of power electronics circuitry is needed between the microcontroller and the actuator.

Transistors provide a way to control a high-powered device with a low power signal. There are many different families of transistors that operate based on different principles. Some examples include Field Effect Transistors and Bipolar Junction Transistors (BJTs). Within each family, there is a wide range of different transistors that have their own characteristics such as maximum voltage and current ratings and switching speed.

Fig. 21.16 shows a schematic of an Arduino Uno controlling the power to an electric resistance heating element using an NPN BJT in what is called a common emitter configuration. BJTs consist of three pins: the **base**, the **emitter**, and the **collector**. The flow of current into the collector of a BJT is controlled by the current that flows into the base of the BJT. When no current flows into the base, no current can flow through the collector. When a small amount of current flows through the base, a large amount of current can flow through the collector. The BJT acts like a switch that is controlled by the current flowing into its base.

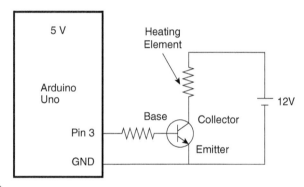

FIGURE 21.16

A BJT controlling a heating element.

In Fig. 21.16, one side of the heating element is connected to 12 V and the other is connected to the collector of the transistor. When the output pin 3 of the Arduino is set to low, no current flows into the base of the transistor so no current can flow into the collector. When there is no current going into the base, the transistor essentially acts as an open switch between the heating element and the ground. When the output pin is set to high, a small amount of current flows from the pin, through the resistor, and into the base of the transistor. This allows for a large amount of current to flow from the collector to the emitter of the transistor. When a small (but sufficient) amount of current is flowing through the base, the transistor essentially acts as a closed switch between the heating element and ground. The power to the heating element can be controlled by having the Arduino Uno output a PWM signal.

It is important to note that for this circuit to work the microcontroller and the power supply for the actuator that is being controlled must share a common ground. But sharing a ground in not ideal since electrical noise generated in the actuator can be transmitted through the ground and affect the operation of the microcontroller.

Mechatronic systems often include direct current (DC) motors. If a given system only requires a DC motor to turn on and off, then a relay may be an appropriate solution for interfacing a microcontroller to the motor. If the speed of a DC motor needs to be controlled, then a transistor may be an appropriate solution. With either of these solutions, a small amount of additional circuitry is often needed since the motor is an inductive rather than load. Any actuator that has a nonnegligible amount of inductance is inductive load. Since motors consist of coils of wire, they have inductance. When inductive loads are switched off, a large voltage can be generated. This can result in arcing between the contacts of a relay or the failure of a transistor. To help reduce the voltage spikes, fly-back diodes are commonly used. Fly back diodes provide a path for current and therefore prevent current from suddenly stopping in an inductive load.

When the speed and direction of a motor needs to be controlled, a single transistor, or a single relay is not enough. Motor drivers, which consist of multiple transistors, diodes, and circuitry on a printed circuit board, are widely available in a wide range of voltage and current ranges. There are many different kinds of motor drivers and each has its own interface, but, in general, a microcontroller can use a motor driver to control the speed and direction of a motor with one or two digital outputs and one or two PWM outputs.

One special type of commonly used DC motors is the radio control (RC) style servo motor. The term **servo** refers to a device that uses some form of feedback to automatically control its output. An RC servomotor typically consists of a DC motor, a motor driver, and potentiometer that is used to measure the angular position of the servo. RC servos have three electrical connections: positive voltage supply (typically between 5 and 9 V), ground, and input signal. The desired angular position of the motor is communicated by sending voltage pulses of varying duration to the input line of the servo. By convention, pulses with a duration of 1 ms will result in the motor moving to its minimum angle and pulses with a duration of 2 ms will result in the motor moving to its maximum angle. Typically, RC servo motors have an angular range of around 180 degrees. Both the angular range and the required pulse durations vary from servo to servo, so it is important to test RC servos before they are incorporated into a system.

The Arduino programming language is distributed with a library, aptly named Servo, that contains the functions needed to control the position of an RC style servo.[5] It is important to note that using Servo library with an Adriano Uno disables PWM on pins 9 and 10. The Servo library uses the internal timers on the microcontroller that are by default used to set up PWM on pins 9 and 10. Fig. 21.17 shows a schematic of a servo connected to an Arduino Uno and Fig. 21.18 shows a program that controls the position of the servo.

---

[5] https://www.arduino.cc/en/Reference/Servo

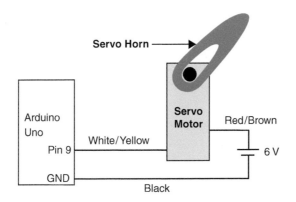

FIGURE 21.17

Schematic of an RC servo motor connected to an Arduino Uno. Wire colors of common RC servos are shown.

```
#include <Servo.h>      // Needed to use the Servo Library.
Servo myservo;          // Creates a servo object (an object is something like a variable).
void setup() {
 myservo.attach(9);  //Pin 9 is where the servo is connected.
}
void loop() {
    myservo.write(60);      // Sets the position of the servo to about 60 degrees.
    delay(1000);            // Waits for one second.
    myservo.write(30);      // Sets the position of the servo to about 30 degrees.
    delay(1000);            // Waits for one second.
}
```

FIGURE 21.18

Program that controls the position of an RC style servo motor.

Summary

As technology advances, the traditional fields of engineering merge and adapt to new challenges.

Mechatronics is a new field of engineering that merges mechanical engineering, electrical engineering, and computer science. It has applications in medicine, industry, military, smart consumer products, and almost every other area of technology. It involves physical computing which is an interactive physical system that can sense and respond to physical circumstances using software and hardware.

In this chapter, the Arduino UNO, an open-source electronics platform that can react to its environment by using sensors and control lights, motors, and other actuators, is discussed. The open-source Arduino Software (IDE) is used to write code which can then be uploaded to the board.

Exercises

1. Define the following abbreviations used in this chapter.

   a) LED b) DAC c) FET d) PWM e) RC f) BJT

2. Some of the following Arduino commands contain errors. Find the errors and write the correct command.
   a) void.loop();
   b) pin.mode();
   c) digital.Write()
   d) Const int y = 3;
   e) Serial.Printi(" ");
   f) Serial, begim(9600);
   g) int Input.Voltage;
   h) analog.Read()
   i) flote(b);

3. The step angle of a digital stepper motor is $\alpha = 3.60$ degrees $= 360.°/n$, where n is the number of integer steps (pulses) needed for one complete revolution of the motor's shaft. The angle through which the motor rotates when given N step pulses is αN and the rotational speed of the motor in rev/min is $\omega = 60f/N$ where f is the number of pulses per second received by the motor.

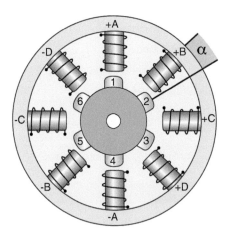

   a) How many pulses are required for the motor to rotate through ten complete revolutions?
   b) What pulse frequency f is required to rotate the motor at 100. rev/min?

4. Alter the code in Example 21.1 so that the LED on the Arduino board blinks on for one second then off for one second three times in a row, and then blinks on for two seconds and off for one second three times in a row. In other words, have the LED send an S–O–S (... – – – ...) in Morse code.

5. What two functions are required in all Arduino programs?
6. What command would you use to tell an Arduino that pin eight will be used as an input?
7. For the circuit and code shown, what is the value of the variable analogIn?

```
int analogIn; void setup() {
        // nothing to set up
}
void loop() {
    analogIn=analogRead(A0);
}
```

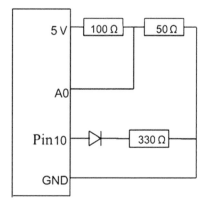

8. Modify the code in Example 21.2 so that the LED connected to pin 13 will turn on when no light is detected by the photo cell and turn off when light is detected by the photocell. Note that code needs to be added to both the setup() and loop() functions.
9. Given the circuit below, write a program that will adjust the brightness of the LED (from fully off to fully on) based on the position of the potentiometer. In this circuit, the voltage of the potentiometer's wiper, which is connected to pin 7, can be varied between 0 and 5 V.

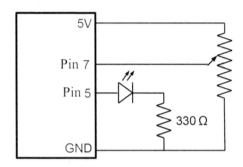

10. Given the circuit below, write a program that will move the servo through its full range of motion (0–90 degrees), based on the position of the potentiometer. In this circuit, the voltage of the potentiometer's wiper, which is connected to pin A0, can be varied between 0 and 5 V.

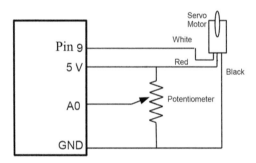

11. In the circuit below, the motor is directly powered by pin 5. Will the motor behave as desired? What could be added to the circuit to make it perform better?

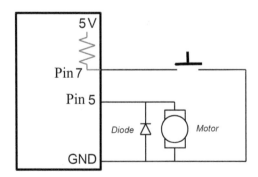

12. Given the code shown below, what will be displayed on the serial monitor when this program runs?

```
int j=2;
void setup() {
      Serial.begin(9600);
      For (int i=0;i<4;i++)
{
      Serial.print("Hi");
      if (j=i)
{
      Serial.println("Bye");
}
}
}

void loop() {
      if (j>10)
{
      j=j+1;
}
      if (j=3){
      Serial.println("Bye Again");
}
}
```

Hands-on

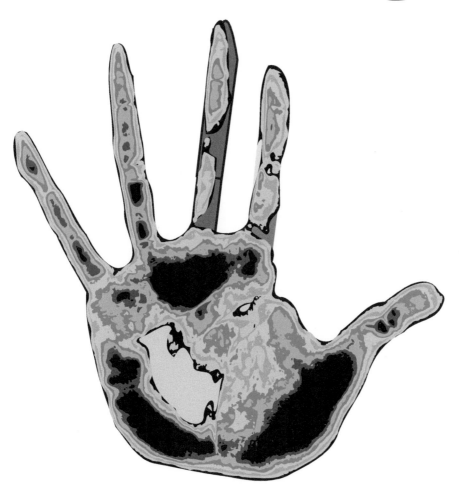

Introduction to engineering design

22

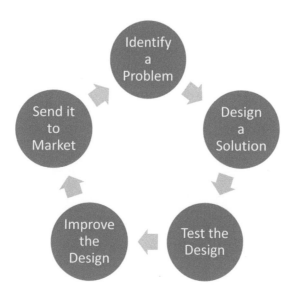

22.1 Introduction

This chapter describes the nature of engineering design, introduces modern design philosophy called **Design for Six sigma (DFSS)**, suggests some benefits of a hands-on design project, indicates the qualities of a good designer, and explains the need for a systematic approach that consists of an eight-step design process.

22.2 Nature of engineering design

In the course of creating new products, engineering design uses available technology to improve performance, lower cost, or reduce risk. For example, if the design of a bridge produces a new structure that is visually stunning with no consideration for its

Exploring Engineering. https://doi.org/10.1016/B978-0-443-13541-5.00005-2

strength, this is design without any engineering. If, on the other hand, the designers of a new concept car use analysis or experiments to evaluate air drag, structural integrity, and manufacturability in addition to style when coming up with a new exterior design, this is engineering design.

Before the 1960s, the preceding description of engineering design might have been adequate, but this is no longer the case. The definition has been broadened to include the systematic thought processes and best practices that define the modern engineering design process. This systematic approach, which has become synonymous with engineering design, forms the heart of these design chapters.

Those of you experienced at tinkering with your own inventions in the basement or garage might say that you got along fine without engineering analysis and without training in the systematic approach. Such a view ignores the realities of engineering design. Usually, designers must search for the best possible design under severe conditions of limited time and limited resources (especially cost). For example, in most design situations, there is barely enough time to produce a single prototype. The idea behind an engineering approach, as embodied in engineering analysis and in the systematic approach, is to minimize the number of design iterations required to achieve a successful final design. Fewer design iterations mean lower costs and shorter development times.

One of the greatest challenges of engineering design is the breadth of knowledge required of the designer. The diversity of topics covered in earlier chapters provides a hint of what a design might entail; some electromechanical designs could conceivably touch on them all. In addition to those topics, there are issues related to manufacturing, economics, aesthetics, ethics, teaming, government regulations, and documentation of the design, to name but a few. In the spirit of these design chapters, we propose that the best way to introduce the multifaceted nature of engineering design is to experience it for yourself through the hands-on design exercise at the end of the chapter.

22.3 Design problems versus homework problems

A design problem is unlike a traditional college homework problem. Homework problems, like those in Parts 1 and 2 ("Lead-On" and "Minds-On," respectively), in this textbook have specific, unique answers; and the student must find those answers using a logical approach such as the need-know-how-solve method. Design problems, on the other hand, do not have unique answers. Several answers (i.e., designs) may satisfy a design problem statement. There may also be a "best" answer based on critical requirements. For example, a design that minimizes manufacturing costs may not produce a very reliable product. If product reliability is a critical requirement of the customer, then minimizing manufacturing costs may not be the "best" design answer.

The challenge of Part 3 of this textbook, which we call "Hands-On," is to transition you from solving well-formulated, single-answer textbook problems to solving what we call *open-ended* engineering design problems. These problems are often complex and sometimes poorly formulated problems. You will find that excellent design work requires considerable creativity beyond that needed to solve textbook problems. In Part 3 of this text, we allow and encourage you to drop your academic inhibitions and explore the joy of engineering creativity.

The problem-solving process we presented in Part 1 of this text (need-know-how-solve) works well for single-answer textbook problems, but it needs to be expanded to encompass open-ended engineering design problems. In recent years, the process of engineering design has been refined to produce more robust and economical designs. The process that most companies embrace today is a multistep process called "Design for Six Sigma."

In Part 3 of this text, we abandon the need-know-how-solve exercise solution method and adopt a systematic eight-step design process.

22.4 Benefits of a hands-on design project

A practicing engineer does not have to be an expert in machining or other basic manufacturing operations. Still, a basic understanding of the challenges involved in manufacturing a product is essential for producing a successful design. The best way to appreciate that fact at an early stage in your career is to manufacture a design yourself.

The lessons to be learned are universal. Do not expect your design to work on the first try. Leave a lot of time for testing. Complicated designs take a lot longer to build and have a lower probability of success, and if you have a choice of manufacturing a part yourself or buying it, buy it. Many such lessons are foretold by the design principles and design for manufacture guidelines of later chapters. The consequences of violating those principles are best understood by experiencing the results of having done so.

Other lessons are to be learned from a hands-on design experience. For electromechanical systems with moving parts, that experience might be the only way to accurately evaluate the design. Also, students gain a sense of accountability by learning that it is not enough for a design to look good on paper. To actually be a good design, it has to lead to an end product that works.

In particular, we recommend that the hands-on design project should be done under a competition format, involving interactions between "machines." It is a natural motivator—the challenge of the task is heightened by having to deal with the unpredictability of your human opponent, and the other designs provide a relative scale against which to assess the quality of performance.

22.5 Qualities of a good designer

These are the qualities of a good designer:

- **Curiosity about how things work.** Seeing other design solutions provides you with a toolbox of ideas that you can draw from when faced with a similar design challenge, so when you come across an unfamiliar device, try to figure out how it works. Take things apart; some companies actually do this and refer to it as *reverse engineering.* Visit a toy store; the products there often demonstrate creative ideas and new technology.
- **Unselfishness.** A key ingredient to effective teaming is suppressing ego and sacrificing personal comfort to serve the best interests of the team.
- **Fearlessness.** It takes a leap of faith to step into the unknown and create something new.
- **Persistence.** Setbacks are inevitable in the course of a design project. Remain resilient and determined in the face of adversity.
- **Adaptability.** Conditions during the design process are constantly evolving. For example, new facts may surface or the rules of the design competition may change in some way. Be prepared to take action in response to those new conditions. In other words, if the ship is sinking, do not go down with the ship—redesign it.

22.6 Using a design notebook

When you are working on a design project and you want to write something down, the design notebook is the place to do it. There is no need for notepads, reams of paper, or sticky notes. The place to record your thoughts is in a permanently bound volume with numbered pages, a cardboard cover, and a label on the front cover identifying its contents. Every college bookstore has them, though they may be called *laboratory notebooks.*

As a starting engineer, now is the best time to start a career-long habit. Just how important the design notebook is can be explained in the case of Dr. Gordon Gould.[1]

On November 9, 1957, a Saturday night just given to Sunday, Gould was unable to sleep. He was 37 years old and a graduate student at Columbia University. For the rest of the ... weekend, without sleep, Gould wrote down descriptions of his idea, sketched its components, projected its future uses.

On Wednesday morning he hustled two blocks to the neighborhood candy store and had the proprietor, a notary, witness and date his notebook. The pages

---

[1] http://inventors.about.com/gi/dynamic/offsite.htm?site=http://www.inc.com/incmagazine/archives/03891051.html

described a way of amplifying light and of using the resulting beam to cut and heat
substances and measure distance ... Gould dubbed the process light amplification
by stimulated emission of radiation, or laser.

It took the next 30 years to win the patents for his ideas because other scientists had filed for a similar invention, although after Gould. Gould eventually won his patents and received many millions in royalties because he had made a witnessed, clear, and contemporaneous record of his invention.

The lesson is that patents and other matters are frequently settled in court for hundreds of millions of dollars by referring to a notebook that clearly details concepts and results of experiments. You must maintain that notebook in a fashion that will expedite your claim to future inventions and patents.

Another more immediate benefit of using a design notebook is that you will know that everything related to the project is in one place. Finding that key scrap of paper in a pile of books and papers on your desk after working for months on the project can be a rather time-consuming endeavor.

Here are some of the most important guidelines for keeping a design notebook.

- Date and number every page.
- *Never* tear out a page.
- Leave no blank pages between used pages. Draw a slash through any such blank pages.
- Include all your data, descriptions, sketches, calculations, notes, and so forth.
- Put an index on the first page.
- Write everything in real time; that is, do not copy over from scraps of paper in the interests of neatness.
- Write in ink.
- Do not use White Out; cross out instead.
- Paste in computer output, charts, graphs, and photographs.
- Write as though you know someone else will read it.
- Document team meetings by recording the date, results of discussions, and assigned tasks.

22.7 The need for a systematic approach

Two main goals of a systematic approach to engineering design are

(1) to eliminate personal bias from the process and
(2) to maximize the amount of thinking and information gathering done up front, before committing to the final design.

These rules result in better design team decisions and fewer costly design changes late in the product development stage.

The engineering design process also provides a blueprint for the design of complex systems. For example, you might be able to get along without a formalized design procedure when designing a new paper clip,[2] but when taking on the daunting task of designing a complex system like the International Space Station, brain gridlock can set in. The design process offers a step-by-step procedure for getting started as well as strategies for breaking down complex problems into smaller manageable parts.

22.8 Steps in the engineering design process

A systematic approach to engineering design that has come to embody the way engineers are supposed to think may be viewed as consisting of eight steps:

1. Define the problem.
2. Generate alternative solutions.
3. Evaluate and select a solution.
4. Detail the design.
5. Defend the design.
6. Manufacture and test the design.
7. Evaluate the performance of the design.
8. Prepare the final design report.

These steps are shown in Fig. 22.1.

The next chapter discusses the importance of effective teaming and how to do it. Subsequent chapters treat each of the preceding eight steps in detail. At the end of each chapter is a suggested milestone for the successful completion of the step in the design process described in the chapter. Milestones are crucial for measuring progress toward the eventual goal of a successful design. After the eight steps have been described, this portion of the book concludes with detailed descriptions of actual student design competitions including representative solutions to the design milestones.

22.9 Hands-on design exercise: "The tower"

Your first design objective is to build the **tallest tower** out of materials to be supplied. Since there are both individual and team winners here, this is both an individual and a team competition.

22.9.1 Setup

- Divide the class into about eight teams of three or more students per team.

---

[2] But see just how difficult it originally was to perfect the paper clip: Henry Petroski, *The Evolution of Useful Things: How Everyday Artifacts—from Forks and Pins to Paper Clips and Zippers—Came to Be as They Are* (New York: Alfred A. Knopf, 1992).

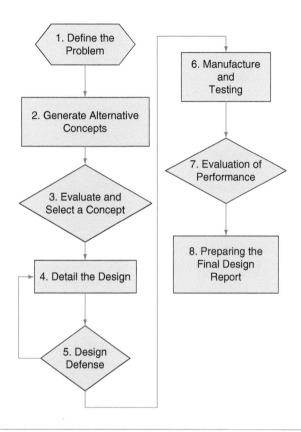

FIGURE 22.1

The design process flowchart.

- Each team should be provided with five sheets of 8.5 × 11 standard copier paper.
- The following material will be distributed **unevenly**:
  - Two teams each receive one roll of Scotch tape.
  - Two teams each receive a roll of duct tape.
  - Two teams each receive one box of paper clips.
  - Two teams each receive one pair of scissors.

22.9.2 Rules

- Each team has just 10 min to build a tower.
- The final tower height measurement must be made by the instructor.
- Teams should indicate to the instructor when they are ready for a measurement.
- If the tower is composed of materials other than the supplied paper, Scotch tape, duct tape, or paper clips, it will be disqualified.
- The tower must be stationary when the measurement is made.

- The tower must be built on a flat surface. The tower cannot lean against or be attached to any other surfaces (wall, table, etc.).
- Any team that intentionally knocks over another team's tower before it has been measured is disqualified.
- After 3 min, **one** individual on each team is offered eight points to join another team.
- There are no other rules.

22.9.3 Scoring

- One point will be awarded for each inch of tower height. (Heights are to be rounded to the nearest inch.)
- The first team that finishes its tower and is ready to be measured receives a bonus of 10 points.
- The "team" winner is the team with the highest point total,
- The "individual" winner is the student with the highest point total.

22.9.4 After the exercise

Discuss the importance of the following issues to the outcome of the competition:

- The "quality of teaming" among different teams: Were any of the extra materials (tape, paper clips, or scissors) shared between teams?
- The "quality of teaming" within the team: Was everyone within the team given the opportunity to contribute ideas or did one person dominate the decision making?
- Ethics: Was it ethical to jump to another team, not share materials, or copy the design of another team?
- Manufacturability: How important was it to have the right materials?

Exercises

1. What is the major difference between the abstract equations used in a mathematics course and the equations used to describe physical objects in design? (**Ans.** The abstract equations used in mathematics do not have units, but those used in design always have units.)
2. Name three additional learning benefits that you feel come from a hands-on design experience.
3. Two best practices for design are to use a design notebook and to team effectively. Which of the following would also make a good best practice for design.
   (a) Work alone.
   (b) Keep the design simple.
   (c) Manufacture all the parts yourself.
   (d) Ignore input from less knowledgeable people.
4. There are eight steps in the design process discussed in this chapter (see Fig. 22.1). Where does the input from the "customer" enter into the process?
5. Who are the "customers" in a design process? (**Ans.** There are many "customers," including those who purchase the final product. Manufacturing is also a customer, as are sales and marketing.)

6. In the design process shown in Fig. 22.1 you are asked to "Generate Alternative Concepts" for the design. Explain why this is a necessary step.
7. In the design process shown in Fig. 22.1, what evaluation criteria would you use in step 3 to evaluate competing ideas for a new consumer product.
8. Define the (a) Steps, (b) Rules, and (c) Scoring for a student design exercise that involves using a single wire coat hanger to make a new functional consumer product.

Design teams

23

Great things in business are never done by one person. They're done by a team.

Steve Jobs, cofounder of Apple Computers.

23.1 Introduction

The design process involves a series of creative steps that lead to a solution to a problem. In engineering, the problem often involves designing a new product that accomplishes a task that meets customer needs. An inventor can sometimes carry out these steps alone, but in large companies, design problems are solved by teams of trained engineers, technicians, marketing personnel, and sometimes lawyers, called **design teams**.

Exploring Engineering. https://doi.org/10.1016/B978-0-443-13541-5.00031-3

While all teams are groups of people, not all groups are teams. Team members must work together toward a common goal, whereas members of a group only share common interests or characteristics. Sometimes it is difficult to tell the difference between a team and a group. For instance, two or more students might meet to discuss homework or watch TV. They may have an agenda and a purpose, but this is a group meeting, not a team meeting. The scope and duration of their meeting is too small to involve the amount of coordination, resources, and effort that teamwork requires.

Engineering students do not always realize that their future success depends on how well they work together in a team. Everyone on a design team comes from a different background. They have different strengths, cultural values, philosophies, and work ethics. Yet they have to join together to form a coherent and productive group. In this chapter, the elements that define a successful design team are discussed. These elements range from project management to team building, team goals, and leadership.

23.2 How to manage a design team project

Design team project management is a carefully planned and organized effort to accomplish a specific project (for example, designing and constructing a robot vehicle to be entered into a competition on a specific date).

Project management includes developing and implementing a plan that defines the project goals, specifies how and by whom the goals will be achieved, identifying needed resources, and developing budgets and timelines. Project management is usually the responsibility of an individual team leader. When the project team members have been identified, the team (or the course instructor) selects one of its members as the project manager. This person has the responsibility for guiding the team design work in a professional, organized, and timely manner. The project manager is also responsible for meeting deadlines and ensuring that the team members are carrying fair workloads. Successful project management involves the following:

- **Understand the project's goals**. You should be able to state the goals of your project in a single sentence.
- **Engage all the team members**. Subdivide the work using functional decomposition to break the project down into individual work assignments, and make sure that everyone knows what he or she is responsible for to meet the project goals.
- **Keep the project moving forward**. Work methodically to meet your benchmarks; do not wait until the last minute and rush to meet a deadline.

23.2.1 Gantt and PERT charts

Design team leaders need to control the timing of all operations, so they develop *schedules* that show specific tasks to be performed during the design process. They assign tasks to work groups, set timetables for the completion of tasks, and make sure that resources will be available when they are needed. There are a number of scheduling techniques, but the most common are **Gantt and PERT charts**.

A **Gantt**[1] **chart** is a horizontal bar chart that provides a graphical illustration of a schedule that helps to plan, coordinate, and track specific tasks in a project. Gantt charts may be simple versions created on graph paper or more complex automated versions created using project management applications such as Microsoft Project or Excel, as in Fig. 23.1.

Gantt chart			Day of Project Work					
Activity	Start day	# of days	1	6	11	16	21	26
Define Problem	1	5	Define Problem					
Brainstorm	5	6		Brainstorm				
Design	10	5			Design			
Construction	14	3				Construction		
Test	16	3				Test		
Improve	19	4					Improve	
Prepare Report	22	4					Prepare Report	

FIGURE 23.1

Gantt chart illustration.

A Gantt chart is constructed with a horizontal axis representing the total time span of the project, broken down into increments (for example, days, weeks, or months) and a vertical axis representing the tasks that make up the project (for example, if the project was designing and constructing a robot vehicle, the major tasks might be: define the problem, brainstorm possible designs, choose a design, construct the design, and so forth).

Horizontal bars of varying lengths represent the time span for each task. Using the same robot vehicle example, you would put "Define Problem" at the top of the vertical axis and draw a bar on the graph that represents the amount of time you expect to spend on defining the problem, and then enter the other tasks below the first one and representative bars at the points in time when you expect them to be undertaken. The bar spans may overlap, as, for example, you may start construction while refining the design during the same time span.

Gantt charts give a clear illustration of project status, but they do not tell how one task falling behind schedule will affect other tasks (however, the PERT chart described below is designed to do this). More complex Gantt charts show more information, such as the individuals assigned to specific tasks and notes about the procedures. Gantt charts should be adjusted frequently to reflect the actual status of project tasks when they diverge from the original plan.

---

[1] Developed as a production control tool in 1917 by Henry L. Gantt, an American engineer and social scientist.

Example 23.1

Tom, Jim, and Sue have a difficult team assignment in their freshman design course. They must design and construct a small robot that will collect Ping-Pong balls scattered on a table and deposit them into a box. Since the project was not well defined regarding the table's size, robot controls, and collection time limit, they decided to take a week (5 days) to define the project with the course instructor.

Since the assignment needed to be completed in 26 days, they created the Gantt chart below to plan the sequence of events. They realized that some of the tasks could overlap. For example, they decided to begin brainstorming solutions 1 day before the problem definition was finished. Construction could also begin a day before the design was completed, and so forth.

They also began to assign responsibilities to team members. Sue was assigned to clarify the problem definition, all three would contribute to the brainstorming effort, Jim and Sue would focus on the design, Tom would be responsible for construction, and they would all participate in the remaining tasks.

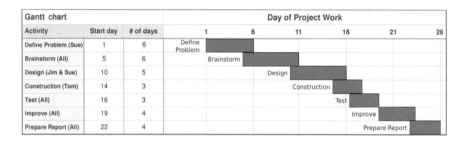

Gantt chart			Day of Project Work					
Activity	Start day	# of days	1	6	11	16	21	26
Define Problem (Sue)	1	6	Define Problem					
Brainstorm (All)	5	6		Brainstorm				
Design (Jim & Sue)	10	5			Design			
Construction (Tom)	14	3				Construction		
Test (All)	16	3				Test		
Improve (All)	19	4					Improve	
Prepare Report (All)	22	4						Prepare Report

This chart was updated throughout the project and the schedule was adjusted as needed to meet the completion deadline. The Gantt chart allowed continuous progress and kept everyone alert to their responsibilities.

Gantt charts are useful when the design and production process is fairly simple. For more complex schedules, PERT charts are used.

A **PERT**[2] **chart** is a project management tool used to schedule, organize, and coordinate tasks within a project. It presents a graphic illustration of a project as a network diagram consisting of numbered *nodes* that represent milestones in the project linked by labeled arrows that represent *tasks* in the project. The direction of the arrows indicates the sequence of tasks. In Fig. 23.2, tasks A and E must be completed in sequence. These are called *dependent* tasks. However, task B does not depend on the completion of task D and thus Tasks D and B can be done simultaneously. These tasks are called *independent* tasks.

---

[2] PERT stands for *Program Evaluation Review Technique*, a methodology developed by the U.S. Navy in the 1950s to manage the Polaris submarine missile program.

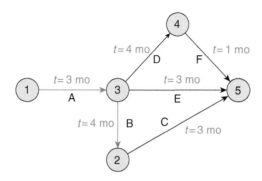

FIGURE 23.2

A PERT chart for a project with five milestones (1 through 5) and six activities (A through F). The critical path is tasks A to B to C that requires a total of 10 months to complete.

The **critical path** is the path that takes the longest amount of time. It is determined by adding the times for the tasks in each sequence and determining which path takes the longest time to complete the project. Thus, the critical path determines the total time required for the project. If you want to produce a product quicker, you have to save time on this path. If you only saved the time on any of the other paths, you still would not produce a product any quicker. You gain efficiency only by improving performance on one or more of the activities along the critical path. The amount of time that tasks along a noncritical path can be delayed without delaying the entire project is called the **slack time**.

Example 23.2

In the PERT chart below, the letters represent tasks and the numbers represent days to complete the task. The label "A,2" means task A requires 2 days to complete.

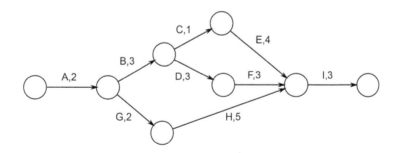

a. Which tasks are on the critical path?
b. What is the slack time for task C?
c. If task C is delayed by 1 day, what effect would this have on the completion date of the project? Why?

Solution

a. Possible paths to completion:

Path 1: A to B to C to E to I = 2 + 3 + 1 + 4 + 3 = 13 days
Path 2: A to B to D to F to I = 2 + 3 + 3 + 3 + 3 = 14 days
Path 3: A to G to H to I = 2 + 2 + 5 + 3 = 12 days

Since Path 2 requires the longest time to complete, it is the critical path.

b. Path D to F, which is along the critical path, takes 6 days and Path C to E takes 5 days. The time difference between Path D to F and C to E is 1 day. This is the slack time.

c. If task C is delayed by 1 day, it would have no effect because task C has 1 day's slack time.

23.3 Effective teaming

Working in teams on a design project is both a joy and a challenge. While there is a sense of security in knowing that others are venturing into the unknown alongside you, the unpredictability of human interactions can be as perplexing as the design itself. To reduce the risk of ineffective teaming, rules of conduct are presented in this section. These are well-accepted best practices based on observations of effective teams and they support two prime objectives of a systematic approach to engineering design:

1) eliminate personal bias from the process and
2) maximize the amount of thinking and information gathering done up front, before committing to the final design.

Several advantages accrue from attacking a design project in teams. First, design requires a wide range of skills and areas of knowledge. No one person is experienced enough to pursue every unfamiliar design challenge in isolation. Teaming provides an opportunity to expand the talents and life experiences brought to bear on the design problem. Second, if done right, teaming serves to keep personal biases in check. Third, more people should mean that more is accomplished in a shorter period of time, although it is puzzling to often see team members standing by politely as one team member does all the work (especially during manufacturing). When best practices are followed, a team is greater than the sum of its parts.

For design projects during the freshman and sophomore years, three students are the ideal team size. Teams of just two students may not experience all of the team dynamics and may not learn as much about teaming. With four student teams, it may be too easy for one team member to hide and not contribute. Student design teams at this level may not be assigned a team leader by the instructor. Leadership typically emerges within the team. If a team leader is assigned, the role is *not* to be the boss but rather to organize and facilitate participation by all team members.

Student teams need to understand these basic rules of teaming:

Assign clear roles and work assignments

A few things are best done as a team, such as brainstorming and evaluation of concepts. Most of the time, it will pay off if everyone has his or her own assigned responsibilities and tasks to which each is held accountable by the team. These tasks should be assigned or updated at the end of each team meeting.

Foster good communication between team members

An atmosphere of trust and respect should be maintained, in which team members feel free to express their ideas without retribution. That trust extends to allowing for civilized disagreement, delicately done so as not to suppress ideas or discourage participation. Everyone should participate in the discussions. Sometimes, this means reaching out with sensitivity to the shy members of the team. If you succeed, you will have a team operating on all cylinders.

Share leadership responsibilities

If there is a designated team leader, that person should empower the other team members with significant leadership responsibilities. This gives those students a strong sense of ownership in the project. At the same time, team members must be willing to step forward to assume leadership roles.

Make team decisions by consensus

Teams can make decisions in one of three ways: (1) the team leader makes the decision, (2) discussions continue until everyone agrees, or (3) after discussions are exhausted, the team takes a vote. Those who disagree with the outcome of the vote are then asked if they can put their opinions aside and move forward in the best interests of the team. In a college-level design project, the best ways to make decisions are (2) and (3), which are examples of decision making by consensus.

23.3.1 How to write a good memo

A memo (or memorandum) is used for internal communications within your design team about meeting schedules, design updates, interactions with the course instructor or other teams, and so forth. A memo is usually sent as an email and can replace the need to have an entire meeting about a small subject which could be explained in a memo. The general format for a good email memo is shown below:

To: Design team member(s) email addresses.

Subject: Give a one-sentence explanation for the memo.

Body: Your memo should convey all the information in the smallest amount of text possible. The last part of the memo should say **exactly what action** you want people to take.

Sender: Give your name and title.

Example of a good memo

To: annay@xxx.edu; genet@xxx.edu; tomasp@xxx.edu

Subject: Schedule for our first design team meeting.

Body: I have reviewed your class schedules and have scheduled our first design team meeting for September 15 at 4:00 p.m. in room 213 of the Kosky engineering building. The agenda for the meeting is attached. Please read it and be prepared to discuss the items.

Sender: William Keat, Team Leader

23.3.2 Team building

Working as a team implies *group accountability* rather than individual accountability, and results in a collective work product. Team building encourages a "team" approach to working on a project. The advantages of this approach include the following:

- Increased flexibility in overall skills and abilities of the team
- More productive than a random student group with individual mindsets
- Encourages both individual and collective team development and improvement
- Focuses on team goals to accomplish tasks
- Improved range of team objectives such as collaboration, communication, and creative or flexible thinking.
- Realization that a necessary part of team building is negotiation, consensus building, and compromise.

Team building centers around various *exercises* designed to bring a group closer together and create an understanding of individual strengths and weaknesses. It increases camaraderie and team morale, which in turn leads to increased productivity. There are many types of team building exercises that range from simple games to challenges that involve complicated tasks designed to improve team performance. These exercises usually involve interpersonal skills like communication, problem solving, and planning, in contrast to technical skills directly involved with the project. Team building tasks can also stimulate interpersonal skills that will increase team performance. Some examples of team building exercises are shown below.

23.3.3 Communication exercise

Communications exercises are geared toward improving communication skills. The issues teams encounter in these exercises are solved by communicating effectively with each other.

A typical communication exercise is to give everyone in a group a sheet of paper and tell them to close their eyes. Then tell everyone to:

1. Fold their paper in half.
2. Fold the lower left corner over the upper right corner.
3. Turn it 90 degrees to your left.
4. Fold it in half again.
5. Turn it 90 degrees to their right.
6. Rip a half-circle from the middle of the right side.

Now tell everyone to open their eyes and unfold their piece of paper. Even though they all received the same instructions and had the same starting material, nearly everyone will have a different result.

A communication problem occurred because they did not all start from the same point (some held their paper vertically and some held it horizontally), some interpreted ripping a half-circle as removing a big piece, some as a small piece, while others found the instructions vague.

23.3.4 Problem-solving exercise

Problem-solving exercises focus specifically on groups working together to solve difficult problems or make complex decisions. Here is a typical exercise: put a mathematical symbol between six and nine to get a number that is greater than six but less than nine (the answer is—a decimal point).

23.3.5 Planning exercise

These exercises focus on the aspects of planning and being adaptable to change. A typical example is to plan a surprise party for a mutual friend. The group is to develop a plan to find a time and place for the party without the friend becoming suspicious, invite trusted guests, arrange for food and entertainment, and do so in an organized and effective manner.

23.3.6 Team leadership

Here are some qualities that a good team leader should have:

Honesty
    If you make honest and ethical behavior a key value, your team will trust you.
Ability to delegate
    Delegation involves identifying the strengths of your team members. If they find a task interesting or enjoyable, they will put more effort into it.
Good communication skills
    Being able to clearly describe what needs to be done is extremely important. If the project's goals cannot be clearly explained to the team, they will not be working toward these goals.
A sense of humor
    Team morale influences productivity, so a team leader can inspire a positive attitude by having a sense of humor.
Confidence
    A team leader should be technically competent and goal-oriented. A team takes cues from its leader, so if the team leader is focused and confident, the team will respond positively.
Commitment
    If a leader expects the team to work hard and produce results, she/he needs to lead by example. Once you have gained the respect of your team, they are more likely to deliver the peak amount of quality work possible.
A positive attitude
    You want to keep your team motivated toward the goals of a project so be enthusiastic about the project.

Creativity

As a team leader, it is important to think outside the box. By utilizing all possible options before making a decision, you can more easily reach the project goals.

23.3.7 Team assessment, feedback, and risks

A team can carry out a self-assessment process to determine its effectiveness and improve its performance. To assess itself, a team needs feedback from group members to find out the teams current strengths and weaknesses. Feedback from the team assessment can be used to identify gaps between the where they are and where they would like to be.

Team assessment can be done in a variety of ways. One of the simplest is to have each team member fill out a questionnaire like the one shown in Table 23.1. Each entry can be given a numerical value, for example, in Table 23.1, we have set Disagree = 0, Undecided = 1, Agree = 2. Then a total score can be computed for each member of the team by summing all the entries. The best possible score would be 20 ("Agree" is selected for every question). A score of 14 or higher typically means this person is a good team member. A score between 8 and 13 means this person's

Table 23.1 Team assessment questionnaire.

	Function	Disagree (enter 0)	Undecided (enter 1)	Agree (enter 2)
1	The team has a clear vision of the project goals.			
2	The team has the skills and resources to achieve its goals.			
3	Everyone on the team has a clear role in the design, construction, and testing.			
4	The team has adequate meeting time, space, and resources.			
5	Team meetings are attended by all team members.			
6	The team members work together effectively.			
7	The team members understand the project requirements and timeline.			
8	Team members are promptly informed of design changes.			
9	Team meetings are run efficiently.			
10	Everyone on the team participates at a satisfactory level.			
			Total score =	

effectiveness as a team member is sporadic. And a score below eight indicates that this person has not yet learned how to be an effective team member.

The major risk of team building is that a team member may become negative and unproductive. This could happen as a result of a teaming event that does not focus on the goals and direction of the design process. For example, if a teaming event seems silly or a waste of time, then it may have a negative impact on team members. This can lead to loss of trust in the leader as well as decreased morale and productivity. Consequently, team building events should be followed by meaningful design project effort.

Exercises

1. The Balloon Tower Challenge

   Equipment: each group receives 10 small uninflated party balloons and a roll of scotch tape.

   Procedure:
   * Split the class into teams of three—five students.
   * Instruct the teams that they need to make the tallest tower possible with the material supplied. The tower cannot be attached to anything except the base (table, etc.) upon which it stands.
   * They must demonstrate a design plan.
   * The teams have 20 minutes to complete the task.
   * They must not use any other equipment and the tower must be free standing.

   Scoring: based on the design plan and the final tower height.

2. Arrange to have your team visit a group of working engineers and have them explain how they function as a team. The engineering team should be one that has worked together for some time. Ask them to explain how a project goes from the idea phase through the production phase. After the presentation, have your student team talk about all the ways the working engineers used teamwork to see their idea through. This activity will give your team a chance to see the value of professional teamwork.

3. Gantt chart exercise: prepare a Gantt chart covering a week of classes that shows how you divide each day into tasks like meals, classes, study time, leisure time, and so forth.

4. Gantt chart exercise: prepare a Gantt chart for the web site development process shown in the table below. Add appropriate dates for task completion.

Item	Task	Duration	Date
Design	Home page design	4 days	
	Product pages design	1 week	
	Shopping cart design—Purchasing	1 week	
	Shopping cart design—Checkout	1 week	
Coding	Shopping cart coding	2 weeks	
	Contact us form coding	1 day	
Testing	Website testing	1 day	
	Shopping cart—Purchasing testing	1 day	
	Shopping cart—Payment testing	1 day	
Delivery	Deliver website to client	1 day	
	Receive feedback and make changes	1 week	

5. Pert chart exercise: what is the critical path on the Pert chart shown below?

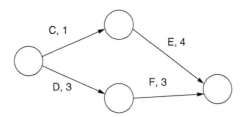

6. Pert chart exercise: develop a Pert chart that shows how you schedule or organize the tasks you follow when you prepare for school in the morning. Start with getting out of bed and end with arriving at your first class.

7. Communication exercise: students write directions for making a peanut butter and jelly sandwich for someone who has never made a sandwich before and will be exactly following your instructions. Discuss what the sandwiches might look like after reading your directions.

8. Problem solving exercise: at a university, all the students are majoring in engineering, business, or both. 55% of the students are engineering majors and 77% are business majors. If there are a total of 500 students, how many of them are majoring in both engineering and business?

(**Answer**: If 55% of the students are engineering majors, we know that 45% are not engineering majors. And if 77% of the students do major in business, then 23% are not business majors. 45% + 23% = 68% of the students are not majoring in both engineering and business; therefore, 100% − 68% = 32% are double majors and 500 × 0.32 = 160 students.)

9. Problem solving exercise: The goal is to build a structure that will prevent an egg from breaking when it is dropped from a height of 8 feet. The only materials allowed are 10 paper drinking straws and 15 inches of ½ -inch wide masking tape.

10. Problem solving exercise: The goal is to transfer a "radioactive" beach ball from a table in one room onto a table in another room (preferably on another floor of the building) without the beach ball touching the ground, any part of the building, or any body part. The only materials allowed are a ball of string and a role of ½ -inch wide masking tape.

11. Planning exercise: You are with a group of 10 students returning to campus after a bus tour. Your bus is scheduled to leave at 6:00 p.m. While waiting, you and your group go to a coffee shop where one of them suddenly becomes seriously ill. While you were helping him, a young girl rushes in and asks for your help with her brother who fell down nearby and broke his leg. It is now 4:00 p.m. and you need to reach campus before 7:00 p.m. Form a plan to deal with this situation.

(**Answer**: Planning priorities—Help your friend and call an ambulance. Find and help the person with a broken leg and call another ambulance. Then go to the bus station and catch the 6:00 p.m. bus back to campus.)

12. Planning exercise: You are in a cave in the mountains of Colorado. It is winter and a terrible blizzard that will last for a week has covered the ground around the cave with five feet of snow. In the cave you have a metal pot, cigarette lighter, dry wood, warm clothing, knife, two candy bars, snowshoes, and a flashlight. You have no food, water, or telephone. Form a plan to survive.

(**Answer**: Planning priorities—You need water to survive a week or more in the cave. Build a fire and melt snow in the metal pot for drinking water, but do not eat snow (why?). Ration the candy bars and keep warm. Clear the opening to the cave and when the blizzard stops, put on the snowshoes and head for home.)

13. Ethics exercise: At a team meeting, you are asked to report on the progress of your project. A team member asks you to report that his/her part of the project is finished even though it is not. He/she tells you that it will be done in three more days. What do you do? Use the engineering ethics matrix.

Design step 1: Defining the problem

24

24.1 Introduction

The design process begins when somebody, whom we will refer to as the *customer*, expresses a need and so enlists the services of an engineer. The customer can be an individual, an organization, a company, or the public. It is up to the engineer to translate the customer's needs into engineering terms. The result is cast in the form of understanding a need, developing a problem definition, and producing a list of design specifications.

24.2 Identifying the need

The basis of a good engineering project depends on identifying a viable "**need**" or "**want**" that can be satisfied. Needs can range from a complaint about an existing

Exploring Engineering. https://doi.org/10.1016/B978-0-443-13541-5.00032-5

consumer product to one of the National Academy of Engineering's Grand Challenges, for example, the global need for new technologies that will make clean water more accessible or capture CO_2 from the atmosphere.

Failure to identify, understand, and validate the need, before designing, is one of the most frequent causes of failure in the entire design process.

The success of the project then depends on whether or not the **customer(s)** are satisfied with the result. The customer should be interviewed to fully understand the needs and the constraints of the project. In many cases, the customer will not have a technical background so care must be taken during the interview to distinguish between the need and a solution, to avoid overconstraining the solution space. Most design projects will have multiple stakeholders, all of whose interests must be satisfied for the project to be considered successful. For example, if developing a consumer product, the needs of the target consumer group, referred to as the **customer's voice,** must be enforced throughout the design process. Examples of other potential stakeholders are representatives from manufacturing, marketing, and government agencies.

Occasionally, as an engineering student, you may be asked to develop your own project idea. The best way to begin is to identify the human need that you wish to address. Here are three methods for identifying needs that could lead to new product ideas:

(1) **Listen to the customer:** Interviews of potential customers can be conducted in person (best) or via surveys. For example, college students seeking new product ideas interviewed the students in Mrs. Harper's third grade class and in two sets of interviews determined each third grader's favorite classic toy and then which of several proposed upgrades the third graders would be most excited to see. Some of the pre-1900 classic games and the corresponding new product ideas are listed in Table 24.1. A few of these product ideas foreshadowed actual products introduced on store shelves years later.

Table 24.1 Classic toys and their updated versions that resulted from two sets of interviews with third graders.

Toy needing an update	Product idea
Checkers	Same game but Harry Potter figurines are jettisoned into the air when captured
Ring toss	Ring posts are mounted to motorized platforms that spin in opposite directions
Kite	Remote-controlled kite
Ball	Soft translucent ball is stuffed with lights, four speakers, and an iPod

(2) Observe customer behaviors: Another way to come up with an innovative product idea is to observe how potential consumer groups (e.g., children) interact with existing products and their environment. Sometimes it takes a fresh pair of eyes to notice how the daily activities of a person's life might be changed to improve their quality of life. For example, based on common experiences observing older parents and relatives, a design team discovered the following needs:

  (a) safe mobility within the home is key to an older person maintaining their independence and health and

  (b) transferring from a chair to a standing position can be a risky maneuver. The product idea of a walker with a powered lift resulted from recognition of this need.

(3) Keep a bug list: Designers are also advised to maintain a "**bug list**," which is a list of annoying flaws that exist in available products and services. This list can be an inspiration for new and innovative product ideas. Online product reviews (a form of a bug list) are often used by consumers to choose between alternative products. The alert designer, like other consumers, is qualified to comment on the shortcomings of existing products, especially in his/her areas of expertise. For example, when asked to come up with a project idea, student athletes will often choose to design a new and improved piece of sports equipment for their respective sport. An experience shared by many students is that of having to deal with the challenges of life in a college dormitory room. This is a dominant theme for students asked to generate their own design project ideas as the needs listed in Table 24.2 might suggest.

Table 24.2 A "bugs list" for life in a college dormitory room and the product ideas those complaints inspired.

Need	Product idea
"Can be annoying to get up to lock or unlock the dormitory room door when half asleep in bed"	Sound activated deadlock bolt
"Alarm clock doesn't always get me out of bed for early morning classes"	An alarm clock that motivates you to get out of bed
"I'm not the best at folding clothes and putting them away"	Automated shirt folder that also drops the shirt in the drawer
"I sometimes forget to charge my phone at night."	A charger that grabs your attention at night
"The room smells bad."	Automated air freshener

24.3 Defining the problem

In this book, "problem definition" refers to a concise 3–4 sentence statement of the design objective that can include mention of some of the most important constraints that are essential to understanding the nature of the desired solution. The problem definition alone is not sufficient for formulating solutions as there are many other design requirements that will first need to be documented in the form of a **list of specifications**.

To develop the problem definition, as well as the list of specifications, information will have to be gathered, beginning with the first interview with the customer. Information is the key to effective decision making. If you are designing a replacement bridge, do you know how many people will use the bridge? When will they use the bridge—all the time or mainly in the morning and evening? If it is mainly in the morning or evening, your problem definition might be "Design a four-lane highway bridge to accommodate 400 cars per hour at peak usage." You might collect several types of information. It will generally fall into one of the following categories:

* Facts (15% of the people do not use the bridge at all)
* Inference (a significant percentage of drivers bypass the bridge)
* Speculation (many trucks use the bridge)
* Opinion (I think a new bridge will be too expensive)

When you are gathering information, you will probably hear all four types of information, and all can be important. Speculation and opinion can be especially important in gauging public opinion. For example, if you are trying to reduce the waiting time to pay for products purchased in a large department store you might find that most people have the opinion that the store should open more checkout lanes. Others may speculate that the store manager does not want to pay for more checkout clerks. However, if you observe the checkout line throughout the day and interview the store manager you may find relevant facts that either support or contradict these opinions and speculations. With the information in front of you, you are ready to write down a problem definition.

The customer's statement of need does not typically take the form of a problem definition. For example, consider the following statement of need from a fictitious client:

> **Need:** *People who work at the Empire State Building are complaining about the long waits at the elevator. This situation must be remedied.*

An engineer might translate this need into the following problem definition:

> **Problem Definition:** *Design a new elevator for the Empire State Building.*

But is this really a good problem definition? Is the main concern of the management at the Empire State Building to reduce average waiting times or to eliminate the complaints? When turning an expressed need into a problem definition, it is important to eliminate assumptions that unfairly bias the design toward a particular solution. A better, less-biased problem definition might be:

> **Improved Problem Definition:** *Increase customer satisfaction with the elevators in the Empire State Building.*

FIGURE 24.1

Blimp returning to base after retrieving a ball.

This would allow such solutions as a mirror on the elevator door or free coffee on the busiest floors.

As another example of an inadequate problem definition, consider the following:

Problem Definition: *Design a device to eliminate the blind spot in a vehicle (the area around the vehicle that cannot be directly observed by the driver while at the controls of the vehicle).*

This proposed problem definition contains an assumption that prematurely limits the designer. The word ***device*** rules out one solution that achieves the design goal of eliminating the blind spot by simply repositioning the front and side mirrors.

A third example occurred in a design competition named Blimp Wars (see Fig. 24.1). The goal was to *design a system to retrieve Nerf balls from an artificial tree and return them to the blimp base.* Inclusion of the word ***blimp*** in the problem definition biased the students toward blimp designs. The alternative of an extendable arm that spans the distance between blimp base and the target balls was not considered.

24.4 List of design specifications

After translating the need into a problem definition, the next step is to prepare a complete **list of design specifications** (or requirements). A typical design requirement has the structure shown in Fig. 24.2.

Design requirements can define performance goals, required features, and constraints on such quantities as time and cost. A checklist of categories that should be addressed by the list of design specifications is provided in Table 24.3. These

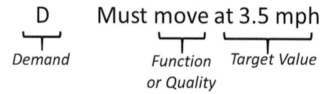

FIGURE 24.2

Typical structure of a design requirement.

Table 24.3 Checklist of design requirement types.

• Performance	• Manufacture
• Geometry	• Standards
• Materials	• Safety
• Energy	• Transport
• Time	• Ergonomics
• Cost	• Environment

categories can be used as headings by which to organize the list of specifications. It also is important to check that the needs expressed by customers and stakeholders are represented in the list.

The list of design specifications includes both "demanded" design characteristics that must be present for the design to be considered acceptable by the customer, and "wished for" design characteristics that are desirable but not crucial to the success of the final design. It is the usual practice to classify each specification as either a demand (**D**) or a wish (**W**). Do not confuse the two. If you treat a wish as if it were a demand, your design may become more complicated than is necessary.

Whenever possible, use numbers to express specifications. For example, instead of merely requiring that weight must be low, state, "Weight must be less than 10 pounds." Sometimes use of numbers is impossible. A quality such as "aesthetically pleasing" is difficult to quantify. However, use numbers wherever possible, even if at this early stage they seem like guesses. The numbers can be refined later on as the design begins to take shape. Ultimately these target values will serve as the metrics by which the success of the design will be measured.

The specifications should be solution independent to avoid bias. For example, if you are designing a small mobile device, requiring that "the wheels must be made of rubber" will bias the design in two respects: in the use of wheels and in the choice of materials. Such decisions are reserved for later in the design process after careful consideration of alternatives.

Example 24.1

The following problem definition was posed to three competing design teams.

Design and build an RC (remote controlled) portable device that will play nine holes of golf at a local golf course with the fewest possible number of strokes.

The instructor also supplied the following list of demands. It was left to the students to develop a complete list of specifications.

Demand (D) specification

Must cost less than $600.

Must be remotely triggered.

Total number of radio-controlled servos[1] is eight.

Device cannot be touching the golf ball before remote triggering of the shot.

Entire device must form a single unit.

Must be portable.

Design must pass a safety review.

Ground supports must fit within a 3-foot circle.

Solution: The first step was to organize the demands under each heading. Then, using the headings as a guide, additional demand (D) or wish (W) specifications were formulated. The results follow.

Performance

D—Must be remotely triggered.

D—Device cannot be touching the golf ball before remote triggering of the shot.

D—Driving distance must be adjustable with a range of between 15 and 250 yards.

D—Putting distance must be adjustable with a range between 0 and 15 yards.

D—Must operate on inclines of up to 45 degrees.

W—Must sink 95% of short putts (less than 3 feet).

W—Driving accuracy of ± 5 yards.

Geometry

D—Total number of radio-controlled servos is eight.

D—Entire device must form a single unit.

D—Ground supports must fit in 3-foot circle.

Materials

W—Materials must not degrade under expected range of weather conditions (including rain, snow, $30°F < T < 90°F$).

Time

D—Must be designed and manufactured in less than 14 weeks.

Cost

D—Must cost less than $600.

Manufacture

D—Must be manufactured using tools available in the machine shop.

D—Must be manufactured using machining skills available within the team.

W—Off-the-shelf parts and materials should be readily available.

Standards

D—Radio must adhere to FAA regulations.

Safety

D—Design must pass a safety review.

Transport

D—Must be portable.

W—Must fit in a car or small truck (for easy transport to golf course).

---

[1] A servo is a control system to amplify a small signal into a large response. Typically it is an electric motor controlled by a small voltage.

24.5 Design milestone #1: Defining the problem

There are two versions of this milestone, depending on the format of the design project. In the case of a **general design project**, unless the project's problem statement is clearly defined by the course instructor, students should meet with the customer (who still may be the course instructor for an ill-defined problem statement) to clarify the customer's needs and convert them into a proper problem statement. Sometimes lay customers are not able to express their needs in technical terms and it is then the responsibility of the students to interpret and clarify these needs into a list of specific technical tasks needed to solve the problem.

If it is a **design competition project**, everyone in the competition must operate under the same set of constraints. Therefore, the responsibility for defining and problem and producing a detailed list of specifications and tasks shifts from the students to the instructor or competition organizer (e.g., NASA Robotic Competitions).

24.5.1 For a general design project

Assignment:

(1) Interview the customer (often the course instructor) to clarify the project and define the design problem you are to solve. In the case of a consumer product, conduct a product survey with likely purchasers of the product.

(2) Prepare a complete list of technical specifications and requirements and develop a clear definition of the problem to be solved.

24.5.2 For design competitions

Assignment:

(1) Review the rules of the competition and ask your instructor (or the organizing committee) for rule clarifications if necessary.

(2) Prepare a complete list of design requirements to supplement those already appearing in the official rules of the competition. For example, set performance goals for your device. You do not have to list requirements already appearing in the rules.

(3) Probe the rules for holes that will allow for concepts not anticipated by the creators of the competition.

(4) Avoid any temptation to bias the requirements toward a particular solution or strategy.

(5) Expect the list of supplemental requirements to be very short if the rules are well defined.

Exercises

1. Develop a bug's list of 10 items long for the experience of going to the airport and taking a long flight. Then, formulate the problem definitions for three different innovative project ideas based on items in your bug list.

2. Develop a bug's list of 10 items long for the choice of using an umbrella on a rainy day. Then, formulate the problem definitions for three different innovative project ideas based on items in your bug list.

3. Develop a bug's list of 10 items long for using a bicycle to commute from home to school. Then, formulate the problem definitions for three different innovative project ideas based on items in your bug list.

4. If you are asked by your manager to design a new medical device that will ultimately be sold to consumers, who are your customers for this device. Include in this list of customers all of the important stakeholders.

5. Translate the following need into a problem definition: ***Students need more parking near campus.***

6. Translate the following need into a problem definition: *Shoppers need a faster checkout process in grocery stores.*

   For the Problem Definitions presented in Exercises 7–12, complete the following:

   (a) Develop a complete List of Design Specifications organized under the headings listed in Table 24.3. Be sure to distinguish between demands (D) and wishes (W).

   (b) What specific information needs to be gathered to finalize the List of Design Specifications?

   (c) List the people you think you would have to interview in the process of gathering information for the List.

7. Design a student backpack that will comfortably hold 10 kg of books, a laptop computer, and lunch.

8. Design an interactive toy for a blind child.

9. Design a device that will launch peanut butter and jelly sandwiches into a crowd of college students.

10. Design a safe, low-cost mode of transport for mall shoppers who prefer to remain vertical while shopping but have difficulty walking.

11. Design a low-cost unmanned underwater vehicle (UUV) to locate a Ford Model T believed to be in pristine condition 120 ft down at the bottom of Ballston Lake, in Ballston NY. Because of high concentrations of iron in the water, visibility is extremely poor at the bottom of the lake.

12. Design an electric-powered Hover Sled for kids who want to experience the thrill of sledding in the summertime. The Hover Sled must ride down grassy slopes on a cushion of air created by an electric fan and carry one person weighing up to 120 lb.

Design step 2: Generation of alternative concepts

25

25.1 Introduction

Once the problem statement is in place and the specifications have been listed, it is time to generate alternative concepts. By *concept*, we mean an idea as opposed to a

Exploring Engineering. https://doi.org/10.1016/B978-0-443-13541-5.00006-4

detailed design. The representation of the concept, usually in the form of a sketch, contains enough information to understand how the concept works but not enough information to build it. By *alternative*, we require that the various proposed ideas must be fundamentally different in some way. The differences must go beyond appearance or dimensions. The usual rule of thumb in design courses is to generate at least three fundamentally different concepts.

In this chapter, four aspects of concept generation are discussed: **brainstorming, concept sketching, research-based strategies, and functional decomposition**.

25.2 Brainstorming

The most common approach for generating ideas is by brainstorming. As the term implies, you rely on your own creativity and memory of past experiences to produce ideas. It can be employed effectively both as an individual or as a member of a team provided some basic rules are observed.

Brainstorming is based on one crucial rule: *Criticism of ideas is not allowed*. This enables each team member to put forth ideas without fear of immediate rejection. For example, a professor once recorded the brainstorming session of a small team of students. At one point, a student offered an idea, and another student referred to it as "stupid". The voice of the first student was never heard from again during the session. Instead of a team of four, it had become a team of three.

It is important to devote some of the brainstorming time searching for bold, un-conventional ideas. At first, such ideas might appear impractical, but with further development they could ultimately lead to very innovative solutions. In the case of a design competition, this could mean searching for holes in the rules that could lead to ideas that the creators of the competition had not anticipated.

When the goal is to brainstorm a list of ideas or trouble shoot a very specific problem, it can be very effective to brainstorm as a team. A facilitator should be selected who writes the ideas on the board in the form of an enumerated list as they are generated, enforces the no criticism rule and also makes sure the voices of all team members are heard. If a slow spot is hit during the brainstorming session, it is also up to the facilitator to redirect the team's thinking in a new more fertile direction.

When the goal is to generate concepts with multiple functional requirements, it is often good practice to have individual team members first brainstorm and sketch ideas on their own, then meet and share ideas as a team. This helps to insure that all team members (even the shy, quiet ones) are fully engaged and contributing.

Only when brainstorming is complete should the team eliminate concepts that are not feasible, not legal, or not fundamentally different. After this weeding-out process, at least three concepts should remain. If not, more brainstorming is in order. Example 25.1 illustrates this step.

Example 25.1

Assuming the alternative concepts in Fig. 25.1 are generated as part of an effort to design a new bat for Major League Baseball, which concepts should be eliminated because they are not feasible, not legal, or not fundamentally different?

Solution:

A is a standard bat to which the new concepts are to be compared.

Not feasible: **E** because it stands no chance of being competitive; **I** because it is too difficult to find in nature.

Not legal: **F**, **G**, **H**, and **J**.

Not fundamentally different from each other: **C** and **D** because the basic shape is the same; only the dimensions differ. Therefore, the condensed list of viable alternatives consists of concepts **B** and **C**.

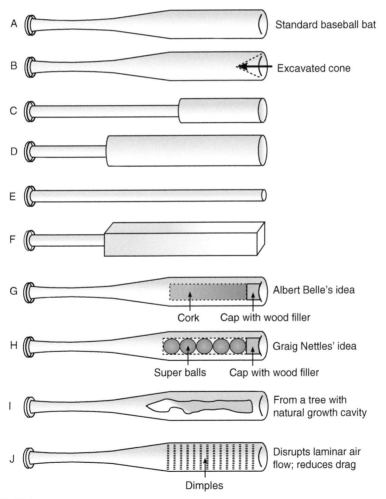

FIGURE 25.1

Alternative concepts for a major league baseball bat.

25.2.1 Mind Mapping

An effective way for an individual to generate ideas is to write the problem you are trying to solve on a piece of paper and draw a circle around it. Then draw lines out from this circle and label each line with a way to solve the problem. This process is called **Mind Mapping**.

Mind mapping is an effective way of getting information in and out of your brain. It is a creative and logical means of sketching that literally "maps out" your ideas. They have a structure that radiates from the center of a page and use lines, symbols, words, color and simple, brain-friendly images. Mind mapping converts a list of information into an organized diagram that works in line with your brain's natural way of doing things.

One simple way to understand a Mind Map is by comparing it to a map of a city. The problem to be solved is in the center of the city and the main streets leading from the center represent your key thoughts and ideas for a solution. The secondary streets represent your secondary thoughts, and so on, as shown in Fig. 25.2. Special images or shapes can represent landmarks of interest or particularly relevant ideas.

FIGURE 25.2

A simple Mind Map.

The essential characteristics of mind mapping are as follows:

- The main idea or problem is in the center.
- The main themes *radiate* from the center as 'branches'.
- Each branch is tagged with a word or pictorial descriptor that represents the theme or idea.
- Topics of lesser importance are represented as "twigs" of a branch.
- The branches form a connected nodal structure.

25.2.2 Ideation

Ideation is a creative method of generating, developing, and communicating new ideas. It can be thought of as a form of "structured brainstorming". The **Ideation Process** is a systematic sequence of actions (vs. brainstorming) that have

- Clearly defined event goals (**Why** are we doing this?),
- Measurable event objectives (**What** will we walk away with?),
- Specific event inputs and outputs (Document **deliverables**),
- "Standard" process actions, inputs, and outputs are **adaptable** to accommodate different or changing event goals and objectives.

If you have ever tried to do anything creative, you know that one of the hardest things to do is to begin. Without structure, it is easy to stall after one or two ideas or simply stare at a blank page. One way of drawing ideas out of the brain is to ask questions. For example, if you are designing a container for an egg that will withstand being dropped 20 feet to the ground without breaking, you might ask:

1. How big can the container be?
2. What container shapes are good at absorbing impact?
3. How fast will the contained be traveling at impact?
4. What other technologies are designed to minimize impact?

Just imagine ideation as a funnel. Insights, views, opinions, data, trends, and knowledge are captured at the top. They are filtered in the middle, and the good stuff—new ideas—come out the bottom.

25.3 Concept sketching

For an idea to be considered a feasible alternative concept, it must be represented in the form of a conceptual sketch. The goal in producing a concept drawing is to convey what the design is and how it works in the clearest possible terms. Any lack of clarity, such as failure to represent one of the subfunctions, translates into doubts about the feasibility of the concept when it comes time to evaluate it.

At the same time, however, this is not a detailed design drawing. Dimensions and other details not relevant to understanding the basic nature of how the concept will work are left out.

It is best to proceed through two phases when generating a concept drawing. First, in the creative phase, hand-sketching is done freestyle and quickly, without regard for neatness or visual clarity. A few simple lines, incomprehensible to others, might be enough to remind you of your idea. Sketching is a means for both storing ideas and brainstorming others. The result is a rough sketch of the concept. Second, in the documentation phase, the concept is neatly redrawn and labeled to facilitate communication with team members and project sponsors.

The final outcome is one or more sketches prepared with the following guidelines in mind:

- Can be hand-sketched or computer generated.
- No dimensions. Remember, this is not a detailed drawing.
- Label parts and main features. If the drawing is hand-sketched, handwritten labeling is acceptable.
- Provide multiple views and close-up views if needed to describe how the design works.

The choice of views is up to you. Isometric views, like those shown in Figs. 25.3 and 25.4, convey a lot of information in a single picture. Most mechanisms can be described effectively using one or more two-dimensional views, as in Fig. 25.5. Despite their apparent informality, the quality of these drawings is crucial to fairly representing the designs during the evaluation process. In some cases, they are the only source of evidence for judging if a design is likely to work.

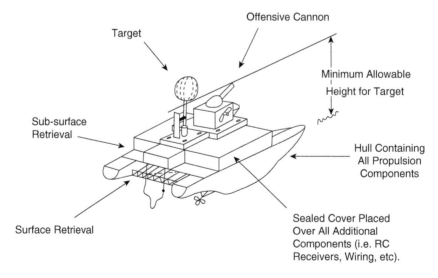

FIGURE 25.3

Concept drawing of a radio controlled (RC) boat for a design competition (hand-drawn isometric).

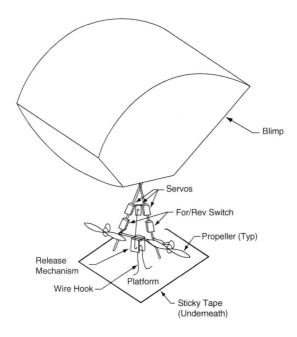

FIGURE 25.4

Concept drawing of a radio-controlled blimp (computer-generated isometric).

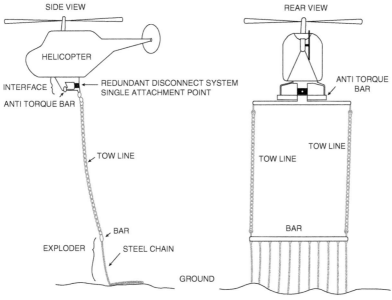

FIGURE 25.5

Concept drawing of an antipersonnel mine clearing system (two hand-drawn views).

25.4 Hands-on design exercise: "The tube"

The design objective is to extract a golf ball from the bottom of a free standing, open-ended mailing tube in the shortest possible time.

25.4.1 Setup

1. Place a mailing tube vertically on the floor and drop a golf ball into the tube.
2. Have a supply of the following materials: string, duct tape, Scotch tape, 8.5 × 11 standard copier paper, and scissors.

25.4.2 Rules

1. Limited to using the supplied materials.
2. Scissors can only be used for manufacturing.
3. Everyone in the group can help in manufacturing, but only one person can extract the ball.
4. Students are not allowed to handle the materials until it is time to test.
5. Must manufacture the design shown on the concept drawing handed to the instructor.
6. Time limit of 3 min to manufacture concept and extract ball.
7. Cannot tip over the tube.
8. Cannot touch the outside of the tube with anything.
9. No forces can be applied to the inside of the tube to hold it vertical; accidental contact with the inside of the tube is okay as long as the tube does not tip over.
10. Violation of any of the preceding rules will lead to immediate disqualification.

25.4.3 Procedure

1. First allow the students 3 min to individually brainstorm (encourage them to draw quick sketches of each of their concepts).
2. Then, divide into teams of four students per team.
3. Allow teams 10 min to collect ideas, brainstorm as a team, select their best concept, and give a sketch of their best concept to the instructor.
4. Instructor should walk around during brainstorming to remind teams to (a) generate multiple solutions before selecting one and (b) try to involve everyone in the process.
5. Allow teams 2 min to assign responsibilities for manufacture and test.
6. One at a time, give each team 3 min to manufacture a concept and attempt to extract the golf ball.
7. The team with the shortest retrieval time wins.

25.5 Research-based strategies for promoting creativity

Some ideas are truly original, but most are drawn from past experience. The following strategies help you to look at old designs to generate new ones.

25.5.1 Analogies

One often-used strategy is to look for analogous design situations in other, unrelated fields. To do this, first translate the design objective into an overall function that is general enough to widely apply. For example, you may want to design a system to "climb a vertical wall" or "walk on two legs" or "move efficiently through the water." Nature is filled with solutions to these problems (but because of their complexity, biological solutions usually have to be simplified and adapted before they can be of practical use). If you are designing a system to "throw an object," a survey of ancient artillery could spark ideas.

25.5.2 Reverse engineering

The basic strategy here is to acquire an existing product that is similar to a design you have in mind, take it apart, figure out how it works, and then either try to improve on it or adapt some of the ideas to your own design. Toy stores are a great place to search for small electromechanical devices that can be reverse engineered.

25.5.3 Literature search

Web-based search engines are very effective at finding existing design solutions. For high-tech applications, you should also search books and technical journals.

25.6 Functional decomposition for complex systems

When confronted with a complex problem, it is frequently advantageous to break it down into smaller, simpler, more manageable parts. In the case of design, those smaller parts usually correspond to the individual functions (or tasks) that must be performed to achieve the overall design objective. This approach, known as **functional decomposition**, is the basis of the procedure described next for generating concept alternatives.

25.6.1 Step 1: Decompose the design objective into a series of functions

Start out by decomposing the overall function into four or five subfunctions. Usually, verbs such as *move*, *lift*, and *control* are used in naming the functions. Fig. 25.6 shows the functional decomposition of the remote-controlled golf machine of earlier examples. It is given in the form of a tree diagram, which is probably the most common form of representation. If more detail is needed, each subfunction could be further broken down into its respective subfunctions.

It is very important that the functional decomposition be general enough to avoid biasing the design solution. For example, the separate drive and chip functions in Fig. 25.6 may cause the design team to overlook the possibility of using the same device to fulfill both functions. If solution bias is unavoidable, introduce multiple functional decompositions.

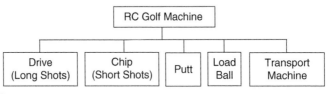

FIGURE 25.6

Functional decomposition for design of a remote-controlled golf machine.

When it is not readily apparent what the subfunctions are, it may help to think in terms of the sequence of tasks that must be performed by the design. The "sequential" functional decomposition for a design to assist disabled people into and out of a bathtub is shown in Fig. 25.7.

FIGURE 25.7

"Sequential" functional decomposition for a design of a system to aid the disabled into and out of a standard bathtub.

25.6.2 Step 2: Brainstorm on alternative concepts for each function and assemble the results in a classification scheme

The classification scheme is a two-dimensional matrix organized as shown in Table 25.1. The first column lists the functions resulting from the functional decomposition. The row of boxes next to each function name contains the corresponding design solutions that have been brainstormed. The design solutions are expressed using a combination of words and pictures, so be careful to draw the boxes large enough to accommodate small illustrations.

Table 25.1 Organization of the classification scheme.

Concepts / Functions	Concept 1	Concept 2	Concept 3	Concept 4
Function A	A1	A2	A3	A4
Function B	B1	B2	B3	B4
Function C	C1	C2	C3	C4
Function D	D1	D2	D3	D4

25.6.3 Step 3: Combine function concepts to form alternative design concepts

Table 25.2 demonstrates how one subfunction concept from each row of the classification scheme is selected to form a total concept. The same subfunction concept can be used with more than one total concept, though keep in mind that the idea is to generate fundamentally different design concepts. The only other rule when deciding on the best combinations is to be sure that the subfunction concepts being combined are compatible.

Table 25.2 Combining of compatible subfunction concepts.

Concepts / Functions	Concept 1	Concept 2	Concept 3	Concept 4
Function A	A1	A2	A3	A4
Function B	B1	B2	B3	B4
Function C	C1	C2	C3	C4
Function D	D1	D2	D3	D4

Total Concept I = A1 + B2 + C2 + D1
Total Concept II = A4 + B2 + C4 + D2

25.6.4 Step 4: Sketch each of the most promising combinations

This is done in accordance with the rules previously presented for concept drawings. Remember that you must end up with drawings for at least three fundamentally different design concepts.

Example 25.2

Use functional decomposition to generate alternative concepts for a proposed radio-controlled blimp, capable of retrieving Nerf balls from an artificial tree and returning them to the blimp base (see Fig. 24.1 of Chapter 24).

Solution: The first step is to produce the functional decomposition shown in Fig. 25.8.

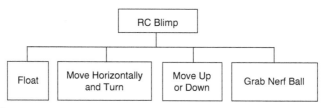

FIGURE 25.8

Functional decomposition for the design of a remote-controlled blimp.

Then, concepts were brainstormed for each of the subfunctions, and the results are assembled in the classification scheme shown in Fig. 25.9.

Concepts / Functions	Concept 1	Concept 2	Concept 3	Concept 4
Float	Hot Air	Helium	Hydrogen	
Move Horizontally and Turn	Prop + Rudder	Two Props	Rotating Turret	2 Props at 90°
Move Up or Down	Pivot Props	Vertical Props	Retract String	Heat or Cool Gas
Grab Nerf Ball	Sticky Tape	Rake	Claw	Pin Cushion

FIGURE 25.9

Classification scheme for remote-controlled blimp.

The total concepts were formed by combining compatible subfunction concepts. The three promising total concepts are as follows:

Total Concept I = Helium + 2 Props + Pivot props + Sticky tape

Total Concept II = Helium + Rotating turret + Vertical prop + Rake

Total Concept III = Helium + Prop with rudder + String + Claw

The final step is to represent each alternative design in the form of a concept drawing. The concept drawing for Total Concept I is shown in Fig. 25.4.

25.7 Design milestone #2: Generation of alternative concepts

This milestone assumes the system to be designed is sufficiently complex (i.e., at least two subfunctions) to warrant the use of functional decomposition. The following assignment is designed for student design competitions:

Assignment

1. For the functional decomposition of your design project, brainstorm **at least five** feasible alternatives for each subfunction and assemble the results in a classification scheme. Include **competition strategy** as one of the items to be brainstormed in the classification scheme. Search the boundaries of the rules for unusual ideas that could potentially dominate the competition.
2. Form **three promising design** concepts by combining compatible subfunction alternatives from your classification scheme. Redraw your concept sketches to enhance clarity and neatness. The quality of the concept drawing, or lack of it, can do much to sway opinions when it comes time to judge the concepts.
3. Firm up your three design concepts by sketching them up in the form of **concept drawings**. Functionality (i.e., how it works) should be clearly indicated in the drawings using labeling and text.

Grading criteria

Technical communication
- Ideas are clearly presented.
- Final concept drawings are neatly rendered.

Technical content
- All concepts are feasible, legal, and fundamentally different.
- Concepts are presented in sufficient detail.
- Requested number of concepts is generated.

Exercises

Brainstorming

Use the technique of **individual** brainstorming for the problems below:
1. Brainstorm by yourself a list of ways you can get to class each day. Do not be afraid to include methods that do not seem currently plausible (e.g., teleportation) since that may trigger other ideas for you.
2. Brainstorm a list of ways you could use to study for an exam. Include when, what, where, and how long to study.
3. Brainstorm ways you can get an A in this course.
4. Brainstorm ways you could create world peace.
5. Brainstorm ways to reduce global warming.

Concept sketching

6. Sketch a bicycle and label all its components (wheels, frame, etc.).
7. Sketch your cell phone and label its operating functions (on/off button, etc.).
8. Sketch your shoe and label all its parts (heel, sole, etc.).
9. Sketch your concept of a space ship and label its components.
10. Sketch a battery powered wheel chair and label all the parts.

Functional decomposition

11. You are designing an electric footstool. It must have an electric heater, move easily on the floor, and have remotely controlled height. Draw a functional decomposition diagram like the one shown in Fig. 25.6 for your footstool.
12. You want to design a hot dog stand for a street vendor. Decide what the stand must be able to do (e.g., heat water, store condiments, buns and wieners, etc.) and prepare a functional decomposition diagram like the one shown in Fig. 25.6 for your design.
13. Examine the toaster in your home and determine all its components. Then draw a functional decomposition diagram like the one shown in Fig. 25.6 for the toaster.
14. Examine a coffee maker at home or in a department store. Then draw a functional decomposition diagram like the one shown in Fig. 25.6 for the coffee maker.
15. You are to design a small robot that will climb a vertical wall. Determine all the subfunctions required of the robot and draw a functional decomposition diagram like the one shown in Fig. 25.6 for the robot.

In Exercises 16 and 17, develop sketches of two different promising designs using the following procedure:

   (a) Brainstorm five alternative concepts for each subfunction in the functional decomposition and assemble the results in a classification scheme.
   (b) Combine compatible subfunction concepts to form two promising alternative design concepts.
   (c) Sketch the two promising design concepts following the guidelines provided in Section 25.3

16. Your goal is to design a spherically shaped remote-controlled robot that moves by rolling and can also jump when needed in order to outperform a rescue dog in a race through an obstacle course. To complete the obstacle course, the robot must (**1**) roll through a tunnel, (**2**) weave its way through eight vertical poles aligned in a straight line and spaced 22 inches apart, (**3**) jump over a 12-inch-high horizontal bar, and (**4**) jump through a hoop like the one shown in figure below. The functional decomposition below has been proposed to guide brainstorming.

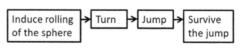

17. Your goal is to design a submersible vehicle to locate an antique Model T Ford coupe believed to be at the bottom of Ballston Lake, NY, at a depth of 120 feet (37 m). The car is expected to be in pristine condition because of lack of oxygen at that depth. However, visibility is very poor at 120 feet because of the high concentrations of particles in the water. The following functional decomposition is proposed to guide your brainstorming:

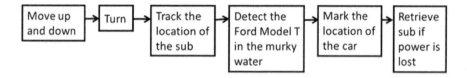

Design step 3: Evaluation of alternatives and selection of a concept

26

Source: Ondrej Pros/iStockphoto.com

Exploring Engineering. https://doi.org/10.1016/B978-0-443-13541-5.00029-5

26.1 Introduction

Suppose you now generated three concepts that meet the problem definition and fulfill the specifications. Which one should you choose as the basis for your final design? There is no magic formula. However, Professor Nam P. Suh of MIT provided two very helpful design principles for evaluating and improving concepts[1]: **(1) minimize information content** and **(2) maintain the independence of functional requirements**.

This chapter adds three additional considerations for evaluating design alternatives: **ease of manufacture**, **robustness**, and **design for adjustability**. It then concludes with a method of pulling together all these ideas in a **decision matrix**.

26.2 Minimize the information content of the design

When choosing among promising alternatives, the best design is often the one that can be uniquely specified using the least amount of information or, alternatively, can be manufactured with the shortest list of directions. This idea is sometimes inelegantly stated as the *KISS* principle: Keep It Simple, Stupid.[2]

A number of design guidelines naturally follow. A few of the most notable ones are as follows:

- Minimize the number of parts.
- Minimize the number of different kinds of parts.
- Buying parts is preferable to manufacturing them yourself.

26.3 Maintain the independence of functional requirements

The functions considered in a functional decomposition provide the basis for Professor Nam P. Suh's second principle. This principle asserts that the functions that a design must accomplish should be independent of each other.

A successful application of this principle is illustrated by the decoupled design in Fig. 26.1. Independence of the functions "lift" and "move" is maintained by designing physically separate mechanisms for each action (scissors jack for lift, wheeled vehicle for move) and by performing the actions in sequence, rather than at the same time. First, the scissors jack lifts the vehicle, the vehicle then slides

---

[1] N. P. Suh, *The Principles of Design* (New York: Oxford University Press, 1990).

[2] The term *KISS* (Keep It Simple, Stupid) was coined by the American design engineer Kelly Johnson to indicate that a military aircraft should be repairable with a limited set of tools under combat conditions. Johnson wrote it as "Keep it Simple Stupid" (without the comma) because it was not meant to imply that an engineer was stupid; just the opposite. A simple solution is better than a complex one even if the solution looks stupid.

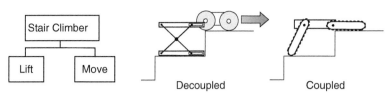

FIGURE 26.1

Two concepts for a stair-climbing machine. The first concept (*center*) decouples the lift and move. Although it does the job, the second concept (*right*) does not.

horizontally onto the next step, and finally the scissors jack closes and is pulled back underneath the vehicle. The coupled design employed four articulated arms, tank-like tracks on each arm, and a complicated motion to both lift and move at the same time. Though both machines performed admirably, the decoupled design has a much higher potential payload and is easier to build, since it requires half as many motors.

The previous example suggests the following design guideline: **Seek a modular design**. A modular design is one in which the design solutions for each function have been physically isolated. The main advantage of a modular design is that the individual modules can be designed, manufactured, and tested in parallel, leading to much shorter product development times.

In looking for opportunities to improve a given design, the situation may arise in which Suh's two principles appear to be in conflict. For example, a design change aimed at increasing the independence of the functional requirements could result in greater complexity. Suh contends that any design change that either increases information or sacrifices the independence of the functional requirements should not be accepted. There always exists a less coupled design with lower information content.

These two design principles are illustrated in the following example.

Example 26.1

A head-to-head student design competition named Davy Jones's Treasure Trove is based on the following problem definition:

Design a system to retrieve surface (ping-pong balls) and subsurface (1 lb mass) objects from a swimming pool.

This is subject to the following major design requirements:
- It must fit in a 2 × 2 × 3 ft volume at the start of the competition.
- It must carry a target that disables the boat if struck by opponent.
  Concept drawings of two of the student designs are shown in Figs. 26.2 and 26.3.

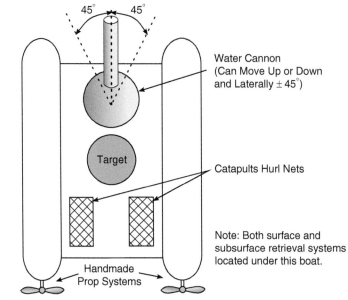

FIGURE 26.2

Water cannon design (*top view*).

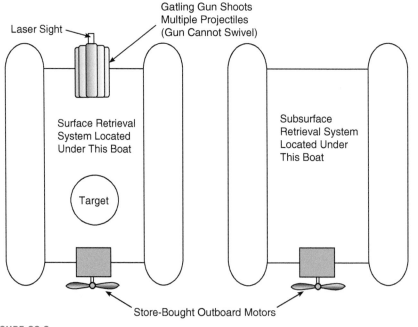

FIGURE 26.3

Twin boat design (*top view*).

Let us look at these figures and evaluate the two student designs by identifying applications and violations of Suh's design principles. (**Hint**: Only relevant features have been labeled in the figures.)

Solution

Water cannon design

- The water cannon might be effective, because aiming and steering are independent.
- Catapults will not be as effective, because aiming is dependent on steering.
- The boat should be very maneuverable, because the two props serve to decouple the move and turn functions; that is, it should be able to turn on a dime.
- Manufacture of the prop systems could be needlessly time consuming.

Twin boat design

- The use of two boats will be very effective, because it decouples the two retrieval functions. One boat can collect the ping-pong balls, while the other collects the 1 l b masses.
- The use of store-bought propellers saves time.
- The use of two nearly identical hull designs simplifies both design and manufacturing, thus saving more time.
- The boats will not be as maneuverable as the Water Cannon Design because the move and turn functions are not independent; that is, the single-prop design needs to be moving forward in order to turn.
- The Gatling gun will not be as effective as the water cannon, because aiming is dependent on steering.

Final note

The preceding characteristics provide accurate insight into how the boats actually performed. The Twin Boat Design won the competition largely on the strength of its dual retrieval system. Although the water cannon was far more effective than the Gatling gun, the Water Cannon Design was ultimately at the mercy of its handmade props, which took a lot of time to manufacture, left little time for testing, and proved to be unreliable.

26.4 **Design for ease of manufacture**

There are clear advantages to going with a design that is easy to manufacture. If, among competing design teams, yours is the first to complete manufacture of your design, the extra time can be used to test, debug, and optimize performance. For a commercial enterprise, first to market can mean a short-term monopoly in a fiercely competitive marketplace. Often, ease of manufacture goes hand in hand with lower costs. So, given the choice of two concepts, both of which satisfy the design requirements to the same degree, where one is more difficult than the other to manufacture, it makes sense to choose the concept that is easier to manufacture.

At this stage in the design process, evaluation of ease of manufacture should be done at a level of abstraction consistent with the concept drawings. The counting of machining operations and assembly steps is reserved for a later time, when the requisite level of detail in the design has been attained. Here, the emphasis is on developing an impression of ease of manufacture as revealed through Suh's design principles.

A student or team with the formidable task of having to build a complex design should be asking the following questions as each concept is evaluated:

Are there a large number of parts?

If there are a lot of parts that need to be made and assembled, it will take a long time to build.

Are there a large number of different kinds of parts?

For parts of comparable complexity, it takes less time to make two of the same part than two different parts because of reduced setup times.

Are there parts with complicated geometry?

These parts take longer to make.

Can some parts be purchased?

This is not always an option in design competitions, but if it is, the time saved and the proven reliability of the prefabricated part usually justifies the purchase.

Do you have the skills to make all of the parts?

The safest thing to do is to choose a concept that you know you can build.

Is it a modular design?

We noted earlier that modular components can be manufactured in parallel by subgroups within the design team, thus saving time.

Are there opportunities to simplify manufacture of the design by:

- Reducing the number of parts?
- Reducing the number of different kinds of parts?
- Simplifying the shape of some parts?
- Purchasing some parts?
- Redesigning the parts that are difficult to make?
- Modularizing the design?

26.5 Design for robustness

Manufacturing errors, environmental changes, and internal wear can cause unexpected variations in performance. When the designed product is insensitive to these three sources of variability, the design is said to be **robust**. Engineers seek a robust design because performance of such a design can be predicted with greater certainty.

The designer must learn to expect the unexpected. All too often, students conceive of a design while assuming ideal operating conditions. Yet, deviations from those ideal conditions can lead to less than ideal performance, as illustrated by the following example.

Example 26.2

For a design competition, students had to design machines that could accurately throw darts at a dartboard. The machines were powered by large falling masses as illustrated in Fig. 26.4. Yet, none of the machines was perfectly repeatable. For example, from 8 feet away, the best that the machine in the following figure could do was to keep the darts within a 1-inch circle. What factors contributed to this loss of dart throwing accuracy?

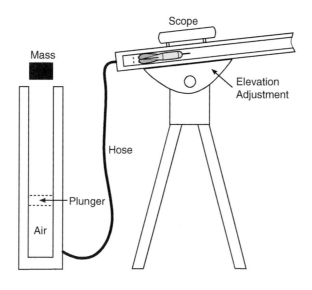

FIGURE 26.4

Concept drawing of a dart throwing machine.

Solution: Relevant manufacturing errors, environmental changes, and internal wear sites were brainstormed. The following list was the result.

Manufacturing errors
- Small dimensional differences between darts in the set of three.

Environmental changes
- Small air currents.
- Inexact repositioning of the plunger.
- Inexact repositioning of the dart within the blow gun.
- Inconsistent releases of the falling mass.

Internal wear
- Damage to the dart fins.
- Blunting of the dart tip.
- Damage to the dartboard.

When evaluating concepts with respect to robustness, you should be asking yourself the following questions:

Will small manufacturing errors dramatically impair performance?
If parts have to be manufactured perfectly for the design to function properly, you should expect to run into problems. You want a design that will work even when part dimensions are a little off. This is one reason why gear sets are such a popular design choice. Small errors in center distance between mating gears do not change the gear ratio.

Will the design function properly over the full range of environmental conditions?
Environmental conditions subject to variation include applied forces, atmospheric conditions, and roughness of surface terrain. The expected range of relevant environmental conditions should be clearly defined in the list of specifications. If they are not there, now is a good time to include them.

In a head-to-head design competition, have the actions of the opposing teams been anticipated?

Those actions can contribute significantly to the variability of the environmental conditions. For example, an opposing machine can apply forces to your machine or alter the roughness of surface terrain by laying obstacles. Therefore, you want to select a strategy-design combination that performs well irrespective of what the opposing teams may do.

26.6 Design for adjustability

In engineering courses, there is usually only enough time and resources to manufacture one design, and that design almost never performs as planned on the first try. Optimizing performance by building several designs is not an option. The only remaining course of action is to design adjustability into the initial implementation.

There are a number of ways to design for adjustability. One way is to design the system with modularity. This can serve to isolate required design changes to a single subsystem. In a mechanical system, dimensional adjustability can be attained by using nonpermanent fastening methods, such as screw joints instead of a permanent method like epoxy.

Design for adjustability can be incorporated into the evaluation process by asking the following questions as each concept is reviewed:

What are the main performance variables?

Usually, only one or two of the most important variables need be considered. Typical performance variables are speed, force, and turning radius.

Can those performance variables be easily adjusted?

Common methods were described previously. Other methods are found by brainstorming and examination of the governing equations.

Example 26.3

A motor-driven moving platform is a common feature of many small-scale vehicle designs. A top view of one such moving platform design follows (Fig. 26.5).

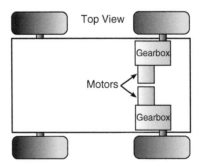

FIGURE 26.5

Moving platform.

Once manufacture of this design is complete, what adjustments can be made to
  a. Increase the speed of the moving platform?
  b. Increase its pushing force?

Solution

  a. Alternative methods for increasing the speed are

   • Decrease the gear ratio.
   • Increase the radius of the tires.
   • Increase the voltage from the power supply.
   • Switch in motors with higher RPM.

  b. Alternative methods for increasing the pushing force are
   • Increase the gear ratio.
   • Decrease the radius of the tires.
   • Increase the voltage from the power supply.
   • Switch to a motor with a higher peak torque.

26.7 Hands-on design exercise: "Waste ball"

26.7.1 Scenario

A company that uses radioactive substances for research sometimes has "spills" of spherical radioactive objects. When this happens, the radioactive objects must be transported to a waste container by the emergency team.

26.7.2 Design objective

Design a method to transfer the radioactive substance (plastic ball) from the site of the spill to a waste container (small refrigerator) at another location.

26.7.3 Setup

• Divide the team, which consists of the entire class, into subfunctional groups of two or three students each. Each subfunctional group is responsible for one leg of the transfer.
• Prior to the class, the instructor has to lay out the course. There must be as many different legs as there are subfunctional groups. The room in which the ball is initially placed and another room that contains the small refrigerator account for two of the legs. Other legs can consist of a corridor, a stairwell, an elevator, or an outdoor excursion. Try to make each challenge a little different to promote development of specialized designs by the subfunctional groups.
• Distribute the following materials to each subfunctional group:
  One daily newspaper (or equivalent).
  One roll of duct tape.
  One foam plate.
  One plastic cup.
  One pair of scissors (for construction only).

  The team also receives three balls of string, which must be shared among the groups.

26.7.4 **Rules**

1. Since the ball is radioactive, no one can be within 8 feet of the ball.
2. Use only the materials provided.
3. For safety purposes, running is not allowed.
4. If necessary, doors must be safely held open by the teams and closed immediately after waste passes through.
5. If, during transport, the ball accidently touches something besides the transport container (e.g., floor), a 30-s penalty is imposed and the group carrying the ball must restart at the location where it received the handoff.
6. The team has 3 min per group to complete each leg of the transfer.
7. The team with the minimum transit time wins.

26.7.5 **After the exercise**

- Assess team performance by comparing times to other sections.
- Discuss the quality of communication between subfunctional groups.
- What were the design considerations that went into selecting your final concept over others?

26.8 **The decision matrix**

The decision matrix promotes a systematic and exhaustive examination of concept strengths and weaknesses. The entire procedure, from selection of evaluation criteria to filling out the matrix, is designed to remove personal bias from the decision-making process. The results give a numerical measure for ranking alternatives and ultimately selecting the best concept.

26.8.1 **Evaluation criteria**

The criteria by which the concepts should be judged are all contained in the list of specifications. Since satisfaction of the design requirements designated as "demands" determines if the design is ultimately successful, "demands" should be prioritized over "wishes" when selecting the evaluation criteria.

The design requirements selected to serve as evaluation criteria are usually reworded to indicate the desired quality. For example, instead of weight, cost, and manufacture, the corresponding evaluation criteria become *low weight, low cost,* and *easy to manufacture.*

Evaluation criteria should be independent of each other to ensure a fair weighting of requirements in the decision matrix. For example, low cost and ease of manufacture are redundant and thus double-counted if the cost of labor is a significant fraction of total cost.

The number of evaluation criteria can vary depending on the situation. We suggest a level of detail consistent with the amount of detailed information available about the concept. For most hands-on student projects, five to seven of the most important evaluation criteria should suffice. *Easy to manufacture* and *low cost* are almost always included in this list.

26.8.2 Procedure for filling out a decision matrix

Step 1. Identify the evaluation criteria

This step is described in the previous Section 26.8.1.

Step 2. Weight the evaluation criteria

Weight values are assigned to each evaluation criterion in proportion to its relative importance to the overall success of the design; the larger the weight, the more important is the evaluation criterion. Though not a mathematical necessity, it is usually a good idea to define the weights such that their sum is equal to 1; that is,

$$W_1 + W_2 + W_3 + \cdots + W_N = \sum_{n=1}^{N} W_n = 1 \qquad (26.1)$$

in which N is the number of evaluation criteria. This constraint instills the view that weights are being distributed among the criteria and, in so doing, helps avoid redundant criteria.

Step 3. Set up the decision matrix

The organization of the decision matrix is illustrated in Table 26.1. The names of the concepts being evaluated are filled in at the top of each column. Likewise, the evaluation criteria and their assigned weights are written in the leftmost columns of the matrix. Scoring and intermediate calculations are recorded within the subcolumns under each concept and totaled at the bottom of the matrix.

Table 26.1 Organization of the decision matrix.

Evaluation criteria	Wt.	Concept 1		Concept 2		Concept 3	
		Val$_1$	Wt × Val$_1$	Val$_2$	Wt × Val$_2$	Val$_3$	Wt × Val$_3$
Criterion 1							
Criterion 2							
Criterion 3							
Criterion 4							
Criterion 5							
Total	1.0		OV$_1$		OV$_2$		OV$_3$

Step 4. Assign values to each concept

Starting in the first row, each concept is assigned a value between 0 and 10 according to how well it satisfies the evaluation criterion under consideration. The values are assumed to have the following interpretation:

$0 = $ *Totally useless* concept in regard to this criterion
$5 = $ *Average* concept in regard to this criterion
$10 = $ *Perfect* concept in regard to this criterion

and are recorded under the first subcolumn of each concept. This process is repeated for each criterion, going row by row to avoid bias. Usually, assignment of values is based on a qualitative assessment, but if quantitative information is available, they can be assigned in proportion to known parameters.

Step 5. Calculate the overall value for each concept

For each concept—criterion combination, the product of the weight and the value is calculated and recorded in the second subcolumn. After these calculations are completed, the overall value (OV) is computed for each concept using the following expression:

$$OV = W_1 \times V_1 + W_2 \times V_2 + W_3 \times V_3 + \cdots W_N \times V_N = \sum_{n=1}^{N} (W_n \times V_n) \qquad (26.2)$$

This is equivalent to summing the second subcolumn under each concept heading. The OVs are recorded at the bottom of the matrix.

Step 6. Interpret the results

The highest overall value provides an indication of which design is best. Overall values that are very close in magnitude should be regarded as indicating parity, given the uncertainty that went into assignment of weights and values. The final result is nonbinding. Thus, there is no need to bias the ratings to obtain the hoped-for final result. Rather, the chart should be regarded as a tool aimed at fostering an exhaustive discussion of strengths and weaknesses.

26.8.3 Additional tips on using decision matrices

1. Every member of the design team should individually fill out a decision matrix prior to engaging in team discussions. This gives everyone a chance to think about the strengths and weaknesses ahead of time and makes it more likely that all will be active participants at the team meeting.
2. Use the matrix to identify and correct weaknesses in a promising design. Give priority to the weaknesses that are most heavily weighted.
3. Feel free to create new alternatives by combining strengths from competing concepts.

Example 26.4

Recalling the golfing machines of earlier examples, the three concepts appearing in Figs. 26.6–26.8 have been proposed as best fulfilling the design requirements. Evaluate these three concepts by using the previously described procedures for filling out a decision matrix.

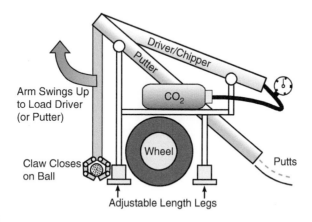

FIGURE 26.6

Concept drawing of the "Cannon".

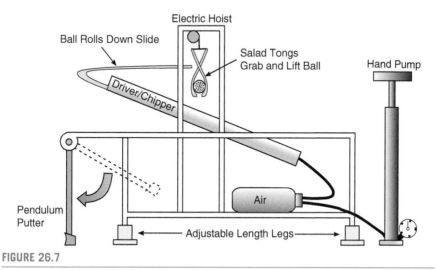

FIGURE 26.7

Concept drawing of the "Original".

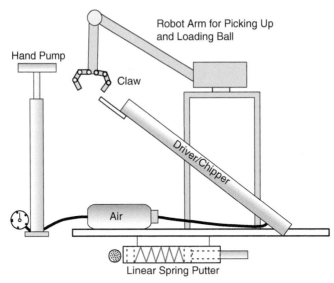

Robot Arm for Picking Up and Loading Ball

Hand Pump

Claw

Driver/Chipper

Air

Linear Spring Putter

FIGURE 26.8

Concept drawing of the "Robogolfer".

Solution: The following design requirements were selected to serve as evaluation criteria:
1. Drives well.
2. Putts well.
3. Ball loader is robust (i.e., picks up ball off of all types of terrain).
4. Easy to transport.
5. Easy to manufacture.

Low cost, which usually appears, was not selected because all three concepts met the cost requirement and cost was not involved in the design competition scoring.

With these in hand, the decision matrix can be drawn up and weights assigned to each criterion. The drives and ball loader were considered equally important, because one cannot work without the other. The drives/chips were weighted slightly higher than the putts, because 43% of all shots taken by golf professionals are putts. Transport is weighted low, because it does not factor into scoring. Ease of manufacture is always important, because of its impact on development times. The resulting weights are listed in the decision matrix of Table 26.2.

Table 26.2 Decision matrix for golf machine concepts.

Evaluation criteria	Wt.	Cannon		Original		Robogolfer	
		Val$_1$	Wt × Val$_1$	Val$_2$	Wt × Val$_2$	Val$_3$	Wt × Val$_3$
Drives well	0.25	9	2.25	8	2.00	8	2.00
Putts well	0.20	4	0.80	4	0.80	8	1.60
Loader is robust	0.25	6	1.50	6	1.50	9	2.25
Easy to transport	0.05	9	0.45	5	0.25	5	0.25
Easy to manufacture	0.25	5	1.25	7	1.75	3	0.75
Totals	1.0		6.25		6.30		6.85

Then, proceeding one evaluation criterion at a time, the team analyzes the strengths and weaknesses of each concept in the context of the given criterion and assigns corresponding values to each concept in the decision matrix. The results of the analysis follow, and the values are recorded in Table 26.2.

1. Drives well

All three drivers appear to be promising, given the effectiveness of the notorious potato gun. However, since the CO_2 tank comes prepressurized and the hand-pumping is subject to a 60 s time limit on preshot preparation, the Cannon is likely to be firing the ball at higher pressures and so should have a greater range.

2. Putts well

The greens at the site of the competition are severely sloped and slow. Therefore, the machines must be capable of executing long putts. Of the three machines, the Robogolfer is the most adjustable, since the springs can be easily replaced if the range proves inadequate. On the other hand, the potential energy of gravity powers the other two putters, and it will be difficult to increase starting heights once these machines are built. Therefore, greater risk is associated with these putters.

3. Loader is robust

A wide range of lies is possible, from severe slopes to sand and divots. The Cannon and the Original address this issue by using legs that are adjustable in length. The robot arm of the Robogolfer is clearly the most flexible design and requires no setup time.

4. Easy to transport

The rules require that only one student from the team may transport the machine to the next shot location. The Cannon is the easiest to transport because only it has wheels.

5. Easy to manufacture

The robot arm of the Robogolfer stands out as easily the most complicated system on any machine. As a three-degree-of-freedom mechanism, it requires three independently controlled motors. The Original's loader should be straightforward to manufacture. The salad tongs and the parts for the electric hoist can be easily purchased.

6. Discussion of results

The Robogolfer is the clear winner on points. But the challenges involved in designing and manufacturing that robot arm should make you nervous (unless you have a robotics expert on your team). The decision matrix also revealed that the putters for the Cannon and the Original are weak concepts. If they are replaced by the Robogolfer's linear spring putter, the Original ends up with the most points.

The concepts in Figs. 26.4–26.6 correspond to actual student designs that were designed, manufactured, and tested. The Original team (so named because they were the first team to develop an air cannon) won the design competition. They compensated for their weak putter by chipping the long putts and adding a ramp to make the short putts. The Robogolfer team took longer to design than the Original and had less time to test. The Cannon had the longest drives and the shortest putts.

26.9 Design milestone #3: Evaluation of alternatives and selection of a concept

Successful completion of this milestone requires three strong design concepts plus an open mind and a lot of careful thought.

Assignment

1. Decide on **five to seven** evaluation criteria that will be used with a decision matrix to evaluate the three concepts from milestone #2.
2. **Assign weights** to the evaluation criteria.
3. Fill out a **decision matrix**. One row at a time, discuss the strengths and weaknesses of all of the concepts in the context of the given criterion, and assign values by consensus before moving on to the next criterion. There is no need to rig the results of the decision matrix to come out to the concept you want as the results are nonbinding.
4. **Analyze the results** of the decision matrix. Use the matrix to look for weaknesses and attempt to correct them by combining ideas from different concepts. Do not blindly obey the results of your decision matrix; the selection of evaluation criteria may have been flawed to begin with.
5. **Select the best concept**. Engage everyone in the decision-making process. Do not shy away from bold designs just because they are different from everyone else's. Those differences could lead to victory at the final competition.
6. **Document your evaluation** process as per Example 26.4.

Grading criteria

- Are weights and values accurate and fully justified?
- Were the results of the decision matrix interpreted thoughtfully when searching for and selecting the best concept?
- Were all three concepts strong designs?
- Is the documentation typed and clearly written?

Exercises

Minimize design information content

1. A company received a complaint from a customer that she bought their product but the box it was supposed to be in was empty. The company then installed an X-ray machine on the assembly line and staffed it with two workers to make sure that no more empty boxes escaped detection. How would you solve this problem using the KISS principle? (**Ans.**: Install a fan on the assembly line that would blow off any empty boxes.)
2. NASA spent $12 million to develop an ink pen that would write in zero gravity and in a wide range of temperatures. How would you solve this problem using the KISS principle? (**Ans.**: Use a pencil.)
3. To test the design ability of an engineering student a professor filled a bathtub with water and then offered the student a tea spoon, a cup, and a bucket to empty the tub. How would you use the KISS principle to empty the tub? (**Ans.**: Pull the plug.)
4. Internet web pages are often confusing. Use the internet to find a web page you find too complex and suggest ways to KISS simplify it.
5. Shakespeare wrote "This life, which had been the tomb of his virtue and of his honor, is but a walking shadow; a poor player, that struts and frets his hour upon the stage, and then is heard no more: it is a tale told by an idiot, full of sound and fury, signifying nothing." Use the KISS principle to simplify this quote.

6. Shakespeare also wrote "Shall I compare thee to a summer's day? Thou art more lovely and more temperate. Rough winds do shake the darling buds of May, and summer's lease hath all too short a date. Sometime too hot the eye of heaven shines, and often is his gold complexion dimmed and every fair from fair sometime declines, by chance, or nature's changing course, untrimmed. But thy eternal summer shall not fade nor lose possession of that fair thou ownest. Nor shall Death brag thou wanderest in his shade, when in eternal lines to time thou grownst; So long as men can breathe or eyes can see, so long lives this, and this gives life to thee." Use the KISS principle to simplify this message to his beloved.

Functional requirement independence

7. Explain whether or not the steering and speed of a large ship are functionally independent? (Ans. they are not because the radius of the turn depends on the speed of the ship.)
8. Explain whether or not the steering and speed of a bicycle are functionally independent?
9. Is there any component in a system that you think should be functionally dependent on another component (think safety)?
10. How could you uncouple the water cannon design steering and catapult aiming shown in Fig. 26.2?
11. How could you uncouple the twin boat design's move and turn functions shown in Fig. 26.3?

Design for ease of manufacturability and robustness

12. What are the seven questions you should ask when designing for ease of manufacture?
13. The Boeing 777 uses three million parts that come from 500 suppliers around the world. Explain why this would be an example of ease of manufacturability.
14. What are the three questions you should ask when designing for robustness?
15. End users often use products for things that the designers did not anticipate. How would you increase the robustness of a ceramic coffee cup so that someone could use it as a hammer?
16. What adjustments can be made to the motor driven moving platform in Example 26.3 to decrease the cost?

Decision matrix

17. Develop a decision matrix for the two student designs in Example 26.1 using the following criteria: (1) It must fit in a $2 \times 2 \times 3$ ft volume at the start of the competition; (2) It must carry a target that disables the boat if struck by opponent; (3) the steering and motion must be independent; (4) it is easy to manufacture.
18. Develop a new decision matrix for Example 26.4 (Table 26.2) by adding "Low cost" to the list of criteria and changing the weighting factors as follows: Drives well 0.15; Putts well 0.20; Loader is robust 0.15; Easy to transport 0.05; Easy to manufacture 0.25; Low cost 0.2. Determine your own values for each of the three golf machine concepts.

Design step 4: Detailed design

27

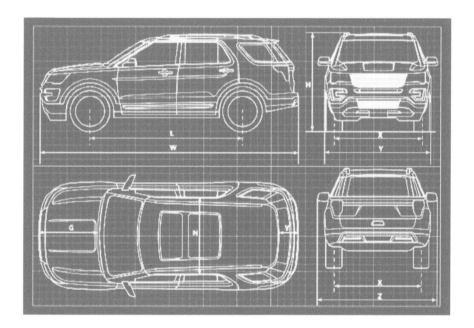

Source: GoldenSkorka/Shutterstock.com

27.1 Introduction

The goal of this step in the design process is to specify the details of the design so that it can be manufactured. Those details are typically the dimensions and material composition of parts, as well as the methods used to join them. The decisions made during detailed design are guided by the following four steps to reduce the "risk" that additional design changes will be needed later:

Step 1 Analysis
Step 2 Experiments

Exploring Engineering. https://doi.org/10.1016/B978-0-443-13541-5.00008-8

Step 3 Models
Step 4 Detailed drawings

27.2 Analysis

Analysis refers to the application of mathematical models to predict performance. The role of analysis in freshman design projects is limited because the analytical capabilities of engineering students are just starting to develop. Calculation of power requirements is the first step in motor or gear box selection for student design projects. If the mechanical power required for a given application is known, then the power ratings for various motors can be reviewed to determine the best motor for use in the design project.

27.2.1 Selection of DC brushed motors

Direct current (DC) brushed motors are commonly used in student design projects due to their availability, price, and performance. They are named for the contacts (brushes) that conduct electricity between the rotating and stationary parts of the motor. Fig. 27.1 shows the motor torque curve for a typical DC brushed motor with a constant applied voltage.

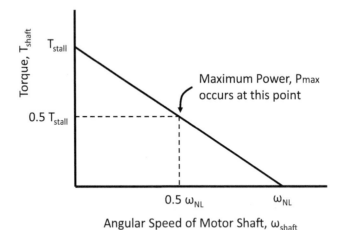

FIGURE 27.1

A typical DC brushed motor torque curve.

In this figure, T_{shaft} is the torque acting on the motor shaft and ω_{shaft} is the resulting angular speed of the motor shaft about its axis measured in revolutions per min (rpm) or radians per second (rad/s). Torque, with units of Force × Length, represents the rotational resistance felt by the motor as it attempts to rotate an element (gear, pulley, or wheel) which has been attached to the motor shaft. The greater the torque on the motor shaft, the slower the shaft will rotate. Thus, the torque on the motor shaft determines the motor speed and not vice versa.

The fact that the motor curve is linear makes the behavior of the motor very predictable and well suited for design applications. It also means that the two motor constants T_{stall} and ω_{NL} are sufficient to completely define the motor curve. The stall torque (T_{stall}) is the applied torque that will cause the motor to stop rotating, or stall.[1] It only takes a few seconds at this torque to burn out the motor. The no load angular speed (ω_{NL}) is how fast the shaft will rotate when there is no torque on the motor's shaft.

In DC motors, the input electrical power, P_{elect}, is converted to mechanical power P_{mech}. However, there are power losses within the motor so, $P_{mech} = P_{elect} - P_{loss}$. For rotational motion, the mechanical power output can be calculated as the product of shaft torque (T_{shaft}) multiplied by the corresponding angular speed (ω_{shaft}), or

$$P_{mech} = T_{shaft} \times \omega_{shaft} \tag{27.1}$$

where P_{mech} has units of Force $\times$ Length/Time (for example, newton-meter/second = watts), and ω_{shaft} has units of 1/Time since it must be in units of radians/Time when substituted into this equation and radians are treated as dimensionless.

If data points along the motor curve in Fig. 22.1 are substituted into Eq. (27.1), and the resulting values of P_{mech} are then plotted against ω_{shaft}, the power curve shown in Fig. 27.2 is obtained.

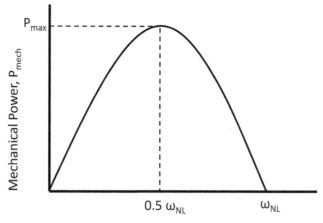

FIGURE 27.2

Shape of the power curve for a DC brushed motor.

---

[1] The motor's stall torque can be measured by attaching a pulley to the output shaft and winding a string around the pulley. With the motor running and winding up the string, slowly add weights to the free end of the string until the motor can no longer lift the weights. The stall torque is then the product of the weight and the pulley radius. Do not stall the motor for a long period of time as that will damage the motor.

According to Fig. 27.2, the maximum mechanical power P_{max} that can be produced by a DC brushed motor occurs when it is running at 50% of its no load speed, or $\omega_{shaft} = 0.5\omega_{NL}$. From Fig. 27.1, we see that this occurs when $T_{shaft} = 0.5T_{stall}$. Therefore, the maximum mechanical power delivered by a DC brushed motor can be calculated from Eq. (27.1) as follows:

$$P_{max} = (0.5T_{stall})(0.5\omega_{NL}) = 0.25T_{stall} \times \omega_{NL} \qquad (27.2)$$

Since the motor-driven mechanism is likely to have internal friction, the DC brushed motor should be selected such that its max mechanical power as defined by Eq. (27.2) is somewhat greater than the required mechanical power that can be estimated from the design requirements using Eq. (27.1), or

$$0.25T_{stall} \times \omega_{NL} > T_{req} \times \omega_{req} \qquad (27.3)$$

where T_{req} and ω_{req} are the torque and angular speed that must be produced by the motor under the operating conditions defined in the list of design requirements.

Example 27.1

Determine the power required to drive a small vehicle for a student design competition. You have determined that the torque on the driving wheels shaft is $T_{shaft} = 3.0$ in-lbf. Your design calls for the wheel rotational speed to be $\omega_{shaft} = 50.$ rpm. What motor power will be required to meet these conditions?

Need: The required motor mechanical power, P_{mech}.

Know: The required torque and shaft speed are $T = 3.0$ in-lbf and $N = 50.$ rpm.

How: The motor power comes from Eq. (27.1), $P_{mech} = T_{shaft} \times \omega_{shaft}$

Solve: First let us convert the units into the metric system

Using the unit conversion equations inside the front cover of this text we get:

$$T_{shaft} = 3.0 \text{ in-lbf} = 0.34 \text{ Nm and } \omega_{shaft} = 50. \text{ rpm} = 5.2 \text{ rad/s}.$$

Then, $P_{mech} = T_{shaft} \times \omega_{shaft} = (0.34 \text{ Nm}) \times (5.2 \text{ rad/s}) = 1.77 \text{ Nm/s} = 1.77 \text{ J/s} = \mathbf{1.8 \text{ W}}$

Considering the unknown losses in the motor, Eq. (27.3) tells us that you will need something more than a 1.8-W motor, say at least a 2.0-W motor.

On the other hand, if you have the situation where the torque acting on the motor shaft is close to zero such that $P_{mech} \approx 0$, then motor selection can be based on matching the no load angular speed (ω_{NL}) of the motor to the angular speed (ω_{req}) that the motor must produce under the operating conditions defined in the list of design requirements.

$$\omega_{NL} \approx \omega_{req} \qquad (27.4)$$

where ω_{req} corresponds to the required angular velocity of a wheel, gear or link that will be attached to the motor shaft.

For most student projects, the torque can be assumed to be small so that motor selection can be based on matching angular speeds as defined in Eq. (27.4). The no load angular speed for small DC brushed motors can be expected to range from 2000 to 20,000 rpm, which is normally too high for your application. However, there are also DC brushed "gear" motors to choose from that are regular DC motors with an integrated gearbox. No load speeds of these motors can vary from 2 to 3400 rpm.

If additional fine-tuning of the angular speed from the selected motor is needed, a gear set with the following gear ratio could be introduced:

$$\textbf{Gear Ratio} = \frac{\omega_{\textbf{NL}}}{\omega_{\textbf{req}}} \qquad (27.5)$$

which can be achieved with a pair of gears or a pair of pulleys with a belt as shown in Fig. 27.3 if the gear ratio is 10 or less. For higher gear ratios, see Section 15.4.2 in the Mechanical Engineering Chapter.

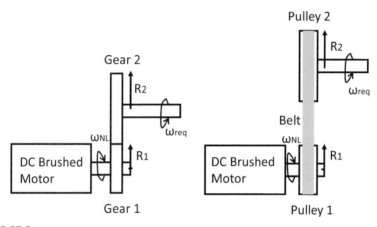

FIGURE 27.3

Two methods for introducing a gear ratio of $GR = R_2/R_1$.

Example 27.2

You need to design a robotic moving platform where the drive wheels have an angular speed of 57.3 rpm on a flat surface. The motor to be used has specifications of $T_{stall} = 0.210$ oz-in and the no-load speed is $\omega_{NL} = 11,600$ rpm. Determine the gear ratio required to achieve the desired angular speed of the drive wheels.

Solution: Since this exercise is a direct use of Eq. (27.5), we can forgo the Need, Know, How, and Solve steps and go directly to Eq. (27.5) to find

$$\textbf{Gear Ratio} = \frac{\omega_{\textbf{NL}}}{\omega_{\textbf{req}}} = \frac{\textbf{11600 rpm}}{\textbf{57.3 rpm}} = 202.$$

27.2.2 Common mechanical linkages

A **mechanical linkage** is a series of rigid *links* connected with *joints* to form a closed chain. A linkage with two or more links movable with respect to a fixed link is called a **mechanism**. Mechanical linkages are designed to perform certain tasks, such as steering an RC vehicle or making a robot climb stairs.

Linkages have different functions. The functions are classified depending on the primary goal of the mechanism. The simplest closed-loop linkage is the 4-bar linkage. It has three moving links, one fixed link, and four pin joints.

Six common 4-bar linkages are shown in Fig. 27.4 and described below the figure.

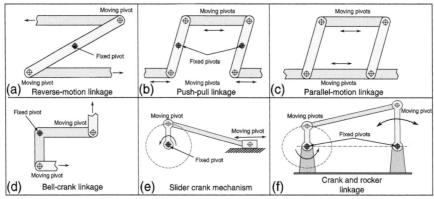

FIGURE 27.4

Common mechanical linkages.

a. **A reverse-motion linkage** makes an object or force move in opposite directions. If the fixed pivot is centered between the moving pivots, output link movement is the same as the input link movement, but in the opposite direction. By varying the position of the fixed pivot, the output link can produce a mechanical advantage.

b. **A push–pull linkage** makes the output link move in the same direction as the input link. It can be rotated through a full 360 degrees.

c. **A parallel-motion linkage** makes links move in the same direction, but at a fixed distance apart. The moving and fixed pivots on the opposing links must be equidistant for this linkage to work correctly. This linkage can also be rotated through 360 degrees.

d. **A bell-crank linkage** changes the direction of a link by 90 degrees. If the pivots are at the midpoints of the cranks, link movements will be equal. However, if those distances vary, a mechanical advantage will occur.

e. **A slider-crank linkage** is used to convert linear motion into rotary motion (and vice versa). When the slider is at its maximum distance from the axis of the crankshaft, it is a *top dead center*; when the slider is at its minimum distance from the axis of the crankshaft, it is at *bottom dead center*.

   f. In a **crank and rocker linkage**, one link rotates continuously while another link rocks back and forth. The windshield wiper on a car uses this mechanism.

27.3 Mechanism control

There are three basic electronic designs that can be used to control the speed and direction of the motors that drive a mechanism:

- **Manual control**—motor speed is controlled by turning a knob on a speed controller.
- **Remote control (RC)**—motor speed and direction are controlled in real time by the operator who, from a distance, transmits radio signals from a hand-held radio to a receiver mounted on the mechanism.
- **Autonomous control**—motor speed and direction are controlled by pre-programmed instructions that are uploaded to a microcontroller mounted on the mechanism.

Each design contains three elements: a DC "**brushed**" motor to move the mechanism, a battery to provide electrical power, and one or more electronic components that translate operator instructions into the desired motor speed and direction. All of these electronic designs are single battery solutions—one battery (or battery pack) powers all of the components in the circuit—but there are other possibilities such as having one battery for each motor.

27.3.1 DC brushed gear motors and batteries

A DC brushed **gear motor** is a regular brushed DC motor with an attached gearbox. It is a popular choice in robotics because the attached gearbox provides lower speeds and higher torque. A DC brushed gear motor normally has a two-part cylindrical shape (see Fig. 27.5). The key word here is "**brushed**" as control systems for DC "**brushless**" motors (such as those used on RC aircraft) are far more complex.

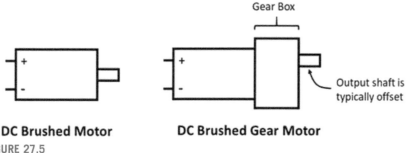

FIGURE 27.5

Recognizing a DC brushed motor and a DC brushed gear motor.

Choosing the right battery is done after selecting a motor, since the voltage of the battery should match the voltage requirement of the motor. In addition to having the right voltage, the selected battery must last long enough to complete the desired task. The battery life can be estimated from its battery capacity C and average current load I_{avg} as follows:

$$\textbf{Battery Life} = \frac{C[\text{mAh}]}{I_{avg}[\text{mA}]} \qquad (27.6)$$

in which typical units have been indicated in brackets with battery capacity C having units of *milli Amp-hours*. I_{avg} can be estimated as $I_{avg} = 0.5\,I_{stall}$ where I_{stall} is the maximum current draw of the motor, also referred to as the stall current. Choice of battery type (see Fig. 27.6) depends on required life and the nature of the application. Attributes and weaknesses of the three most commonly used battery types are listed below:

FIGURE 27.6

Battery types (from left to right): lead acid, lithium polymer, four AA alkaline battery pack.

- **Alkaline batteries:** Are readily available; small and lightweight but also low battery capacity because of their small size; multiple batteries can be conveniently housed in series in a plastic battery pack (a typical 6 V battery pack holds four 1.5-volt AA batteries). A disadvantage is that alkaline batteries cannot be recharged.
- **Lead acid**: Have high capacity but are relatively large and heavy. However, low cost 6 and 12 V battery chargers are easily available.
- **Lithium Polymer (LiPo)**: Have high battery capacity per unit weight relative to lead acid. But they require a "smart" charger, are a potential fire hazard, and can be difficult to dispose of.

Use alkaline batteries if the operating voltage is 6 V or less. For higher operating voltages, use lead acid batteries if weight is not of consequence; otherwise use LiPo batteries.

27.3.2 Manual control of motor speed

In the controls setup of Fig. 27.7, the motor and battery have been wired to an adjustable speed controller. The battery powers both the motor and the speed controller. Turning the control knob changes the voltage applied to the motor, and consequently the motor speed. Applied voltage can be varied from 0 up to the rated voltage of the battery. The direction of the motor speed cannot be changed by turning the knob. This is a low-cost, effective solution when motor speed must be adjusted either to attain a desired speed or to optimize performance of the mechanism being driven by the motor. These speed controllers typically work for a wide range of battery voltages and currents.

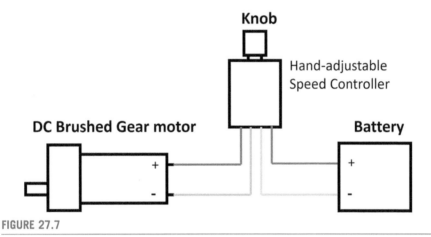

FIGURE 27.7

Motor speed control with a hand-adjustable speed controller.

27.3.3 Remote control of motor speed

RC enables the operator to vary motor speed in real time while being situated some distance away from the mechanism. The speed and direction of a motor can be varied by moving a control stick on the hand-held RC radio controller where the speed is proportional to the movement of the control stick. The number of motors that can be controlled by one radio controller is limited by its number of channels (typical radio channels are 2, 4, and 6). The signal from the radio controller is sent wirelessly to an RC receiver mounted on the mechanism. The receiver has ports for plugging in RC cables, one port for powering the receiver plus one additional port for each channel of the radio controller. Each RC cable consists of three wires—two for power and one for sending a signal.

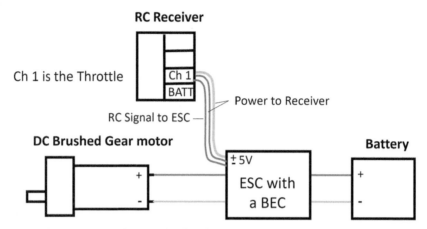

RC Receiver

Ch 1 is the Throttle

Ch 1

BATT

Power to Receiver

RC Signal to ESC —

DC Brushed Gear motor

Battery

± 5V

+

ESC with a BEC

+

-

-

Electronic Speed Controller (**ESC**) with Battery Elimination Circuit (**BEC**)
(Note that there is no connection to the BATT port on the RC receiver)

FIGURE 27.8

Remote control of motor speed using an RC electronic speed controller.

The electronics design in Fig. 27.8 employs only standard RC components—radio controller, receiver, and an **electronic speed controller (ESC)** with built-in **battery elimination circuit (BEC)**. The BEC eliminates the need to supply power to the receiver through its battery port. Instead, a single RC cable plugged into one channel port is sufficient to both power the receiver and pass on the signal from the radio controller to the ESC. The RC signal from the receiver, like the turning of the knob in the manual design, communicates to the ESC how the motor speed should be adjusted. This is probably the easiest of the RC designs to implement; however with each motor requiring its own ESC, the cost of the ESCs can quickly escalate.

The alternative RC design shown in Fig. 27.9 is a more practical design than the one just discussed. In this design, the ESC has been replaced by a **motor controller**. Motor controllers distribute power from the battery, some of it at a lower voltage to provide power to an RC receiver or to a microcontroller. They can also interpret signals from an RC receiver or microcontroller to modify the speeds of up to two motors at a time. Unlike ESCs, motor controllers are compatible with microcontrollers which often make them a more practical purchase. The wiring of this design is slightly different from the one in Fig. 27.8. Since the motor controller lacks a BEC, the motor controller has to divert some of the power from the battery to the battery port of the RC receiver.

27.3.4 Autonomous control of motor speed

Autonomous control, which means varying a motors speed without real-time human interaction, can be achieved by replacing the RC radio and receiver combination of the previous design by a **microcontroller** such as the Arduino Uno (see Fig. 27.10).

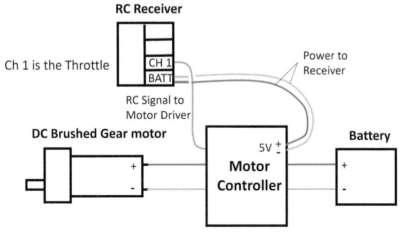

FIGURE 27.9

Remote-control of motor speed using a motor controller.

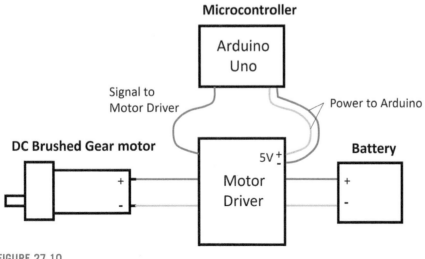

FIGURE 27.10

Autonomous control of motor speed using a microcontroller with a motor driver.

User-defined instructions for varying motor speed are first coded into a computer then uploaded to the microcontroller. Since there is no need for an RC interface here, a low cost **motor driver** is used in place of a motor controller to distribute power to the Arduino and motor at different voltages. Code-generated signals from the microcontroller are interpreted by the motor driver as voltages that are applied to the motor.

This setup is especially useful when the voltage of the motor exceeds the voltage required to power the microcontroller. When this is not the case, there is always the option of eliminating the motor driver from the circuit.

Use of a microcontroller offers the additional advantage that it can process input signals received from one or more sensors and use the results to modulate motor speeds and directions. Details are discussed in Chapter 21.

27.3.5 Control of servo-actuated functions

A **servo** is a special form of gear motor with its own built-in control system that allows angular displacement of its shaft to be easily controlled. The maximum rotation of a typical servo motor shaft is 180 degrees. Because its gears and outer casing are made of plastic, servos are not as well suited for high torque applications as a metal gear motor. Thus, applications tend to be limited to the servo acting as a mechanical switch or driving one of the links of a low torque mechanism. Servos are however easily implemented in microcontrollers and RC systems. For example, servos plug directly into one of the channel ports of the receiver. See Chapter 21, Section 21.5.5 for details on how to control the servo with an Arduino Uno microcontroller.

27.4 Experiments

Physical **experiments** are a particularly effective way to reduce risk when working with small robotic systems. Because of the small scale, materials needed for the experiments can be scavenged or at least obtained at low cost, and realistic forces can easily be applied. Also, physical experiments are often more accurate than idealized mathematical models at this scale.

Since the actual design has not been built yet, the subfunction being investigated may have to be idealized for the purposes of the experiment. For example, you might use cheaper materials or use your hands to create the motion. The errors introduced by these approximations are tolerable if they are much smaller than the changes in performance being observed.

Knowing when to use experiments requires a keen awareness of the sources of risk[2] in a design. This is no time for overconfidence; you can safely assume that *if something can go wrong, it will*. Therefore, it is vital that you be able to distinguish between the aspects of the design that you are sure about and those aspects that you are not so sure about. The four steps for formulating an experimental plan are as follows:

1. Identify aspects of the design and its performance about which you are uncertain.

---

[2] Risk is the possibility that something bad or unpleasant (such as having to redesign your robot) will happen.

2. Associate the aspects in step 1 with one or more physical variables that can be varied by means of simple experiments.
3. Carry out the experiments that will do the most to reduce risk of redesign within the available time frame.
4. If possible, document the results in the form of graphs or tables.

Example 27.4

A concept for a robotic design competition named Dueling Duffers has been proposed and is shown in Fig. 27.11. The object of this head-to-head competition is to be the first to deposit up to 10 golf balls into a hole in the center of the tabletop playing field shown in Fig. 27.12.

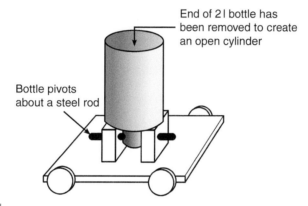

FIGURE 27.11

Proposed concept for the "Dueling Duffers" design competition.

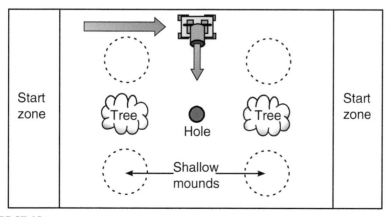

FIGURE 27.12

Playing field for the "Dueling Duffers" design competition.

These are the main features of the proposed design:
- It holds 10 golf balls in the top half of a 2-L bottle.
- It uses the side rail to steer.
- When it is even with the hole, it dumps the golf balls in the general direction of the hole.

The path of the vehicle and the direction in which the balls are dumped are indicated in Fig. 27.12. For this example, you are asked to (a) identify the main sources of risk and (b) propose experiments to address those sources of risk.

Solution

Sources of risk

The sources of risk are presented here in the form of questions, the answers to which are currently unknown. These are the same type of questions that will be asked by the jury at the oral design defense:

1. Will all 10 balls fit in the top half of the 2-L bottle?
2. What is the optimal height from which to dump the balls?
3. Will all 10 balls drop into the hole when dumped in this manner?
4. Is it important to have the vehicle perfectly positioned before dumping the balls?

Proposed experiments

The experimental setup in Fig. 27.13 can be used to address each of the first four sources of risk just listed. The last item in the list is best handled by computing the overall gear ratio using Eq. (27.5).

The bottle, the steel rod, and the two blocks of wood are all easily obtained. Assume that the actual playing field is available for testing. The bottom of the plastic bottle will need to be cut off, and the steel rod will need to be inserted through the top of the bottle. There are no other parts that need to be joined. You can use your hands to keep the rod in place above the blocks while taking care to let the bottle and balls fall under their own weight.

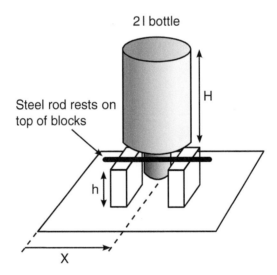

FIGURE 27.13

Experimental setup for establishing key dimensions and reducing risk.

The numbers of the experiments described next correspond to the preceding numbered sources of risk:

1. The control variable is H (refer to Fig. 27.13). Put 10 golf balls into the bottle and measure the minimum H required to hold 10 balls.

2. The control variable h determines how fast the balls roll when they pass the hole. Increase h by inserting books under the blocks; decrease h by sawing the ends of the blocks. For each value of h, dump the balls three times and record the number of balls that drop into the hole. Document the results by plotting h versus the average number of balls that dropped.
3. Here, the concern is less with the speed and more with the distribution of the balls as they pass the hole. Control variables might be H or the manner in which the balls are packed within the bottle (e.g., a vertical divider could be used to keep the balls on the side of the bottle facing the hole). Again, use three trials for each value of the control variable, and plot the control variable versus the average number of balls that dropped.
4. The control variable is X. Perform three trials for each value of X, and plot X versus the average number of balls that drop. The shape of this graph provides the answer to question 4.

27.5 **Models**

Models are scaled replicas constructed out of inexpensive, readily available materials. In the case of small electromechanical devices, they are often constructed out of cardboard or foamboard. Models are used to check geometric compatibility, establish key dimensions of moving parts, and visualize the overall motion.

Typical examples are shown in Figs. 27.14 and 27.15. The foam board model of the stair-climbing device in Fig. 27.14 is used to prove the feasibility of the design. Several other stair-climbing concepts, mainly wheel and track designs, were regarded as feasible until models proved otherwise. In Fig. 27.15, a dart-throwing mechanical linkage is modeled in 3D using cardboard. All the links move, including the grip mechanism that releases the dart when the arm strikes a stopper.

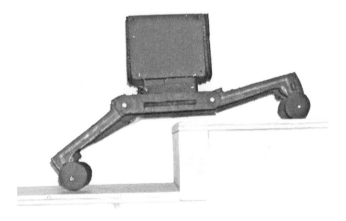

FIGURE 27.14

Model of a stair-climbing device.

FIGURE 27.15

Model of a dart-throwing machine.

27.6 Detailed drawings

By definition, a **detailed drawing** contains all the information required to manufacture the design. The drawings should be so complete that if you handed them off to someone unfamiliar with the design, that person would be able to build it. The various elements of engineering drawing, dimensioning, and sketching are discussed in Chapter 2. Table 2.1 illustrates the five common types of drawing projections.

The usual practice is to specify dimensions on multiple orthogonal views of the design. An isometric view is also sometimes provided to assist with visualization. In all, six orthogonal views are possible: front, back, left, right, top, and bottom. Three views, however, are the most common. Fig. 27.16 shows a detailed drawing with five orthogonal views.

Additional information such as material specification, part type, and assembly directions are conveyed through written notes on the drawings. Close-up views can be employed to clarify small features.

While practicing engineers generate drawings like Fig. 27.16 using computer-aided design (CAD) software, first-year engineering students may or may not be taking a CAD course (or had taken one in high school). Therefore, we recommend that

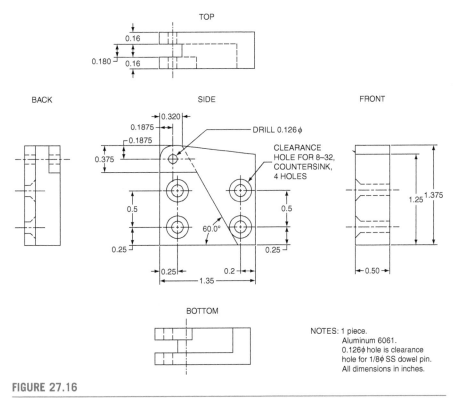

FIGURE 27.16

Five views of the right toe of an animatronic eastern gray squirrel.

the usual standards for preparation of detailed drawings should be relaxed somewhat and replaced by the following set of guidelines:

- Drawings can be neatly hand-drawn using a ruler and compass.
- Drawings must be drawn to scale, though not necessarily full scale.
- Drawings of at least two orthogonal views of the design should be prepared. An isometric view is not required, but close-up views should be used to clarify small features.
- Show hidden lines only when they enhance clarity. These are dashed lines that are used to show edges not visible from the viewer's perspective.
- It is acceptable to show only essential dimensions, that is, key dimensions that either have a direct impact on performance or are needed to demonstrate that geometry constraints are satisfied.
- Use notes or labels to indicate material specification, part type, and assembly directions.
- It is acceptable if the manufacturing details are incomplete. Students who lack the experience to fully specify them have to discover those details by trial and error during building.

• If an electric circuit was designed, show it in the form of a neatly hand-drawn but fully specified circuit diagram.

An example of a detailed drawing prepared in accordance with these guidelines is shown in Fig. 27.17.

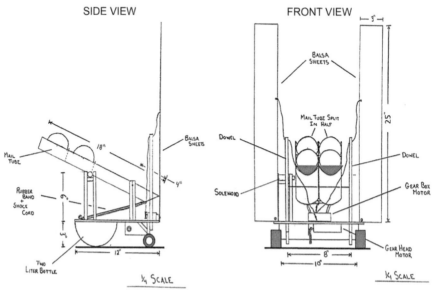

FIGURE 27.17

Hand-drawn detailed drawing of a competition vehicle.

Although the freedom to leave out some dimensions and manufacturing details has been allowed, keep in mind that missing details amount to higher risk in the minds of those being asked to provide resources to the project. Thus, a design with fewer missing details may be viewed by the instructor as having lower risk.

27.7 Design milestone #4: Detailed design

This milestone is all about reducing risk, not only in your own mind but in the minds of the jury at the upcoming oral design defense.

Assignment

1. Use **analysis, experiments, and models** to help establish dimensions and proof of concept. Time spent now on analysis, experiments, models, and detailed drawings pays off later in fewer design changes during manufacturing and testing.
2. Prepare **detailed drawings** of the design concept you selected. Of the four methods used in reducing design risk (analysis, experiments, models, and

detailed drawings), good, detailed drawings reap the most rewards in a freshman design project.

Documentation

- Write the analysis details in the usual format, stating all assumptions.
- For each experiment, (a) state the purpose of the experiment, (b) describe the experimental procedure, (c) present results, and (d) state conclusions.
- Summarize useful information yielded by models and turn in the models constructed.
- Attach hand-drawn detailed drawings.

Grading criteria

- From examination of the detailed drawings, does the design have a chance of working?
- Have opportunities to reduce the level of risk (through analysis, experiments, and models) been fully exploited?
- Is there enough information in the detailed drawings to manufacture the design?
- What is the overall quality of the detailed drawings?

Exercises

Analysis

1. If the rotational speed of a shaft is 250 rpm, what is its rotational speed in radians per second?
2. What is the mechanical power output of a shaft rotating at 1500 rpm that has a torque of 200 Nm?
3. If a small DC motor has a power output of 3 W at 700 rpm, what is its torque?
4. A small DC brushed motor has a torque load of 350 Nm acting on its motor shaft. If the motor has a stall torque of 500 Nm and a no-load angular speed of 6000 rpm, what is the angular speed of the motor shaft?
5. If the no-load motor speed in Example 27.2 was 3600 rpm, what overall gear ratio is required for the drive wheels of the moving platform to have an angular speed of 38.2 rpm?
6. If the overall gear ratio in Example 27.2 was 50, what is the angular speed of the drive wheels of the platform?
7. If the motor in Example 27.2 was replaced with a similar motor that had a torque of 3.42 ounce-inches and a no-load speed of 1759 rpm, what overall gear ratio is required for the drive wheels of the moving platform to have an angular speed of 65.8 rpm?
8. If the moving platform in Example 27.2 had a constant acceleration of 0.056 ft/s^2, how long would it take for the platform to reach its top speed of 6 inches per second?

Experiments

9. The Dueling Duffers concept shown in Fig. 27.11 requires the bottle to be tipped over far enough to spill all the golf balls out of it. How would you develop an experiment to determine how far the bottle needs to be tipped so that all the golf balls escape?
10. In the Dueling Duffers concept shown in Fig. 27.11, it was decided to tip the bottle by pushing on it halfway up its surface at H/2. How would you develop an experiment to determine the force required at this height to tip the bottle far enough to spill the golf balls?
11. During the development of the Dueling Duffers concept shown in Fig. 27.11, it was decided to glue the bottle to the steel pivot rod and use a gear motor to turn the rod to tip the bottle and

spill the golf balls. How would you develop an experiment to determine how much torque would be required by the motor?

12. In Example 26.2 of the previous chapter, the concept of a dart throwing machine was presented. How would you develop an experiment to determine how much mass the plunger should have to propel the dart into a target 8 feet away?

13. In the three concept drawings shown in Example 26.4 of the previous chapter, how would you develop an experiment to determine how much air or CO_2 pressure is required to drive the golf ball 250 yards?

Models and drawings

14. Make a hand-drawn detailed sketch of the Dueling Duffers concept shown in Fig. 27.11. Choose your own dimensions for the components.

15. Make a hand-drawn detailed sketch of the stair climbing device shown in Fig. 27.14. Choose your own dimensions for the components.

Design step 5: Design defense

Source: Maksym Drozd/Shutterstock.com

28.1 Introduction

Engineers must convince customers that a design is worth expenditures of money and the time of skilled people. In a student design project, this process of convincing customers is simulated by an **oral design defense**. The goal of this oral presentation is to win the confidence of the project sponsors, henceforth referred to as the **jury**.

28.2 How to prepare an oral defense

In assessing a team's chances for future success, the jury is searching for answers to the following questions:

- Did the team adhere to the systematic approach?
- How does the final concept work?

Exploring Engineering. https://doi.org/10.1016/B978-0-443-13541-5.00027-1

- What level of risk is associated with this design?
- Do the students appear to be teaming effectively?

The jury's concerns suggest some strategies that should be effective. First, the organization of the presentation should parallel the steps in the design process, as shown in Table 28.1. This is your way of saying that you followed a systematic approach. Second, you should try to get the jury to understand how your final concept works as quickly as possible. This frees up more time during questions for alleviating concerns about the design. Third, you should anticipate that the jury will ask questions about potential sources of risk and prepare evidence in advance to quell those concerns. This evidence should be in the form of high-quality detailed drawings, results of calculations and experiments, and models. If you built models, bring them; if you conducted experiments, try to bring some evidence that indeed you did them. The strategies that serve to reduce risk, and their counterparts that do not, are summarized in Table 28.2.

Meanwhile the jury also is evaluating your teaming. It will base its impressions on the quality of your design and oral presentation (see Table 28.3 for some tips on delivery and visual aids). There are other telltale signs. For example, did everyone

Table 28.1 Suggested organization of the oral design defense.

Organization	Slides
Title	1
Outline of presentation	1
Problem definition	1
Important design requirements	1
Alternative concepts not selected	2
Final concept	1–2
Describe main features	
Explain why you selected it	
Detailed design	
Show main drawings	2
Explain how the design works	
Explain how you will construct it	
Zero in on special features with closeup views	1–2
Present results of analyses, experiments, and models	1–2
Summary	1
Summarize strengths of the design	
Quantify performance expectations (e.g., top speed)	
Describe strategy at the final competition	
Total slides	12–15

Table 28.2 Strategies that either reduce or amplify risk.

Risk reducers
1. High-quality concept drawings and detailed drawings.
2. Calculations, experiments, and models that establish proof of concept.
3. Explanation of manufacturing details.
4. Quality visual aids.
Risk amplifiers that could delay manufacture
1. Poorly detailed drawings.
2. One or more subfunctions obviously will not work.
3. No thought given to manufacturing.

Table 28.3 Oral presentation tips.

Tips on delivery
1. Do not read sentences directly off the slides.
2. Look at the audience.
3. Stand next to the screen when speaking.
4. Practice the presentation.
5. Be positive and dynamic.
Tips on visual aids
1. Avoid using too many words.
2. Use a font size that can be seen easily.
3. Include concept or detailed drawings.
4. Put a heading on each slide.
5. Show results of calculations or experiments.
6. View the projected images in advance to make sure all the words and pictures are visible to the audience.

contribute equally to the presentation? Was everyone involved in answering questions? Did team members refer to themselves as "we" or "I" when citing accomplishments?

When answering questions, be forthright and honest. Failure to do so will lead to an unending chain of questions. If your response is an opinion and not a fact, state so, because one erroneous answer can damage your credibility and thus elevate the risk associated with your design.

28.3 Design milestone #5: Oral design defense

To qualify to receive parts and materials for the manufacturing phase of a hands-on design project, a majority of the jurors (or your instructor) must be convinced that your design will work. In the event, such a consensus is not achieved, teams will be asked to revise their designs and resubmit at a later date.

Assignment

1. Prepare the **visual aids** for the oral design defense. You must use PowerPoint or an equivalent software package. Relevant drawings should be scanned. Grading tends to be proportional to the quality of your drawings.
2. **Practice the presentation.**
3. Deliver the oral presentation to your instructor or a jury of evaluators. Evidence of the use of **calculations, experiments, and models** does much to reduce the level of risk in the minds of the jurors. Visual representations of the results are more effective than just saying you did them.

Typical format

- Eight minutes for the oral presentation; 4 minutes for questions.
- All team members participate in the presentation and in responding to questions.

Grading Criteria

- What level of risk is associated with the design?
- What was the quality of the drawings and the other visual aids?

Exercises

Preparing an oral defense

1. A team of students prepared an oral defense for their design project that contained only the following items: a statement of the problem and the final detailed design. What else should they have included?
2. When you are called upon in class to make an oral defense for your design project, you suddenly realize that you do not have your presentation slides with you. What can you do to make a successful oral defense without them?
3. While making their oral defense, a student group begins by telling a joke and showing a cartoon slide. This approach is often effective for certain types of presentations. Do you think it is appropriate for a design project oral defense? Why or why not?
4. When you are making your oral defense, your instructor does not want you to use notes of any kind (3 by 5 cards, etc.). What is the best way to move through your defense without reading your notes? (**Ans.:** use your slides to prompt your memory.)
5. One of the tips in Table 28.2 is to look at the audience. Some students are shy and find this very difficult to do. How can you look like you are looking at the audience without actually looking at them? (**Ans.:** look at the back wall of the room.)

Design step 6: Manufacturing and testing

29

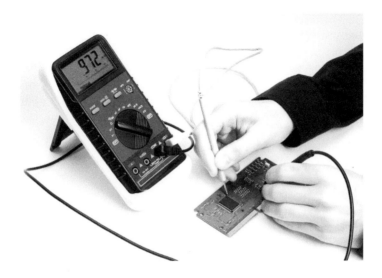

29.1 Introduction

Manufacturing (i.e., the construction of your design)[1] begins once the detailed design has been approved and ends when your device is placed in the starting zone of the final competition. In between, your device may undergo numerous modifications. Few new designs work on the first try. Manufacturing and testing tend to take much longer than expected—**probably three to five times as long**. This chapter begins with a summary of good manufacturing and testing strategies, then moves on to describe materials, joining methods, and hand tools.

---

[1] "Manufacturing" is the same as "construction" in this chapter.

Exploring Engineering. https://doi.org/10.1016/B978-0-443-13541-5.00001-5

29.2 Manufacturing and testing strategies

You can employ several strategies during the construction of your design to minimize the time it takes to get the first prototype ready for initial tests. The extra time freed up for testing and design refinements can prove decisive in a design competition.

The following time-saving **manufacturing strategies** have been observed in successful teams:

- **Talk to a machinist**. Professional machinists are considered partners in the design process. What they lack in knowledge of the engineering design process and analysis, they make up for in manufacturing and practical experience. As a practicing engineer, you will be required to consult with a machinist before finalizing your detailed design.
- **Do not delay in getting started**. Take the leap of faith and begin construction as soon as possible. Only then does the team gain a realistic sense of the manufacturing timeline.
- **Divide up responsibilities so that team members can work in parallel on different subassemblies**. Otherwise, you may find the entire team standing around waiting for the glue joint to dry.
- **Keep detailed drawings up to date**. If they are not up to date, only one person knows what the actual design looks like. That one person ends up doing most of the manufacturing, while the other team members watch. With accurate drawings, team members can work in parallel.
- **Set and enforce intermediate deadlines**. Manufacturing can span several weeks. The instructor sets the big deadlines through the required milestones and the teams should set little ones in between.

Testing is as important as manufacturing in preparing for a design competition. For example, three design teams were assigned the task of designing electromechanical machines capable of playing 18 holes of miniature golf at a local course. The first team was stocked with experienced machinists, so it built its machine out of thick steel parts. The second team had an expert welder, so it welded together its machine out of steel beams and plates. The third team chose to make its machine out of wood. Three very different manufacturing skill sets, yet all three machines failed at the final competition for the same reason—not enough testing.

The first team did not test its machine on synthetic grass before the competition. Its steel machine was heavy, which created large friction forces between its tank treads and the synthetic grass. When it attempted to turn, the treads broke, immobilizing the machine.

The second team did not have time to test their steering mechanism due to last-minute modifications. As a result, it could not consistently maneuver into position for putts within the time constraint.

The third team completed manufacture and testing of its moving platform 2 weeks before the final competition. Over the next 2 weeks, while the putting mechanism was being made, it did not test the moving platform again until about 30 minutes before the start of the competition. The machine never moved.

The lessons learned by these three teams apply to all design competitions. They are summarized in the following testing tips:

- Always leave a lot of time for testing.
- When conducting tests, try to simulate as closely as possible the conditions at the final competition. If these conditions are not known, test under a variety of conditions to insure robustness.

29.3 Materials

When possible, the designs should be made of wood to facilitate manufacture and keep costs down. Manufacturing can be simplified still further by constraining the designs to be small (for example, less than 1 ft^3) and lightly loaded. Under these conditions, balsa can be used as the main structural material.

Recommended **materials** for a small, lightly loaded electromechanical device are listed in Table 29.1, along with their attributes relative to balsa. Balsa is listed as easy to use with hand tools, because it can be cut and shaped easily with a sharp knife. On the other hand, plastic tends to deform rather than shear cleanly under the action of cutting tools, and so it is listed as harder to work with.

When selecting a material from Table 29.1, the strength and stiffness requirements of the given part also need to be considered. For example, if there are concerns about a part breaking under load, strength considerations override ease of manufacture, leading to the use of plywood instead of balsa. If a small-diameter axle requires high stiffness so that gears can remain engaged, then a steel rod may be the best choice.

Table 29.1 List of recommended materials for a small electromechanical design project.

Material	Relative strength	Relative stiffness	Ease of manufacture	Useful forms
Balsa	1	1	Very easy	Sheets, beams, blocks
Wood	5	4	Easy	Plywood sheets and wooden dowels
Plastic	5	1	Hard	Plastic sheets and prefabricated gears
Steel	50	80	Medium	Thin metal sheets and rods
Rubber	0.5	Very small	Very easy	Long strands

The strength and stiffness are normalized with respect to values for balsa wood.

29.4 Joining methods

For small-scale balsa and wood structures, the preferred **joining methods** for parts are adhesives, wood screws, and machine screws with nuts. Typical joint configurations employing these methods are shown in Fig. 29.1. Use of tapes, especially duct tape, is frowned upon for their nonpermanence and poor aesthetics. Nails are not particularly compatible with balsa because of the large impact forces involved and the possibility of wood splitting.

Adhesives, in particular hot glue, are the method of choice when balsa is the predominant structural material. Hot glue cures quickly, and though essentially permanent for wood-to-wood bonds, metal-to-wood bonds can be adjusted or broken by heating. It is so general purpose that it can be used to mount a motor to a plywood base. The main drawback of hot glue is its low strength, but this is usually not an issue for lightly loaded balsa structures. When it does become an issue—for example, when the surface area available for the glue joint is very small—one of the higher strength adhesives listed in Table 29.2 may be substituted.

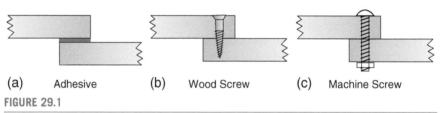

(a)   Adhesive   (b)   Wood Screw   (c)   Machine Screw

FIGURE 29.1

Different joining methods.

Table 29.2 Common adhesives.

Adhesive	Typical uses	Setting time[a]	Curing time[b]	How to apply	Relative strength
Wood glue	Wood, paper	8 hours	24 hours	Apply in liquid form directly from bottle.	13
Epoxy	Wood, metal	5 minutes to 12 hours	3 hours to 3 days	Comes in 2 tubes; mix equal amounts and apply with stick.	14
Hot glue	Almost anything	1 minutes	2 minutes	Place glue sticks in heated gun and apply with gun.	1

[a] Setting time = Time to harden.
[b] Curing time = Time to reach maximum strength.

On balance, wood screws are a less popular alternative for balsa-to-balsa joints. To reduce the chances of wood splitting, a pilot hole equal in diameter to the screw without the threads should be drilled prior to inserting the screw. This material removal plus the persistent possibility of wood splitting reduces the effective strength of the balsa members. In situations where adjustability is needed, the benefits of easy screw removal may override these issues.

Wood-to-wood joints are a different matter. Wood is much stronger than balsa, so wood could be used when a higher strength joint is needed. In such cases, use of screws is often preferable to the long curing times of the higher-strength adhesives.

29.5 **Useful hand tools**

Small-scale electromechanical devices can be crafted using common **hand tools**, provided balsa or wood is the primary structural material. In this section, we offer a short compendium of those tools for your easy reference. Our goal in presenting them is to make you aware of the alternative manufacturing operations at your disposal. We do not go into much detail on how to use them. Instead, we advise you to observe your classmates, talk to a machinist, or just give it a try.

For each tool, we will (1) show a picture of it (so that you can find it), (2) name it (so that you can ask for it), (3) describe its use, and (4) provide additional comments on its usage.

29.5.1 Tools for measuring

Name: Tape measure; steel ruler (Fig. 29.2).
Use: For measuring length.
Comments: A steel ruler can also be used as a straight edge when making straight cuts in balsa with the X-Acto knife.

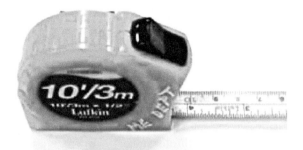

FIGURE 29.2

Tape measure/ruler.

29.5.2 Tools for cutting and shaping

Name: X-Acto knife (Hunt Manufacturing Co.) (Fig. 29.3).
Use: For cutting and shaping balsa and some plastics.
Comments: Blades come in different shapes and are replaceable.

FIGURE 29.3

X-Acto knife.

Name: Coping saw (Fig. 29.4).
Use: For cutting curves in wood and plastic.
Comments: The blade can be mounted in the frame to cut on either the push or pull stroke. Unscrewing the handle relieves tension on the blade so that it can be removed, rotated up to 360 degrees, or inserted through a small drilled hole to cut out a larger hole.

FIGURE 29.4

Coping saw.

Name: Hacksaw (Fig. 29.5).
Use: For making straight cuts in metal.
Comments: It cuts on the push stroke, when the blade is mounted correctly. Though designed for metal, it can also be used with wood. Blades are detachable and can be rotated to cut in four directions.

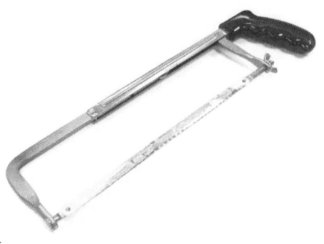

FIGURE 29.5

Hacksaw.

Name: Dremel (Robert Bosch Tool Corporation) (Fig. 29.6).
Use: For shaping and drilling holes in wood, thin metal, and plastic.
Comments: This is an electric hand drill without the handle. Because of its shape, forces can be applied more easily when grinding or sanding.

FIGURE 29.6

Dremel tool with bits.

29.5.3 Tools for drilling holes

Name: Cordless hand drill (Fig. 29.7).
Use: For drilling holes in metal, wood, and plastic.
Comments: Release trigger lock and pull trigger to start drilling. Power switch controls the direction of spin.

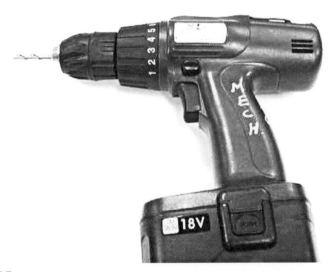

FIGURE 29.7

Cordless hand drill.

29.5.4 Tools for joining parts

Name: Hot glue gun (Fig. 29.8).
Use: For applying hot glue.
Comments: Insert a glue stick in the back, wait 3–5 minutes for the gun to heat up, then apply the glue by pulling the trigger. Handle with care, as the glue can reach temperatures of 400°F.

FIGURE 29.8

Hot glue gun.

Name: Spring clamps (Fig. 29.9).

Use: For holding parts together when waiting for an adhesive to set.

Comments: If the grips do not open wide enough for your application, C-clamps are a good alternative.

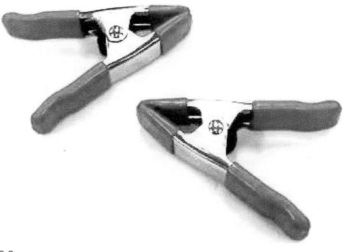

FIGURE 29.9

Spring clamps.

Name: Screwdriver (Fig. 29.10).

Use: For inserting and removing screws.

Comments: The longer and thicker the handle, the easier it is to apply torque. The screwdriver in Fig. 29.10 has a removable tip. You can pull out the tip, rotate it 180 degrees, and reinsert it to switch from a regular screwdriver ($\ominus$) to a Phillips head screwdriver ($\otimes$).

FIGURE 29.10

Screwdrivers.

Name: Adjustable wrench (Fig. 29.11).
Use: For holding nuts when tightening a screw or bolt.
Comments: Thumb adjustment changes the distance between jaws. Unlike pliers, there is no gripping force, so parts being held must have flat surfaces.

FIGURE 29.11

Adjustable wrench.

Name: Claw hammer (Fig. 29.12).
Use: For driving and removing nails.
Comments: Nails should not be used to join balsa parts.

FIGURE 29.12

Claw hammer.

29.5.5 Tools for wiring

Name: Soldering iron (Fig. 29.13).
Use: For attaching or connecting wires with solder.
Procedure: (1) Create a mechanical connection (e.g., by twisting wires together); (2) let iron heat up; (3) apply solder to tip of iron (called *tinning*); (4) heat wires with iron; (5) bring solder in contact with heated wires until solder melts and flows; and (6) let wires cool.

FIGURE 29.13

Soldering iron.

Name: Long-nosed pliers (Fig. 29.14).
Use: For bending wires, grasping parts, cutting wires, and reaching into tight places.
Comments: Clamping forces are highest at the wire cutters and smallest at the tip of the nose.

FIGURE 29.14

Long-nosed pliers.

Name: Wire cutters (Fig. 29.15).

Use: For cutting and stripping wires.

Comments: You may not need this tool if your long-nosed pliers have a good wire cutter.

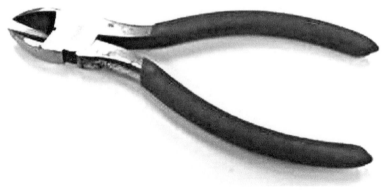

FIGURE 29.15

Wire cutters.

Name: Wire strippers (Fig. 29.16).

Use: For stripping insulation from wires, cutting wires, and shearing small diameter bolts.

Comments: To strip insulation, lay wires across jaw in notch of same diameter as metal wire, and then pull wire while holding jaws compressed.

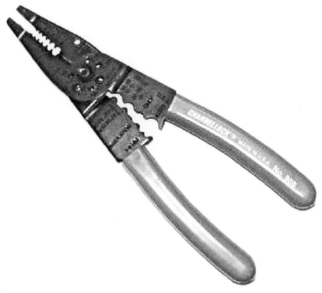

FIGURE 29.16

Wire strippers.

29.6 Design milestone #6A: Manufacturing and testing assessment I

The **Design for Manufacture** (DFM) assessments are so named because the teams that gave serious consideration to DFM principles when first formulating their designs have the best chance of doing well on these milestones. This first DFM assessment should occur about halfway through the first manufacturing iteration. For a design project of the scale proposed in these chapters, that is about 1 week after receiving the box of parts.

Example 29.1

Assignment

1) Make as much progress as possible toward the completion of construction and testing of your design by the due date. Team members should work in parallel on the different aspects of the design to accelerate progress.

2) From here on, the design project is a race to begin testing. Teams with the most testing time tend to win competitions.

Grading Teams should bring their materials and a detailed drawing of their design to class on the due date. Grading will be based on the amount of progress made:

100 points = Manufacturing is more than 50% completed

80 points = Manufacturing is between 25% and 50% completed

60 points = Some progress made (but less than 25%)

0 points = No progress

29.7 Design milestone #6B: Manufacturing and testing assessment II

By the time this milestone is reached, teams should have completed their first manufacturing iteration and begun testing. For a freshman project, this could be as soon as 1 week after Design Milestone #6A described above.

The instructor evaluates progress by conducting one or more well-defined tests to determine whether or not the various subfunctions work. Evaluation of how well they work is reserved for the next milestone (Performance Evaluation, Chapter 29). Subfunctions to be tested must be carefully selected, so as not to be biased toward a particular design or strategy.

Example 29.2

Assignment Get all subfunctions working by the assigned due date.

Subfunction Test To be defined by the instructor.

Grading Team grades are based on the level of functionality demonstrated in the best of three trials and computed as follows:

If manufacturing is 95% completed, then

$$\text{Grade} = B + \sum_{i=1}^{I}(W_i \times s_i)$$

where

Grade = Assigned grade on a scale of 100 points

B = Base grade (typically $B = 70$ points)

W_i = Number of grade points associated with the ith subfunction and defined such that

$$W_1 + W_2 + W_3 + \cdots + W_I = \sum_{i=1}^{I} W_i = 100 - B$$

I = Total number of subfunctions being evaluated

$s_i = 1$ (if the subfunction worked)

$s_i = 0$ (if the subfunction did not work)

If manufacturing is less than 95% completed, then

$$\text{Grade} = f \times B$$

where

Grade = Assigned grade on a scale of 100 points

B = Base grade (typically, $B = 70$ points)

f = Fraction of manufacturing that was completed

Exercises

Manufacturing and testing strategies

1. In Chapter 23 (Design Teams) we discussed Gantt charts as a way of planning your design activities. It can also be used to develop manufacturing and testing strategies. Suppose you are in a design team with three other students (Sara, Jim, and Leslie) and are beginning to construct (i.e., manufacture) your design. Develop a Gantt chart that will show the responsibilities and deadlines for each student (including yourself) for the following activities: (1) talk to a machinist for advise on making unique components of your design, (2) search for components on the Internet or in hardware stores (wheels, bolts, etc.), (3) collect receipts for purchases, (4) keep drawings up to date, (5) assemble the components to make a working model of the design.

2. Now that you have built you design it's time to test it. You only have 2 weeks left in the semester and during your first test you determine that you prototype does not function properly. How do you salvage your design before the semester ends?

Materials and joining methods

3. When you begin testing the prototype of your design you find that an important component you made from balsa wood breaks under load. You need to fabricate a new version of the component and replace it quickly. What material do you choose for its replacement?

4. Your team's final design calls for gluing two pieces of metal together. During testing, the glue does not hold, and the prototype fails to function properly. How do you fix this problem quickly and securely?

5. One of your team members wants to assemble the main components of the design with bolts. It turns out that the bolts add a lot of weight to the prototype and are difficult to keep from coming loose during testing. What would you recommend using instead of bolts at this late date and why?

Design step 7: Performance evaluation

30

Source: Nonwarit/iStockphoto.com

30.1 Introduction

Once the design has been manufactured, it is time to evaluate its performance. If the student devices are required to interact with each other, the performance should be measured in two stages:

- **Stage 1** is the individual performance evaluation. The manufactured device is tested alone, under controlled conditions, to verify that it is capable of doing what the problem definition requires.
- **Stage 2** is the head-to-head class competition performance. Each device is tested against other devices in a series of head-to-head matches to determine the best overall design. The student grade resulting from this performance evaluation typically constitutes up to 50% of the project grade.

Exploring Engineering. https://doi.org/10.1016/B978-0-443-13541-5.00025-8

30.2 Stage 1 performance evaluation tests

Performance of a given device in the second stage head-to-head matches may vary with the opponents. For example, an offensive or defensive strategy that works well against one opponent may not work against another. So, the only way to test all of the machines under the same set of conditions is to test them individually, in isolation, with no opponent.

The basic approach is to measure one or more quantities, referred to as "**metrics**" that are good predictors of success in the second stage head-to-head evaluation matches. Typical metrics are time, speed, pushing force, or number of points scored against a stationary obstacle representing the opponent. Some basic rules when selecting metrics are (1) they should be easily measurable, (2) they should be continuously variable to maximize information content, and (3) they should not be biased toward particular design solutions.

The number of different physical tests required to measure all of the metrics depends on the choice of metrics. Ideally, you want to be able to design a single test that measures all of the metrics. Sometimes, each metric requires its own test.

The individual performance evaluation grade is the overall measure of performance expressed on a scale of 100. We recommend that it be computed as a weighted sum of the metric values as expressed in design milestone #7 in the next section of this chapter.

The specifics of the individual performance evaluation tests are often not revealed to the students until a week before they take place. This is to prevent students from tailoring their machines to the performance evaluation tests instead of winning the final competition, which is the real design objective.

30.3 Design milestone #7: Stage 1 performance evaluation

Gradewide, this is the most important milestone. The performance evaluation enforced by this milestone and the previous one prepares the machines for the final competition.

Assignment

Optimize performance of your machine in preparation for performance testing.

Performance Evaluation Test

To be defined by the instructor.

Grading:

Team grades are based on the quality of performance demonstrated in the best of three trials and are computed as follows:

$$\text{Grade} = B + \sum_{i=1}^{I} W_i \left(\frac{m_i}{M_i}\right)^n - \sum_{j=1}^{J} P_j$$

where

Grade = Assigned grade on a scale of 100 points

B = Base grade (typically, $B = 70$ points)

W_i = Number of grade points associated with the ith metric and defined such that

$$W_1 + W_2 + W_3 + \cdots + W_I = \sum_{i=1}^{I} W_i = 100 - B$$

I = Total number of metrics

m_i = Measured value of the ith metric

M_i = Best value of m_i recorded by any team in the class

$n = 1$ (if performance is directly proportional to m_i)

$n = -1$ (if performance is inversely proportional to m_i)

P_j = Number of grade points associated with the jth penalty

J = Total number of penalties assessed for rules violations

If the grade calculated is less than or equal to the Base grade (B), then

$$\text{Grade} = f \times B$$

where

    Grade = Assigned grade on a scale of 100 points

    f = Fraction of manufacturing that was completed

    B = Base grade (typically, $B = 70$ points)

30.4 Stage 2 performance evaluation—The final competition

The final competition pits the machines against each other, usually in a series of head-to-head matches that may involve direct interactions between the machines. This is the ultimate test of the machines, as it is the only way to accurately evaluate the effectiveness of offensive and defensive strategies, robustness, durability, and the wisdom of past design choices.

However, it is probably best that the results of head-to-head competition not be linked to the performance grade, since each machine faces a different set of challenges. For example, one machine may not match up well against a particular opponent, or a prefabricated part may fail unexpectedly due to an accidental collision.

Exercises

1. What metrics are normally used to predict the performance of your entry in a head-to-head competition?
2. What additional metrics beyond those listed for exercise 1 would you use to predict your design's performance in a head-to-head RC model boat competition?
3. How would you measure the Stage 1 pushing force of your entry in a head-to-head competition?
4. If you had designed and constructed an entry in a head-to-head RC walking robot competition, what Stage 1 performance evaluation tests would you carry out before the competition?
5. You are to design and construct a small autonomous vehicle that will seek out a lit candle and blow it out. What Stage 1 performance evaluation tests would you implement before the competition?

Design step 8: Design report

31

31.1 Introduction

The design report documents the final design. It enables someone unfamiliar with a design to figure out how it works, evaluate it, and reproduce it. It is the final step in the design process. This chapter summarizes the organization of the report and provides some writing guidelines.

31.2 Organization of the report

Like the oral design defense, the organization of a design report follows the steps of the design process (see Table 31.1). The report begins with a concise statement of the "Problem Definition," in the student's own words.

The "Design Requirements" section does not need to repeat the competition rules. Instead, it should begin with a short paragraph describing competition strategy. Then, list all performance requirements, for example, "Must have a top speed of 1 ft/s"; "Must deposit at least six ping-pong balls"; "Must be able to steer."

Almost all the content of the "Conceptual Design" section should be available from previous milestones. In Section 3.1, present the sketches of your alternative

Exploring Engineering. https://doi.org/10.1016/B978-0-443-13541-5.00020-9

Table 31.1 Suggested organization of the design report.

Organization	Pages
Title and authors	1
Table of contents (with page numbers)	1
List of individual contributions to the report	1
1. Problem definition	0.5
2. Design requirements	1
3. Conceptual design	
3.1 Alternative concepts	2
3.2 Evaluation of alternatives	1
3.3 Selection of a concept	0.5
4. Detailed design	
4.1 Main features and how they work	3
4.2 Results of analysis, experiments, and models	1
4.3 Manufacturing details	1
5. Performance evaluation	1
6. Lessons learned	1–2
Total	**15–16**

design concepts and briefly describe how each one works. In Section 3.2, present your decision matrix and use the matrix as a vehicle to discuss the strengths and weaknesses of each concept. In Section 3.3, indicate the concept you selected and give your rationale.

The "Detailed Design" section describes the design that appeared at the final performance evaluation. New detailed drawings have to be prepared. Place these new drawings in Section 4.1, along with text describing the operation and main features of the final design. Describe the overall design first, and then zero in on the details of special features. If possible, include digital photographs of your device. To create Section 4.2, retrieve the results presented at the oral design defense and add text. Section 4.3 is primarily a summary of the joining methods used.

The results of the performance tests are summarized in the "Performance Evaluation" section. Describe how your machine fared, both during individual performance testing and at the final competition, being as quantitative about it as you can. Also compare performance predictions to actual results.

The "Lessons Learned" section is an opportunity to reflect back on the design experience. Write it in the form of three paragraphs, with each paragraph dedicated to answering one of the following questions. (1) How would you redesign your machine to improve performance? (2) What general lessons did you learn about the design process? (3) What general lessons did you learn about teaming?

31.3 Writing guidelines

Use double spacing to leave room for instructor comments. Use the section and subsection headings of Table 31.1 and boldface them so that they stand out. Finally, figures should be embedded within the text (rather than placed at the end of the report) for ease of reference. In addition,

1. Be concise.
2. Begin each paragraph with a topic sentence that expresses the theme or conclusion of the entire paragraph. The reader should be able to overview the entire report by reading just the topic sentences.
3. Generate high-quality concept drawings and detailed drawings to pass the "flip test." The first thing the instructor does before reading the report is to flip through the pages to examine the figures. The figures are the instructor's first impression of the report.
4. Give each figure (i.e., drawing, graph, etc.) a figure number and a self-explanatory figure caption. For example, **Fig. 3: Decision matrix**.
   Figures should be numbered consecutively and should be referenced from the text using the figure numbers. For example, "The results of the comparison are summarized in the decision matrix of Figure 3."
5. Use a spelling checker. Spelling mistakes cause the reader to question technical correctness.
6. Employ page numbers both in the text and in the table of contents.

31.4 Technical writing is "impersonal"

Technical writing is always done using an impersonal writing style. This style is limited to using the passive voice, third-person pronouns, and impersonal "things" rather than people as subjects of sentences.

In a sentence with an action verb, the subject of the sentence performs the action indicated by the verb, and the sentence is in the **active voice**. If the subject of a sentence is *acted upon* by the verb, the subject is *passive*, and the sentence is in the **passive voice**. So in a sentence using active voice, the subject of the sentence *performs* the action expressed in the verb. In the passive voice, the subject *receives* the action of the verb:

In technical writing, we do not use the active voice. The passive voice is always used in writing technical and laboratory reports because the person writing is not important (passive) but the process or principle being described is important. Examples of appropriate passive and inappropriate active voice sentences in technical writing are shown in Table 31.2.

Table 31.2 Examples of passive and active voice writing.

Use passive voice	Do *not* use active voice
The angle was observed to be …	I observed the angle to be …
It is suggested …	We suggest …
A graph was used to …	We used a graph …

Technical writing also only uses third-person pronouns. Examples of the correct third-person and incorrect first-person pronouns are shown in Table 31.3.

Table 31.3 Examples of third-person and first-person writing.

Use third person	Do *not* use first person
It was found that …	I found …
It was assumed that …	I assumed that …

Finally, impersonal items rather than people are subjects of sentences in technical writing. Examples using impersonal rather than personal (people) as sentence subjects are shown in Table 31.4.

Table 31.4 Examples of impersonal and personal sentence subjects.

Use impersonal sentence subjects	Do *not* use people as sentence subjects
Analysis of the data indicates …	During the tests I noticed that …
This report presents …	In this report I show …

31.5 Design milestone #8: Design report

This is the time both for documentation and for reflection.

Assignment

Prepare a design report in accordance with the guidelines of this chapter.
   Grading criteria
- Is the report complete?
- Does it pass the flip test?
- Is the writing style solid?
- Is it clear how the design works?
- Could someone unfamiliar with your machine manufacture it from the drawings and information in the report?

Exercises

1. While writing a design report, a team member copies material from the Internet and pastes it into their section of the report. He/she feels that this is important information relevant to the operation of your design. How do you deal with this situation?

2. In reviewing your team's design report, you come across this statement in the section on "Conceptual design": "We brainstormed three alternate solutions and then we used the design matrix to decide on which one we should use for our project." How would you restate this sentence to make it sound more professional?

3. One of your team members didn't contribute very much to the design but spent a lot of time constructing the competition entry. When writing the design report section on "Manufacturing details," she/he wrote the following sentence: "I made the base out of cardboard and then I glued the motor and gear wheel thing on it and I hoped it would work out ok." How would you restate this sentence to make it sound more professional?

4. Sometimes you can be too concise in writing your design report. Consider the concise statement "we made it out of wood". How would you change this to still be concise and yet give the reader more relevant information?

5. A team member wants to put several cartoons in the design report to entertain the instructor who will be reading it. The cartoons have nothing to do with the design process but they do inject some humor into the performance evaluation. List several reasons why you think this is or is not acceptable.

Examples of design competitions

32

32.1 Introduction

In this chapter, we present typical design competitions along with one team's solutions to the first few milestones. Design competitions like the ones described in this chapter have proven to be very successful with first- and second-year engineering students. The rules, tabletop playing field, and list of parts provided may be used as a template in defining similar competitions. This chapter has three design competitions for students to study: "**A Bridge Too Far**," the "**Mars Meteorite Retriever Challenge**," and the "**Automatic Air Freshener**."

Exploring Engineering. https://doi.org/10.1016/B978-0-443-13541-5.00016-7

32.2 Design competition example 1: A Bridge Too Far

A design project begins on the day that the instructor distributes the rules governing the design competition. This package of rules typically consists of a statement of the design objective, a list of design requirements, a drawing of the playing field, and a list of parts and materials available.

In the design competition named **A Bridge Too Far**, the objective is *to design and build a device to outscore your opponent in a series of head-to-head matches.* The playing field is shown in Fig. 32.1 with an elevated bridge separating the two scoring pits. A team receives +1 point for every ball resting in its scoring pit at the end of play. In all, there are 17 scoring balls. Six of the balls start out in the possession of the two teams; the remaining 11 balls have starting positions on the playing field as indicated in Fig. 32.1. Other key rules are as follows:

- The device must fit within a 1-ft by 1-ft by 2-foot high volume.
- Parts and materials are limited to those listed in Table 32.1.
- Each device can have up to three independent tethered controls.
- There is a 1-min setup time.
- One game lasts 30 s.
- If there is a tie, both teams lose.

The parts and materials of Table 32.1 are not supplied to the teams until they successfully defend their designs at the oral design defense. The complete set of rules is given in Section 32.4 of this chapter.

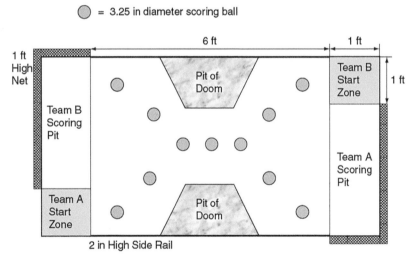

FIGURE 32.1

A Bridge Too Far tabletop playing field.

Table 32.1 The list of parts and materials for "A Bridge Too Far."

Quantity	Item
1	Motor with adjustable gear box
1	Gearhead motor
1	Electric relay (for use with gearhead motor)
1	Electric diode (for use with gearhead motor)
1	Electric pull solenoid
1	Set of six gears
1	3/8 in. by 36 in. wooden dowel
2	$1/16$ in. by 3 in. by 36 in. balsa wood sheets
2	$3/8$ in. by $3/8$ in. by 36 in. balsa wood beams
1	¼ in. by 12 in. by 12 in. plywood sheet
10	Large craft sticks (tongue depressors)
1	$1/8$ in. diameter by 20 in. long steel rod
1	10 in. piece of string
2	1 in. long metal hinges
1	¼ in. diameter by 10 in. long elastic cord with hooks
1	12 in. long rubber-band strip
4	2 in. diameter wheels with hubs (for $1/8$ in. diameter shaft)
1	2 in. diameter mailing tube
1	2-L plastic soda bottle (provided by the team)

32.3 Some design milestone solutions for "A Bridge Too Far"

32.3.1 Design milestone #1: Clarification of the task

In the case of a design competition, clarification of the task is mostly the responsibility of the instructor. There were just two things left for the design team to do to complete this milestone.

First, the team directed the following questions to the rules committee:

1. Can a machine score from the Pit of Doom?
2. Can you drop obstacles to interfere with the other machine?
3. Can you score points by driving a machine loaded with balls into a Scoring Pit?
4. Can a machine attach itself to the other machine?
5. Can a machine expand itself beyond the 1-ft by 1-ft by 2-ft high dimensions once the game begins?

The answer to all these questions is "yes," since these actions do not violate any of the stated rules. The motivation behind the questions is to fully understand the design constraints and probe for omissions that could lead to a design advantage.

Second, the team compiled a list of performance requirements that were specific to their design and their competition strategy.

D—Must score at least four points.
W—Must score the first point within 10 s.
D—Must hinder the opponent's ability to score.

The list is short to avoid solution bias. Later, after a design has been selected, other performance requirements can be added.

32.3.2 Design milestone #2: Generation of alternative concepts

The functional decomposition settled on is shown in Fig. 32.2. Consideration of offensive strategy is done through the subfunction "Deliver Balls."

The results of brainstorming each subfunction were collected in the classification scheme of Fig. 32.3.

```
                    "A Bridge Too Far" Machine
```

| Hold Balls | Move and Steer | Pick Up Balls | Deliver Balls | Defensive Strategies |

FIGURE 32.2

Functional decomposition for "A Bridge Too Far."

Concepts / Functions	Concept 1	Concept 2	Concept 3	Concept 4	Concept 5
Hold balls	Place on machine	Keep under machine	Push balls in front	Container to put balls in	Drag behind machine
Move and steer	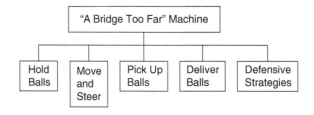 2 Motors	2 Motors	1 Motor steers	No steering	Don't move
Pick up balls	Don't pick up balls	Wedge	Scoop	Plow	Claw
Deliver balls	Push them into pit	Throw them into pit	Drive into pit with balls	Roll them down a ramp	Push opponent into pit
Defensive strategies	Block bridge	Ignore opponent	Throw extra balls into Pit of Doom	Pick up opponent	Throw obstacles

FIGURE 32.3

Classification scheme for "A Bridge too Far."

Compatible subfunction concepts were combined to form the four promising designs shown in the concept drawings of Figs. 32.4–32.7.

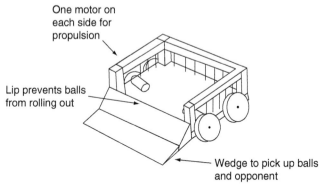

One motor on each side for propulsion

Lip prevents balls from rolling out

Wedge to pick up balls and opponent

FIGURE 32.4

Concept drawing of the "wedge."

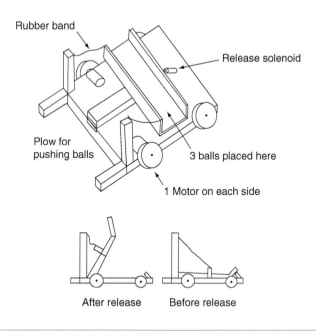

Rubber band

Release solenoid

Plow for pushing balls

3 balls placed here

1 Motor on each side

After release   Before release

FIGURE 32.5

Concept drawing of the "catapult."

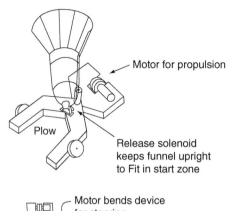

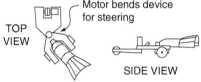

FIGURE 32.6

Concept drawing of the "snake."

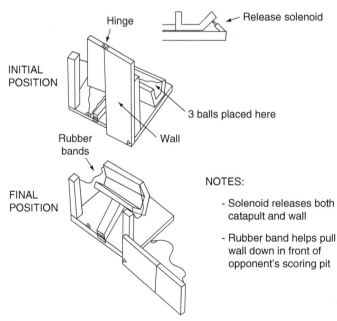

FIGURE 32.7

Concept drawing of the "wall."

32.3.3 Design milestone #3: Evaluation of alternative concepts

The four alternative concepts developed in the previous milestone are illustrated in Fig. 32.8 for easy reference.

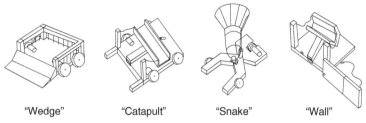

"Wedge"   "Catapult"   "Snake"   "Wall"

FIGURE 32.8

Alternative design concepts from "A Bridge Too Far."

The selection of evaluation criteria was tricky, because a high-scoring machine that consistently gathers 9 of the 17 balls should be ranked on the same level as a defensive machine that consistently wins by the score of 1 to 0. The evaluation criteria and their respective weights are shown in Table 32.2. Both of the first two criteria are needed to describe scoring potential, because it is not enough that you can transport a lot of balls; you must be able to score with them with an opponent in your way. Their combined weighting of 0.4 is slightly higher than the weight of 0.3 assigned to defensive capabilities (third criterion), since you need to score to have a chance of winning. "Easy to manufacture" has its usual importance, but "low cost" was not included because it is not a factor in the competition. "Easy to control" appears because of the potential challenges involved in maneuvering across the bridge. The discussion of concept strengths and weaknesses is organized by evaluation criterion.

1) **Has a large payload**

   The "Wall" is rated low because it can score with only the three original balls. For the other three concepts, ratings are proportional to carrying capacity. Since the funnel design of the snake expands to be much larger than the start zone, it has a much larger carrying capacity than the other two.

2) **Robust scoring capability**

   The two designs that launch the balls score highest here, because it is much harder to block a ball that is thrown than one that is pushed. Also, these machines can score without having to cross the bridge.

3) **Can disrupt opponent**

   The "Wall" is an obvious favorite for this category, since it is the only one that actively blocks the opponent's goal. The "Wedge" is designed to be able to pick up the opponent and drop it in the appropriate goal to gain more points, so it too can theoretically disrupt the opponent well.

4) **Easy to manufacture**

   The "Wedge" is rated higher than the "Catapult," because it has fewer functions to build. The "Wall" also does well in this category, because it does not involve any motors or corresponding drivetrains. In fact, it would be rated even higher were it not for the difficulties anticipated with sequencing the release of the wall and the catapult arm.

Table 32.2 Decision matrix for "A Bridge Too Far."

Evaluation criteria	Wt	Wedge		Catapult		Snake		Wall	
		Val_1	$Wt \times Val_1$	Val_2	$Wt \times Val_2$	Val_3	$Wt \times Val_3$	Val_4	$Wt \times Val_4$
1) Large payload	0.2	6	1.2	5	1.0	8	1.6	2	0.4
2) Robust scoring	0.2	5	1.0	9	1.8	3	0.6	7	1.4
3) Disrupts opponent	0.3	7	2.1	5	1.5	4	1.2	8	2.4
4) Easy to manufacture	0.2	7	1.4	5	1.0	5	1.0	7	1.4
5) Easy to control	0.1	7	0.7	6	0.6	3	0.3	9	0.9
Total	1.0		6.4		5.9		4.7		6.5

5) Easy to control

The "Wall" simply requires flipping one switch, which is almost as easy as doing nothing. The "Snake" is slightly alien to most things people will have controlled and, as a result, would be hard to drive.

32.3.3.1 Discussion of results

All the machines demonstrated strength in some areas. Observed weaknesses cannot be corrected without diminishing the attributes that made them strong. For example, the "Snake's" funnel, which gives it the largest payload, also makes it vulnerable to being pushed around by the opponent.

The "Wall" is the highest-rated design and also the boldest, in that it dares to remain stationary. Students tend to shy away from designs like this one, because they are different. This design was not selected because it was felt that its defensive capabilities were overrated. The design may not be able to resist pushing by an opponent at the end of the wall.

The "Wedge" was finally selected as the best design. It is interesting that it is rated high, even though it is not a clear winner in any of the categories. It compares very closely with the "Catapult." The simplicity, lifting potential, and large payload of the "Wedge" were the decisive factors.

32.3.4 Design milestone #4: Detailed design

As you may recall, the "Wedge" (shown on the left in Fig. 32.9) was selected as the final concept. The original wedge design was dual purpose: (1) to lift and push the opponent's machine and (2) to allow balls to roll up into the holding bin above the moving platform.

32.3.4.1 Experiments

There were concerns that the forward speed of the vehicle might not be sufficient to allow the balls to roll up high enough into the bin. Experiments (see Fig. 32.10) were

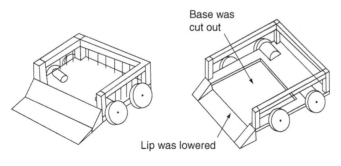

Base was
cut out

Lip was lowered

ORIGINAL DESIGN   MODIFIED DESIGN

FIGURE 32.9

Design modifications in response to experiments.

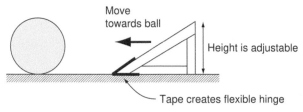

FIGURE 32.10

Experiment used to determine optimal wedge geometry.

conducted to check this out. The results showed that the concerns were justified. Acceptable dimensions for the wedge and holding bin were established, based on the results of the experiments. The modified concept is shown on the right in Fig. 32.9.

32.3.4.2 Analysis

Two identical gearhead motors were available for creating the drivetrains for the left and right sides of the vehicle. The no-load angular speed (N_{noload}) at 24 V was listed in the catalog as 145 RPM. To maintain a speed of 1.00 ft/s or 12.0 inches/s on the flat with tire diameters of 2.00 inches, the overall gear ratio needs to be (see Eq. (27.5) of Chapter 27—Design Step 4):

$$\text{GR} = \frac{2\pi \ N_{noload} \ R_{tire}}{60 \ V_{mp}} = \frac{(2\pi \ \text{rad/rev}) \ (145 \ \text{rev/min}) \ (1.00 \ \text{in})}{(60 \ \text{s/min})(12.0 \ \text{in/s})} = 1.26$$

where N_{noload} was converted to rad/s using 1 rad/sec = $2\pi/60$ RPM and for a moving platform it is known that $\omega_{tire} = V_{mp}/R_{tire}$ in which ω_{tire} is the angular velocity of the tire, V_{mp} is the linear velocity of the moving platform and R_{tire} is the radius of the tire. With the pushing requirement of this design, it might seem that the actual gear ratio needs to be much higher. However, this particular motor is quite powerful and the motor shaft could not be visibly slowed by manually gripping the ends of the shaft. Given the power of the motor and the low weights of the vehicles, this gear ratio was deemed to be reliable, even with the pushing requirement. The resulting drivetrain geometry is shown in Fig. 32.11.

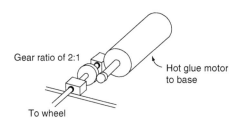

FIGURE 32.11

Close-up view of the drivetrain.

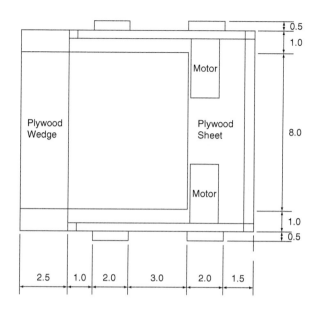

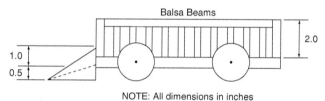

NOTE: All dimensions in inches

FIGURE 32.12

Detailed drawing (two views).

32.3.4.3 Detailed drawing
Two views of the final design are shown in the hand-drawn detailed drawing of Fig. 32.12. Some manufacturing details are provided.

32.4 Official rules for the "A Bridge Too Far" design competition

32.4.1 Objective
Design and build a device to outscore your opponent in a series of head-to-head matches.

32.4.2 Constraints

1. The parts and materials that may be used in the construction of the device are limited to those listed in Table 32.1. Each team is provided with a box containing all legal parts and materials.
2. The devices must be constructed entirely by the members of the team (i.e., if you lack the required expertise or tools to manufacture a certain part of the device, redesign it).
3. Each device can be loaded with up to three scoring balls. If a team chooses not to load all three balls, the discarded balls are removed from the playing field.
4. When placed on the table at the start of the game, each machine must fit completely within its assigned 1-ft by 1-ft by 2-foot high start zone.
5. An external power source is made available to each device for use during the matches. It consists of
   - Overhead wires that connect to the device. (Note: These wires are considered to be part of the playing field.)
   - A control box with two forward-reverse-off switches and one on-off switch.

32.4.3 The game

1. Just prior to the game, there is a 1-min setup time, during which each team should:
   - Place its device in the starting zone.
   - Attach and check the electrical connections.
   - Load its device with up to three scoring balls.
2. Except for manipulation of the electrical control boxes located at one end of the playing field, no human interaction with the device (or playing field) is allowed once the game begins.
3. The game begins when indicated by the referee and ends 30 s later when the power is switched off.
4. The game, and all scoring, ends as soon as one of the following occurs:
   - All movement stops as a result of power being switched off.
   - Five seconds have elapsed since power was switched off.

32.4.4 Scoring

1. At the end of the game, each team receives one point for every ball in its scoring pit, irrespective of which team caused the ball to fall into the pit. The team that scores the most points wins.
2. In all, there are 17 scoring balls; 6 of the balls start out in the possession of the two teams, and the remaining 11 balls have starting positions on the playing field as indicated in Fig. 32.1.
3. A ball is counted as lying in a scoring pit if an imaginary vertical line through the center of the ball lies within the boundary of the scoring pit. The referee can

ascertain the status of a scoring ball at the end of the game by removing other balls from the scoring pit; if the ball in question then falls into the pit, it counts as lying in the scoring pit.

4. In the event of a tie,
   - If neither team has scored, both teams lose.
   - If both teams have scored, the winner is the machine that has advanced farthest down the field as based on final positions. If this criterion proves indecisive, both teams advance to the next round.

32.4.5 Other rules

1. If the tethers (i.e., electrical connections) should entangle as a result of the machines passing each other, time stops and both machines are returned to their respective starting zones. Play then continues at the signal of the referee. This situation can occur, because each machine is tethered to the nearest of two overhead rods running parallel to the length of the table. To avoid entanglement, keep to the right of the opposing machine when passing.
2. Devices that permanently damage the playing field or the balls are disqualified.
3. Any attempt to intentionally inflict permanent damage on an opponent's device results in immediate disqualification. However, devices should be designed to hold up under expected levels of nonaggressive contact. For example, some pushing should be expected, but that pushing should not occur at significant ramming speeds.
4. Any device deemed unsafe is not allowed to participate in the matches.
5. Implementation of any strategies that are not directly addressed in the rules but that are clearly against the spirit of the rules (e.g., intentionally interfering with the person at the controls) leads to disqualification.
6. The rules committee has the final word on any interpretation of the rules.

32.5 Design competition example 2: Mars Meteorite Retriever Challenge

Meteorites from Mars occasionally land on Earth. Those that land on Antarctica are particularly easy to spot because of their color contrast with ice. NASA is developing automated rovers to retrieve these meteorites. Your job is to implement such a rover on a simulated Antarctica: a wooden board 96″ long and 48″ wide. The meteorite is simulated by a small object containing a light source. NASA identified the most promising "landing zone," simulated by a rectangle on the board 18″ by 24″. The meteorite can be anywhere within that landing zone.

Your *autonomous* vehicle begins its trip at a Robot Depot, simulated by a 12″ by 12″ square on the board. Your challenge is to locate the meteorite, travel to it, pick it up, carry it without touching the ground at any point (to avoid contamination), and deposit the meteorite at the Meteorite Lab (another 12″ by 12″ square), see Figs. 32.13 and 32.14. Two completed student designs are shown in Fig. 32.15. The vehicle starts at the Robot Depot and the meteorite is placed randomly within the landing zone.

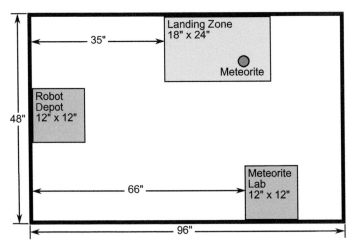

FIGURE 32.13

Layout of simulated Antarctica.

FIGURE 32.14

The actual Mars Meteor Retriever Challenge game board.

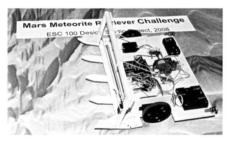

FIGURE 32.15

Student vehicles.

32.6 Some design milestones for the "Mars Meteorite Retriever Challenge"

32.6.1 Design milestone #7: Performance evaluation

32.6.1.1 Grading

Team grades are in proportion to level of functionality attained in the best of two tests, as defined by the following increasing levels of performance: The vehicle

- Is completely manufactured, 60%.
- Moves when powered, 70%.
- Moves with program control, 75%.
- Navigates to the loading dock, 80%.
- Navigates to the loading dock and touches meteorite under program control, 85%.
- Lifts the meteorite from the table surface and leaves the loading dock completely, 95%.
- Lifts the meteorite from the table surface and deposits the light source in the laboratory area, 100%.

32.6.2 Design milestone #8: The design report

32.6.2.1 Purpose

The purpose of the report is to provide a concise, accurate, and informative record of your design efforts. The report must be no more than 12 pages long, including the cover page, table of contents, and drawings.

32.6.2.2 Organization

1. Cover page.
2. Table of contents.
3. Introduction: State the design objective and cite some of the design constraints.
4. Strategy: Describe your competition strategy.
5. Conceptual design:

- Discussion of alternatives with clearly labeled concept drawings. Discuss the strengths and weaknesses of each.
- Include your functional decomposition, classification scheme, decision matrix, and the results.

6. Detailed design:
   - How it works (i.e., details of the operation of each subfunction).
   - Include detailed drawings of the final design, and explain how it is put together.
   - Include the results of any calculations or experiments.
7. Performance evaluation: Summarize the results of performance tests.
8. Conclusions and recommendations: Discuss the strengths and weaknesses of your design and your execution, and conclude with a discussion of lessons learned.

32.6.2.3 Grading
Technical communications, 50%

- The report is neat, concise, well organized, and includes all the items just listed.
- The report is free from spelling and grammatical errors.
- The report is presented attractively.

Technical content, 50%

- The report shows clear evidence that you understand and have followed the design methodology learned in the design studio.
- The report describes how the design works clearly enough that a person who has never seen the device could produce a working machine from the descriptions and drawings in the report.

32.6.2.4 Oral presentation
During the competition, each team gives a 2-min presentation to the judges. This presentation must include a detailed sketch or picture of the final design and a brief description of the design features, such as the steering mechanism, pickup mechanism, and the like that make the team's design unique. You should also provide a brief description of a problem encountered and the way in which the problem was solved using teamwork. If you did not encounter any difficulties in your design, discuss how teamwork made this possible.

Each team must practice its presentation and time it to ensure that it does not exceed 2 min. Your presentation should consist of two PowerPoint slides: a title page that includes the team members' names and team name (if you have no team name, select one) and a slide with a detailed drawing of the final design.

32.7 Official rules for the "Mars Meteorite Retriever Challenge" design competition

32.7.1 Objective

The objective of the competition is to construct an autonomous vehicle that can retrieve a (simulated) Mars meteorite from the (simulated) icy wastes of Antarctica.

32.7.2 Constraints

1. Parts and materials that may be used in the construction of the device are limited to those listed in Table 32.3. Each team will be provided with a box containing all legal parts and materials.
2. The devices must be constructed entirely by the members of the team (i.e., if you lack the required expertise or tools to manufacture a certain part of the device, redesign it).
3. When placed on the table at the start of the game, each machine must fit completely within its assigned 1-cubic foot start zone.

32.7.3 Scoring

The overall challenge is to accumulate points for design quality and vehicle performance. The teams with the most points will be designated "superteams." Design

Table 32.3 List of parts and materials for the "Mars Meteorite Retriever Challenge."

Quantity	Item
1	$7/8$ in. by $11^{7}/_8$ in. by ¼ in. plywood
2	$3/8$ in. by $3/8$ in. by 36 in. bass wood beam
1	$3/8$ in. by 36 in. hardwood dowel
6	Large craft sticks (tongue depressors)
1	$3/_{32}$ in. by 3 in. by 36 in. balsa sheet
2	Parallax continuous rotation servos
2	Boe-Bot wheels and tires
1	Tamiya 4-speed crank axle gearbox and motor
1	Plastic ball caster
2	20 in. long 6–32 threaded rod with eight nuts and washers
2	Battery holders for four AA batteries
1	Parallax microcontroller board with a Basic Stamp processor
1	Box of paper clips
2	Light sensors
2	Micro switches

quality points (up to 30) are awarded by the judges for creativity, construction, and presentation. Vehicle performance points (up to 70) are awarded for accomplishing the tasks of the challenge in the demonstration.

1. Teams give a 2-min presentation about their design for the design quality judging.
2. Teams demonstrate their vehicles. Vehicles start from a designated Robot Depot.
3. At the judges' "go" signal, the teams immediately operate any necessary switches and release their vehicles. No team member may subsequently make either direct or indirect contact with the vehicle or playing field until the "stop" signal is given.
4. Such contact, even if accidental, results in immediate disqualification.
5. The vehicle operates until it fully completes the task or until 3 min have elapsed. If 3 min have elapsed, the judges give a "stop" signal, and the team terminates the vehicle's operation.
6. Points are awarded for the demonstration of tasks as follows:
   - 5 points, the vehicle completely departs the Robot Station.
   - 5 points, the vehicle at least touches the Landing Zone.
   - 10 points, the vehicle touches the meteorite.
   - 10 points, the vehicle picks up the meteorite.
   - 10 points, the vehicle moves the meteorite outside the landing area (5 points for moving it but letting it touch the surface).
   - 10 points, the vehicle delivers the meteorite to the Meteorite Lab.
   - 10 points, the vehicle deposits the meteorite completely within the Meteorite Lab.
   - 10 points, the vehicle navigates completely outside the Meteorite Lab.

32.7.4 Rules

1. Vehicles must be a single unit.
2. Vehicles must at all times fit within a cube 12″ on a side
3. No vehicle can damage or be attached to the playing field. All vehicles must move in a way that does no harm to the field.
4. All vehicles using batteries must include in their circuits a length of fuse wire between one pole of each battery pack and the next connected component. No vehicle may use more than eight AA batteries or any other kind of battery.
5. The microcontroller board must be mounted in a way that can be removed after the competition and returned undamaged.
6. Imaginative strategies in the spirit of the game are encouraged.
7. However, any strategies determined by the judges to be contrary to the spirit of the game are excluded. Contestants have the responsibility of clearing with the judges before the competition any strategies that might possibly violate this rule.

32.7.5 Additional supplies

A circuit design and parts that allow the Tamiya motor to be reversed are provided to those who would like to implement them. Teams must purchase AA batteries for the microcontroller. Teams may purchases simple and inexpensive connectors, both mechanical (e.g., screws, nails) and electrical (e.g., simple contact switches). The simplicity and inexpensive nature of such additional supplies must be cleared with your instructor and made known to and available to other contestants.

Teams that, in the opinion of the judges, are attempting to win the competition by exclusive use of expensive or otherwise difficult-to-access outside technology (i.e., components not made by the team itself from approved parts) are disqualified.

32.8 Design competition example 3: Automatic Air Freshener

This example is a consumer product design that differs from the design projects in Examples 1 and 2 in two important respects. First, the students are not handed the design objective along with most of the design requirements. Here, it is left to the student to **identify the customer need**, translate this need into a **design objective**, and then brainstorm the **design requirements**, some of which must now be defined by future consumers of the product. Second, this project requires that all proposed designs must be operated autonomously using an Arduino microprocessor. This significantly increases the challenge of the electronics component of the design and introduces the need to learn some computer programming.

This type of **open-ended** design project makes it difficult to define a universal list of parts and materials that will meet the needs of all the proposed designs by student teams in a class. We propose using a more flexible strategy that leverages the fast delivery times made possible by the internet. Students are allowed a fixed budget (e.g., $65) with which to order parts and are steered to a small set of on-line suppliers to reduce ordering costs and time. This budget does not include the cost of the microprocessor kits (e.g., Arduino) which can be provided by the school or purchased separately by the students. Table 32.4 lists the building materials that are made available by the school in unlimited supply subject to a size constraint imposed on the designs.

The following design milestone solutions 1 through 4 were adapted from an actual design developed and successfully implemented by a team of three first-year students. The remaining design milestones #5 through #8 are not presented here.

32.8.1 Design milestone #1: Defining the problem

Prior to the start of the design, the instructors imposed a few design constraints to insure an equitable distribution of resources and a common set of learning objectives. Some of the most noteworthy ones are as follows:

- Maximum footprint of the device at the start of operation is 30 in. by 30 in.
- Spending limit of $65
- Must use at least one Arduino microprocessor for control

Table 32.4 Parts and materials in unlimited supply.

3/16 in. foam board
1/4 in. plywood sheets
1/8 in. balsa wood sheet
3/4 × 1/2 × 36 in. wood beams
8 ½ × 1/8 in. rubber bands, 7 × 5/8 in. rubber bands
6 × 11/16 in. tongue depressors
6/32 × 6 in. threaded rods with nuts and washers
Wire
Wood screws

- Must employ some combination of audio/visual sensors and electric actuators
- Must be activated by pressing a single start button

The burden of identifying "the need" then fell to the teams. Three methods were offered as ways to discover needs that might inspire future products: (1) observe potential customers in action with a truly open mind, (2) interview potential customers (e.g., children if designing a toy, or people with a disability if designing an assistive technology), and (3) keep a "bug" list in which you record aspects of products, services, or student life that annoy you. For example, the teams were asked to think about conditions in their dormitory rooms and brainstorm a bug list. From these origins, one student team came up with the following **Statement of Need**:

Everyone has had the experience of walking into someone's dormitory room where it smells horrible. Right as the door is opened, a wave of body odor assaults the nose. Most of the time, the occupant of the room doesn't even notice the smell because they live in it.

Translating, the team arrived at the following **Design Objective**:

Design a device to help keep a college dormitory room odor free.

Among the essential constraints cited by the students were (i) that it must activate automatically upon the students entering the room, (ii) that it must work with an off-the-shelf air freshener dispenser and (iii) that it must rotate the dispenser about a vertical axis as it sprays. The full list of **Design Requirements** developed by the team appears in Table 32.5.

32.8.2 Design milestone #2: Generation of alternative concepts

The **functional decomposition** presented in Fig. 32.16 shows the sequence in which the individual functional requirements are performed. Control signals to and from the Arduino microprocessor are shown as dashed lines.

Table 32.5 List of design specifications.

Performance
D—Must have movement that is visible from 10 ft away
D—Must have an audio or visual sensor and electric actuators
D—Must be still at beginning of operation
D—Must be activated by the press of a single button
D—Must have consistent and repeatable operation with 1 min of setup time in between runs
D—Must spin and spray at the same time
D—Must spin and spray when motion is sensed by sensor
D—Must hold air freshener in a stable upright position
W—Must be able to easily replace air fresher can when empty
D—Must be able to push button of can without the device slipping
D—Must operate for between 20 and 60 s
Geometry
D—Must have a smaller footprint than 30 in. by 30 in.
W—Sensor can be placed up to 10 ft from the device
Energy
D—Device uses 9 V battery for each red board, and up to 8 AA batteries
Controls
D—Uses at least one red board
Cost
D—Must be at or below $65
Schedule
D—Must be complete by November 9th
Safety
D—Must be safe for all
D—Must be nondamaging to the room
Aesthetics
W—Must be aesthetically pleasing

Solutions for each functional requirement were brainstormed and then documented in the form of a **classification scheme**, as shown in Fig. 32.17. An extra row was added to the classification scheme to display the various motor options. This was to ensure that each motor type is fully considered when selecting mechanisms to depress the nozzle of the air freshener can and to rotate it.

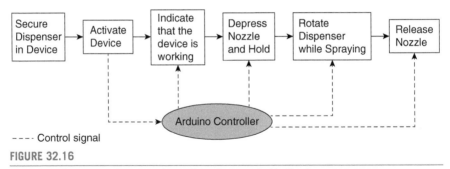

---- Control signal

FIGURE 32.16

Sequential functional decomposition of the Automatic Air Freshener.

Concepts / Functions	Concept 1	Concept 2	Concept 3	Concept 4	Concept 5
Secure dispenser	Drop into cup-like holder	Push into metal spring clips	Velcro to structure	Hold with rubber bands	Hold with Magnets
Activate device	Motion sensor on device or door	Timer	Push button	Sound sensor	Smell sensor
Indicate device is working	LED light	Let movement of device indicate this	Liquid crystal display (LCD)	Emit Sound	Raise a flag when complete
Depress/ hold/ release nozzle	Motor-driven cam or arm	Motor-driven lever	Solenoid	Use motor to lower large weight onto nozzle	Slide wedge
Rotate dispenser	Motor-driven rotating platform	Platform rotates on driven wheels	Wind and unwind wrapped cord	————	————
Motor options	DC motor	DC Gear motor	Servomotor	Continuous rotation servomotor	Stepper motor

FIGURE 32.17

Classification scheme for the Automatic Air Freshener.

Finally, compatible solutions appearing in the classification scheme were combined and sketched to form the three alternative concepts shown in Fig. 32.18.

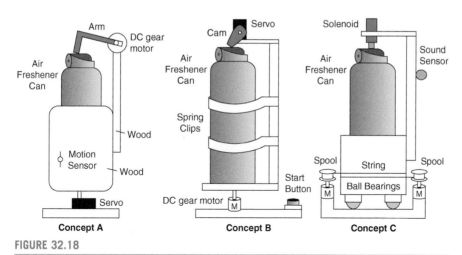

FIGURE 32.18

Concept sketches of the three alternative designs.

32.8.3 Design milestone #3: Evaluation of alternative concepts

The **decision matrix** shown in Table 32.6 was developed to encourage an exhaustive discussion of the strengths and weaknesses of each concept.

Table 32.6 Decision matrix for the three concepts.

	Evaluation criteria	Wt	Concept A		Concept B		Concept C	
			Val1	Wt × Val1	Val2	Wt × Val2	Val3	Wt × Val3
1	Easy to build	0.2	9	1.8	7	1.4	6	1.2
2	Easy to program	0.3	6	1.8	7	2.1	4	1.2
3	Reliable/robust operation	0.3	8	2.4	8	2.4	5	1.5
4	Easy to insert and replace air dispenser	0.1	8	0.8	6	0.6	8	0.8
5	Sophisticated electronics	0.1	7	0.7	6	0.6	8	0.8
	Total	1.0		7.5		7.1		5.5

Evaluation criteria prioritized functionality and ease of fabrication, where the latter encompasses the degree of difficulty associated with both implementing the control system (easy to code) and building the structure and mechanical systems (easy to build). The easier the design is to implement, the greater the amount of time that will be available for testing and debugging later. Students usually consider implementation of the controller to be the greater of the two challenges, thus its slightly higher weighting. While "Reliable/robust" was a requirement of the project, both it and "easy to insert and replace dispenser" enforce the consumer's voice, consistent with this being a product design. "Sophistication of electronics" reflects the point totals that would have been assigned to the electronic component selections during grading of the project and thus served as a counterbalance to ease of fabrication during the decision-making process.

Values were assigned in row-wise order to the three concepts based on the following assessment of their respective attributes:

(1) Easy to build

Concept A has the fewest parts and thus the highest rating.

(2) Easy to code

Both A and B have similar ratings because they each involve one servo and one DC gear motor. A slight edge goes to Concept B because the push button was more familiar to the students than the motion sensor. Concept C has the lowest rating for two reasons: (i) it has the most the most electronic components to deal with and (ii) a typical solenoid operates at relatively high voltages, requiring either a transistor or motor driver to interface the high voltage circuit with the 5 V circuit powered by the Arduino.

(3) Reliable/robust operation

In Concepts A and B, the mechanisms used to depress the nozzle and rotate the dispenser are very similar and so should exhibit similar reliabilities. Concept C received a lower rating because (i) it has more parts and thus more opportunities for something to go wrong and (ii) the holder may tend to drift to one side as both motors wind in unison, possibly leading to unforeseen issues.

(4) Easy to insert and replace dispenser

All three concepts require the user to align the dispenser with respect to the upper structure, leading to reductions in their respective ratings. Lowest rating goes to Concept B because the force required to insert the dispenser into the spring clips may require two hands to prevent the device from tipping over.

(5) Sophisticated electronics

Ratings correspond to the point totals associated with each set of electronic components (e.g., $+2$ = reversible DC motor, $+2$ = motion sensor, $+1$ = servo, $+2$ = solenoid, $+1$ = button)

Although the decision matrix was decisive in eliminating Concept C from further consideration, the final concept shown in Fig. 32.19 proved to be a strategic combination of all three concepts. It was decided to deploy high torque servos both to depress the nozzle and rotate the dispenser, as servos are specifically designed to produce finite rotations. By adding shaft encoders, DC gear motors could have worked too, but at the cost of extra coding. The start button was added to stay consistent with the rules governing the project; however, once the device is turned on, it is a motion sensor on the door frame that determines when to begin spraying.

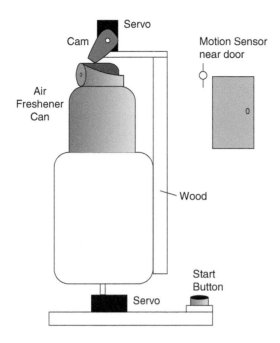

FIGURE 32.19

Final concept for the automatic air freshener. Not shown are the LED and the liquid crystal display (LCD) that were selected to indicate that the device has been activated.

32.8.4 Design milestone #4: Detailed design

In this application, both the mechanical and electronic systems must undergo a detailed design phase. There is some coupling between these two systems. For example, the length of the cam must be small enough that it can still function within the torque limits of the servo, and consideration must be given to how the motors attach to the structure. But for the most part, the mechanical and electronic designs can be developed in parallel.

32.8.4.1 Mechanical design

Detailed drawings comparable to those in Fig. 32.12 were prepared for the Automatic Air Freshener prior to the build. Another way to document the mechanical design, especially when the original design has undergone significant changes, is through photos of the final product. Figs. 32.20 and 32.21 provide both far-field and close-up views of the Automatic Air Freshener in its final form.

32.8.4.2 Electronics design with a microcontroller

The detailed design documented here had four deliverables that were developed in the following order: (1) circuit diagram, (2) Fritzing diagram[1] showing the implementation of the circuit diagram on an Arduino and breadboard, (3) flowchart of

---

[1] Fritzing is an open-source free software tool that can be used to create circuit diagrams (see http://fritzing.org).

FIGURE 32.20

Final mechanical design.

FIGURE 32.21

Close-up view of the top mounted servo and clear plastic cam used to depress the nozzle.

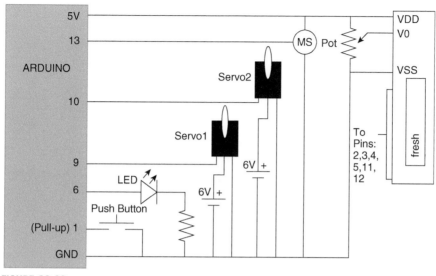

FIGURE 32.22

Circuit diagram.

the control logic, and (4) the computer code that implements the control logic and is uploaded to the Arduino.

The **circuit diagram** of the Arduino-based control system is shown in Fig. 32.22 as a series of independently controlled loops (one loop per component) with a common ground. There are three independent power sources. Each servo is powered by its own 6 V battery pack. The push button, the LED, the motion sensor, and the liquid crystal display (LCD) are all powered by the 5 V from the Arduino. A pull-up resistor will be declared at Pin 1 to avoid a short circuit when the start button is pressed. The potentiometer (labeled as POT) is used to control the brightness of the LCD.

The implementation of the circuit diagram on an Arduino and breadboard was documented in Fig. 32.23 using the **Fritzing** open-source software. The potentiometer and LCD are not shown in the interest of clarity. This software has some limitations (e.g., cannot change the size of components) and so finishing touches such as adding battery packs were done from within PowerPoint by exporting the Fritz image as a .jpg file.

The **flowchart** shown in Fig. 32.24 was developed to organize the sequencing of actions to be taken by the Arduino, as preparation for writing the computer code. These actions consist mainly of turning on and off the electronic components at the appropriate times and controlling the angular position of the servos' output shafts. The arrows indicate the direction of flow of the logic. The diamond-shaped boxes are decision points which are coded using "if" or "while" statements. The void loop repeats until the Arduino is either reset or powered down. Delays are used to provide extra time for processing or to slow the angular rotation of the servos' shafts.

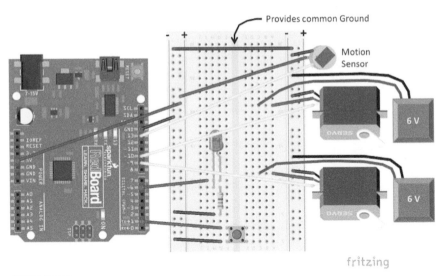

FIGURE 32.23

Implementation of the circuit diagram on an Arduino and breadboard, documented using the Fritzing software.

The final deliverable for this milestone is the code itself. The working **computer code** developed by the students is presented in Table 32.7 along with plenty of **descriptive comments**.

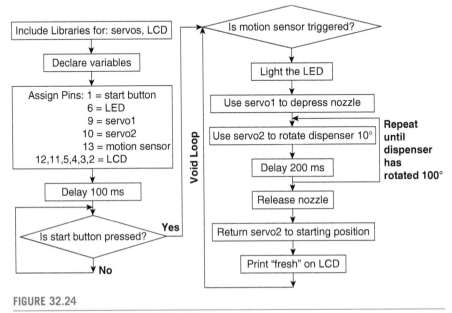

FIGURE 32.24

Flowchart of the control logic.

Table 32.7 Computer code uploaded to the Arduino.

Computer code	Descriptive comments
`#include <Servo's>`	//Include the servo library
`#include <LiquidCrystal.h>`	//Includes the LCD library
`LiquidCrystal lcd(12, 11, 5, 4, 3, 2);`	//Assigns pins to the LCD
`Servo servo1;`	//Names the first servo as servo1
`Servo servo2;`	//Names the second servo as servo2
`int pos = 0;`	//Declares "pos" to be an integer and sets it equal to 0
`void setup()`	
`{`	
`lcd.begin(16, 2);`	
`pinMode(13,INPUT);`	//Introduces pin 13 as an input (motion sensor)
`pinMode(6,OUTPUT);`	//Introduces pin 6 as an output (LED)
`pinMode(1,INPUT_PULLUP);`	//Introduces pin 1 as an input (button) with a pull-up resistor
`servo1.attach(9);`	//Attaches servo1 to pin 9
`servo2.attach(10);`	//Attaches servo2 to pin 10
`servo1.write(0);`	//Turns servo1 to 0 degrees to reset it
`servo2.write(0);`	//Turns servo2 to 0 degrees to reset it
`delay(100);`	//Delay to give the Arduino time to process
`while (digitalRead(1)==HIGH){}`	//Holds the code here until the button is pressed
`}`	

Continued

Table 32.7 Computer code uploaded to the Arduino.—*cont'd*

Computer code	Descriptive comments
`void loop()`	
`{`	
`if (digitalRead(13)==LOW)`	//Holds the code here until the motion sensor is triggered
`{`	
`digitalWrite(6,HIGH);`	//Lights the LED
`servo1.write(68);`	//Turns servo1 to 70 degrees to press the button
`delay(1);`	//A small delay to give the Arduino time to process
`for (int pos = 0; pos <= 100; pos += 10)`	//Increase "pos" by 10° until it reaches 100°
`{`	
`servo2.write(pos);`	//Turns servo2 to an angle equal to "pos"
`delay(200);`	//Delays servo2 to slow the overall rotation
`}`	
`servo1.write(0);`	//Turns servo1 to 0 degrees to stop spraying
`delay(100);`	//Delay to give the Arduino time to process
`digitalWrite(6,LOW);`	//Turns the LED off
`servo2.write(0);`	//Turns servo2 back to its starting position
`lcd.print("fresh ");`	//Prints "fresh" on the LCD screen
`}`	

Exercises

1. Other examples of first year engineering student design competitions can be found on the Internet. An excellent list of well over 100 first year competitions from a wide variety of universities can be found at the Directory of Projects for First-Year Engineering Students (http://www.discovery-press.com/discovery-press/studyengr/projects.asp).

2. The Rube Goldberg design projects are especially interesting for first year engineering students (http://www.rubegoldberg.com).

3. **TryEngineering** offers a variety of lesson plans that allow teachers and students to apply engineering principles in the classroom (http://www.tryengineering.org/lesson-plans).

4. Over 200 Arduino projects with source code, schematics, and complete DIY instructions can be found at https://circuitdigest.com/arduino-projects. Other Arduino projects can be found at https://playground.arduino.cc/Projects/Ideas.

Closing remarks on the important role of design projects in engineering education

33

Source : Malcom Romain/iStockphoto.com

If you ask professors or students why they do design projects in engineering courses, you can expect to hear responses like these:

- "They are motivational tools."
- "They apply the analytical methods taught in courses."
- "They help develop written and oral communication skills."
- "They teach teaming."

Indeed, these are all valuable outcomes of a design project, but each one can be achieved by some other means. The answer must lie elsewhere.

Exploring Engineering. https://doi.org/10.1016/B978-0-443-13541-5.00002-7

Part of the answer is found in the view that every engineering endeavor is ultimately about finding a solution to an expressed need. The analytical methods, the teaming skills, and all the rest are just tools for achieving that goal. They are the means to the end, not the end itself. Each design project offers a rare opportunity for students who spend most of their time deeply immersed in learning theoretical methods to see the bigger picture.

The rest of the answer has to do with the real purpose behind these design chapters. Engineering design is at its core an unbiased and structured methodology for dissecting and solving complex problems. It is the way engineers should and must think. In contrast to theoretical methods that are limited to their own special class of analytical problems, design methodology has universal applicability—to design, to research, to all fields of study. Design projects are the best way we know to exercise and develop this most fundamental of all engineering methods.

Hands-on design projects come closest to fully realizing these goals in part because they do not end with the detailed design; there will be design modifications to be made during manufacturing, testing, and the final performance evaluation. Students learn the importance of design principles by experiencing the results of having failed to follow them. They also gain a sense of accountability by learning that it is not enough for a design to look good on paper—it has to *work*.

Index